D1764953

3 0116 00423 5592

This book is due for return not later than the
last date stamped below, unless recalled sooner.

Lecture Notes
in Computational Science
and Engineering

9

Editors
M. Griebel, Bonn
D. E. Keyes, Norfolk
R. M. Nieminen, Espoo
D. Roose, Leuven
T. Schlick, New York

Research and Technology Organization of NATO:
Educational Notes RTO-EN-5

Springer
Berlin
Heidelberg
New York
Barcelona
Hong Kong
London
Milan
Paris
Singapore
Tokyo

Timothy J. Barth
Herman Deconinck
(Eds.)

High-Order Methods for Computational Physics

Springer

Editors

Timothy J. Barth
Information Sciences Directorate
NASA Ames Research Center
Moffett Field, CA 94035, USA
e-mail: barth@nas.nasa.gov

Herman Deconinck
von Karman Institute for Fluid Dynamics
Waterloosesteenweg, 72
1640 Sint-Genesius-Rode, Belgium
e-mail: deconinck@vki.ac.be

Cataloging-in-Publication Data applied for
Die Deutsche Bibliothek – CIP-Einheitsaufnahme

High order methods for computational physics / Timothy J. Barth; Herman Deconinck (ed.). –
Berlin; Heidelberg; New York; Barcelona; Hong Kong; London; Milan; Paris; Singapore; Tokyo: Springer, 1999
(Lecture notes in computational science and engineering; 9)
ISBN 3-540-65893-9

This book has been published jointly by Springer-Verlag and NATO's Research and Technology
Organization (RTO), both of whose logos appear on the front cover.

The cover image depicts iso-density contours of a time evolving double Mach reflection generated
from a Mach 10 shock wave impinging upon an inclined ramp geometry (calculation courtesy
B. Cockburn).

Mathematics Subject Classification (1991): primary: 65-06
                                      secondary: 65M60, 65M70, 65J15

ISBN 3-540-65893-9 Springer-Verlag Berlin Heidelberg New York

Cover Design: Friedhelm Steinen-Broo, Estudio Calamar, Spain
Cover production: *design & production* GmbH, Heidelberg
Typeset by the authors using a Springer TeX macro package

SPIN 10700458   46/3143 – 5 4 3 2 1 0 – Printed on acid-free paper

Preface

The development of high-order accurate numerical discretization techniques for irregular domains and meshes is often cited as one of the remaining challenges facing the field of computational fluid dynamics. In structural mechanics, the advantages of high-order finite element approximation are widely recognized. This is especially true when high-order element approximation is combined with element refinement (h-p refinement). In computational fluid dynamics, high-order discretization methods are infrequently used in the computation of compressible fluid flow. The hyperbolic nature of the governing equations and the presence of solution discontinuities makes high-order accuracy difficult to achieve. Consequently, second-order accurate methods are still predominately used in industrial applications even though evidence suggests that high-order methods may offer a way to significantly improve the resolution and accuracy for these calculations.

To address this important topic, a special course was jointly organized by the Applied Vehicle Technology Panel of NATO's Research and Technology Organization (RTO), the von Karman Institute for Fluid Dynamics, and the Numerical Aerospace Simulation Division at the NASA Ames Research Center. The NATO RTO sponsored course entitled "Higher Order Discretization Methods in Computational Fluid Dynamics" was held September 14-18, 1998 at the von Karman Institute for Fluid Dynamics in Belgium and September 21-25, 1998 at the NASA Ames Research Center in the United States. During this special course, lecturers from Europe and the United States gave a series of comprehensive lectures on advanced topics related to the high-order numerical discretization of partial differential equations with primary emphasis given to computational fluid dynamics (CFD). Additional consideration was given to topics in computational physics such as the high-order discretization of the Hamilton-Jacobi, Helmholtz, and elasticity equations.

This volume consists of five articles prepared by the special course lecturers. These articles should be of particular relevance to those readers with an interest in numerical discretization techniques which generalize to very high-order accuracy. The articles of Professors Abgrall and Shu consider the mathematical formulation of high-order accurate finite volume schemes utilizing essentially non-oscillatory (ENO) and weighted essentially non-oscillatory (WENO) reconstruction together with upwind flux evaluation. These formulations are particularly effective in computing numerical solutions of conservation laws containing solution discontinuities. Careful attention is given by the authors to implementational issues and techniques for improving the overall efficiency of these methods. The article of Professor Cockburn discusses the discontinuous Galerkin finite element method. This method naturally extends to high-order accuracy and has an interpretation as a finite volume method. Cockburn addresses two important issues associated with the discontinuous Galerkin method: controlling spurious extrema near solution discontinuities via "limiting" and the extension to second order advective-diffusive equations (joint work with Shu). The articles of Dr. Henderson and Professor Schwab consider the mathematical formulation and implementation of the h-p finite element methods using hierarchical basis functions and adaptive mesh refinement. These methods are particularly useful in computing high-order accurate solutions containing perturbative layers and corner singularities. Additional flexibility is obtained using a mortar FEM technique whereby nonconforming elements are interfaced together. Numerous examples are given by Henderson applying the h-p FEM method to the simulation of turbulence and turbulence transition.

The organizers gratefully acknowledge the special course lecturers for their substantial effort in preparing the articles for publication. The organizers also acknowledge the generous support of NATO RTO, the von Karman Institute, and the NASA Ames Research Center for sponsoring and holding the special course. Additional thanks is also given to RTO and Springer-Verlag for publishing the articles in a quality book form and so making them available to a wide readership.

Timothy Barth (NASA Ames Research Center)

Herman Deconinck (von Karman Institute for Fluid Dynamics)

March 1999

Table of Contents

High Order Approximations for Compressible Fluid Dynamics on Unstructured and Cartesian Meshes

Rémi Abgrall[1], Thomas Sonar[2], Oliver Friedrich[2] and Germain Billet[3]

[1] Université Bordeaux I, 351 Cours de la Libération 33 405 Talence Cedex
[2] Institut für Angewandte Mathematik, Universität Hamburg
Bundesstraße 55, D-20146 Hamburg, Germany
[3] ONERA, DSNA, Avenue de la Division Leclerc, 92 230 Châtillon sous Bagneux,
France

Abstract. We describe in detail some techniques to construct high order MUSCL type schemes on general meshes : the ENO and WENO type schemes. Special attention is given to the reconstruction step. Extesio to Hamilton Jacobi equations is sketched. We also present some hybrid techniques that use simple modifications of classical TVD schemes yielding in a very clear improvements of the accuracy. We discuss means of improving the efficiency using Harten's multiresolution analysis. We provide several numerical examples and comparisions with more conventional schemes.

Table of Contents

1   Introduction

During the past few years, a growing interest has emerged for constructing
high order accurate and robust schemes for simulations of compressible fluid
flow. One of the difficulties is the appearance of strong discontinuities that
may interact even for smooth initial data. To get rid of this difficulty, a
possible solution is to use a TVD (Totally Variation Diminishing) scheme.
Such a scheme has the property, at least for 1D scalar equations, not to
create new extrema, and hence to provide a nice treatment of discontinuities.
They have been successfully and widely used with any type of meshes (see for
example, [48] for a review and, among many others, [25] for simulations on
finite element type meshes). They are now of common use even for industrial

simulations of flows in complex geometries. Nevertheless, one of their main weaknesses is that the order of accuracy boils down to first order in regions of discontinuity and at extrema, leading to excessive numerical dissipation.

Various methods have been proposed to overcome this difficulty (adaptation of the mesh, for example see [27,32]) but one promising way – this is not the only possibility as we show in this paper – may also be the class of Essentially Non-Oscillatory schemes (ENO for short) introduced by Harten, Osher and others [18,19,34,15] and their WENO modifications.

The basic idea of ENO schemes is the use of a Lagrange type interpolation with an adapted stencil: when a discontinuity is detected, the procedure looks for a region around this discontinuity where the function is smooth and least oscillatory. This reconstruction technique may be applied either to the nodal values [34] or to a particular function constructed from cell averages in control volumes [18,19]. In this latter case, the approximation is conservative. This enables one to approximate any piecewise smooth function with any desired order of accuracy.

One of the purposes of this paper is twofold: first to provide a description of the basic ideas of the "classical" ENO/WENO methods, and second to show how it is possible to adapt them to general geometries. It is not possible to provide a complete overview of ENO methods. Hence, this paper will concentrate on finite volume ENO methods and triangular meshes. Nevertheless, we believe that it will provide information on the typical features of ENO methods on general geometries, the difficulties and problems associated with them.

This paper is organized as follows: In §2, we first recall the principle of finite volume schemes of MUSCL-type [52]. This enables us to describe the three steps of a finite volume scheme: the reconstruction, the evolution and the projection step. In general, the last two steps are merged by means of a Riemann solver and an appropriate temporal discretization scheme. Before entering the main topic of this paper, the reconstruction step, we describe Runge-Kutta methods due to Shu and Osher that allow to keep the TVD or TVB (Total Variation Boundedness) features of the first order approximation of a generalized Riemann problem with non-constant data. Then we move to the reconstruction problem. First, in §3, we detail the "classical" methods that are applied to real valued functions and show why it cannot be applied on general geometries.

To overcome this difficulty, we introduce a new reconstruction procedure in §4. We show its properties that appear to be the same than those of a more classical Lagrange reconstruction. The practical calculation of this polynomial is discussed in detail in §5, and we show that it can be determined by an algorithm very similar to the Newton algorithm for divided differences. We also discuss the impact of the choice of the polynomial expansion in the calculation from the point of view of stability. The ENO reconstruction is then described in §6. We also discuss other types of expansions using splines in §8. In §11, we briefly discuss how all these methods can be extended to first order

Hamilton Jacobi equations. A third order ENO method for CFD problems is applied in §9 to several several flow problems, and we show that the accuracy is improved considerably in comparison with 2nd order computations.

However, these methods are by nature quite costly. This is why we also discuss (§10) a particular technique aimed at reducing the CPU cost of ENO methods. Toward the same goal (getting very high order schemes with the lowest computational cost), we also show (§12) how to modify the now very classical second order MUSCL type schemes (on cartesian meshes) so that numerical diffusion is reduced a lot.

2   An overview of finite volume schemes

2.1   The Euler equations

Let us quickly recall elementary facts about the Euler equations of a calorically perfect gas:

$$\frac{\partial W}{\partial t} + \frac{\partial F(W)}{\partial x} + \frac{\partial G(W)}{\partial y} = 0 \tag{2.1}$$

As usual, in equation (2.1), W stands for the vector of conserved quantities and F (respectively G) is the flux in the x direction (resp. y direction):

$$W = \begin{pmatrix} \rho \\ \rho\, u \\ \rho\, v \\ E \end{pmatrix}, F(W) = \begin{pmatrix} \rho\, u \\ \rho\, u^2 + \mathsf{p} \\ \rho\, uv \\ u(E + \mathsf{p}) \end{pmatrix},$$

$$G(W) = \begin{pmatrix} \rho\, v \\ \rho\, uv \\ \rho\, v^2 + \mathsf{p} \\ v(E + \mathsf{p}) \end{pmatrix}, \tag{2.2}$$

with initial and boundary conditions. In equation (2.2), ρ is the density, u, v are the components of the velocity, E is the total energy and p the pressure, related to the conserved quantities by the equation of state:

$$\mathsf{p} = (\gamma - 1)\left(E - \frac{1}{2}\rho(u^2 + v^2)\right) \tag{2.3}$$

The ratio of specific heats γ is kept constant.

It is well known that the system defined by equations (2.1), (2.2) and (2.3) is hyperbolic: for any vector $n = (n_x, n_y)$, the matrix:

$$A_n = n_x \frac{\partial F}{\partial W} + n_y \frac{\partial G}{\partial W} \tag{2.4}$$

is diagonalizable and has a full set of real eigenvalues and eigenvectors. Let us describe now the construction of a high order scheme.

2.2   Finite volume formulation

We consider a mesh $\mathcal{M}$ consisting of control volumes $\{C_i\}$, for example the triangles of a conforming triangulation or the boxes of a dual mesh, see [1]. The semi discrete finite volume formulation of (2.1) is:

$$\frac{\partial}{\partial t}\overline{W}_i(t) = -\frac{1}{|C_i|}\int_{\partial C_i} \mathcal{F}_{\boldsymbol{n}}[W(x,t)]dl = \mathcal{L}_i(t) \qquad (2.5)$$

Here, $W_i(t)$ is the (spatial) mean value of $W(x,t)$ at time t over C_i, $\boldsymbol{n} = (n_x, n_y)$ is the outward unit normal to ∂C_i, and $\mathcal{F}_{\boldsymbol{n}} = n_x F + n_y G$. We first describe the spatial approximation of (2.5), then the temporal discretization of the resulting set of ordinary differential equations. Last, we give details concerning the boundary conditions.

Spatial discretization The first step is to discretize $\mathcal{L}_i(t)$ up to k^{th} order. We define the integer number p such that either $k = 2p$ or $k = 2p + 1$. We can rewrite $|C_i|\mathcal{L}_i(t)$ as:

$$\int_{\partial C_i} F_{\boldsymbol{n}}[W(x,t)]dl = \sum_{\Gamma_s}\int_{\Gamma_s} F_{\boldsymbol{n}}[W(x,t)]dl \qquad (2.6)$$

where, as in Figure 13, the set of the Γ_s's is that of the edges of C_i. On each Γ_s, $\boldsymbol{n}$ is constant. We consider, on any Γ_s, the p Gaussian points $\{G_l\}_{1\leq l\leq p}$ associated to the Gaussian formula of order $2p+1$. The integral $\int_{\Gamma_s} F_{\boldsymbol{n}}[W(x,t)]dl$ is approximated by

$$\sum_{l=1}^{p} \omega_l \mathcal{G}_{\boldsymbol{n},l}(t), \qquad (2.7)$$

where the term $\mathcal{G}_{\boldsymbol{n},l}(t)$ has to be defined. Let C_j be another control volume of which Γ_s is a boundary part. In C_i and C_j, compute approximations of W at time t, as well as so–called reconstruction functions (recovery functions) $R_i[W(.,t)]$ and $R_j[W(.,t)]$. The ENO reconstruction described in this paper (see §4) is applied to the physical variables, then the conserved ones are derived from them. We define

$$\mathcal{G}_{\boldsymbol{n},l}(t) = F_{\boldsymbol{n}}^{\text{Riemann}}\left\{R_i[W(.,t)](G_l), R_j[W(.,t)](G_l)\right\}. \qquad (2.8)$$

In equation (2.8), $F_{\boldsymbol{n}}^{\text{Riemann}}$ may be any of the available Riemann solvers. In all the examples below, we have chosen Roe's Riemann solver with the Harten-Hyman entropy correction. The boundary conditions are implemented as in [25].

We see that the only remaining degree of freedom is the evaluation of $R_j[W(.,t)]$ which should be a "good" approximation of W. It is natural to ask for the following properties [18,19] for the reconstruction $R[u]$ of a function u:

P1- If u is of class C^r with $r \geq k$, then the l-th order derivative of $u - R[u]$, $l \leq k$ satisfies $u^{(l)} - R[u]^{(l)} = O(h^{k+1-l})$,

P2- $TV(R[u]) \leq TV(u) + O(h^r)$,

P3- The average of $R[u]$ over $[x_{i-1/2}, x_{i+1/2}]$ is equal to that of u.

Roughly speaking, one may say that the property **P1** guaranties the accuracy of the approximation, the property **P2** guarantes (for reasonable flux functions) that the scheme is Total Variation Bounded (TVB) and hence converges to the correct solution for any $t \in [0, T]$ while the property **P3** states the consistency of the scheme. Before entering into the details of the reconstruction step, let us briefly comment on the evolution operator.

Approximation of the evolution operator   A classical way of solving a set of ordinary differential equations like (2.5) is to use a Runge–Kutta scheme. Among these schemes, some of them have the property not to increase the total variation [34]. They are built as follows. The set of equations to be solved are supposed to be written in the form

$$\frac{d\overline{u}}{dt} = \mathcal{L}(\overline{u}),$$

where the operator $\mathcal{L}$ contains spatial derivatives.

1. **Second order scheme**: This is the classical Heun's method. It is TVD under CFL=1.
$$\overline{u}^{(1)} = \overline{u}^{(0)} + \Delta t \mathcal{L}(\overline{u}^{(0)})$$

$$\overline{u}^{(2)} = \tfrac{1}{2}\overline{u}^{(0)} + \tfrac{1}{2}\overline{u}^{(1)} + \tfrac{1}{2}\mathcal{L}(\overline{u}^{(1)})$$

2. **Third order scheme**, TVD under CFL=1 :
$$\overline{u}^{(1)} = \overline{u}^{(0)} + \Delta t \mathcal{L}(\overline{u}^{(0)})$$

$$\overline{u}^{(2)} = \tfrac{3}{4}\overline{u}^{(0)} + \tfrac{1}{4}\left(\overline{u}^{(1)} + \Delta t \mathcal{L}(\overline{u}^{(1)})\right)$$

$$\overline{u}^{(3)} = \tfrac{1}{3}\overline{u}^{(0)} + \tfrac{2}{3}\left(\overline{u}^{(2)} + \Delta t \mathcal{L}(\overline{u}^{(2)})\right)$$

Until the end of this paper, we will not give more details on the time stepping since our purpose is to concentrate on the reconstruction step of a finite volume scheme.

3   The reconstruction step : classical methods

3.1   An essentially non oscillatory Lagrange interpolation

The classical ENO reconstruction methods are derived using two well known properties of the Lagrange interpolation of a function u: consider an increasingly finer subdivision of $\mathbb{R}$, (y_i), and P a polynomial of degree r such that $P(y_l) = u(y_l)$, $m \leq l \leq m + r$ Then

1. if u admits k continuous derivatives on $[y_m, y_{m+r}]$ then for any $x \in [y_m, y_{m+r}]$,

$$|P^{(k)}(x) - u^{(k)}(x)| \leq C\max_{m \leq l \leq m+r-1}|y_l - y_{l+1}|^{r-k+1}$$

2. if the k-th derivative of u, $k < r$, has a jump $[u^{(k)}]$ in $x_0 \in]y_m, y_{m+r}[$, then a_k, the leading coefficient of P in its Taylor expansion around x_0, behaves like $\max_{m \leq l \leq m+r}|y_l - y_{l+1}|^{-r+k}$.

Although the first property stated is a well known result in numerical analysis the second property is (up to our knowledge) nowhere clearly stated and proved explicitly.

These two facts explain that if u is smooth then the coefficients of P remain *bounded* when the mesh size decreases but *blow up* if one of the derivatives of u of order smaller than the degree of P has a jump.

From these two remarks one may construct the following essentially non–oscillatory Lagrange reconstruction in a neighborhood of a mesh point y_i. The idea is to construct an "adaptive" stencil $\mathcal{S}_i^k$, the points of which are used to compute the Lagrangian interpolant P_i^k, where P_i^k is a polynomial of degree k.

First, we recall that the divided differences table can easily be recursively constructed:

$k = 0 : [y_i]u = u(y_i)$

$k > 0 : [y_i, y_{i+1}, \cdots, y_{i+k-1}, y_{i+k}]u =$
$$\frac{[y_{i+1}, \cdots, y_{i+k}]u - [y_i, y_{i+1}, \cdots, y_{i+k-1}]u}{y_{i+k} - y_i}.$$

We can now describe the classical ENO interpolation algorithm of Harten [19], in which a polynomial is constructed which does not interpolate through a discontinuity of u.

ENO interpolation algorithm:

1. Start with $\mathcal{N}_i^0 = \{y_i\}$
2. for $l \leq k$, consider $\mathcal{N}_i^{l-1} = \{y_{j_0} < \cdots < y_{j_0+l-1}\}$ as the stencil for P_i^{l-1}. Compute the divided differences $[y_{j_0-1}, y_{j_0}, \cdots, y_{j_0+l-1}]u$ and $[y_{j_0}, \cdots, y_{j_0+l-1}, y_{j_0+l}]u$.
   - if $|[y_{j_0-1}, y_{j_0}, \cdots, y_{j_0+l-1}]u| < |[y_{j_0}, \cdots, y_{j_0+l-1}, y_{j_0+l}]u|$ then $\mathcal{N}_i^l = \mathcal{N}_i^{l-1} \bigcup \{y_{j_0-1}\}$,
   - else , $\mathcal{N}_i^l = \mathcal{N}_i^{l-1} \bigcup \{y_{j_0+l+1}\}$.

end

When one applies this algorithm to the Heaviside function on a grid with uniform mesh size Δx, one gets

$$R(x) = \begin{cases} 1 & \text{if } x < \frac{\Delta x}{2} \\ -1 & \text{if } x > \frac{\Delta x}{2} \end{cases}$$

It is conjectured in [19] that if the function u is smooth, say of class larger than the degree k of the reconstruction, then the derivative of $u - R$ satisfy $\max |u^{(l)} - R^{(l)}| \leq Ch^{k+1}$ where C is a constant that depends on the mesh and u, and $TV(R) \leq TV(u) + O(h^k)$.

3.2   Application to finite volume schemes

In finite volume schemes, the variables are not known at nodes but only their mean values on control volumes are given. There are two ways of using the above ENO Lagrange interpolation in order to get an ENO reconstruction: the so–called *reconstruction via primitive functions* and *reconstruction via deconvolution*. Since the reconstruction via deconvolution can work only for regular meshes [19], we concentrate on the reconstruction via primitive functions.

Let us consider a mesh $(x_i)_{i \in \mathbb{N}} \subset \mathbb{R}$ and a real valued function u. The averages u_i of u are given on the control volumes $[x_{i-1/2}, x_{i+1/2}]$. It is possible to know the values of the primitive W of u defined by

$$W(x) = \int_{x_{-1/2}}^{x} u(t)dt$$

at the nodes $x_{i+1/2}$, because

$$W(x_i) = \int_{x_{i-1/2}}^{x_{-1/2}} u(t)dt = \sum_{j=0}^{j=i}(x_{j+1/2} - x_{j-1/2})u_i.$$

The choice $i = 0$ in the definition of W is arbitrary. One now determines a local ENO Lagrange interpolant of W up to degree $k + 1$, say P_i^{k+1}, on the interval $]x_{i-1/2}, x_{i+1/2}[$. This is obtained by constructing the ENO interpolant using the above ENO interpolation algorithm. The nodal values $W(x_{i+1/2})$ are taken as interpolation data. The interpolant is finally restricted to the intervall $]x_{i-1/2}, x_{i+1/2}[$, i.e.

$$R[u] = \frac{dP_i^{k+1}}{dx} \qquad \text{on }]x_{i-1/2}, x_{i+1/2}[.$$

It is clear that if u is smooth enough, then property P1 is satisfied. If u has only *isolated* discontinuities itself or in one of its derivatives, property P2 is also satisfied, see [1]. The last property P3 is a consequence of the construction:

$$\int_{x_{i-1/2}}^{x_{i+1/2}} R[u]dx = P_i^{k+1}(x_{i+1/2}) - P_i^{k+1}(x_{i-1/2}) =$$
$$(x_{i+1/2} - x_{i-1/2})u_i$$

3.3   Possible extensions to higher dimensions and their weaknesses

The extension of these method to higher dimensions has been carried out, for example by Casper et al. [10] for *regular structured* grids. The basics of their method are the following. They first assume their mesh is a Cartesian product $\{(x_i, y_j), 1 \le i \le N, 1 \le j \le M\}$ and they take rectangular control volumes. With the notation $\Delta_k \xi = \xi_{k+1} - \xi_k$, the data are

$$\overline{\overline{w}}_{ij} = \frac{1}{\Delta_i x \, \Delta_j y} \int_{x_{i-1/2}}^{x_{i+1/2}} \int_{y_{j-1/2}}^{y_{j+1/2}} w(x, y) dx \, dy. \tag{3.1}$$

For $y_{j-1/2} < y < y_{j+1/2}$, they consider the primitive function $\overline{W}_j(x)$ associated with w defined by:

$$\overline{W}_j(x) = \int_{x_0}^{x} \frac{1}{\Delta_j y} \left[\int_{y_{j-1/2}}^{y_{j+1/2}} w(\xi, y) dy \right] d\xi \tag{3.2}$$

From (3.1), they notice that

$$\Delta_i x \overline{\overline{w}}_{ij} = \overline{W}_j(x_{i+1/2}) - \overline{W}_j(x_{i-1/2})$$

so that they can consider the reconstruction via primitive function of degree k of $\overline{W}_j$: $v_j(x) = R(x, w)_j$.

Then, the procedure (3.2) is performed for any j. Since

$$\frac{d}{dx} \overline{W}_j(x) = \frac{1}{\Delta_j y} \int_{y_{j-1/2}}^{y_{j+1/2}} w(x, y) dy,$$

$R(x, w)_j$ can be interpreted as a one-dimensional cell average on $[y_{j-1/2}, y_{j+1/2}]$ of some function $v(x, y)$. For a fixed x, one considers the set $\{R(x, w)_j\}$ and a primitive $V(x, y)$ associated to v,

$$V(x, y) = \int_{y_0}^{y} v(x, y) dx dy,$$

whose pointwise values are known at the interfaces:

$$V(x, y_{j+1/2}) = \sum_{k=j_0}^{j} \Delta_k y \overline{v}_k(x).$$

We can once more apply the same reconstruction via primitive function to v of degree k and construct a reconstruction $R^2(x, y, w) = R(y; R(x; \overline{\overline{w}}))$ of w . It is clear that this new reconstruction will have the conservation property, essentially non–oscillatory and precision properties: they are directly inherited from the one-dimensional reconstruction properties.

When the mesh consists of quadrilaterals that are *not* a Cartesian product, one has to assume the existence of a smooth transformation from the physical $x - y$ plane the rectangular $\xi - \nu$ plane: $x = x(\xi, \nu)$, $y = y(\xi, \nu)$ The Jacobian determinant $J(\xi, \nu)$ should *never* vanish. The control volume C_{ij} of the physical plane are the control volumes $D_{ij} =]\xi_{i-1/2}, \xi_{i+1/2}[\times]\nu_{i-1/2}, \nu_{i+1/2}[$ mapped by the transformation. The averaged values are:

$$\overline{\overline{u}} = \frac{1}{a_{ij}} \int_{\xi_{i-1/2}}^{\xi_{i+1/2}} \int_{\nu_{i-1/2}}^{\nu_{i+1/2}} u(x(\xi, \nu), y(\xi, \nu)) d\xi \; d\nu$$

where a_{ij} is the area of C_{ij}. Then one uses the above reconstruction on the $\xi - \nu$ mesh. The reconstruction $\widetilde{R}$ is:

$$\widetilde{R}(\xi, \nu, \overline{\overline{u}}) = \frac{1}{J(\xi, \nu)} R^2(\xi, \nu, a\overline{\overline{u}})$$

The scaling factors are introduced so that $\widetilde{R}$ satisfies the conservation property.

From this, it is clear that this kind of reconstruction algorithm is very dependent on the structure of the mesh. For example, for the reconstruction via primitive function, one needs to gather control volumes into subsets so that their collection is a square. For a reconstruction of degree k, one should be able to gather them into subsets containing k^2 control volumes. This is in general not possible, see Figure 13.

For all these reasons, one needs other algorithms to handle more general geometries. In Section 4, we show how this can be done in the context of general unstructured meshes.

4   The reconstruction problem on unstructured meshes

4.1   Preliminaries

In the sequel, the symbol $\Pi_n[x, y]$ denotes the set of polynomials P in the variables x and y of degree less or equal than n:

$$P(x, y) = \sum_{l=0}^{n} \sum_{i+j=l} a_{ij} x^i y^j \tag{4.1}$$

The set $\Pi_n[x, y]$ is a vector space of dimension $N(n) = \frac{(n+1)(n+2)}{2}$, a basis of which is the set of monomials $\{(x - x_0)^i (y - y_0)^j\}_{i+j \leq n}$ where (x_0, y_0) is any point in $\mathbb{R}^2$. The degree of P does not depend on the choice of (x_0, y_0). As we will show later, this kind of basis is not well suited for practical calculations.

Let a set of points be given. Associated with this set we also consider a triangulation $\mathcal{T}$. We may consider several kinds of control volumes, for example the triangles of $\mathcal{T}$ themselves or the dual mesh. The dual mesh with its control volumes is constructed as follows: For each point M_i the control volume is obtained by connecting the midpoints of the segments adjacent to it and the center of gravity of the triangles of which it is a vertex. Let us denote by $\{C_i\}$ the set of control volume. We only require the following properties:

- For any $i \neq j$, $C_i \cap C_j$ is of empty interior,
- C_i is connected,
- There is an algebraic dependency of the C_i's in terms of the points of $\mathcal{M}$, i.e. the points of $\mathcal{M}$ are within a specific location inside the boxes, or the node points of the triangles, respectively.
- The boundary of C_i is a polygonal line with at most N_0 vertices.

We consider the following problem (problem $\mathcal{P}$ or *approximation in the mean* for short):

Let u be regular enough (say in L^1). Given two integers N and n, a set of control volumes $\mathcal{S} = \{C_{i_l}\}_{1 \leq l \leq N}$, find an element $P \in \mathbb{R}_n[x, y]$ such that for $1 \leq l \leq N$,

$$\bar{u}_l := \langle \, A \, (C_{i_l}), u \rangle := \frac{\int_{C_{i_l}} u \, dx}{|C_{i_l}|} = \langle \, A \, (C_{i_l}), P \rangle. \qquad (4.2)$$

For that problem to have a unique solution, one must satisfy two conditions: $N = \frac{(n+1)(n+2)}{2} = N(n)$ and the following Vandermonde type matrix must be non singular

$$\mathcal{V} = (\langle \, A \, (C_l), x^i y^j \rangle)_{i+j \leq n, 1 \leq l \leq N(n)}. \qquad (4.3)$$

If $\det \mathcal{V} \neq 0$, then we will say that this stencil $\mathcal{S}$ is *admissible*. In that case, there is a unique solution to problem (4.2).

A similar problem was first considered by Barth et al. [8] for smooth functions, then by Harten et al. [17], Vankeirsbilck et al. [39,38], Abgrall [1] and Sonar [20]. In the four first references [8,17,39,38], the authors consider overdetermined systems for two reasons: first, the problem $\mathcal{P}$ has not always have a unique solution, second they claim that the condition number of the overdetermined system is better than that of problem $\mathcal{P}$. In [1], the same approach as here was adopted. To support this choice, we note that (4.3) is generally not singular. Second, the condition number of the linear system mainly depends on the basis used for the polynomial expansion, as it is shown in Section 4.4. For these two reasons, we have prefered this approach which also has the advantage of simplifying the coding of the global scheme.

4.2 Some general results about problem $\mathcal{P}$

In this section, we give two results on the reconstruction (4.2) of a given function u if either it is smooth or not. They generalize well-known properties of the Lagrange interpolation of 1D real-valued functions that have been used as a building block by Harten and his coauthors to design an essentially non-oscillatory reconstruction. Throughout this section, if $\mathcal{S}_n$ is an admissible stencil for degree n, the symbol $K(\mathcal{S}_n)$ denotes the convex hull of the union of the elements of $\mathcal{S}_n$.

The case of a smooth function In [1], we show the following result. Its proof follows easily from Ciarlet and Raviart's proof [11] on Lagrange and Hermite interpolation:

Theorem 1. *Let S be an admissible (for degree n) stencil of $\mathbb{R}^2$, let h and ρ be the diameter of $K(\mathcal{S})$ and the supremum of the diameters of all circles contained in $K(\mathcal{S})$, respectively. Let u be a function that has everywhere in $K(\mathcal{S})$ a derivative $D^{n+1}u$ with*

$$M_{n+1} = sup\{\|D^{n+1}u(x)\|; x \in K(\mathcal{S})\} < +\infty.$$

If P_u is the solution to problem $\mathcal{P}$, then for any integer m, $0 \le m \le n$,

$$sup\{\|D^m u(x) - D^m P_u(x)\|; x \in K(\mathcal{S})\} \le C M_{n+1} \frac{h^{n+1}}{\rho^m}$$

for some constant $C = C(m, n, \mathcal{S})$. Moreover, if S' is obtained from S by an affine transformation (i.e. there exists $x_0 \in \mathbb{R}^2$ and an invertible matrix A such that

$$C'_k \in \mathcal{S}' \text{ iff there exists } C_k \in \mathcal{S} \text{ such that } C'_k = A \ C_k + x_0,$$

then
$$C(n, m, \mathcal{S}) = C(n, m, \mathcal{S}').$$

This result basically expresses that if the stencil $\mathcal{S}$ is not too flat, i.e. the ratio h/ρ is not too large, then P_u will be a good approximation of u. Let us turn now to the case of unsmooth functions.

4.3   The case of a nonsmooth function

We begin with some notations. Let $\mathcal{S}_n$ be an admissible stencil. For the sake of simplicity, we assume that (x_0, y_0) is any point in $K(\mathcal{S}_n)$. Throughout this subsection, we adopt the lexicographic ordering for polynomials: if i and j are two indices such that $i + j = p \le n$, we set $l = N(p-1) + i + 1$ (with the convention $N(-1) = 0$) and denote by P_l the monomial $(x - x_0)^i(y - y_0)^j$. We also set

$$R_l = \left(\langle A \ (C_1), P_l \rangle, \cdots, \langle A \ (C_{N(n)}), P_l \rangle \right)^T.$$

Given a set of $N(n)$ real numbers u_i, the solution $P = \sum_{j=1}^{N(n)} a_j P_j$ of problem (4.2) may be seen as the solution of the linear system $M_n \left(a_1, \cdots, a_{N(n)} \right)^T =: U = \left(\overline{u}_1, \cdots, \overline{u}_{N(n)} \right)^T$, where the lth column of M_n is denoted by R_l.

The aim of this section is to give the asymptotic behavior of the leading coefficients of P_l. By scaling arguments, we see that if the data u_i are the average of some u that is smooth up to order $p < n$, one should have

$$|a_l| \simeq h^{p-n},$$

where h is the scaling factor (a typical size of the cells). Unfortunately, this kind of argument assumes that we work with a very particular set of stencils: all stencils are obtained by a similarity transformation from a "mother" stencil. This is far from what we need, namely an estimate involving the typical size of admissible stencil of size h small. Moreover, it is possible that for some stencils

$$\sum_{N(n-1)+1 \leq l \leq N(n)} |a_l| = 0.$$

This means that P is of degree $n - 1$ at most. With this kind of stencil, no information at all is obtained from the leading coefficients.

We need to work with admissible stencils such that the polynomial P, solution of problem (4.2) (for degree n) with data that are either 0 or taken among linear combination of terms belonging to $\{\langle A (C), P_l \rangle\}_{C \in \mathcal{S}_n}$ is *exactly* of degree n. More precisely, we define the set $\mathcal{P}_{n,p}^{\alpha,\beta} \subset \mathbb{R}^M$ for α, $\beta > 0$, $p \in \{1, \cdots n\}$ and $M = (N(p) - N(p-1)) \times N_0 \times N(n)$ by

Definition 2. Given any $p \in \{1, \cdots n\}$, and two real number $\alpha > 0$, $\beta > 0$, we define the set $\mathcal{P}_{n,p}^{\alpha,\beta}$ as follows: $\mathcal{S}_n \in \mathcal{P}_{n,p}^{\alpha,\beta}$ if and only if

1. The diameter h of K, the convex hull of $\mathcal{S}_n$, is 1 and $(0,0) \in K$;
2. The stencil $\mathcal{S}_n$ is admissible and
   $$\left| \det(R_1, \cdots, R_{N(n)}) \right| \geq \beta$$
3. For any polynomial Q of degree exactly p, $Q = \sum_{l=0,p} \sum_{i+j=l} \lambda_{ij} (x - x_0)^i (y - y_0)^j$ with $\max_{l,i+j=l} |\lambda_{ij}| = 1$, and for any partition $\mathcal{S}_0$, $\mathcal{S}_1$ of $\mathcal{S}_n$, with $\#\mathcal{S}_0 > 0$ and $\#\mathcal{S}_1 > 0$, the polynomial $P \in P_n(\mathbb{R}^2)$ defined by $\langle A (C_k), P \rangle = 0$ if $C_k \in \mathcal{S}_0$ and $\langle A (C_k), P \rangle = \langle A (C_k), Q \rangle$ if $C_k \in \mathcal{S}_1$ satisfies $\sum_{i+j=n} |a_{ij}| \geq \alpha$.

In the above definition, the polynomial Q cannot identically vanish. The polynomial P is of degree exactly n and its leading coefficients cannot be very small. It can be shown the inequality $\left| \det(R_1, \cdots, R_{N(n)}) \right| \geq \beta$ implies that the stencils are not too flat, i.e. the ratio $\frac{h}{\rho} = \frac{1}{\rho}$ is not too large. Algebraic arguments indicate that for any n and p, we can find $\alpha, \beta > 0$ such that the set $\mathcal{P}_{n,p}^{\alpha,\beta}$ is not empty.

To motivate Definition 2, we give a counterexample in $\mathbb{R}$; counterexamples in higher dimensions can also be obtained [1]. Consider the stencil $\{x_0 =$

$0, x_1 = 1, x_2 = 2\}$, and P, polynomial of degree two, such that $P(0) = 1$, $P(1) = 0$, $P(2) = 0$. We then have $P(3) = 1$. But the stencil $\{x_0 = 0, x_1 = 1, x_2 = 2, x_3 = 3\}$ is admissible for degree three, and hence does not satisfy the analogue of Definition 2 in $\mathbb{R}$.

We have the following result.

Theorem 3. *Let n, p, be integers and α, β and δ real numbers be integers and real numbers, as in Definition 2 and $\mathcal{S}_n$ be an admissible stencil such that there exists an affine invertible transformation ϕ for which $\phi^{-1}(\mathcal{S}_n) \in \mathcal{P}_{n,p}^{\alpha,\beta}$ and $\|\phi^{-1}\| \leq \delta$. Let u be a real-valued function defined on an open subset of Ω in $\mathbb{R}^2$, $u \in C^p, p \leq n$, except on a locally C^1 simple curve C. The intersection of the convex hull K of $\mathcal{S}_n$ and C is assumed to be nonempty. We also assume that the pth-order derivative of u has a jump such that $\min_{(x,y) \in C} \|[D^p u](x,y)\| \geq \gamma > 0$.*

Then there exists a constant $C(n, p, \alpha, \beta, \delta) > 0$, invariant by affine transformations, such that the coefficients in the Taylor expansion of u around any point (x_0, y_0) satisfy

$$\sum_{i+j=n} |a_{ij}| \geq C(n, p, \alpha, \beta, \delta) \frac{\gamma}{h^{n-p}} \tag{4.4}$$

if h is small enough.

This result enables us to distinguish between regions of smoothness and those where a jump in one of the derivatives occurs.

4.4   Three polynomial expansions

In this section, we intend to study the numerical system that has to be solved in order to get P from the data. We will consider three kinds of expansions of P:

1. the "natural" expansion: for any point $(x_0, y_0) \in \mathbb{R}^2$,

$$P = \sum_{0 \leq i+j \leq n} a_{ij}(x - x_0)^i (y - y_0)^j, \tag{4.5}$$

2. an expansion using scaling. Define a local scaling factor $s := 1/\sqrt{|C_1|}$ which should be read as an approximation for $1/h$ and change the "natural" expansion into

$$P = \sum_{0 \leq i+j \leq n} a_{ij} s^{i+j} (x - x_0)^i (y - y_0)^j, \tag{4.6}$$

3. an expansion using barycentric coordinates. Let $\mathcal{S}_n = \{C_1, C_2, C_3, \ldots, C_{N(n)}\}$ be an admissible stencil. Hence, at least one subset of three elements of $\mathcal{S}_n$ is an admissible stencil for $n = 1$. We may assume that the

set $\{C_1, C_2, C_3\}$ is admissible. We consider the three polynomials Λ_i of degree 1 defined by $A(C_j)\Lambda_i = \delta_i^j$, $1 \leq i \leq 3$, $1 \leq j \leq 3$. Clearly, we have $\Lambda_1 + \Lambda_2 + \Lambda_3 = 1$. These polynomials are the barycentric coordinates of the triangle constructed on the barycenters of C_1, C_2, and C_3. In order to get expansion (4.5), a strategy may be to look first for the expansion of the polynomial P in terms of powers of Λ_2 and Λ_3:

$$P = \sum_{i+j \leq n} a_{ij} \Lambda_2^i \Lambda_3^j \qquad (4.7)$$

and then to get the Taylor expansion of P around the center of gravity of C_1 from (4.7) (the Theorems 1 and 3 give the behaviour of the leading coefficients of P whatever point chosen in the convex hull of $\mathcal{S}$).

In order to get the expansions (4.5), (4.6) or (4.7), one has to solve linear $N(n) \times N(n)$ systems:

$$\mathcal{B}(a_{00}, \cdots, a_{0n})^T = \left(\langle A(C_{i_1}), u \rangle, \cdots, \langle A(C_{i_{N(n)}}), u \rangle \right)^T \qquad (4.8)$$

where the matrix $\mathcal{B}$ is obtained by taking the average of $(x - x_0)^i (y - y_0)^j$ for (4.5), the same average times s^{i+j} for (4.6) and $\Lambda_2^i \Lambda_3^j$ for (4.7). Let us now study the properties of these linear systems.

The case of the natural expansion A very easy consequence of the inequality (4.4) is:

Proposition 4. *Let us assume that the conditions of* Theorem 3 *hold and let h be the supremum of the diameters of the spheres containing $K(\mathcal{S}_n)$. Then the condition number of system (4.8) is at least $O(h^{-n})$ for h small enough.*

This fact is well known for 1D Lagrange interpolation and has motivated the search of more efficient algorithms, such as the Newton algorithm. There exist algorithms that generalizes it [31,30]. In §5, we propose a completely algebraic algorithm that we show, to be equivalent to the generalization of [31,30] for the cell average recovery problem (4.2). These method makes use of the barycentric coordinate expansion (4.7).

The case of the barycentric expansion In the case of expansion (4.7), we have the following result:

Proposition 5. *Under the assumption of* Theorem 3 *the condition number of the system (4.8) for the expansion (4.7) is bounded from above and below by constants independent of h, the supremum of the diameters of the circles containing $K(\mathcal{S}_n)$.*

For this reason, the barycentric expansion is more suitable in practical calculations.

The case of the scaled expansion In [21] it is shown:

Proposition 6. *The condition number of system (4.8) for expansion (4.6) is invariant with respect to isotropic grid scaling. In particular it does not depend from h.*

This means that system (4.8) can be solved stably even on fine grids. Which is a requirement for practical computations.

The scaling technique is simpler than using the barycentric expansion but for unisotropically scaled grids it is not clear whether it is sufficient or not.

5   The explicit calculation of the reconstruction: Mühlbach expansions, Tschebyscheff systems and divided differences

In this section, we sketch the main results of [6] and provide the link between the previous section and divided differences. In fact, when computing the reconstruction polynomial by the reconstruction-via-primitive technique, it is surprising to see that the formulae look very similar to divided differences formula, even though an additional derivative has been taken. In this section we explain this fact and show that the algorithm is indeed the same as the one based on divided differences.

We begin with a definition and stay as close as possible to the notations introduced by Mühlbach in [31]. The functional space V is in practice the space of continuous functions on Ω or $L^1(\Omega)$.

Definition 7. The functions $\varphi_1, \ldots, \varphi_n \in V$ form an **I-Tschebyscheff system** on Ω, if the condition

$$V \begin{pmatrix} \varphi_1, \ldots, \varphi_n \\ \lambda_1, \ldots, \lambda_n \end{pmatrix} := \begin{vmatrix} \langle \lambda_1, \varphi_1 \rangle & \cdots & \langle \lambda_1, \varphi_n \rangle \\ \vdots & \ddots & \vdots \\ \langle \lambda_n, \varphi_1 \rangle & \cdots & \langle \lambda_n, \varphi_n \rangle \end{vmatrix} \neq 0$$

holds for the set of linear forms (information) $\mathbf{I} = (\lambda_1, \ldots, \lambda_n)^T$.

We refer to $V \begin{pmatrix} \varphi_1, \ldots, \varphi_n \\ \lambda_1, \ldots, \lambda_n \end{pmatrix}$ as the **generalized Van der Monde determinant**.

If $\{\underline{x}_1, \ldots, \underline{x}_n\}$ denotes a set of n distinct points in Ω and $\lambda_i = \delta_{\underline{x}_i}$ then we are back at the classical interpolation condition. In the type of applications we are most interested in, $\{C_1, \ldots, C_n\}$ denotes a set of pairwise disjoint control volumes and $\mathbf{I}$ is the information about cell averages $\langle \lambda_i, \Phi \rangle := A(C_i)\Phi$ of $\Phi \in V$. In §4.4, the functions $\phi_1, \cdots, \phi_n$ where either of the type $(x - x_0)^i(y - y_0)^j$ for (4.5) or $\Lambda_1^i \Lambda_2^j$ for (4.7). The Van der Monde condition has already been introduced in (4.3). However, the results we present here can be applied to more general problems.

The simple rules on linear systems enable us to get the following

Lemma 8. *The following three statements are equivalent.*

1. $\varphi_1, \ldots, \varphi_n$ *constitute an* **I**-*Tschebyscheff system on* Ω.
2. *For every function* $\Phi : \Omega \to \mathbb{R}$ *and* $\mathbf{I} = (\lambda_1, \ldots, \lambda_n)^T$, *there exists a unique linear combination*

$$
p_n\Phi := p\Phi \begin{bmatrix} \varphi_1, \ldots, \varphi_n \\ \lambda_1, \ldots, \lambda_n \end{bmatrix} := \sum_{i=1}^{n} \alpha_i \varphi_i
$$

of $\varphi_1, \ldots, \varphi_n$ *satisfying the* **recovery conditions**

$$
\langle \lambda_i, p_n\Phi \rangle = \langle \lambda_i, \Phi \rangle, \quad i = 1, \ldots, n. \tag{5.1}
$$

Note that the conditions (5.1) are, in the case of cell averages, the conditions (4.2). We are now ready to define the generalized divided differences.

Definition 9. The coefficients α_i in the representation

$$
p_n\Phi = \sum_{i=1}^{n} \alpha_i \varphi_i,
$$

i.e. the coefficients corresponding to the φ_i's, are called **generalized divided differences** of Φ with respect to the **I**-Tschebyscheff system $\varphi_1, \ldots, \varphi_n$ and will be denoted by

$$
\alpha_i := \begin{bmatrix} \varphi_1, \ldots, \varphi_n \\ \lambda_1, \ldots, \lambda_n \end{bmatrix} \begin{matrix} \Phi \\ i \end{matrix} \Bigg].
$$

The function

$$
r_n\Phi := r\Phi \begin{bmatrix} \varphi_1, \ldots, \varphi_n \\ \lambda_1, \ldots, \lambda_n \end{bmatrix} := \Phi - p\Phi \begin{bmatrix} \varphi_1, \ldots, \varphi_n \\ \lambda_1, \ldots, \lambda_n \end{bmatrix}
$$

is called the **recovery error function**.

Lemma 10. *With the notation of* Definition 7 *the representation*

$$
\begin{bmatrix} \varphi_1, \ldots, \varphi_n \\ \lambda_1, \ldots, \lambda_n \end{bmatrix} \begin{matrix} \Phi \\ i \end{matrix} \Bigg] =
$$

$$
\frac{V \begin{pmatrix} \varphi_1, \ldots, \varphi_{k-1}, & \Phi, & \varphi_{k+1}, \ldots, \varphi_n \\ \lambda_1, \ldots, \lambda_{k-1}, & \lambda_k, & \lambda_{k+1}, \ldots, \lambda_n \end{pmatrix}}{V \begin{pmatrix} \varphi_1, \ldots, \varphi_n \\ \lambda_1, \ldots, \lambda_n \end{pmatrix}}
$$

for all $k = 1, \ldots, n$ *as well as*

$$
r_n\Phi(\underline{x}) = \frac{V \begin{pmatrix} \varphi_1, \ldots, \varphi_n, & \Phi \\ \lambda_1, \ldots, \lambda_n, & \delta_{\underline{x}} \end{pmatrix}}{V \begin{pmatrix} \varphi_1, \ldots, \varphi_n \\ \lambda_1, \ldots, \lambda_n \end{pmatrix}}
$$

hold.

We now generalize Mühlbach's Newton-type interpolation formula constructed in [31] to the case of the recovery problem.

Theorem 11. *Let $m < n$ be natural numbers. Suppose that $\varphi_1, \ldots, \varphi_n$ constitute an $\mathbf{I}$-Tschebyscheff system with $\mathbf{I} = (\lambda_1, \ldots, \lambda_n)^T$ such that its subsystem $\varphi_1, \ldots, \varphi_m$ is again an $\mathbf{I}$-Tschebyscheff system with respect to $\lambda_1, \ldots, \lambda_m$. Then, for every function $\Phi \in V$ and every $\underline{x} \in \Omega$ there holds the* **Mühlbach expansion**

$$\Phi \begin{bmatrix} \varphi_1, & \ldots, & \varphi_n \\ \lambda_1, & \ldots, & \lambda_n \end{bmatrix} (\underline{x}) = p\Phi \begin{bmatrix} \varphi_1, & \ldots, & \varphi_m \\ \lambda_1, & \ldots, & \lambda_m \end{bmatrix} (\underline{x}) +$$
$$\sum_{k=m+1}^{n} \begin{bmatrix} \varphi_1, & \ldots, & \varphi_n & \Phi \\ \lambda_1, & \ldots, & \lambda_n & k \end{bmatrix} r\varphi_k \begin{bmatrix} \varphi_1, & \ldots, & \varphi_m \\ \lambda_1, & \ldots, & \lambda_m \end{bmatrix} (\underline{x}).$$

It would be desirable for numerical purposes to compute the generalized divided differences not by means of the clumsy determinant formula given in Lemma 10 but only from previously calculated divided differences as in the Newton polynomials in $\mathbb{R}$.

5.1   A linear system for divided differences

Before presenting Mühlbach's recurrence relation for generalized divided differences in the recovery case we present a related result which already allows the computation of the divided differences as solutions of linear systems of equations.

Theorem 12. *For any $\Phi : \Omega \to \mathbb{R}$, $\Phi \in V$, the generalized divided differences*

$$\alpha_k = \begin{bmatrix} \varphi_1, & \ldots, & \varphi_n & \Phi \\ \lambda_1, & \ldots, & \lambda_n & k \end{bmatrix}, k = m+1, \ldots, n,$$

are uniquely determined as solution of the system of $n - m$ linear equations

$$\left\langle \lambda_i, \sum_{k=m+1}^{n} \alpha_k \cdot r\varphi_k \begin{bmatrix} \varphi_1, & \ldots, & \varphi_m \\ \lambda_1, & \ldots, & \lambda_m \end{bmatrix} \right\rangle =$$
$$\left\langle \lambda_i, r\Phi \begin{bmatrix} \varphi_1, & \ldots, & \varphi_m \\ \lambda_1, & \ldots, & \lambda_m \end{bmatrix} \right\rangle,$$

for $i = m+1, \ldots, n$.

5.2   A recurrence relation for divided differences

The method of computing the generalized divided differences described in Theorem 12 does not use previously computed divided differences exclusively but requires the computation of recovery error terms. In transfering Mühlbach's generalized recurrence relation for divided differences in interpolation problems to the recovery case we finish the description of Mühlbach expansions.

Theorem 13. *Let $m < n$ be integers and suppose that $\varphi_1, \ldots, \varphi_n$ constitute an $\mathbf{I}$-Tschebyscheff system, $\mathbf{I} = (\lambda_1, \ldots, \lambda_n)^T$, such that its subsystem $\varphi_1, \ldots, \varphi_m$ is also an $\mathbf{I}$-Tschebyscheff system with regard to $\lambda_1, \ldots, \lambda_m$. For $\Phi \in V$ let $\underline{a}$ denote the vector*

$$
\underline{a} = \begin{bmatrix} \alpha_{m+1} \\ \vdots \\ \alpha_n \end{bmatrix} = \begin{bmatrix} \begin{bmatrix} \varphi_1, \ldots, \varphi_n \\ \lambda_1, \ldots, \lambda_n \end{bmatrix}\begin{array}{c} \Phi \\ m+1 \end{array}\end{bmatrix} \\ \vdots \\ \begin{bmatrix} \varphi_1, \ldots, \varphi_n \\ \lambda_1, \ldots, \lambda_n \end{bmatrix}\begin{array}{c} \Phi \\ n \end{array}\end{bmatrix} \in \mathbb{R}^{n-m}
$$

of generalized divided differences. Then $\underline{a}$ is uniquely determined as solution of the linear system

$$
\underline{\underline{C}}\,\underline{a} = \overline{u},
$$

where $\underline{\underline{C}} \in \mathbb{R}^{m(n-m) \times (n-m)}$. If $\underline{\underline{C}} = (\underline{c}_1, \ldots, \underline{c}_{n-m})$ is the representation of $\underline{\underline{C}}$ in terms of the column vectors $\underline{c}_k$, then the k-th column vector is given by

$$
\underline{c}_k = \begin{bmatrix} \begin{bmatrix} \begin{bmatrix} \varphi_1, \ldots, \varphi_m \\ \lambda_2, \ldots, \lambda_{m+1} \end{bmatrix}\begin{array}{c} \varphi_{m+k} \\ 1 \end{array}\end{bmatrix} - \begin{bmatrix} \varphi_1, \ldots, \varphi_m \\ \lambda_1, \ldots, \lambda_m \end{bmatrix}\begin{array}{c} \varphi_{m+k} \\ 1 \end{array}\end{bmatrix} \\ \vdots \\ \begin{bmatrix} \begin{bmatrix} \varphi_1, \ldots, \varphi_m \\ \lambda_{n-m+1}, \ldots, \lambda_n \end{bmatrix}\begin{array}{c} \varphi_{m+k} \\ 1 \end{array}\end{bmatrix} - \begin{bmatrix} \varphi_1, \ldots, \varphi_m \\ \lambda_{n-m}, \ldots, \lambda_{n-1} \end{bmatrix}\begin{array}{c} \varphi_{m+k} \\ 1 \end{array}\end{bmatrix} \\ \vdots \\ \begin{bmatrix} \begin{bmatrix} \varphi_1, \ldots, \varphi_m \\ \lambda_2, \ldots, \lambda_{m+1} \end{bmatrix}\begin{array}{c} \varphi_{m+k} \\ m \end{array}\end{bmatrix} - \begin{bmatrix} \varphi_1, \ldots, \varphi_m \\ \lambda_1, \ldots, \lambda_m \end{bmatrix}\begin{array}{c} \varphi_{m+k} \\ m \end{array}\end{bmatrix} \\ \vdots \\ \begin{bmatrix} \begin{bmatrix} \varphi_1, \ldots, \varphi_m \\ \lambda_{n-m+1}, \ldots, \lambda_n \end{bmatrix}\begin{array}{c} \varphi_{m+k} \\ m \end{array}\end{bmatrix} - \begin{bmatrix} \varphi_1, \ldots, \varphi_m \\ \lambda_{n-m}, \ldots, \lambda_{n-1} \end{bmatrix}\begin{array}{c} \varphi_{m+k} \\ m \end{array}\end{bmatrix} \end{bmatrix},
$$

while the right hand side $\overline{u} \in \mathbb{R}^{m(n-m)}$ is given by

$$
\overline{u} = \begin{bmatrix} \begin{bmatrix} \begin{bmatrix} \varphi_1, \ldots, \varphi_m \\ \lambda_2, \ldots, \lambda_{m+1} \end{bmatrix}\begin{array}{c} \Phi \\ 1 \end{array}\end{bmatrix} - \begin{bmatrix} \varphi_1, \ldots, \varphi_m \\ \lambda_1, \ldots, \lambda_m \end{bmatrix}\begin{array}{c} \Phi \\ 1 \end{array}\end{bmatrix} \\ \vdots \\ \begin{bmatrix} \begin{bmatrix} \varphi_1, \ldots, \varphi_m \\ \lambda_{n-m+1}, \ldots, \lambda_n \end{bmatrix}\begin{array}{c} \Phi \\ 1 \end{array}\end{bmatrix} - \begin{bmatrix} \varphi_1, \ldots, \varphi_m \\ \lambda_{n-m}, \ldots, \lambda_{n-1} \end{bmatrix}\begin{array}{c} \Phi \\ 1 \end{array}\end{bmatrix} \\ \vdots \\ \begin{bmatrix} \begin{bmatrix} \varphi_1, \ldots, \varphi_m \\ \lambda_2, \ldots, \lambda_{m+1} \end{bmatrix}\begin{array}{c} \Phi \\ m \end{array}\end{bmatrix} - \begin{bmatrix} \varphi_1, \ldots, \varphi_m \\ p\lambda_1, \ldots, \lambda_m \end{bmatrix}\begin{array}{c} \Phi \\ m \end{array}\end{bmatrix} \\ \vdots \\ \begin{bmatrix} \begin{bmatrix} \varphi_1, \ldots, \varphi_m \\ \lambda_{n-m+1}, \ldots, \lambda_n \end{bmatrix}\begin{array}{c} \Phi \\ m \end{array}\end{bmatrix} - \begin{bmatrix} \varphi_1, \ldots, \varphi_m \\ \lambda_{n-m}, \ldots, \lambda_{n-1} \end{bmatrix}\begin{array}{c} \Phi \\ m \end{array}\end{bmatrix} \end{bmatrix}.
$$

5.3   Some examples

An interpolatory example We explain the notions introduced above by means of a simple example corresponding to $\mathbf{I}\Phi = (\Phi(\underline{x}_1), \Phi(\underline{x}_2), \Phi(\underline{x}_3))^T$, $\underline{x}_1, \underline{x}_2, \underline{x}_3$ forming a non-degenerated triangle in $\mathbb{R}^2$. The Mühlbach expansion is sought with respect to the system $\varphi_1, \varphi_2, \varphi_3$ defined by $\varphi_i(\underline{x}) := a_{00}^i + a_{10}^i x_1 + a_{01}^i x_2, \varphi_i(\underline{x}_k) = \delta_i^k, i, k = 1, 2, 3$. Thus, we consider the linear finite element functions on the triangle spanned by $\underline{x}_1, \underline{x}_2, \underline{x}_3$ interpolating Φ at these points, i.e. the three linear functions taking the value 1 on one node of the triangle while vanishing on the remaining two nodes. We would expect the interpolant to be the function $\Phi(\underline{x}_1)\varphi_1(\underline{x}) + \Phi(\underline{x}_2)\varphi_2(\underline{x}) + \Phi(\underline{x}_3)\varphi_3(\underline{x})$.

The Mühlbach expansion can be written in the form

$$p\Phi \begin{bmatrix} \varphi_1, & \varphi_2, & \varphi_3 \\ \delta_{\underline{x}_1}, & \delta_{\underline{x}_2}, & \delta_{\underline{x}_3} \end{bmatrix} (\underline{x}) =$$
$$p\Phi \begin{bmatrix} \varphi_1 \\ \delta_{\underline{x}_1} \end{bmatrix} (\underline{x})$$
$$+ \begin{bmatrix} \varphi_1, & \varphi_2, & \varphi_3 & \Phi \\ \delta_{\underline{x}_1}, & \delta_{\underline{x}_2}, & \delta_{\underline{x}_3} & 2 \end{bmatrix} r\varphi_2 \begin{bmatrix} \varphi_1 \\ \delta_{\underline{x}_1} \end{bmatrix} (\underline{x})$$
$$+ \begin{bmatrix} \varphi_1, & \varphi_2, & \varphi_3 & \Phi \\ \delta_{\underline{x}_1}, & \delta_{\underline{x}_2}, & \delta_{\underline{x}_3} & 3 \end{bmatrix} r\varphi_3 \begin{bmatrix} \varphi_1 \\ \delta_{\underline{x}_1} \end{bmatrix} (\underline{x}).$$

According to Lemma 8 the function $p\Phi \begin{bmatrix} \varphi_1 \\ \delta_{\underline{x}_1} \end{bmatrix} = p_1\Phi$ satisfies the recovery condition (i.e. interpolation condition) $p_1\Phi(\underline{x}_1) = \Phi(\underline{x}_1)$ and can be written in the form $p_1\Phi(\underline{x}) = \alpha_1\varphi_1(\underline{x})$. Thus, $p_1\Phi(\underline{x}_1) = \Phi(\underline{x}_1) = \alpha_1\varphi_1(\underline{x}_1) = \alpha_1$ and therefore $p_1\Phi(\underline{x}) = \Phi(\underline{x}_1)\varphi_1(\underline{x})$. According to Lemma 10 we furthermore have

$$\begin{bmatrix} \varphi_1, & \varphi_2, & \varphi_3 & \Phi \\ \delta_{\underline{x}_1}, & \delta_{\underline{x}_2}, & \delta_{\underline{x}_3} & 2 \end{bmatrix} = \Phi(\underline{x}_2),$$

Analogously,

$$\begin{bmatrix} \varphi_1, & \varphi_2, & \varphi_3 & \Phi \\ \delta_{\underline{x}_1}, & \delta_{\underline{x}_2}, & \delta_{\underline{x}_3} & 3 \end{bmatrix} = \Phi(\underline{x}_3).$$

Also according to Lemma 10 we have $r\varphi_2 \begin{bmatrix} \varphi_1 \\ \delta_{\underline{x}_1} \end{bmatrix} (\underline{x}) = \varphi_2(\underline{x})$ and $r\varphi_3 \begin{bmatrix} \varphi_1 \\ \delta_{\underline{x}_1} \end{bmatrix} (\underline{x}) = \varphi_3(\underline{x})$. Therefore, the Mühlbach expansion results in $p_3\Phi(\underline{x}) = \sum_{i=1}^{3} \Phi(\underline{x}_i)\varphi_i(\underline{x})$, which indeed is the required interpolant.

An efficient algorithm for quadratic polynomial recovery We show that the problem of recovering a quadratic polynomial from cell averages can be broken up into two 3×3-systems instead of solving one 6×6-system of equations. A quadratic polynomial is sought on the triangulation shown in Figure 13. With each of the nodes $\underline{x}_i, i = 1, \ldots, 6$, we associate the linear functional $\lambda_i = \langle A(C_i), \cdot \rangle$. As already explained, a direct computation of a recovery polynomial $p(\underline{x}) := \sum_{|\alpha|\leq 2} a_\alpha \underline{x}^\alpha$ satisfying the recovery conditions $\langle$ A $(C_i), p\rangle = \langle$ A $(C_i), \Phi\rangle$ for $i = 1, \ldots, 6$, would require the solution of a 6×6-system for the unknown coefficients a_α. The determinant of the coefficient matrix of this system is of generalized Van der Monde-type, thus we have high condition numbers resulting in numerical problems during the solution process. Anyway, it would be desirable to break down the computation into smaller subproblems. We now show that Mühlbach expansions can accomplish this task. We assume that the triangulation is chosen such that polynomials of degree not exceeding two form an **I**-Tschebyscheff system. Note that this can always be assured in practice.

On the triangle $T_{\min}$ shown in Figure 13 we compute three linear finite element functions $\Lambda_1, \Lambda_2, \Lambda_3$ according to the recovery conditions A $(C_i)\Lambda_j = \delta_i^j$, $i, j \in \{1, 2, 3\}$. According to our notations these three functions comprise the recovery function

$$p_3\Phi = p\Phi \begin{bmatrix} \Lambda_1, & \Lambda_2, & \Lambda_3 \\ A\ (C_1), & A\ (C_2), & A\ (C_3) \end{bmatrix}$$

which can be thought of as the linear part in a Mühlbach expansion. Defining the remaining functions $\Lambda_4(\underline{x}) = \Lambda_1\Lambda_2$, $\Lambda_5(\underline{x}) = \Lambda_1^2$, $\Lambda_6(\underline{x}) = \Lambda_2^2$, the complete Mühlbach expansion is then given as

$$p\Phi \begin{bmatrix} \Lambda_1, & \cdots, & \Lambda_6 \\ A\ (C_1), & \ldots, & A\ (C_6) \end{bmatrix} (\underline{x}) =$$
$$p\Phi \begin{bmatrix} \Lambda_1, & \Lambda_2, & \Lambda_3 \\ A\ (C_1), & A\ (C_2), & A\ (C_3) \end{bmatrix}$$
$$+ \sum_{k=4}^{6} \alpha_k \cdot r\Lambda_k \begin{bmatrix} \Lambda_1, & \Lambda_2, & \Lambda_3 \\ A\ (C_1), & A\ (C_2), & A\ (C_3) \end{bmatrix},$$

where the α_k's denote the generalized divided differences again. Due to the recovery properties of the linear functions $\Lambda_i, i = 1, 2, 3$, it follows that

$$V \begin{pmatrix} \Lambda_1, & \Lambda_2, & \Lambda_3 \\ A\ (C_1), & A\ (C_2), & A\ (C_3) \end{pmatrix} = 1.$$

Thus, according to Lemma 10 we obtain

$$r_3\Lambda_k(\underline{x}) = \Lambda_k(\underline{x}) - \sum_{\ell=1}^{3} \Lambda_\ell(\underline{x})\langle$ A $(C_\ell), \Lambda_k\rangle,$$

for $k \in \{4, 5, 6\}$, and $r_3\Phi = \Phi(\underline{x}) - \sum_{\ell=1}^{3} \Lambda_k(x)\langle$ A $(C_\ell), \Phi\rangle$. Following Theorem 12 the divided differences can be computed as solutions of a 3×3-system. In our case it is easy to verify that the system has the form $\underline{\underline{A}}\,\underline{a} = \bar{u}$ with $\underline{\underline{A}} = (a_{ij})_{1 \leq i, j \leq 3}$ given by

$$a_{ij} = A\ (C_{i+3})\Lambda_{j+3} - \sum_{\ell=1}^{3} A\ (C_{i+3})\Lambda_\ell \cdot \langle$ A $(C_\ell), \Lambda_{j+3}\rangle,$$

the right hand side $\bar{u} = (b_1, b_2, b_3)^T$ is

$$b_i = A\ (C_{i+3})\Phi - \sum_{\ell=1}^{3} A\ (C_{i+3})\Lambda_\ell \cdot \langle$ A $(C_\ell), \Phi\rangle,$$

and $\underline{a} = (\alpha_1, \alpha_2, \alpha_3)^T$. Thus, the process of recovery can be conveniently broken down into smaller subproblems by using Mühlbach expansions.

Remark 14. In [3], the computation of the polynomial expansion Φ was carried out by using the error $r_3\Phi$, as here. Instead of introducing the error functions $r_3\Lambda_4$, $r_3\Lambda_5$ and $r_3\Lambda_6$, the error $r_3\Phi$ was expanded in terms of the Λ_i's, $i = 1, \ldots, 6$. Then the 6×6-system is reduced to a 3×3 one. It turns out that the method of [3] exactly reduces to the one presented here for degree 2. For higher degree, say degree r, the method of [3] needs the solution of two linear systems. One of them is a $(r + 1) \times (r + 1)$ system, the other one is a $\frac{(r+1)(r+2)}{2} \times \frac{(r+1)(r+2)}{2}$ system. It is clear that the present method is much more efficient in general.

6   The ENO reconstruction

6.1   ENO on general meshes

In [1], we have found that only a few stencils were indeed necessary to achieve an essentially non-oscillatory reconstruction of a piecewise smooth function on a triangular mesh. This set has to be as isotropic as possible. Moreover, the ENO reconstruction was found to achieve the expected order of accuracy for smooth functions, even on very irregular meshes. In what follows, a_{ij} always stands for any of the coefficients of the reconstruction P in the natural basis, $\{(x - x_0)^i(y - y_0)^j\}$.

Then we can apply the procedure of [1] in a straightforward manner. Let us describe our procedure for reconstruction up to third order: (i) We start from a given cell, C_0, assigned to a point of $\mathcal{M}$, say (x_0, y_0) ; (ii) Consider all the triangles having (x_0, y_0) as a vertex, and choose the one, say T_{min}, that minimize $\sum_{i+j=1} |a_{ij}|$. Here, S_1 is the set of control volumes located around the vertices of T_{min}, (see Figure 13). For a regular unstructured mesh, there are six possible triangles. (iii) Consider T_{min}. For each of its edges, consider the three triangles, T_1, T_2, T_3 as in Figure 13-a. We choose the configuration that minimize the sum

$$\sum_{i+j=2} |a_{ij}|.$$

What can be done for fourth (and higher) order reconstruction is explained in [1].

6.2   Numerical examples

We have performed several tests on the second, third and fourth order ENO interpolation and ENO reconstruction, but we only report the third order results since they are *a priori* more computationally interesting. In particular, we intend to check numerically that the expected order of accuracy is in fact reached for smooth functions.

In all these examples, we have assumed that the control volumes are elements of the dual mesh. The practical calculations of the averages in these control volumes have been performed with a 5-th order quadrature formula ([37], Table 4.1, p.184).

The tests on smooth functions are performed on:

$$u(x,y) = \cos(2\pi(x^2 + y^2)). \tag{6.1}$$

All the error estimates have been obtained on irregular meshes as the one presented in Figure 13. The main difference between such a mesh and the regular structured one is that the number of triangles each node belongs to is different. We also have done the same tests with regular meshes, and we have not seen any degradation of the convergence.

The locally smooth function we have chosen is obtained by a modification of the one used by Harten in [15] for example : if ϕ is any angle, let a function f_ϕ be defined by:

$$f_\phi(x,y) = \begin{cases} -r\sin\left(1.5\pi r^2\right) & ; r \leq -\frac{1}{3} \\ 2r - 1 + \frac{1}{6}\sin\left(3\pi r\right) & ; r \geq \frac{1}{3} \\ |\sin\left(2\pi r\right)| & ; |r| < \frac{1}{3} \end{cases}, \tag{6.2}$$

where $r = x - \frac{y}{\tan(\phi)}$. From f_ϕ we finally define u to be:

$$u(x,y) := \begin{cases} f_{\sqrt{\pi/2}}(x,y) & ; x \leq \frac{1}{2}\cos\pi y \\ f_{-\sqrt{\pi/2}}(x,y) + \cos\left(2\pi y\right) & ; x > \frac{1}{2}\cos\pi y \end{cases} \tag{6.3}$$

The function defined by (6.2)-(6.3) has discontinuities in the function itself and its first order derivatives; some of the discontinuities are straight lines (never aligned to the mesh), one is a curved line where the jump changes from one point to another. Last, the behavior of u is basically one-dimensional on the left of the curve $x = \cos\pi y/2$ and really two-dimensional on the right.

A plot of this function is given in Figure 13-(B). One should obtain straight lines and smooth transitions at discontinuities contrary to what is shown in the Figure: this is an effect of the plotting procedure in which linear finite element hat functions are used for interpolation purposes.

Results on the smooth function We have displayed in Table 13.1 the L^∞-error of the third-order ENO reconstruction. The experiments have been done in two different contexts. The column "(a)" of Table 13.1 corresponds to different meshes that have been generated independently. In this case, the constant C of Theorem 1 is different for each mesh, so that the slope -3 has to be expected in the mean only. The column "(b)" of Table 13.1 corresponds to meshes that have been successively refined: the same constant C appears, and the slope -3 is recovered much better. Here h is the maximal radius of the circumscribed circles of the triangles, r_c is a number such that the error is proportional to h^{r_c}.

Results on the nonsmooth function In Figure 13-(A), we have displayed the nodal values of the third-order ENO reconstruction for the mesh shown in Figure 13. There are no oscillations in the reconstruction. In [1], the same representation is given for the fourth order reconstruction, and the only visible difference is a better resolution of the area surrounding the triple points.

Where the function is smooth, we should recover the asymptotic order of convergence obtained for a smooth function. In order to check this, we have computed the error between the reconstruction and the exact function u at different points of the line $y = 0$, namely at $x = 0.4, 0.2, 0.1, 0, -0.2, -0.5, -0.75$. The results are presented in Table 13 for third-order accuracy. We get what is expected. In particular, the point $x = 0$ is on the line where u is continuous but its first-order derivative has a jump, so that only a first-order approximation is obtained in any case. Elsewhere, a third order accuracy is recovered.

In [1], another selection procedure has also been proposed. It includes a much richer set of stencils but no real improvement has been noticed. From all our experiments, we can conclude that the choice discussed here is indeed sufficient.

7   Weighted ENO reconstruction

Although the results for ENO reconstruction are reasonable, critical investigations show two weak points of the approach:

1. In smooth regions the accuracy is not as good as for TVD schemes.
2. Convergence for steady state flows is usually not achieved.

In this section we are facing these problems and show how they can be circumwented.

7.1   Motivation of WENO reconstruction

The main idea of ENO schemes is to compute several candidates P_i for a reconstruction P and to choose that one with the lowest oscillation. If we assume that we are in a smooth region of the flow then non of the P_i does really oscillate. In this context choosing that candidate P_i with the lowest oscillation means to choose the flattest reconstruction or to maximize dissipation. This explains why the accuracy is not that good in smooth regions.

Furthermore it is obvious that if there are enough candidates P_i then there will be more than one candidate with a comparable low oscillation.

Thus, even small changes of the data will force a switch from one candidate to another. This digital switching prevents convergence of the scheme for steady state flows.

Both drawbacks can be removed or at least reduced by modifying the ENO scheme in the following way: Instead of digitally selecting the least oscillating reconstruction we use a weighted sum:

$$P := \sum_i w_i P_i.$$

The positive weights w_i with $\sum_i w_i = 1$ are choosen such that w_i is small if the oscillation of P_i is high and w_i is larger for less oscillating P_i. This scheme is then called weighted ENO scheme (WENO). It was introduced for the one-dimensional case in [40,24] and applied to the case of unstructured grids in two dimensions in [21].

7.2   Choice of weights

For the computation of weights, it has to be clarified how *the oscillation* of P is measured. From theorem 3 one comes to the conclusion that $\sum_{i+j=n} |a_{ij}|$ should be used. However, numerical tests (see [21]) have demonstrated that this oscillation measure is not well suited as a base for the weights. Much better results are obtained using the following quantity:

$$osc(P) := \left(\sum_{1 \le i+j \le n} \int_C h^{2(i+j)-4} \|D^{i+j} P(x)\|^2 dx \right)^{\frac{1}{2}},$$

where C is the cell P has to computed for and $h = \sqrt{|C|}$.

The oscillation measures $osc(P_i)$ are then used to compute the weights as follows:

$$\tilde{w}_i := (\epsilon + osc(P_i))^{-r},$$

where r is positive, and

$$w_i := \frac{\tilde{w}_i}{\sum_i \tilde{w}_i}.$$

Note, that if there is exactly one P_i with a maximum oscillation then the WENO scheme tends to the classical ENO scheme if r tends to infinity. On the other hand, if r tends to zero then the oscillation will not be taken into account for the weights, which means that the scheme will become an unstable central scheme.

Numerical tests showed that $r = 4$ is large enough to hold the scheme stable even for flows with strong discontinuities and small enough to obtain a significant improvement over the ENO scheme for smooth flows.

7.3   Required modifications of the reconstruction algorithm

It would be a violation of the main idea to use the hierachical recovery algorithm described in §5 for third order reconstruction which leads to a number of only three stencils for polynomial degree two. For WENO reconstruction the linear part of the Mühlbach expansion can not be fixed. Instead, the stencils for all the triangles and not only one have to be taken into account. For each of the triangles Mühlbach expansions are used.

7.4   A stencil selection algorithm that does not need triangles

In the last sections we have described the ENO and WENO reconstruction algorithm using a triangulation to select stencils. This is no principal restriction on the kind of used control volumes as was stated before because a triangulation of the control volumes' centers can always be constructed to obtain the required topological information.

However, this technique may be impractical and one may wish to select stencils for a finite volume grid without the need of a triangulation.

Grids like the dual mesh of a triangulation and also grids obtained from a dual mesh by fusing together cells own a nice topological property: If the boundaries of two cells have a common point then they already have a common edge. This property is not given for primary triangular grids and also not for quadrilateral grids where cells can touch at single points.

In the following we call cell b a *neighbour* of cell a if their boundaries have have a common edge. We say cell b is *touching* cell a if their boundaries have a common edge. With this definition the topological property described above means that for this kind of grids touching cells are already neighbours.

For this kind of grids the stencil selection algorithm described in [21] is used:

Polynomial degree 1: For a polynomial degree $n = 1$ the required stencil size is three. We select all that sets of three cells $\{C_\ell, C_a, C_b\}$ as stencils for cell C_ℓ which have the following properties:

- C_a is a neighbor of C_ℓ, and
- C_b is a neighbor of C_ℓ and of C_a.

Polynomial degree 2: For a polynomial degree $n = 2$ the required stencil size is six. We select all that sets of six cells $\{C_\ell, C_a, C_b, C_c, C_d, C_e\}$ as stencils for cell C_ℓ which have the following properties: First, $\{C_\ell, C_a, C_b\}$ has to be a selected stencil for polynomial degree $n = 1$. Second, C_c, C_d and C_e have to fulfill one of the following three conditions (see figure 13 for an example of each type. C_ℓ is dark shaded):

1. Central stencil:
    - C_c is a neighbor of C_ℓ and of C_a, and

- C_d is a neighbor of C_ℓ and of C_b, and
- C_e is a neighbor of C_ℓ and of either C_c or C_d.
2. Almost central stencil:
   - C_c is a neighbor of C_ℓ and of C_a, and
   - C_d is a neighbor of C_ℓ and of C_b, and
   - C_e is a neighbor of C_a and of C_b.
3. One-sided stencil:
   - C_c is a neighbor of C_a and of C_b, and
   - C_d is a neighbor of C_a and of C_c, and
   - C_e is a neighbor of C_b and of C_c.

Note, that apart from the *central stencils* this results to the same stencils as those described in §5.3.

8   Other recovery techniques

In all what preceded, we have worked with piecewise polynomial reconstructions. This is quite standard thanks to the ease of computing polynomials. This is also accurate since in the regular case we have error estimates. However, one might wonder whether this is optimal in the sense of minimizing the error between the reconstruction and the function u to be reconstructed. Since the latter is known only through its average values on the control volume, it is better to ask that the distance between the reconstruction and the space in which u lives, is minimized provided some linear constraints are added.

Let us give a simple example. It is well known that the Lagrange interpolation is not optimal when we want to interpolate data while minimizing other quantities, like a norm of derivatives. Let $a = x_0 < x_1 < \cdots < x_{n-1} < x_n = b$ and $y_i, i = 1, \cdots, n$ and if one wishes to minimize $\int_a^b [f'']^2(x)dx$ in the space of continuously twice differentiable functions with the constraints $f(x_i) = y_i$, the answer is given by cubic splines.

Since accuracy as well as robustness of such approximations applied to hyperbolic conservation laws depend mainly on the recovery algorithm it makes sense to ask for recovery algorithms satisfying an optimality condition. It turns out that the solution to this class of problems can be found in an abstract setting in the papers by Golomb and Weinberger [13] and Micchelli and Rivlin [29], in which a theory of optimal recovery is developed. Within this theory one is able to show that polynomial recovery is only optimal in a trivial sense.

The idea of applying the theory of optimal recovery to numerical approximations of differential equations goes back to Morton and his co-workers in 1988, see [7]. They considered finite element approximations and used piecewise polynomial recovery to get information about point values and derivatives of the unknown solution.

The details of these recovery techniques can be found in [20,36].

An example is the following. Instead of taking polynomial expansion (4.1), the following expansion is considered (this corresponds to a spline in a Beppo–Levi space, the so-called thin-plate spline)

$$R(x) = \sum_{j=0}^{M-1} \lambda_j \left\langle A(C_{i_j})^y, \left(|x-y|^2 \log(|x-y|)\right)\right\rangle$$

$$+ a_{00} + a_{10}x + a_{01}y$$

(8.1)

where the averaging process is done with respect to the variable y, and the integrals are computed with a quadrature formula. The cells C_{i_j} belong to a stencil S_i around the node M_i of the mesh. It is constructed in the same spirit as before. The constraints are

$$\left\langle A\left(C_{i_j}\right), R\right\rangle = \left\langle A\left(C_{i_j}\right), u\right\rangle, \quad k = 1, \cdots, M-1,$$

where C_{i_0} is the control volume associated with M_i, and

$$\text{for all } p \in \{1, x, y\}: \quad \sum_{j=1}^{M-1} \lambda_j \left\langle A\left(C_{i_j}\right), p\right\rangle = 0.$$

This gives a $(M+3) \times (M+3)$ linear system that is solvable if the stencil S_i contains an admissible stencil of 3 elements for the linear reconstruction. The expansion (8.1) minimizes

$$J(u) = \int_{\mathbb{R}^2} \sum_{i=0}^{2} \binom{2}{i} \left(\frac{\partial^2 u}{\partial x^i \partial y^{2-i}}\right)^2 dx dy.$$

If $M = \#S_i = 3$, R is a linear polynomial. Thus, in practical applications, a stencil of 4 elements is taken. For ENO applications, the reconstruction is performed on the stencil that has the smallest total variation, computed once more by a quadrature formula.

Several numerical applications have been tried with this technique, in particular in [36], on rotating cone problems and the Collela and Woodward test case of supersonic flow in a channel with foreward facing step. Improvements with respect to linear reconstruction technique are reported there, they are particularly pronounced for the rotating cone problem.

9   A class of high order numerical schemes for compressible flow simulations

We have applied the polynomial reconstruction method to various test cases, with polynomial of degree 2, on various test cases. Here, as said before, the physical variables are approximated in the setting of §2.

We have reduced the order of accuracy of the reconstruction for cells that are too close to the boundary. For them, a proper calculation of the ENO stencil may be impossible because the set of possible stencils is biased in one direction due to the boundary. For the third-order scheme, these cells are those related to a mesh point that belongs to a triangle having at least one point on the boundary.

9.1   Numerical tests

All the examples we propose now have been computed with the second and third-order ENO schemes. The ratio of specific heats, γ is always set to 1.4. We present numerical computations of the reflection of a shock on a wedge. Other calculations, including the Collela and Woodward test case and 2D shock tube problems can be found in [3,2].

In these two examples, the post shock conditions are $\rho = \gamma$, $u = v = 0$ and $p = 1$. The preshock conditions are determined from the Rankine-Hugoniot relations with a shock Mach number of 5.5. The only difference between the two cases is the angle of the wedge, $\theta = 30^0$ in one case and $\theta = 45^0$ in the other one. The kind of mesh we use is also different. In the first case, it is a triangular mesh with 8569 nodes and 16806 triangles, in the second one it is made of squares and triangles on the boundary. It has 23990 nodes and 23771 elements (triangles and quadrangles).

In the first case, we have a double Mach reflection, [9]. By comparing the density displayed in Figures 13.6-A and 13.6-B, it is clear that our 3rd order ENO scheme improves the resolution of the various features of the flow. In particular, the slip line coming out of the triple point is clearly visible on Figure 13.6-B while barely existing on Figure 13.6-A.

The second example is even more interesting. First it shows that our methodology is easily extendable to more general meshes, see Figure 13.8. Second the test case itself demonstrates the improvement between accuracy of first order (Figure 13.9-A), second order (Figure 13.9-B) and third order (Figure 13.9-C). Following Ben Dor [9], Figure 2.42-c, page 102, we see that $\theta = 45^o$ and $M = 5.5$ corresponds to a double mach reflection very close to the regular reflection transition. On Figure 13.9-A, we see a regular reflection. On Figures 13.9-B and 13.9-C, we see double mach reflection, but the details of the internal shock and the slip line coming out of the triple point are much better resolved in the third order simulation.

9.2   Some remarks on the formal accuracy of the scheme

We would like to point out some difficulties of these high order finite volume schemes that have been apparently unnoticed yet. Following many authors, we have recovered the density, the $x-$ and $y-$ component of the velocity and the pressure. The choice of the last three variables is dictated by the fact that (i) the density should remain positive, (ii) the pressure is a Riemann invariant and should remain positive, (iii) in a Riemann problem, the normal component of the velocity is also a Riemann invariant, hence the choice of the velocity component may be wise. Nevertheless, all we have said on the recovery problem assumes that the variable to be reconstructed is a conserved one which is true for the density only. So one may question the validity of our approach.

We first discuss the case of the velocity. In fact, the starting point of the reconstruction procedure is the averaged velocity:

$$\bar{u}_C = \frac{\langle A(C), (\rho u) \rangle}{\langle A\ (C), \rho \rangle} = \frac{\int_C \rho u\, dx}{\int_C \rho\, dx}$$

and

$$\bar{v}_C = \frac{\langle A(C), (\rho v) \rangle}{\langle A\ (C), \rho \rangle} = \frac{\int_C \rho v\, dx}{\int_C \rho\, dx}.$$

Thus $\bar{u}_C$ and $\bar{v}_C$ appear to be *true* averages, not with respect to the measure $\langle A\ (C), \cdot \rangle = \frac{\int_C \cdot\, dx}{\int_C dx}$ but with respect to the measure $\langle \cdot \rangle_C = \frac{\int_C \rho\, dx}{\int_C \rho\, dx}$. Now, one can easily convince onself that all that has been done with $\langle A\ (C), \cdot \rangle$ is also true for $\langle \cdot \rangle_C$, and things become clear.

This is no longer true for the pressure, since the "averaged" pressure is

$$\frac{\bar{p}_C}{(\gamma - 1)} = \langle A\ (C), E \rangle - \frac{1}{2\langle A\ (C), \rho \rangle} \left\{ \langle A(C), (\rho u) \rangle^2 + \langle A(C), (\rho v) \rangle^2 \right\}$$

$$= \langle A\ (C), p \rangle + \mathcal{R}$$

where

$$2\mathcal{R} = \langle A\ (C), (\rho u)^2 \rangle + \langle A\ (C), (\rho v)^2 \rangle - \frac{\langle A\ (C), (\rho u) \rangle^2 + \langle A\ (C), (\rho v) \rangle^2}{\langle A\ (C), \rho \rangle}.$$

If there exists a measure μ such that $\bar{p}_C = \int_C p\, d\mu$, a necessary condition is $\mathcal{R} = 0$. Unfortunately, this can not be expected in general.

An alternative to these problems is to work directly on the conserved variables, but then there is no control on the positivity of the pressure. However, the improvement in accuracy is obvious, despite all these problems, as it can be seen from Figure 13.6-(A), 13.6-(B) and 13.9.

We end this set of remarks by noticing that in the second order case, there is no problem because one can interpret the averaged quantities as their values at the centroid of the control volume with second order accuracy. Then the "averaged" pressure has to be understood as the pressure at the centroid, with second order accuracy.

10   Multiresolution Analysis

10.1   Introduction

The simulation of engineering problems requires more and more sophisticated numerical models, finer and finer meshes discretizing complex geometries.

Even with the most powerful computers, these tasks are very challenging and cheaper computing techniques are needed.

The modern numerical methods, such as the TVD or ENO schemes, use many switches that are essential only in a small part of the flow. To reduce their CPU cost, the use of the solution structure appears as an appealing guide to a better distribution of the computer resources. This goal can be achieved via multiresolution (MR) analysis. Recently, A. Harten has developed a framework that is general enough to contain some of the wavelets families [12] on $\mathbb{R}$ but can also be applied when the data are represented on unstructured meshes by cell averages, the natural output of finite volume schemes.

Here, we first describe a technique to represent data which originate from discretizations of functions in unstructured meshes in terms of their local scale components and give some numerical applications. Then, we show how to exploit a particular version of Harten's multiresolution analysis to reduce their CPU cost. Last, we provide some numerical illustrations.

10.2   Harten's multiresolution analysis on general meshes

This section is a very compact résumé of [5]. We consider a domain Ω, with a triangulation $\mathcal{T}(\Omega)$.

We construct a set of control volumes $(C_i)_{i=1,N}$ as before ; they should exactly cover Ω such that if $i \neq j$, $C_i \cap C_j = \emptyset$. In all the numerical examples, the control volumes are the elements of a dual mesh. If f belongs to $L^1(\Omega)$, we can represent f by its average values A $(C_i)f$. The idea is to represent f not by the set (A $(C_i)f)_{i=1\cdots N}$ but by an *equivalent* representation made of the cell averages on a coarser mesh *and* a set of scale coefficients that measure the difference in information between the representation of f on coarser and coarser meshes.

The method needs three ingredients: (i) an *agglomeration procedure* to construct levels of decreasing resolution, (ii) a *discretization mapping* from each level of resolution onto a finite dimensional vector space and (iii) a *reconstruction mapping* that is a right inverse of the discretization. We detail each item of the above list. The definition of the first two items is very closely related. For a complete set of details, the reader is referred to [5]

Discretization We assume that we are given a sequence $\{C_i\}$ of control volumes that are non overlapping. We set $\mathcal{D} : L^1(\Omega) \longmapsto \mathbb{R}^{N_l}$ defined by $\mathcal{D}_i(f) = \langle$ A $(C_i), f\rangle$. In the next paragraph, the sequences of control volume clusters $\{\{C_i^l\}\}_{l=1\cdots L}$ are labeled by l, and D^l will refer to the discretization defined with the cells $\{C_i^l\}$ for one level l.

For numerical purposes, it is essential that if one knows the representation of f on one level l, one will know its representation on the coarser levels, i.e. for indices smaller that l. This nestedness defined by Harten can be stated formally as:

$$D^l(f) = 0 \quad \text{implies} \quad D^{l-1}(f) = 0. \tag{10.1}$$

Since the discretization operator is known, everythings will rely on the way the cells C_i^l are constructed.

An agglomeration procedure We wish to construct $L > 0$ levels of discretization. We rename the control volumes C_i defined on $\mathcal{T}(\Omega)$ by C_i^L, there are $N_L \equiv N$ such control volumes. We set $C^L = \{C_i^L\}_{i=1,N_L}$. Assume that C^l, $1 \le l < L$ is known. If $\{\mathcal{I}_1^l, \cdots, \mathcal{I}_{N_{l-1}}^l\}$ is a partition of $\{1, \cdots, N_l\}$, we set

$$C_k^{l-1} = \cup_{j \in \mathcal{I}_k^l} C_j^l \tag{10.2}$$

and $C^{l-1} = \{C_1^{l-1}, \cdots, C_{N_{l-1}}^{l-1}\}$. Clearly, $\overline{\cup_{k=1}^{N_{l-1}} C_k^{l-1}} = \Omega$ and $C_i^{l-1} \cap C_j^{l-1} = \emptyset$ if $i \ne j$, since the C_i^l are assumed to be open. Thanks to (10.2), the nestedness property (10.1) is true. In fact an explicit calculation shows that

$$\mathcal{D}_i^{l-1}(f) = \sum_{j \in \mathcal{I}_i^l} \frac{|C_j^l|}{|C_i^{l-1}|} \mathcal{D}_i^l(f).$$

This obvious equality enables one to get knowledge of the discretization of f at any level $l \le L$ without knowing f explicitly, provided $\mathcal{D}^L(f)$ is known. Now the key issue is the definition of the $\mathcal{I}_j^l$s.

If the control volumes were squares, like on a cartesian mesh, the obvious procedure would be to gather four cells provided they share a common corner. This is what is done in multigrid, or in domain decomposition methods (except here we are likely to have many subdomains, depending on the level of resolution).

In the present context, the same principles have been applied. In [5], we have used the agglomeration procedure described in [14] initially derived for multigrid acceleration; it has been used for the numerical examples of this section. For the flow simulations, we have preferred to use a recursive domain decomposition algorithm [26] because it enabled us to have a better control of the number of agglomerated cells, and a better control of their shape.

Reconstruction Once the discretization of f is known at level l, we need a reconstruction of f, i.e. we need to find a function $R_l(f) \in L^1(\Omega)$ such that $\mathcal{D}^l[R_l(f)] = \mathcal{D}^l(f)$. We have chosen to look for a piecewise polynomial function of degree r ($= 2$ in practice) that is defined locally, for each cell C_i^l. Since R_l is a right inverse of $\mathcal{D}^l$, the particular choice of the discretization imposes $\langle \text{ A } (C_i), R_l(f) \rangle = \langle \text{ A } (C_i), f \rangle$ but this is the only constraint. Any other recovery procedure might have been employed, provided this conservation constraint is true. In particular, the recovery procedure might be non linear, or it might have used non polynomial functions as suggested by [36] and sketched in §8.

This shows that the suitable reconstruction technique should be the same as the one we have used for the ENO methods in §4. The only remaining question is how to define the stencil $\mathcal{S}_i^l$. This is achieved by an heuristic procedure inspired from [1,3] as described in §6. However, we only need one stencil per cell C_i^l instead of several as in §6.

First we identify each cell C_i^l with its center of gravity. Thanks to this, the method is not restricted to control volumes generated by triangular meshes, because at this level we may forget the origin of the control volumes. Second, we build a Delaunay mesh[1] on these points, and remove the spurious triangles that lie outside the original domain Ω. More precisely, we say that a triangle lies outside the domain if its centroid is not in the domain. This can be checked in practice with the help of an efficient sorting tool. Then, we also remove the triangles that are on the boundary of this new mesh and are too flat. Once this is done, we construct the stencils: for each triangle (A, B, C) of this mesh (i.e. for a set of 3 cells at level l), we add the three other points/cells shown in Figure 13.10. More details can be found in [3].

Data compression Once all this is done, we can consider the N_{l+1} errors, defined for each cell C_i^l, by $e_i^{l+1} = \langle$ A $(C_i^{l+1}), R_l(f) \rangle - \mathcal{D}_i^{l+1}(f)$. By construction, we have $\sum_{i \in \mathcal{I}_j^{l+1}} |C_j^{l+1}| e_j^{l+1} = 0$.

We have N_l such linear relations between the errors at level $l + 1$, thus we can define $N_{l+1} - N_l$ *independent* scale coefficients d_i^{l+1}, by the following computation: for each $j \le N_l$, we set $d_i^{l+1} = e_i^{l+1}$ for all $i \in \mathcal{I}_j^{l+1}$ except the last index, and we set $d^{l+1} = (d_1^{l+1}, \cdots, d_{N_{l+1}-N_l}^{l+1})$. It can be shown that

$$\mathcal{D}^L(f) = (\langle \text{ A } (C_1^L), f \rangle, \cdots, \langle \text{ A } (C_{N_L}^L), f \rangle) \\ \longleftrightarrow [\mathcal{D}^1(f), d^2, \cdots, d^L] \tag{10.3}$$

is a (linear) one-to-one mapping.

Moreover, from Theorem 1, if f admits a p-th continuous derivative, then $d_i^l = O(h_l^{p+1})$ (h_l is a characteristic size of the C_i^l), provided the mesh is regular enough. This remark enables one to represent $\mathcal{D}^L(f)$ within a given tolerance ϵ, with less than N_L degree of freedom. To do so, we replace the scale coefficients in (10.3) by truncated ones,

$$\tilde{d}_i^l = \begin{cases} d_i^l & \text{if } |d_i^l| > \frac{\epsilon}{2^k} = \epsilon_k \\ 0 & \text{else.} \end{cases}$$

More sophisticated expressions for ϵ_k can be used, but it does not affect that much the compression factor μ defined as the ratio between N_L and N_1 plus the number of non-zero $\tilde{d}_i^l$,

$$\mu = \frac{N_L}{N_1 + \sum_{l=2,N_L} \#\{|d_j^l| > \epsilon_l\}}.$$

---

[1] A triangulation is a Delaunay triangulation if no point of the triangulation lies within the outer circles of each of the triangles.

10.3   Numerical examples

The power of the method is demonstrated by means of a piecewise smooth function on a complex geometry which looks like what is shown in the upper left part of Figure 13.11. In the upper part of the figure one sees cell averages on a sequence of coarser grids. The two plots in the lower part are isolines of truncated scale coefficients, one plot for the restriction from the fine to the medium, one for the restriction from the medium to the coarse grid. If the reconstruction starts from the coarsest grid and uses only the non-zero scale coefficients, then the cell averages on the finest grid as shown in the upper left part are recovered within plot accuracy.

In Table 13, with the entry f_2, we represent the tolerance ϵ, the compression factor μ, the L^∞- and L^1-error for this particular function. The same information is given for $f_1 = \cos 2\pi(x^2 + y^2)$. In [5], other examples and details are presented. In particular, we try to quantify what we loose by using unstructured meshes, compared to Cartesian ones.

They all indicate that our method is stable and has the same accuracy on structured and unstructured meshes.

10.4   Multiresolution analysis and ENO schemes

In the ENO method of §9, the key point is the use of a piecewise polynomial reconstruction, the same as here, and a stencil selection procedure. Then, the MUSCL method is applied, with an ENO reconstruction on the physical variables. Nevertheless, this is costly. The previous concepts can help to reduce significantly the number of ENO reconstructions. The idea is to use a two-level multiresolution scheme. Only one set of scale coefficients is produced and we modify the ENO reconstruction as follows: for each fine cell C_i^2 and physical variable f_i, if $\frac{|d_i^2|}{|f_i|} \leq \epsilon$, we use the reconstruction of f on the coarse level in the MUSCL method, else, we apply the ENO algorithm. The other details of the scheme remain the same.

10.5   A numerical experiment

We have applied this simplified ENO scheme to various configurations: the interaction of a shock with a 90° wedge, of a shock with a ramp and of a shock and a vortex with a ramp. Here, we illustrate the method on the interaction of a shock and a 30^0 ramp. The shock Mach number, evaluated from the post–shock conditions, is 5.5. The fine mesh has 33943 points and 67224 triangles, the coarse level is made of 10000 cells. Figure 13 shows the Mach number isolines. Most of the known structures of this double Mach reflection are well represented, in particular the slip line out of the triple point, as well as the vortex that results in the interaction of the weak shock out of the reflected one and this slip line.

Figure 13.13 shows the isolines of the ENO–indicator for the density (all the error indicators look similar, except for the slip lines where nothing is detected for the pressure). Here, $\epsilon = 10^{-2}$, and only 14% of pure ENO reconstruction was done at this stage of the computation. The compression factor is always larger than $3/4$ of its maximum possible value N_L/N_1. The simplified ENO scheme runs, in this case, 2.5 times faster than the pure ENO one. On the other examples, the ratio was about 2–2.5. This ratio is clearly problem dependent. The most expensive part of the scheme is the flux evaluations (Roe's scheme here), while the reconstruction cost, overhead included, becomes almost negligible.

11   Other applications : Hamilton Jacobi equations

Eno schemes have been applied to other problems, in particular the approximation of the Hamilton Jacobi equations

$$
\begin{aligned}
&\frac{\partial u}{\partial t} + H(x, u, \nabla u) = 0 \quad x \in \Omega, t > 0 \\
&u(x, 0) = u_0(x) \qquad\qquad x \in \Omega, t = 0 \\
&\text{Boundary conditions .}
\end{aligned}
\tag{11.1}
$$

In (11.1), $u : \Omega \times \mathbb{R}^+ \longrightarrow \mathbb{R}$ where Ω is an open subset of $\mathbb{R}^N$. Here, $N = 2$ but this does not change anything to the discussion. The boundary conditions can be of the Dirichlet type for example, but this point will not be discussed here, see [45,41] for details.

The existence and uniqueness of the viscosity solution of the Cauchy problem (11.1) is discussed in [46,45] and the reference therein. Our purpose is to discuss some elements the numerical approximation of

$$
\frac{\partial u}{\partial t} + H(\nabla u) = 0 \; x \in \mathbb{R}^2, t > 0
$$
$$
u(x, 0) = u_0(x) \quad x \in \mathbb{R}^2, t = 0
\tag{11.2}
$$

with a triangular unstructured mesh. More details are given in [42], generalisation to (11.1) is rather obvious via [41].

In [44,47], several numerical upwind schemes have been constructed. They rely on the strong formal analogy between the viscosity solutions of (11.2) and the weak solutions of

$$
\frac{\partial W}{\partial t} + \frac{\partial H(W)}{\partial x} = 0 \; x \in \mathbb{R}^2, t > 0
$$
$$
W(x, 0) = \nabla_x u_0(x) \; x \in \mathbb{R}^2, t = 0.
\tag{11.3}
$$

From (11.3), any reasonable numerical scheme for conservation law should give rise to a numerical scheme for (11.2) : Godunov, Lax Friedrichs, ect. In

[47], only the case of regular Cartesian grid was considered. Any of the proofs could be applied even to a non orthogonal structured mesh. Their work has been generalised in [42] and error estimates are provided.

We consider schemes writting like

$$u_i^{n+1} = u_i^n - \Delta t \mathcal{H}(\nabla_{T_1} u^n, \cdots, \nabla_{T_{k_i}} u^n). \tag{11.4}$$

In (11.4), the set $\{T_1, \cdots, T_{k_i}\}$ is the set of triangles that share M_i as vertex, the quantities are $\nabla_{T_l} u^n$ are numerical gradients of the node values u_j^n. The scheme is formally first order when $\nabla_{T_l} u^n$ is the gradient of the (continuous) piecewise linear interpolation at the vertices of the mesh.

It is shown that a first order monotone scheme that has the additional property of beeing "intrinsic" is convergent, and one has the following error estimate

$$\max_{M_i, n \in \mathbb{N}} |u_i^n - u(M_i, t_n)|| \le C(\text{ Mesh}, T, H, u_0) \sqrt{h}$$

where M_i is a generic mesh point, $t_n = n\Delta t$, the constant C depends on standard regularity properties of the mesh, the Lipschitz constant of H and u_0. The parameter h is the maximum of the diameters of the circumscribed circles of the triangles. The proof does not depend on $N = 2$, the dimension of N. By saying that a numerical Hamiltonian is "intrinsic", we mean the

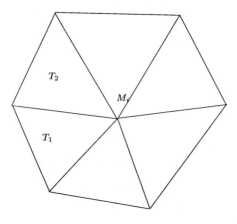

Fig. 11.1. The neighboring triangles of M_i

following property : take any triangle T, cut it in two parts as on Figure 11.2, then, since u is assumed to be linear in T, the gradient of u is the same in T_{cut} and T'_{cut}. The number of arguments in $\mathcal{H}$ is increased by one, but two among them are the identical : they are $\nabla_T u$. The numerical Hamiltonian is intrinsic if the value of $\mathcal{H}$ is not modified. This is obviously true for the Godunov solver, and appropriate weights enable to have the same property for the

Lax–Friedrichs one. This property is usefull in the proof : the fundemental difference between an unstructured mesh and a Cartesian one is that the mesh is not invariant by translation. The "intrinsic" property, in some sense, replace the invariance by translation one. Note that a structured mesh is not, in general, invariant by translation.

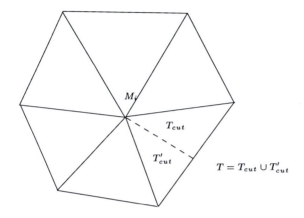

Fig. 11.2. Illustration of the intrinsic property

To increase the accuracy of the scheme, any of the previous ENO or WENO techniques can be applied. The algorithm is :

1. given $(u_i^n)_{M_i}$, compute an ENO/WENO Lagrange reconstruction within each triangle with the algorithm of §6. We call $\pi_T u^n$ the reconstruction in a generic triangle T,
2. Given any node M_i, consider the set $\{T_1, \cdots, T_{k_i}\}$ and compute

$$\nabla_{T_1} \pi_{T_1} u^n(M_i), \cdots, \nabla_{T_{k_l}} \pi_{T_{k_l}} u^n(M_i)$$

   the gradients node values at the vertex M_i
3. Compute the numerical Hamiltonian with these arguments,
4. Take any Runge Kutta scheme, for example the TVD ones above, to update the solution.

The ENO algorithm has been successfully used in [42] to compute the ray paths through a lense, or for a Geophysical problem [43].

12   Schemes with adaptive limiters and fluxes

The simulations of severe flow conditions, such as in the reactive flows, require robust numerical methods. Many computations use a class of algorithms based either on flux vector splitting (FVS) or on flux difference splitting (FDS). Liou-Steffen [49] have proposed a remarkably simple upwind FVS. This splitting, called AUSM, treats the convective and pressure terms separately. The convective quantities are upwind-biased extrapolated to the cell interface using a properly defined cell face advection Mach number. AUSM keeps the qualities of FVS (robustness and efficiency) and recovers the accuracy attributed to FDS. Radelspiel-Kroll [50] proposed several modifications in order to improve the scheme's ability to solve viscous flows correctly. In particular, this includes a switch from AUSM to van Leer flux splitting (VL) through strong shock waves. In this Vsection, we retain this idea but we use it differently [59]. The switch, here, is related with the local accuracy of the scheme. When the scheme degenerates into a first-order one (outside a shock wave), it is convenient to use AUSM; and when the scheme is a second- or third-order, VL is better in order to minimize the error terms. To capture strong and (or) rapid physical fluctuations accurately, the local variation of each quantity has to be incorporated as much as possible in the writing of the scheme. ENO schemes choose the stencil which provides the most regular solution in order to minimize numerical over and undershoots. In this method, we take the stencil which minimizes the numerical error terms (dissipative and dispersive terms). These terms have different expressions following the local evolution of quantities. To improve efficiency, the equivalent system (ES) needs to be studied, including the expression for the slope limiters. Their expressions are controlled by the local but also by the environing physical variation of the quantities. For each quantity, six different cases are considered, each associated with a different physical variation. A triad of limiters is defined which minimizes or cancels the second-order truncation errors. From this study, a new explicit scheme is written. Compared with a standard TVD-MUSCL scheme, it is no more complicated and it gives a more precise solution. It is applied to the 1-D test case proposed by Shu-Osher [51] to simulate the interaction between a moving shock wave and a turbulent flow. The results show that we obtain the same precision as with their ENO scheme. The improved accuracy is also demonstrated by the computations of a 2-D supersonic jet.

12.1   MUSCL Approach and Flux Splittings

The hyperbolic part of the conservation form of the 1-D Navier-Stokes equations is classically written:

$$W_t + F_x = 0 \quad \text{with} \qquad W = [\rho, \ \rho u, \ \rho E]^T ,$$
$$W(x,0) = W^0(x), \quad -\infty < x < +\infty, \quad t \geq 0 \tag{12.1}$$

where ρ, u and E are the density, the velocity and the total energy. In the discrete form, (12.1) is expressed as:

$$W^{n+1} = W_j - \sigma \left(\mathcal{F}_{j+\frac{1}{2}} - \mathcal{F}_{j-\frac{1}{2}} \right) \tag{12.2}$$

with $W_j = W_j^n = W(V_j^n)$, $\sigma = \Delta t / \Delta x$, and $\mathcal{F}_{j+\frac{1}{2}} \equiv \mathcal{F}(V_{j-1}^n, V_j^n, V_{j+1}^n, V_{j+2}^n)$. Note that W are the conserved variables, the flux $\mathcal{F}$ may not be expressed in terms of W but in terms of new variables V. This method is devoted to improve the numerical simulation of compressible mixing layers, with possibly chemical reaction. because of that, we have chosen $V = (\rho, u, T)^T$ where T is the temperature. Note that all the limitation procedures are performed on the variables V. The numerical $\mathcal{F}$ verifies $\mathcal{F}(V, ..., V) = F(V)$ (Note we write it in termes of the V-variables). Δx is assumed to be constant and Δt is related to Δx by a CFL condition. With the MUSCL approach [52], the backward and forward extrapolated values of $V_{j+\frac{1}{2}}$ at the interface $j + \frac{1}{2}$ can be written as:

$$V_{j+1/2}^L = L(V_j, V_{j+1}, \varphi_1) \quad \text{and} \quad V_{j+1/2}^R = R(V_j, V_{j+1}, \varphi_2^R).$$

At the interface $j - 1/2$, we have:

$$V_{j-1/2}^L = L(V_{j-1}, V_j, \varphi_2^L) \quad \text{and} \quad V_{j-1/2}^R = R(V_{j-1}, V_j, \varphi_1),$$

where φ_1 and $\varphi_2^{L,R}$ are non-linear functions of r_j with $r_j = \frac{V_j - V_{j-1}}{V_{j+1} - V_j}$ (component by component). The non-linear interpolations L and R have to verify the following properties: homogeneity, translation invariance, left-right symmetry, monotonicity and convexity. The flux $F_{j+\frac{1}{2}}$ is written in the general form $F_{j+\frac{1}{2}} = F(V_{j+\frac{1}{2}}^L, V_{j+\frac{1}{2}}^R) - \Phi \Delta G$ where $\Phi \Delta G = \Phi \left[G(V_{j+\frac{1}{2}}^R) - G(V_{j+\frac{1}{2}}^L) \right]$ is a dissipation term. We are more particularly interested in FVS_{VL} and AUSM schemes. As in [50] we couple both schemes; but our coupling is different. It is based on an analysis of the ES and takes account of the properties of each scheme at the first and second-order. This coupling is advantageous because the expression for the fluxes is both very similar and yet exhibits different properties. In the case of a perfect gas, with the constant-pressure specific heat Cp and the specific heat ratio γ assumed to be constant, if we define

$$F_M^{FS} = M = F_M^+ + F_M^- = \frac{(M^L+1)^2}{4} - \frac{(M^R-1)^2}{4}$$

$$F_a^{FS} = \begin{bmatrix} 0 \\ p \\ 0 \end{bmatrix} = F_a^+ + F_a^- = \begin{bmatrix} 0 \\ p^L \frac{1+M^L}{2} + p^R \frac{1-M^R}{2} \\ 0 \end{bmatrix},$$

$$F_c^{DS} = \begin{bmatrix} \rho c \\ \rho c u \\ \rho c H \end{bmatrix},$$

then these both splittings can be written, at the grid point $j + \frac{1}{2}$ and for $-1 \le M \le 1$, as follows

- The FVS$_{VL}$ scheme:

$$F = F_M^{FS} F_c^{DS} + F_a^{FS}$$
$$=$$
$$F_M^+ F_c^{DS}(V^L) + F_M^- F_c^{DS}(V^R) + F_a^+ + F_a^- \qquad (12.3)$$

$$\Phi = 0$$

- The AUSM scheme:

$$F = (F_M^+ + F_M^-)\frac{F_c^{DS}(V^L)+F_c^{DS}(V^R)}{2} + F_a^+ + F_a^- \qquad (12.4)$$
$$\Phi = |M| \quad \text{and} \quad \Delta G = \tfrac{1}{2}\left[F_c^{DS}(V^R) - F_c^{DS}(V^L)\right]$$

where c, p and H represent, respectively, the sound speed, the static pressure, and the total enthalpy.

The analysis of (ES), obtained from Taylor expansions, quantifies the truncation error of the discrete form as Δx and $\Delta t \to 0$. V and F are assumed to be analytic functions. For each component V_i, the expansions reflect the environing physical behaviour associated with the specific approach used here. Six different cases are considered for each component W_i (Fig 13.14):

- No extremum at j
    - case 1 : monotonic evolution,

    - case 2 : extremum at the nodes j-1   and   j+1,

    - case 3 : extremum at the node j-1   or   j+1,
- Extremum at j
    - case 4 : no extremum at the nodes j-1   and   j+1,

    - case 5 : extremum at the nodes j-1   and   j+1,

    - case 6 : extremum at the node j-1 or j+1.

12.2   First-order Error Terms in Space

After calculated the Taylor expansions of $r(V)$, $\varphi_\alpha(r(V))$ ($\alpha = 1,2$) and of the fluxes $\Psi = \Psi(\varphi_\alpha(r(V)))$ with $\Psi = F_M^-, F_M^+, F^{DS}$,... at the node j for both cases $V_x \neq 0$ and $V_x = 0$, (1) is transformed into:

$$W_t + F_x + \Delta x\,[A]\,V_{xx} + O(\Delta x^2) = 0 \qquad (12.5)$$

where$[A]$ is a $(3,3)$ matrix. The first-order error term in space, for the kth equation ($k = 1,...,3$) of the system (12.5) and for the splitting (12.3), can be expressed:

$$Er_k = \sum_{i=1}^{3} A_{ki}V_{i_{xx}}$$
$$= \frac{1}{2}\sum_{i=\rho,u,T}\left\{\begin{array}{l}(F^{DS}F_i^{+\prime} + \frac{F^{FS}}{2}F_i^{L\prime} + \frac{\Phi}{2}G_i^{L\prime})\,[g_i^L] \\ -(F^{DS}F_i^{-\prime} + \frac{F^{FS}}{2}F_i^{R\prime} - \frac{\Phi}{2}G_i^{R\prime})\,[g_i^R]\end{array}\right\}V_{i_{xx}}.$$

For the AUSM approach (12.4), the expression is a little more complicated:

$$Er_k =$$

$$\frac{1}{2}\sum_{i=\rho,u,T}\left\{\begin{array}{l}(F_c^{DS}F_{M_i}^{+\prime}+\frac{F_M^{FFS}}{2}F_{C_i}^{L\prime}+F_{a_i}^{+}+\frac{\Phi}{2}G_i^{L\prime})\,[g_i^L]\\-(F_c^{DS}F_{M_i}^{-\prime}+\frac{F_M^{FFS}}{2}F_{C_i}^{R\prime}+F_{a_i}^{-}-\frac{\Phi}{2}G_i^{R\prime})\,[g_i^R]\end{array}\right\}V_{i_{xx}}$$

where $g^{L,R} = g^{L,R}(\varphi) = \frac{\varphi_1(-1)}{2} + \frac{\varphi_2^{L,R}(3)}{2} - 1$ if $V_x = 0$ $g^{L,R} = g^{L,R}(\varphi') = \varphi'_2^{L,R}(1) - \varphi'_1(1)$ if $V_x \neq 0$, $\varphi' = \frac{d\varphi}{dr}$, $F^{+\prime} = \frac{dF^+}{dV^L}$, $(F^-)' = \frac{dF^-}{dV^R}$, $(F^{L,R})' = \frac{dF^{DS}}{dV^{L,R}}$, $(G^{L,R})' = \frac{dG}{dV^{L,R}}$, etc.

In order to develop a simple analysis, we assume at node j: $|V_i^R - V_i^L| <<$ $\max(|V_i^R|, |V_i^L|)$, $(i = 1, ..., 3)$. This says the jump at the interface j is considered to be weak (the strong discontinuities are excluded of the analysis). The case $|V_i^R - V_i^L| \approx \max(|V_i^R|, |V_i^L|)$ is not considered in this paper, although it may exist in the velocity under certain circumstances, such as when this quantity has strong fluctuations around zero. The expansions are calculated for positive values of M. The expressions for $M < 0$ are obtained by symmetry (g_i^L is replaced by g_i^R and reciprocally). The first-order term cancels if $g_i^{L,R} = 0$. In general, we assume that

$$\text{for } r < 0, \quad \varphi_\alpha = \varphi'_\alpha = 0 \quad \text{and} \quad \text{then} \quad \varphi_1 = \varphi_1(-1) = 0. \tag{12.6}$$

Therefore, the first-order error term cancels if

$$\varphi_2^L = \varphi_2^R = 2; \text{ for } r = 3 \tag{12.7}$$

(if there is an extremum at node j or if

$$\varphi'_2^L = \varphi'_2^R = \varphi'_1 \text{ for } r = 1 \tag{12.8}$$

if there is no extremum at node j.

The Taylor expansions at node j include the presence of one extremum (cases 4-6) or none (cases 1-3) at this point. On the other hand, they do not say whether one extremum exits or not at the neighbors $j - 1$ and $j + 1$. If there is no extremum associated with $j - 1$ and $j + 1$ (cases 1 and 4), no additional constraint appears; but if an extremum is present at these points, then either a different definition of φ (cases 2 and 3) is required in order to preserve the second-order accuracy or the scheme accuracy automatically degenerates (cases 5 and 6).

If $V_{i_x} \neq 0$ at node j, the condition (12.8) is easily met if the nodes $j - 1$, j and $j + 1$ have no extremum for component V_i (case 1). In this case, it is sufficient to take the same function in the second-order TVD domain for each point $j - 1$, j, $j + 1$. If one extremum exits for one or both neighbors of node j (case 2 or 3), the condition is more restrictive. Since we have $\varphi'_2^L = 0$ and/or , the second-order accuracy is ensured only if

$$\varphi'_1(1) = 0. \tag{12.9}$$

If $V_{i_x} = 0$ at point j, the condition (12.7) is obtained if $j - 1$ and $j + 1$ are not associated with an extremum (case 4). But this condition is no longer met if there exists at least one extremum at one of the neighbors of j. For these cases (cases 5 and 6), the first-order error term does not cancel. Case 5 corresponds to local phenomena of wave length $2\Delta x$, and case 6 to phenomena of wave length $2\Delta x$ or $3\Delta x$. Therefore, when we have physical variations with wave length fluctuations greater than $3\Delta x$, the scheme is second-order in space if the expression of φ is well-defined. When wave length fluctuations are smaller than or equal to $3\Delta x$ (cases 5 and 6) the first-order error term is still present. In this case, the scheme has strong dissipative properties that can eliminate the numerical instabilities.

For $M > 1$, the error terms have the same expression whatever the splitting; but for $M \leq 1$, their expression, $[A] = [A]_c^{AUSM}$ for Liou-Steffen and $[A] = [A]_c^{VL}$ for van Leer, depends on the splitting chosen. For case 5, where $g_i^{L,R} = -1$, and for case 6 where $g_i^L = -1$ and $g_i^R = 0$, the error terms are written:

$$
\begin{aligned}
[A]^{AUSM} &= [A]_c^{AUSM} + [A]_a^{VL}, \\
[A]^{VL} &= [A]_c^{VL} + [A]_a^{VL}, \quad \text{with} \\
[A]_c^{AUSM} &= [A]_c^{VL}
\end{aligned}
$$

and $[A]_c^{VL}$ given by

$$
[A]_c^{VL} = \begin{bmatrix} A_{11} & A_{12} & A_{13} \\ uA_{11} & uA_{12} + \rho A_{11} & uA_{13} \\ HA_{11} & HA_{12} + \rho u A_{11} & HA_{13} + \rho C_p A_{11} \end{bmatrix}
$$

where

$$
A_{11} = \tfrac{c}{2}M, \; A_{12} = \tfrac{\rho}{2}(M\delta_{l5} + b\delta_{l6}),
$$

$$
A_{13} = \tfrac{\rho c}{2T}Md\delta \qquad \text{for } [A]_c^{AUSM},
$$

$$
A_{11} = \tfrac{c}{2}(a^2\delta_{l5} + b^2\delta_{l6}),
$$

$$
A_{12} = \tfrac{\rho}{2}(M\delta_{l5} + b\delta_{l6}),
$$

$$
A_{13} = \tfrac{\rho c}{2T}bd\delta \quad \text{for } [A]_c^{VL}.
$$

$$
a^2 = \tfrac{1+M^2}{2}, b = \tfrac{1+M}{2},
$$

$$
d = \tfrac{1-M}{2},
$$

$$
\delta = \delta_{l5} + \tfrac{\delta_{l6}}{2},
$$

$$
[A]_a^{VL} = -\tfrac{c^2}{2\gamma}\begin{bmatrix} 0 & 0 & 0 \\ \tfrac{2}{\rho}A_{12} & \tfrac{\rho}{c}\delta & \tfrac{1}{T}(A_{12} + \tfrac{\rho}{4}\delta_{l6}) \\ 0 & 0 & 0 \end{bmatrix}.
$$

δ_{l5} and δ_{l6} are the Kronecker symbol, $l = 5$ (case 5) or 6 (case 6).

The error term induced by AUSM splitting is always smaller. The differences become greater when $M \to 0$. For case 5, for example, when $M \to 0$, many components of $[A]$ cancel with AUSM splitting (Fig. 13.15). From this study, we deduce

Condition 1: the AUSM splitting (12.4) is chosen when the scheme degenerates into a first-order scheme (cases 5 and 6).

12.3   Second-order Error Terms in Space

In this section, we see that the first order terms of (ES) cancel if the limiters satisfy given properties at some specific points only. The second order terms only remain and we specify explicitly which limiters should be taken so that these term also cancel. The system (12.2) has the following expression:

$$W_t + F_x + \Delta x^2 (BV_{xxx} + CV_{xx} + DV_{xx} + EV_x) = O(\Delta x^3) \qquad (12.10)$$

where

$$B = B(\chi_1, V),$$

$$C = C(\varphi''_1, \varphi''^R_2, \varphi''^L_2, V, V_x) \text{ if } V_{i_x} \neq 0 \text{ at node } j$$
$$(\text{ cases } 1-3),$$

$$C \equiv 0 \text{ if } V_{i_x} = 0 \qquad \text{at node } j \text{ (case 4)},$$

$$D = D(\chi_2, V \qquad , V_x),$$

$$E = E(V, V_x)$$

with

- If $V_{i_x} \neq 0$ at j, $\chi_1 = 1 - 3\varphi'_2$ and $\chi_2 = 1 - \varphi'_1 - \varphi'_2$.
- If $V_{i_x} = 0$ at j, $\chi_1 = 2 + \varphi_1 - \varphi_2 + 2\varphi'_1 - 4\varphi'_2$ and $\chi_2 = 2 + \varphi_1 - \varphi_2$

The matrices B, C, D, E are provided in annex A.

By homogeneity, the cancellation of C for cases 1-3 gives the following condition on φ'': $\varphi''_1(1) = \varphi''^R_2(1) = \varphi''^L_2(1)$ .The terms $E_i V_x$ ($i = 1$ (continuity eq.), 2 (momentum eq.), 3 (energy eq.)) come from fluxes that contain products of at least three primary quantities (for example $\rho u^2, ...$). For each equation, the error terms are expressed in the annex A. With AUSM and VL splittings, the error terms are the same, excepted for $E_i V_x$ and the dissipation term in the energy equation :

$$(E_1 V_x)^{AUSM} = \frac{(LogT)_x^2}{32}\left[-3\rho u_x - cM\rho_x + 2\rho cM(LogT)_x\right],$$

$$(E_1 V_x)^{VL} = 0$$

$$(E_2 V_x)^{AUSM} =$$
$$\frac{1}{32}\left[\begin{array}{l}16\rho_x u_x u_x + 4cM\rho_x u_x(LogT)_x - 4\rho u_x u_x(LogT)_x + \\ c^2 M^2 \rho_x(LogT)_x^2 + 2\rho c^2 M^2(LogT)_x^3\end{array}\right],$$

$$(E_2 V_x)^{VL} = \tfrac{1}{12}(\rho_x u_x u_x),$$

$$(E_3 V_x)^{AUSM} =$$
$$\frac{1}{32}\left[\begin{array}{l}20cM\rho_x u_x u_x + 16c^2(C_2 - \frac{M^2}{4})\rho_x u_x(LogT)_x + \\ 2\rho cM u_x u_x(LogT)_x - c^3 M(3C_2 - \frac{M^2}{2})\rho_x(LogT)_x^2 - \\ \rho c^2(3C_2 + \frac{M^2}{2})u_x(LogT)_x^2 + \tfrac{3}{2}\rho c^3 M^3(LogT)_x^3\end{array}\right],$$

$$(E_3 V_x)^{VL} = \tfrac{1}{8}\left[cM\rho_x u_x u_x + 3C_p\rho_x u_x T_x + \rho u_x u_x u_x\right],$$

$$(D_3 V_{xx})^{VL} = (D_3 V_{xx})^{AUSM} + \tfrac{\chi_2}{4}\left[C_p cM(\rho_x T_x)_x\right],$$

where $C_2 = \frac{C_p}{R} - 1$. In order to avoid the appearance of numerical oscillations corresponding to case 5 or 6, it is better to eliminate the dispersive error term BV_{xxx}. Although these oscillations are damped by the scheme, as we have seen in the previous paragraph, it is harmful to drop the scheme accuracy artificially if this is not necessary Therefore, for cases 1 and 4, we let $\chi_1 = 0$. Applying conditions (12.7–12.9), we have

$$\varphi_2'(1) = \varphi_1'(1) = \frac{1}{3} \quad \text{if} \ \ V_{i_x} \neq 0 \tag{12.11}$$

at nodes $j-1$, j and $j+1$ (case 1) and

$$\varphi_2'(3) = 0 \quad \text{if} \ \ V_{i_x} = 0 \tag{12.12}$$

at node j (case 4).

For cases 2 and 3, as the constraints (12.8–12.9) are already imposed, we have: $\chi_1 = 1$. Therefore, BV_{xxx} does not cancel for these cases. But in fact, it is possible to eliminate the dispersive term if we use a multi-time stepping scheme and if we apply different expressions of φ at every time step (not presented here). From conditions (12.7–12.9) and (12.11–12.12), it is possible to define an adequate limiter under the form of a triad of limiters, each adapted to the local variation of the physical quantities. If at node j we have:

– case 1, we take ([53])

$$\varphi_1 = \varphi_2 = \varphi = \begin{cases} \frac{r+2}{3} & \text{if } \frac{2}{5} \leq r \leq 4 \\ 2 & \text{if } r > 4 \\ 2r & \text{if } 0 \leq r \leq \frac{2}{5} \\ 0 & \text{if } r \leq 0 \end{cases} \tag{12.13}$$

– cases 2 and 3, we choose

$$\varphi_1 = \varphi_2 = \varphi = \begin{cases} 1 & \text{if } r \geq \frac{1}{2} \\ 2r & \text{if } 0 \leq r \leq \frac{1}{2} \\ 0 & \text{if } r \leq 0 \end{cases} \tag{12.14}$$

– case 4, we define

$$\varphi_1 = \varphi_2 = \varphi = \varphi_{superbee}, \tag{12.15}$$

– cases 5 and 6, φ has to verify only the constraint (12.6). It is easy to see that the triad (12.13–12.15) verifies this condition.

From (12.13–12.15), $\chi_2 = \frac{1}{3}$ for case 1, $\chi_2 = 1$ for cases 2 and 3 and $\chi_2 = 0$ for case 4. If all the components V_i have an isolated extremum at j (case 4) at the same time, the second-order dissipative error terms vanish. For this case, $E_i V_x$ vanishes too. The scheme is then third-order accurate in space. For the second-order error terms, the main difference between AUSM and VL splittings is in the expression of terms $E_i V_x$. As long as the temperature gradients are weak, these terms can be neglected. But for reactive flows, their values become unnegligible and the choice of the splitting becomes important. For this kind of flow, it is better to take VL splitting because, with it, the expressions for $E_i V_x$ are much simpler and remain the same as those associated with the supersonic flow. Their values are weaker too. Therefore

Condition 2: when the scheme remains second-order or third-order (cases 1-4), it is recommended to use VL splitting (12.3).

Although simpler with this splitting, the $E_i V_x$ terms are not automatically negligible in particular when the temperature variations become high. So, it is to our advantage to eliminate these terms, which appear as additional transport terms in the conservation equations:

$$\rho_t + (\rho u)_x + \Delta x^2 (B_1 V_{xxx} + D_1 \qquad\qquad V_{xx}) = O(\Delta x^3)$$

$$(\rho u)_t + (\rho u^2 + p)_x + (\delta \rho \delta u)\tfrac{u_x}{12} + \Delta x^2 (B_2\ V_{xxx} + D_2 V_{xx}) = O(\Delta x^3)$$

$$(\rho E)_t + (\rho u H)_x + \delta(\rho H)\tfrac{u_x}{8} + \Delta x^2 (3 \qquad\qquad V_{xxx} + D_3 V_{xx} = O(\Delta x^3)$$

where

$$\delta(\rho H) = 3C_P \delta \rho \delta T + u \delta \rho \delta u + \rho \delta u \delta u$$

and δV_i represents the variation of V_i on the mesh size Δx. Formally, the residual error terms can be corrected by adding the opposite value to the expression for the fluxes. For example, by defining the flux at the interfaces $j + \frac{1}{2}$ and $j - \frac{1}{2}$ in the following form:

$$F_{j+1/2}^{FS} = F^-(V_{j+1/2}^R) + F^+(V_{j+1/2}^L) - \delta Q_j V_{j+1/2},$$
$$F_{j-1/2}^{FS} = F^-(V_{j-1/2}^R) + F^+(V_{j-1/2}^L) - \delta Q_j V_{j-1/2},$$
$$V_{j+1/2} = cM = \frac{c^L + c^R}{2}(F_M^+ + F_M^-), \quad c^{L,R} = c(U^{L,R}),$$

$$\delta Q_j = \begin{bmatrix} 0 \\ \frac{\delta \rho \delta u}{12} \\ \frac{3 C_P \delta \rho \delta T + u \delta \rho \delta u + \rho \delta u \delta u}{8} \end{bmatrix}_j,$$

where $\delta V_j = V_{j+1/2} - V_{j-1/2}$, the terms $E_i V_x$ disappear. This correction is activated only if the scheme remains a second or third-order scheme in space.

12.4   Multi-time stepping algorithm

The analysis of the previous sections was based on the hypothesis of a single time step. When we use a multi-time stepping scheme, two questions come to the mind:

- What is the effect of a multi-stepping scheme on the spatial error terms and on the CFL criterion?
- What is the minimum number of time steps needed to achieve sufficient accuracy?

For the particular second-order scheme in time:

$$\tilde{V}_j = V_j - \sigma(\mathcal{F}_{j+1/2} - \mathcal{F}_{j-1/2})$$
$$V_j^{n+1} = \frac{1}{2}\left[(V_j + \tilde{V}_j) - \sigma(\tilde{F}_{j+1/2} - \tilde{F}_{j-1/2})\right] \tag{12.16}$$

we show that if we choose the same limiters ($\varphi = \varphi_1 = \varphi_2$ for the predictor stage and $\tilde{\varphi} = \tilde{\varphi}_1 = \tilde{\varphi}_2$ for the corrector one) for each case considered, the spatial error terms are the same as those generated by a single-time stepping scheme. But, a more restrictive condition on the CFL number is introduced in order to the scheme (12.16) verifies (12.6) and remains second-order in space at a node j where, for one or several components, $V_{i_x} = 0$ and $V_{i_{xx}} \neq 0$ (equation 12.7):
$\tilde{\varphi}(-a) = \tilde{\varphi}(-1/a) = 0$ and $\tilde{\varphi}(b) = \tilde{\varphi}(c) = 2$ with $a = 1 + 4\sigma F_{i_{xx}}/(V_{i_{xx}} - 2\sigma F_{i_{xx}})$, $1/a = 1 - 4\sigma F_{i_{xx}}/(V_{i_{xx}} + 2\sigma F_{i_{xx}})$, $b = 1 + 2V_{i_{xx}}/(V_{i_{xx}} - 2\sigma F_{i_{xx}})$,

$c = 1 + 2V_{i_{xx}}/(V_{i_{xx}} + 2\sigma F_{i_{xx}})$, then

$$\sigma \leq \frac{1}{2}\left|\frac{V_{i_{xx}}}{F_{i_{xx}}}\right|.$$

This inequality associated with the classical condition gives a new condition on the local time step Δt_j :

$$\Delta t_j \leq \min\left[\frac{2\Delta x}{3\max|\lambda_i|}, \frac{\Delta x}{2\max\left|\frac{F_{i_{xx}}}{V_{i_{xx}}}\right|}\right] \quad \text{where the values of the second deriva-}$$

tives act. λ_i represent the eigenvalues of the Jacobian A of F. Δt has to verify $\Delta t \leq \min_j \Delta t_j$.

For example, with the scalar non-linear equation (inviscid Burgers equation), the condition on Δt_j, associated with an extremum ($V_x = 0$ and $V_{xx} \neq 0$), is the most restrictive since $|\lambda_i| = u$ and $\left|\frac{F_{i_{xx}}}{V_{i_{xx}}}\right| = u$ and therefore $\Delta t_j \leq \frac{\Delta x}{2u}$.

This result has not been generalized to a scheme with three-time stepping; but it would seem that the spatial error terms would be still the same with probably, a new condition on the CFL number. Knowing $W_{ttt} = -\left(F'^2_W(F'_W W_x)\right)_{xx}$ [54], we can write the condition (ES):

$$W_t + F_x - \frac{\Delta t^2}{6}(A^3 V_x)_{xx}$$

$$+ 2^{nd} \text{ order spatial error terms } = O(\Delta x^3, \Delta t^3).$$

The time error terms are similar and of the same order as the spatial error terms when the CFL number is close to unity and, what is essential, there is theoretically no way of canceling or even controlling them. That is to say, if we keep this accuracy in time, all the effort devoted to the spatial discretization in order to control the error terms will become useless. Since all these terms cannot be eliminated easily (they are still more complicated than the spatial error terms), it is recommended that a higher order scheme in time (at least third-order) be applied in order to remove them automatically. The second-order and, even more, the first-order error terms are then controlled by the spatial discretization. For example, we can use the third-order scheme in time defined in [55].

12.5 Applications

In 1-D, the example proposed by Shu and Osher [51] is interesting because it uses the Euler equations to simulate the interaction between a moving Mach 3 shock and a turbulent flow represented by sine waves in density. The initial conditions are described as:

- $\rho = 3.857143$, $u = 2.629369$, $p = 10.33333$ if $x < -4$,
- $\rho = 1 + 0.2 \sin 5x$, $u = 0$, $p = 1.$ if $x \geq -4$.

As in [51], the CFL number is equal to 0.5 and the final time is $t = 1.8$. Since the exact solution for this problem is unknown, the solid line representing the numerical solution with 1600 cells is assumed to be the exact solution.

Figs. 13.16a, b and c show the solution of the density field with 400 cells and the limiters *minmod*, *superbee*, and $\varphi = (r + 2)/3$, applied separately. The limiters *minmod* and *superbee* give middling solutions. If AUSM-VL splittings (12.3–12.4) with the selected triad (12.13–12.15) are applied (Fig.13.16d), the solution is comparable with that of the third-order ENO scheme. In particular, the high frequencies are well-represented and the compression waves and the shock are well-captured. If the nodes where (outside the shock wave) the scheme degenerates to a first-order scheme in space are plotted (Fig. 13.17), we see that it degenerates, not in the regions of strong fluctuations but rather in the relatively quiet regions; that is to say, to eliminate essentially numerical micro-oscillations.

The second test case is a 2-D axisymmetric supersonic mixing layer. The inlet conditions are:

- central jet: $M = 1.74$, $p = 8 \; 10^3$ Pa, $T = 200$ K
- peripheral jet: $M = 2.$, $p = 8 \; 10^3$ Pa, $T = 580$ K.

We solve the Euler equations using a splitting method. The 2-D finite difference operator is split into a product of simpler operators $U^{n+2} = (L_r L_z L_z L_r) \, U^n$ where L_r and L_z are hyperbolic 1-D difference operators in directions r and z. The computations have been done on a grid mesh of 351 nodes in z direction and 93 nodes in r direction. CFL number is equal to 0.5.

Fig. 13.18 a shows an instantaneous view of the temperature field, with the scheme using AUSM splitting and the limiter $\varphi = (r + 2)/3$, and Fig. 13.18b shows the same view, with the same time-stepping scheme but using the conditions 1 and 2 for AUSM-VL splittings and the triad of limiters. The transitional zone (A) shows a greater sensitivity of the scheme proposed here to the physical instabilities. We can also see the very weak diffusion of the scheme in the shear layer. In the growth of the large eddies (B), the mixing in the core of the eddies is more detailed with the method presented here.

12.6   Summary

This section shows it is possible to improve the accuracy of TVD-MUSCL approach if:

- the accuracy in time is greater than the accuracy in space,

- the non-linear functions φ are expressed in a triad (13-15) taking into account the local variations of each quantity,

- AUSM splitting (12.4) is used when the scheme degenerates into a first-order, and VL splitting (12.3) is applied when the scheme remains second or third-order.

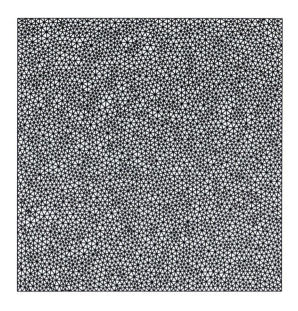

Fig. 13.4. Zoom of a typical mesh, 4545 nodes, 8848 triangles

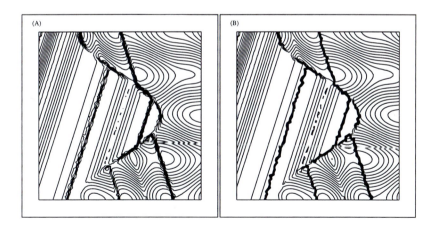

Fig. 13.5. Reconstructed function (A) and exact function (B)

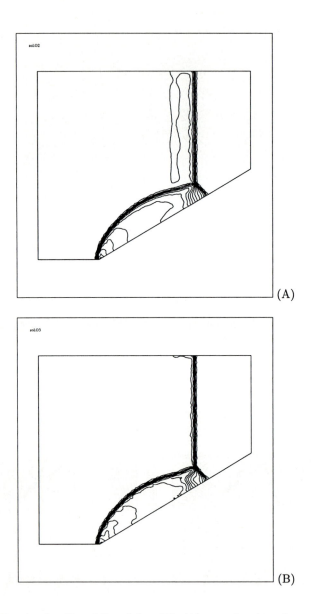

(A)

(B)

Fig. 13.6. Density for $M = 5.5$ and $\theta = 30^o$, (A):second order scheme, (B): third order scheme

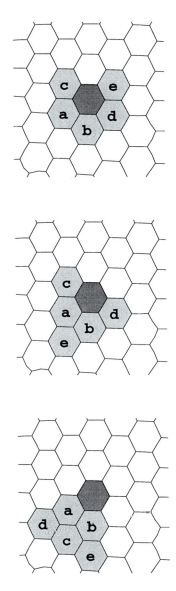

Fig. 13.7. Different types of stencils for quadratic reconstruction.

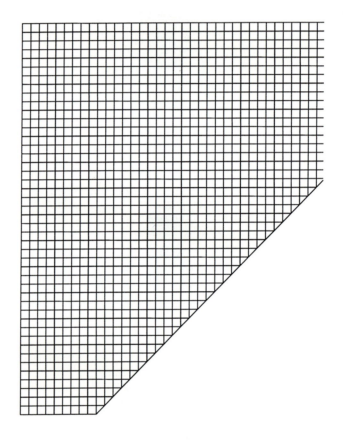

Fig. 13.8. Mesh for $M = 5.5$ and $\theta = 45^o$

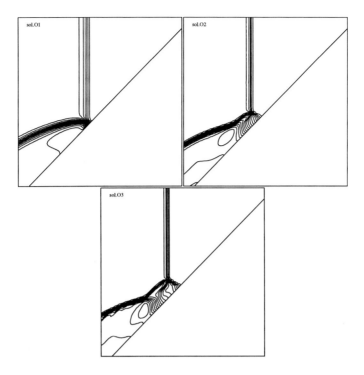

Fig. 13.9. Density for $M = 5.5$ and $\theta = 45°$, (A) : first order, (B) : second order, (C): third order. Zoom around the triple point

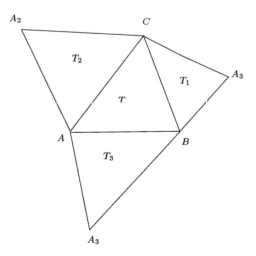

Fig. 13.10. The stencil for the reconstruction.

Fig. 13.11. Encoding/Decoding procedure: Fine, medium and coarse discretizations and the truncated scale coefficients.

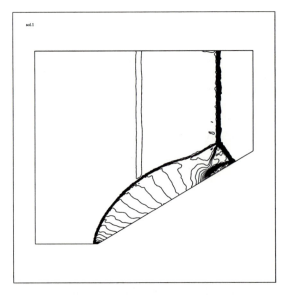

Fig. 13.12. 40 Isolines of the Mach number

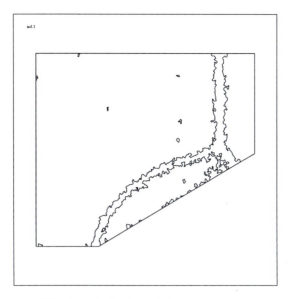

Fig. 13.13. Isolines of the error sensor

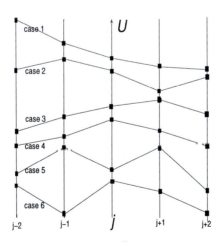

Fig. 13.14. Different evolutions of W.

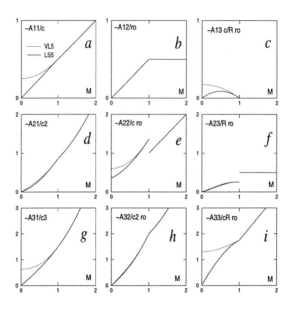

Fig. 13.15. Evolution of the first-order error terms.

A   Details of (36)

The detailed expression of (37) for each equation is written:

A.1   Continuity equation

$$\rho_t + (\rho u)_x + \Delta x^2 \{ \tfrac{\chi_1}{6} (cM\rho_{xxx} + \rho u_{xxx}) + \tfrac{\chi_2}{4} (\rho_x u_x)_x$$

$$+ \tfrac{(LogT)_x^2}{32} [-(3\rho u_x + cM\rho_x) + 2\rho cM(LogT)_x] I_{ausm} \}$$

$$= O(\Delta x^3)$$

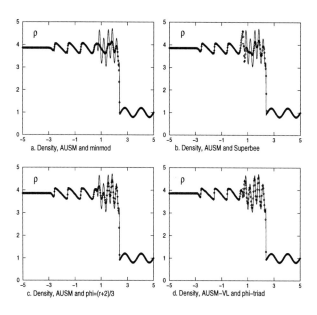

Fig. 13.16. Moving shock in a sinusoidal density field.

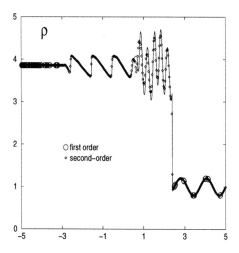

Fig. 13.17. Location of degeneration into first-order.

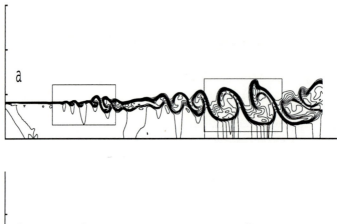

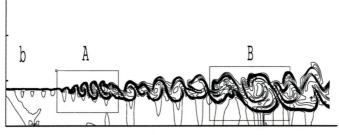

Fig. 13.18. Instantaneous temperature field in a supersonic jet.

A.2   Momentum equation

$$(\rho u)_t + (\rho u^2 + p)_x$$

$$+\Delta x^2 \{ \tfrac{\chi_1}{6} [c^2(M^2 + \tfrac{1}{\gamma})\rho_{xxx} + 2\rho c M u_{xxx} + \rho R T_{xxx}]$$

$$+\tfrac{\chi_2}{4} [2cM(\rho_x u_x)_x + \rho(u_x u_x)_x + R(\rho_x T_x)_x]$$

$$+(\tfrac{I_{ausm}}{2} + \tfrac{I_{vl}}{12})\rho_x u_x u_x$$

$$+\tfrac{(LogT)_x}{32} [4cM\rho_x u_x - 4\rho u_x u_x + c^2 M^2 \rho_x (LogT)_x$$

$$+2\rho c^2 M^2 (LogT)_x^2] I_{ausm} \}$$

$$= O(\Delta x^3)$$

A.3   Energy equation

$$(\rho E)_t + (\rho u H)_x$$

$$+\Delta x^2 \{ \tfrac{\chi_1}{6}[c^3 M(\tfrac{M^2}{2} + C_2)\rho_{xxx}$$
$$+\rho c^2(\tfrac{3M^2}{2} + C_2)u_{xxx} + \tfrac{\rho C_2 c^3 M}{T}T_{xxx}]$$

$$+\tfrac{\chi_2}{4}[c^2(\tfrac{3M^2}{2} + C_2)(\rho_x u_x)_x + \tfrac{3\rho c M}{2}(u_x u_x)_x + \tfrac{\rho C_2 c^2}{T}(u_x T_x)_x]$$

$$+\tfrac{\chi_2}{4}[C_p c M(\rho_x T_x)_x]I_{vl}$$

$$+(\tfrac{5 I_{ausm}}{8} + \tfrac{I_{vl}}{8})c M \rho_x u_x u_x + (\tfrac{3}{8}C_p \rho_x u_x T_x + \tfrac{1}{8}\rho u_x u_x u_x)I_{vl}$$

$$+\tfrac{(LogT)_x}{32}[16c^2(C_2 - \tfrac{M^2}{4})\rho_x u_x + 2\rho c M u_x u_x$$

$$-c^3 M(3C_2 - \tfrac{M^2}{2})\rho_x(LogT)_x - \rho c^2(3C_2 + \tfrac{M^2}{2})u_x(LogT)_x$$

$$+\tfrac{3}{2}\rho c^3 M^3(LogT)_x^2]I_{ausm}\}$$

$$= O(\Delta x^3)$$

where $I_{ausm} = 1$ and $I_{vl} = 0$ for AUSM splitting, and $I_{ausm} = 0$ and $I_{vl} = 1$ for VL splitting.

B   A piece of code

For cases 1-4, the calculation of the right- and left-values of U (MUSCL approach) is done. The lines written with the capital letters already exist in codes with the classical limiters. The ten lines written with the small letters correspond with the adding of the triad of limiters. The writing is very simple and the additional consuming time for a complete code is very small.

```
c The right and left values are calculated as
C follows
c
c UR(L+1/2)=U(L+1)-PHI(1/R(L+1))
c                       *( U(L+1)-U(L) )
c
c UL(L+1/2)=U(L)   +PHI(R(L))
c                       *( U(L+1)-U(L) )
c
c R(L)=( U(L)-U(L-1) )/( U(L+1)-U(L) )
c
c if R <= 0
C PHI(R)=0
```

```
C if R>0
c PHI(R)= ( (1-ETA)* MIN( R, (3-ETA)/(1-ETA))
c               + (1+ETA) *
c            MIN( 1, (3-ETA)*R/(1-ETA)) )/4
c
c DELSIGN(L) = sign of variation U(L+1)-U(L)
c               (+1 if U(L+1)-U(L)=0)
c
c if is1234=1 and is23=1 we have case 1
c if is1234=0 and is23 we have cases 2 and 3
c if is1234=0 and is23=0 we have case 4
c
c case 1        : PHI(R)=(R+2)/3 if 2/5 < R < 4
c                  2        if R>4
c                  2*R      if 0<R<2/5
c
c case 2 and 3 : PHI(R)=1    if R>1/2
c                        2*R if 0 < R< 1/2
c
c case 4        : PHI(R)= SUPERBEE(R)
c
c ************ Phi-triad *************

DO L = 1, LMAX-1

DELU(L) = U (L+1) - U (L)

DELSIGN(L) = SIGN (1., DELU(L))

ENDDO

DO L = 3, LMAX-2

DEL = DELU (L) - DELU (L-1)

c*** automatic choice of the limiter ***

etasbee = dim(sign(1., del), 0.)
            - dim(0., sign(1., del))

is1 = delsign (L-2)

is2 = delsign (L-1)

is3 = delsign (L )
```

```
is4 = delsign (L+1)

is1234 = iabs (is1+is2+is3+is4) /4

eta13 = is1234 / 3. + (1-is1234)

is23 = iabs (is2+is3) /2

eta(L) = eta13 * is23 + etasbee * (1-is23)

omega = is23 * (3. - eta(L))
              / (1. - eta(L)) + 2.* (1-is23)

c************

A = DELU (L ) * DELSIGN (L-1)

B = DELU (L-1) * DELSIGN (L )

ABMIN = MIN (A,OMEGA*B)

DELUP (L) = MAX(0.,ABMIN) * DELSIGN (L )

ABMIN = MIN (B,OMEGA*A)

DELUM (L) = MAX(0.,ABMIN) * DELSIGN (L-1)

ENDDO

c*U-right and U-left at the interface L+1/2 *

DO L = 3 , LMAX-3

UR (L) = U (L+1)
         - 0.25*( (1.-ETA (L+1))* DELUP (L+1)
         + (1.+ETA (L+1))*DELUM (L+1) )

UL (L) = U (L )
         + 0.25*( (1.-ETA (L))* DELUM (L)
         + (1.+ETA (L))* DELUP(L) )

ENDDO
```

Discontinuous Galerkin Methods for Convection-Dominated Problems

Bernardo Cockburn[1]

School of Mathematics, University of Minnesota,
206 Church Street S.E., Minneapolis, MN 55455, USA

Abstract. We present and analyze the Runge Kutta Discontinuous Galerkin method for numerically solving nonlinear hyperbolic systems. The basic method is then extended to convection-dominated problems yielding the Local Discontinuous Galerkin method. These methods are particularly attractive since they achieve formal high-order accuracy, nonlinear stability, and high parallelizability while maintaining the ability to handle complicated geometries and capture the discontinuities or strong gradients of the exact solution without producing spurious oscillations. The discussed methods are readily applied to the Euler equations of gas dynamics, the shallow water equations, the equations of magneto-hydrodynamics, the compressible Navier-Stokes equations with high Reynolds numbers, and the equations of the hydrodynamic model for semiconductor device simulation. As a final example, consideration is given to the application of the discontinuous Galerkin method to the Hamilton-Jacobi equations.

Table of Contents

1   Introduction

1.1   The purpose of these notes

In these notes, we study the Runge Kutta Discontinuous Galerkin method for numerically solving nonlinear hyperbolic systems and its extension for convection-dominated problems, the so-called Local Discontinuous Galerkin method. Examples of problems to which these methods can be applied are the Euler equations of gas dynamics, the shallow water equations, the equations of magneto-hydrodynamics, the compressible Navier-Stokes equations with high Reynolds numbers, and the equations of the hydrodynamic model for semiconductor device simulation; applications to Hamilton-Jacobi equations is another important example. The main features that make the methods under consideration attractive are their formal high-order accuracy, their nonlinear stability, their high parallelizability, their ability to handle complicated geometries, and their ability to capture the discontinuities or strong gradients of the exact solution without producing spurious oscillations. The purpose of these notes is to provide a short introduction to the devising and analysis of these discontinuous Galerkin methods. Most of the material of these notes has been presented in [17].

Acknowledgements The author would like to thank T.J. Barth for the invitation to give a series of lectures in the NATO special course on ' Higher Order Discretization Methods in Computational Fluid Dynamics,' the material of which is contained in these notes. He would also like to thank F. Bassi and S. Rebay, and I. Lomtev and G. Karniadakis for kindly suplying several of their figures, and to C. Hu and Chi-Wang Shu for provinding their material on Hamilton-Jacobi equations. Thanks are also due to Rosario Grau for fruitful discussions concerning the numerical experiments of Chapter 2, to J.X. Yang for a careful proof-reading the appendix of Chapter 6, and to A. Zhou for bringing the author's attention to several of his papers concerning the discontinuous Galerkin method.

1.2   A historical overview

The original Discontinuous Galerkin method The original discontinuous Galerkin (DG) finite element method was introduced by Reed and Hill [76] for solving the neutron transport equation

$$\sigma\, u + div(\, \overline{a}\, u) = f,$$

where σ is a real number and $\overline{a}$ a constant vector. A remarkable advantage of this method is that, because of the linear nature of the equation, the approximate solution can be computed element by element when the elements are suitably ordered according to the characteristic direction.

LeSaint and Raviart [58] made the first analysis of this method and proved a rate of convergence of $(\Delta x)^{k}$ for general triangulations and of $(\Delta x)^{k+1}$ for Cartesian grids. Later, Johnson and Pitkaränta [52] proved a rate of convergence of $(\Delta x)^{k+1/2}$ for general triangulations and Peterson [75] numerically confirmed this rate to be optimal. Richter [77] obtained the optimal rate of convergence of $(\Delta x)^{k+1}$ for some structured two-dimensional non-Cartesian grids. In all the above papers, the exact solution is assumed to be very smooth. The case in which the solution admits discontinuities was treated by Lin and Zhou [60] who proved the convergence of the method. The issue of the interrelation between the mesh and the order of convergence of the method was explored by Zhou and Lin [93], case $k = 1$, and later by Lin, Yan, and Zhou [59], case $k = 0$, and optimal error estimates were proven under suitable assumptions on the mesh. Recently, several new results have been obtained. Thus, Falk and Richter [39] obtained a rate of convergence of $(\Delta x)^{k+1/2}$ for general triangulations for Friedrich systems; Houston, Schwab and Süli [42] analyzed the *hp* version of the discontinuous Galerkin method and showed its exponential convergence when the solution is piecewise analytic; and, finally, Cockburn, Luskin, Shu, and Süli [22] showed how to exploit the translation invariance of a grid to *double* the order of convergence of the method by a simple, local postprocessing of the approximate solution.

Nonlinear hyperbolic systems: The RKDG method The success of this method for linear equations, prompted several authors to try to extend the method to nonlinear hyperbolic conservation laws

$$u_t + \sum_{i=1}^{d} (f_i(u))_{x_i} = 0,$$

equipped with suitable initial or initial–boundary conditions. However, the introduction of the nonlinearity prevents the element-by-element computation of the solution. The scheme defines a nonlinear system of equations that must be solved all at once and this renders it computationally very inefficient for hyperbolic problems.

• **The one-dimensional scalar conservation law.**

To avoid this difficulty, Chavent and Salzano [13] constructed an explicit version of the DG method in the one-dimensional scalar conservation law. To do that, they discretized in space by using the DG method with piecewise linear elements and then discretized in time by using the simple Euler forward method. Although the resulting scheme is explicit, the classical von Neumann analysis shows that it is unconditionally unstable when the ratio $\frac{\Delta t}{\Delta x}$ is held constant; it is stable if $\frac{\Delta t}{\Delta x}$ is of order $\sqrt{\Delta x}$, which is a very restrictive condition for hyperbolic problems.

To improve the stability of the scheme, Chavent and Cockburn [12] modified the scheme by introducing a suitably defined 'slope limiter' following the ideas introduced by van Leer in [88]. They thus obtained a scheme that was proven to be total variation diminishing in the means (TVDM) and total variation bounded (TVB) under a fixed CFL number, $f' \frac{\Delta t}{\Delta x}$, that can be chosen to be less than or equal to $1/2$. Convergence of a subsequence is thus guaranteed, and the numerical results given in [12] indicate convergence to the correct entropy solutions. However, the scheme is only first order accurate in time and the 'slope limiter' has to balance the spurious oscillations in smooth regions caused by linear instability, hence adversely affecting the quality of the approximation in these regions.

These difficulties were overcome by Cockburn and Shu in [26], where the first Runge Kutta Discontinuous Galerkin (RKDG) method was introduced. This method was constructed (i) by retaining the piecewise linear DG method for the space discretization, (ii) by using a special explicit TVD second order Runge-Kutta type discretization introduced by Shu and Osher in a finite difference framework [80], [81], and (iii) by modifying the 'slope limiter' to maintain the formal accuracy of the scheme extrema. The resulting explicit scheme was then proven linearly stable for CFL numbers less than 1/3, formally uniformly second order accurate in space and time including at extrema, and TVBM. Numerical results in [26] indicate good convergence behavior: Second order in smooth regions including extrema, sharp shock transitions (usually in one or two elements) without oscillations, and convergence to entropy solutions even for non convex fluxes.

In [24], Cockburn and Shu extended this approach to construct (formally) high-order accurate RKDG methods for the scalar conservation law. To device RKDG methods of order $k + 1$, they used (i) the DG method with polynomials of degree k for the space discretization, (ii) a TVD $(k + 1)$-th order accurate explicit time discretization, and (iii) a generalized 'slope limiter.' The generalized 'slope limiter' was carefully devised with the purpose of enforcing the TVDM property without destroying the accuracy of the scheme. The numerical results in [24], for $k = 1, 2$, indicate $(k+1)$-th order order in smooth regions away from discontinuities as well as sharp shock transitions with no oscillations; convergence to the entropy solutions was observed in all the tests. These RKDG schemes were extended to one-dimensional systems in [21].

- **The multidimensional case.**

The extension of the RKDG method to the multidimensional case was done in [20] for the scalar conservation law. In the multidimensional case, the complicated geometry the spatial domain might have in practical applications can be easily handled by the DG space discretization. The TVD time discretizations remain the same, of course. Only the construction of the generalized 'slope limiter' represents a serious challenge. This is so, not only because of the more complicated form of the elements but also because of inherent accuracy barriers imposed by the stability properties.

Indeed, since the main purpose of the 'slope limiter' is to enforce the nonlinear stability of the scheme, it is essential to realize that in the multidimensional case, the constraints imposed by the stability of a scheme on its accuracy are even greater than in the one dimensional case. Although in the one dimensional case it is possible to devise high-order accurate schemes with the TVD property, this is not so in several space dimensions since Goodman and LeVeque [41] proved that any TVD scheme is at most first order accurate. Thus, any generalized 'slope limiter' that enforces the TVD property, or the TVDM property for that matter, would unavoidably reduce the accuracy of the scheme to first-order accuracy. This is why in [20], Cockburn, Hou and Shu devised a generalized 'slope limiter' that enforced a **local** maximum principle only since they are not incompatible with high-order accuracy. No other class of schemes has a proven maximum principle for general nonlinearities **f** and arbitrary triangulations.

The extension of the RKDG methods to general multidimensional systems was started by Cockburn and Shu in [25] and has been recently completed in [28]. Bey and Oden [10], Bassi and Rebay [4], and more recently Baumann [6] and Baumann and Oden [9] have studied applications of the method to the Euler equations of gas dynamics. Recently, Kershaw *et al.* [56], from the Lawrence Livermore National Laboratory, extended the method to arbitrary Lagrangian-Eulerian fluid flows where the computational mesh can move to track the interface between the different material species.

- **The main advantages of the RKDG method.**

The resulting RKDG schemes have several important advantages. First, like finite element methods such as the SUPG-method of Hughes and Brook [44], [49], [45], [46], [47], [48] (which has been analyzed by Johnson *et al.* in [53], [54], [55]), the RKDG methods are better suited than finite difference methods to handle complicated geometries. Moreover, the particular finite elements of the DG space discretization allow an extremely simple treatment of the boundary conditions; no special numerical treatment of them is required in order to achieve uniform high order accuracy, as is the case for the finite difference schemes.

Second, the method can easily handle adaptivity strategies since the refining or unrefining of the grid can be done without taking into account the continuity restrictions typical of conforming finite element methods. Also, the degree of the approximating polynomial can be easily changed from one element to the other. Adaptivity is of particular importance in hyperbolic problems given the complexity of the structure of the discontinuities. In the one dimensional case the Riemann problem can be solved in closed form and discontinuity curves in the (x, t) plane are simple straight lines passing through the origin. However, in two dimensions their solutions display a very rich structure; see the works of Wagner [90], Lindquist [62], [61], Tong and Zheng [86], and Tong and Chen [85]. Thus, methods which allow triangulations that can be easily adapted to resolve this structure, have an important advantage.

Third, the method is highly parallelizable. Since the elements are discontinuous, the mass matrix is block diagonal and since the order of the blocks is equal to the number of degrees of freedom inside the corresponding elements, the blocks can be inverted by hand once and for all. Thus, at each Runge-Kutta inner step, to update the degrees of freedom inside a given element, only the degrees of freedom of the elements sharing a face are involved; communication between processors is thus kept to a minimum. Extensive studies of adaptivity and parallelizability issues of the RKDG method have been performed by Biswas, Devine, and Flaherty [11], Devine, Flaherty, Loy, and Wheat [32], Devine and Flaherty [31], and more recently by Flaherty *et al.* [40]. Studies of load balancing related to conservation laws but not restricted to them can be found in the works by Devine, Flaherty, Wheat, and Maccabe [33], by deCougny *et al.* [30], and by Özturan *et al.* [74].

Convection-diffusion systems: The LDG method The first extensions of the RKDG method to nonlinear, convection-diffusion systems of the form

$$\partial_t \mathbf{u} + \nabla \cdot \mathbf{F}(\mathbf{u}, D\,\mathbf{u}) = 0, \text{ in } (0, T) \times \Omega,$$

were proposed by Chen *et al.* [15], [14] in the framework of hydrodynamic models for semiconductor device simulation. In these extensions, approximations of second and third-order derivatives of the discontinuous approximate solution were obtained by using simple projections into suitable finite elements spaces. This projection requires the inversion of global mass matrices, which in [15] and [14] were 'lumped' in order to maintain the high parallelizability of the method. Since in [15] and [14] polynomials of degree one are used, the 'mass lumping' is justified; however, if polynomials of higher degree were used, the 'mass lumping' needed to enforce the full parallelizability of the method could cause a degradation of the formal order of accuracy.

Fortunately, this is not an issue with the methods proposed by Bassi and Rebay [3] (see also Bassi *et al* [4]) for the compressible Navier-Stokes equations. In these methods, the original idea of the RKDG method is applied to *both u and D u* which are now considered as *independent* unknowns. Like the RKDG methods, the resulting methods are highly parallelizable methods of high-order accuracy which are very efficient for time-dependent, convection-dominated flows. The LDG methods considered by Cockburn and Shu [27] are a generalization of these methods.

The basic idea to construct the LDG methods is to *suitably rewrite* the original system as a larger, degenerate, first-order system and then discretize it by the RKDG method. By a careful choice of this rewriting, nonlinear stability can be achieved even without slope limiters, just as the RKDG method in the purely hyperbolic case; see Jiang and Shu [51]. Moreover, error estimates (in the linear case) have been obtained in [27]. A recent analysis of this method is currently being carried out by Cockburn and Schwab [23] in the one dimensional case by taking into account the characterization of the viscous boundary layer of the exact solution.

The LDG methods [27] are very different from the so-called Discontinuous Galerkin (DG) method for parabolic problems introduced by Jamet [50] and studied by Eriksson, Johnson, and Thomée [38], Eriksson and Johnson [34], [35], [36], [37], and more recently by Makridakis and Babuška [68]. In the DG method, the approximate solution is discontinuous only in time, not in space; in fact, the space discretization is the standard Galerkin discretization with *continuous* finite elements. This is in strong contrast with the space discretizations of the LDG methods which use *discontinuous* finite elements. To emphasize this difference, those methods are called **Local** Discontinuous Galerkin methods. The large amount of degrees of freedom and the restrictive conditions of the size of the time step for explicit time-discretizations, render the LDG methods inefficient for diffusion-dominated problems; in this situation, the use of methods with continuous-in-space approximate solutions is recommended. However, as for the successful RKDG methods for purely hyperbolic problems, the extremely local domain of dependency of the LDG methods allows a very efficient parallelization that by far compensates for the extra amount of degrees of freedom in the case of convection-dominated flows. Karniadakis *et al.* have implemented and tested these methods for the compressible Navier Stokes equations in two and three space dimensions with impressive results; see [64], [65], [63], [66], and [91].

Another technique to discretize the diffusion terms have been proposed by Baumann [6]. The one-dimensional case was studied by Babuška, Baumann, and J.T. Oden [2] and the multidimensional case has been considered by Oden, Babuška, and Baumann [70]. The case of convection-diffusion in multidimensions was treated by Baumann and Oden in [7]. In [8], Baumann and Oden consider applications to the Navier-Stokes equations.

Finally, let us point bring the attention of the reader to the non-conforming staggered-grid Chebyshev spectral multidomain numerical method for the solution of the compressible Navier-Stokes equations proposed and studied by Kopriva [57]; this method is strongly related to discontinuous Galerkin methods.

1.3   The content of these notes

In these notes, we study the RKDG and LDG methods. Our exposition will be based on the papers by Cockburn and Shu [26], [24], [21], [20], and [28] in which the RKDG method was developed and on the paper by Cockburn and Shu [27] which is devoted to the LDG methods. We also include numerical results from the papers by Bassi and Rebay [4] and by Warburton, Lomtev, Kirby and Karniadakis [91] on the Euler equations of gas dynamics and from the papers by Bassi and Rebay [3] and by Lomtev and Karniadakis [63] on the compressible Navier-Stokes equations. Finally, we also use the material contained in the paper by Hu and Shu [43] in which the application of the RKDG method is extended to Hamilton-Jacobi equations.

The emphasis in these notes is on *how the above mentioned schemes were devised*. As a consequence, the chapters that follow reflect that development. Thus, Chapter 2, in which the RKDG schemes for the one-dimensional scalar conservation law are constructed, constitutes the core of the notes because it contains all the important ideas for the devising of the RKDG methods; chapter 3 contains its extension to one-dimensional Hamilton-Jacobi equations. In chapter 4, we extend the RKDG method to multidimensional systems and in Chapter 5, to multidimensional Hamilton-Jacobi equations. Finally, in chapter 6 we study the extension to convection-diffusion problems.

We would like to emphasize that the guiding principle in the devising of the RKDG methods for scalar conservation laws is to consider them as *perturbations of the so-called monotone schemes*. As it is well-known, monotone schemes for scalar conservation laws are stable and converge to the entropy solution but are only first-order accurate. Following a widespread approach in the field of numerical schemes for nonlinear conservation laws, the RKDG are constructed in such a way that they are high-order accurate schemes that 'become' a monotone scheme when a piecewise-constant approximation is used. Thus, to obtain high-order accurate RKDG schemes, we 'perturb' the piecewise-constant approximation and allow it to be piecewise a polynomial of arbitrary degree. Then, the conditions under which the stability properties of the monotone schemes are still valid are sought and enforced by means of the generalized 'slope limiter.' The fact that it is possible to do so without destroying the accuracy of the RKDG method is the crucial point that makes this method both robust and accurate.

The issues of parallelization and adaptivity developed by Biswas, Devine, and Flaherty [11], Devine, Flaherty, Loy, and Wheat [32], Devine and Flaherty [31], and by Flaherty *et al.* [40] (see also the works by Devine, Flaherty, Whea, and Maccabe [33], by deCougny *et al.* [30], and by Özturan *et al.* [74]) are certainly very important. Another issue of importance is how to render the method computationally more efficient, like the quadrature rule-free versions of the RKDG method recently studied by Atkins and Shu [1]. However, these topics fall beyond the scope of these notes whose main intention is to provide a simple introduction to the topic of discontinuous Galerkin methods for convection-dominated problems.

2   The scalar conservation law in one space dimension

2.1   Introduction

In this section, we introduce and study the RKDG method for the following simple
model problem:

$$u_t + f(u)_x = 0, \quad \text{in } (0,1) \times (0,T), \tag{2.1}$$

$$u(x,0) = u_0(x), \quad \forall \, x \in (0,1), \tag{2.2}$$

and periodic boundary conditions. This section has material drawn from [26] and
[24].

2.2   The discontinuous Galerkin-space discretization

The weak formulation To discretize in space, we proceed as follows. For each
partition of the interval $(0,1)$, $\{x_{j+1/2}\}_{j=0}^{N}$, we set $I_j = (x_{j-1/2}, x_{j+1/2})$, $\Delta_j = x_{j+1/2} - x_{j-1/2}$ for $j = 1, \ldots, N$, and denote the quantity $\max_{1 \le j \le N} \Delta_j$ by Δx .

We seek an approximation u_h to u such that for each time $t \in [0, T]$, $u_h(t)$
belongs to the finite dimensional space

$$V_h = V_h^k \equiv \{v \in L^1(0,1) : \quad v|_{I_j} \in P^k(I_j), \; j = 1, \ldots, N\}, \tag{2.3}$$

where $P^k(I)$ denotes the space of polynomials in I of degree at most k. In order to
determine the approximate solution u_h, we use a weak formulation that we obtain
as follows. First, we multiply the equations (2.1) and (2.2) by arbitrary, smooth
functions v and integrate over I_j, and get, after a simple formal integration by
parts,

$$\int_{I_j} \partial_t \, u(x,t) \, v(x) \, dx - \int_{I_j} f(u(x,t)) \, \partial_x \, v(x) \, dx$$

$$+ f(u(x_{j+1/2}, t)) \, v(x_{j+1/2}^-) - f(u(x_{j-1/2}, t)) \, v(x_{j-1/2}^+) = 0, \tag{2.4}$$

$$\int_{I_j} u(x,0) \, v(x) \, dx = \int_{I_j} u_0(x) \, v(x) \, dx. \tag{2.5}$$

Next, we replace the smooth functions v by test functions v_h belonging to the finite
element space V_h, and the exact solution u by the approximate solution u_h. Since
the function u_h is discontinuous at the points $x_{j+1/2}$, we must also replace the
nonlinear 'flux' $f(u(x_{j+1/2}, t))$ by a *numerical* 'flux' that depends on the two values
of u_h at the point $(x_{j+1/2}, t)$, that is, by the function

$$h(u)_{j+1/2}(t) = h(u(x_{j+1/2}^-, t), u(x_{j+1/2}^+, t)), \tag{2.6}$$

that will be suitably chosen later. Note that *we always use the same numerical flux
regardless of the form of the finite element space*. Thus, the approximate solution
given by the DG-space discretization is defined as the solution of the following weak

formulation:

$$\forall\ j = 1, \ldots, N, \qquad \forall\ v_h \in P^k(I_j):$$

$$\int_{I_j} \partial_t\, u_h(x,t)\, v_h(x)\, dx - \int_{I_j} f(u_h(x,t))\, \partial_x\, v_h(x)\, dx$$

$$+h(u_h)_{j+1/2}(t)\, v_h(x_{j+1/2}^-) - h(u_h)_{j-1/2}(t)\, v_h(x_{j-1/2}^+) = 0, \qquad (2.7)$$

$$\int_{I_j} u_h(x,0)\, v_h(x)\, dx = \int_{I_j} u_0(x)\, v_h(x)\, dx. \qquad (2.8)$$

Incorporating the monotone numerical fluxes To complete the definition of the approximate solution u_h, it only remains to choose the numerical flux h. To do that, we invoke our main point of view, namely, that *we want to construct schemes that are perturbations of the so-called monotone schemes*. The idea is that by *perturbing* the monotone schemes, we would achieve high-order accuracy while keeping their stability and convergence properties. Thus, we want that in the case $k = 0$, that is, when the approximate solution u_h is a piecewise-constant function, our DG-space discretization gives rise to a monotone scheme.

Since in this case, for $x \in I_j$ we can write

$$u_h(x,t) = u_j^0,$$

we can rewrite our weak formulation (2.7), (2.8) as follows:

$$\forall\ j = 1, \ldots, N:$$
$$\partial_t\, u_j^0(t) + \{h(u_j^0(t), u_{j+1}^0(t)) - h(u_{j-1}^0(t), u_j^0(t))\}/\Delta_j = 0,$$
$$u_j^0(0) = \frac{1}{\Delta_j} \int_{I_j} u_0(x)\, dx,$$

and it is well-known that this defines a monotone scheme if $h(a,b)$ is a Lipschitz, consistent, monotone flux, that is, if it is,

(i) locally Lipschitz and consistent with the flux $f(u)$, i.e., $h(u, u) = f(u)$,
(ii) a nondecreasing function of its first argument, and
(iii) a nonincreasing function of its second argument.

The best-known examples of numerical fluxes satisfying the above properties are the following:

(i) The Godunov flux:

$$h^G(a,b) = \begin{cases} \min_{a \le u \le b}\, f(u), & \text{if } a \le b \\ \max_{b \le u \le a}\, f(u), & \text{otherwise.} \end{cases}$$

(ii) The Engquist-Osher flux:

$$h^{EO}(a,b) = \int_0^b \min(f'(s), 0)\, ds + \int_0^a \max(f'(s), 0)\, ds + f(0);$$

(iii) The Lax-Friedrichs flux:

$$h^{LF}(a,b) = \frac{1}{2}\left[f(a) + f(b) - C(b-a)\right],$$

$$C = \max_{\inf u^0(x) \le s \le \sup u^0(x)} |f'(s)|;$$

(iv) The local Lax–Friedrichs flux:

$$h^{LLF}(a,b) = \frac{1}{2}\left[f(a) + f(b) - C(b-a)\right],$$

$$C = \max_{\min(a,b) \le s \le \max(a,b)} |f'(s)|;$$

(v) The Roe flux with 'entropy fix':

$$h^{R}(a,b) = \begin{cases} f(a) & \text{if } f'(u) \ge 0 \quad \text{for} \quad u \in [\min(a,b),\ \max(a,b)], \\ f(b) & \text{if } f'(u) \le 0 \quad \text{for} \quad u \in [\min(a,b),\max(a,b)], \\ h^{LLF}(a,b) & \text{otherwise.} \end{cases}$$

For the flux h, we can use the Godunov flux h^G since it is well-known that this is the numerical flux that produces the smallest amount of artificial viscosity. The local Lax-Friedrichs flux produces more artificial viscosity than the Godunov flux, but their performances are remarkably similar. Of course, if f is too complicated, we can always use the Lax-Friedrichs flux. However, numerical experience suggests that as the degree k of the approximate solution increases, the choice of the numerical flux does not have a significant impact on the quality of the approximations.

Diagonalizing the mass matrix If we choose the Legendre polynomials P_ℓ as local basis functions, we can exploit their L^2-orthogonality, namely,

$$\int_{-1}^{1} P_\ell(s)\, P_{\ell'}(s)\, ds = \left(\frac{2}{2\ell+1}\right)\delta_{\ell\ell'},$$

to obtain a *diagonal* mass matrix. Indeed, if, for $x \in I_j$, we express our approximate solution u_h as follows:

$$u_h(x,t) = \sum_{\ell=0}^{k} u_j^\ell\, \varphi_\ell(x),$$

where

$$\varphi_\ell(x) = P_\ell(2\,(x - x_j)/\Delta_j),$$

the weak formulation (2.7), (2.8) takes the following simple form:

$$\forall\ j = 1,\ldots,N \text{ and } \ell = 0,\ldots,k:$$

$$\left(\frac{1}{2\ell+1}\right)\partial_t u_j^\ell(t) - \frac{1}{\Delta_j}\int_{I_j} f(u_h(x,t))\,\partial_x\varphi_\ell(x)\,dx$$

$$+ \frac{1}{\Delta_j}\left\{ h(u_h(x_{j+1/2}))(t) - (-1)^\ell\, h(u_h(x_{j-1/2}))(t)\right\} = 0,$$

$$u_j^\ell(0) = \frac{2\ell+1}{\Delta_j}\int_{I_j} u_0(x)\,\varphi_\ell(x)\,dx,$$

where we have use the following properties of the Legendre polynomials:

$$P_\ell(1) = 1, \qquad P_\ell(-1) = (-1)^\ell.$$

This shows that after discretizing in space the problem (2.1), (2.2) by the DG method, we obtain a system of ODEs for the degrees of freedom that we can rewrite as follows:

$$\frac{d}{dt} u_h = L_h(u_h), \qquad \text{in } (0,T), \tag{2.9}$$

$$u_h(t = 0) = u_{0h}. \tag{2.10}$$

The element $L_h(u_h)$ of V_h is, of course, the approximation to $-f(u)_x$ provided by the DG-space discretization.

Note that if we choose a different local basis, the local mass matrix could be a full matrix but it will always be a matrix of order $(k+1)$. By inverting it by means of a symbolic manipulator, we can always write the equations for the degrees of freedom of u_h as an ODE system of the form above.

Convergence analysis of the linear case In the linear case $f(u) = c\,u$, the $L^\infty(0,T;L^2(0,1))$-accuracy of the method (2.7), (2.8) can be established by using the $L^\infty(0,T;L^2(0,1))$-stability of the method and the approximation properties of the finite element space V_h.

Note that in this case, all the fluxes displayed in the examples above coincide and are equal to

$$h(a,b) = c\,\frac{a+b}{2} - \frac{|c|}{2}(b-a). \tag{2.11}$$

The following results are thus for this numerical flux.

We state the L²-stability result in terms of the jumps of u_h across $x_{j+1/2}$ which we denote by

$$[u_h]_{j+1/2} \equiv u_h(x_{j+1/2}^+) - u_h(x_{j+1/2}^-).$$

Proposition 1. (L²-stability) *We have,*

$$\tfrac{1}{2}\|u_h(T)\|^2_{L^2(0,1)} + \Theta_T(u_h) \le \tfrac{1}{2}\|u_0\|^2_{L^2(0,1)},$$

where

$$\Theta_T(u_h) = \tfrac{|c|}{2} \int_0^T \sum_{1 \le j \le N} [u_h(t)]^2_{j+1/2}\, dt.$$

Note how the jumps of u_h are controlled by the L²-norm of the initial condition. This control reflects the subtle built-in dissipation mechanism of the DG-methods and is what allows the DG-methods to be more accurate than the standard Galerkin methods. Indeed, the standard Galerkin method has an order of accuracy equal to k whereas the DG-methods have an order of accuracy equal to $k+1/2$ for the same smoothness of the initial condition.

Theorem 2. (First L^2-error estimate) *Suppose that the initial condition u_0 belongs to $H^{k+1}(0,1)$. Let e be the approximation error $u - u_h$. Then we have,*

$$\| e(T) \|_{L^2(0,1)} \leq C \, | u_0 |_{H^{k+1}(0,1)} (\Delta x)^{k+1/2},$$

where C depends solely on k, $|c|$, and T.

It is also possible to prove the following result if we assume that the initial condition is more regular. Indeed, we have the following result.

Theorem 3. (Second L^2-error estimate) *Suppose that the initial condition u_0 belongs to $H^{k+2}(0,1)$. Let e be the approximation error $u - u_h$. Then we have,*

$$\| e(T) \|_{L^2(0,1)} \leq C \, | u_0 |_{H^{k+2}(0,1)} (\Delta x)^{k+1},$$

where C depends solely on k, $|c|$, and T.

Theorem 2 is a simplified version of a more general result proven in 1986 by Johnson and Pitkäranta [52] and Theorem 3 is a simplified version of a more general result proven in 1974 by LeSaint and Raviart [58]. To provide a simple introduction to the techniques used in these general results, we give *new* proofs of Theorems 2 and 3 in an appendix to this chapter.

The above theorems show that the DG-space discretization results in a $(k+1)$th-order accurate scheme, at least in the linear case. This gives a strong indication that the same order of accuracy should hold in the nonlinear case when the exact solution is smooth enough, of course.

Now that we know that the DG-space discretization produces a high-order accurate scheme for smooth exact solutions, we consider the question of how does it behave when the flux is a nonlinear function.

Convergence analysis in the nonlinear case To study the convergence properties of the DG-method, we first study the convergence properties of the solution w of the following problem:

$$w_t + f(w)_x = (\nu(w) \, w_x)_x, \tag{2.12}$$
$$w(\cdot, 0) = u_0(\cdot), \tag{2.13}$$

and periodic boundary conditions. We then mimic the procedure to study the convergence of the DG-method for the piecewise-constant case. The general DG-method will be considered later after having introduced the Runge-Kutta time-discretization.

The continuous case as a model. In order to compare u and w, it is *enough* to have (i) an entropy inequality and (ii) uniform boundedness of $\| w_x \|_{L^1(0,1)}$. Next, we show how to obtain these properties in a formal way.

We start with the entropy inequality. To obtain such an inequality, the basic idea is to multiply the equation (2.12) by $U'(w-c)$, where $U(\cdot)$ denotes the absolute value function and c denotes an arbitrary real number. Since

$$U'(w-c)\,w_t = U(w-c)_t,$$
$$U'(w-c)\,f(w)_x = \left(U'(w-c)\,(f(w)-f(c))\right)$$
$$\equiv F(w,c)_x,$$

and since

$$U'(w-c)\,(\nu(w)\,w_x)_x = \left(\int_c^w U'(\rho-c)\,\nu(\rho)\,d\rho\right)_{xx} - U''(w-c)\,\nu(w)\,(w_x)^2$$
$$\equiv \Phi(w,c)_x x - U''(w-c)\,\nu(w)\,(w_x)^2,$$

we obtain

$$U(w-c)_t + F(w,c)_x - \Phi(w,c)_x \leq 0,$$

which is nothing but the entropy inequality we wanted.

To obtain the uniform boundedness of $\|\,w_x\,\|_{L^1(0,1)}$, the idea is to multiply the equation (2.12) by $-(U'(w_x))_x$ and integrate on x from 0 to 1. Since

$$\int_0^1 -(U'(w_x))_x\,w_t = \int_0^1 U'(w_x)\,(w_x)_t\frac{d}{dt}\|\,w_x\,\|_{L^1(0,1)},$$
$$\int_0^1 -(U'(w_x))_x\,f(w)_x = -\int_0^1 U''(w_x)\,w_{xx}\,f'(w)\,w_x = 0,$$

and since

$$\int_0^1 -(U'(w_x))_x\,(\nu(w)\,w_x)_x = -\int_0^1 U''(w_x)\,w_{xx}\,(\nu'(w)\,(w_x)^2 + \nu(w)\,w_{xx})$$
$$= -\int_0^1 U''(w_x)\,\nu(w)\,(w_{xx})^2$$
$$\leq 0,$$

we immediately get that

$$\frac{d}{dt}\|\,w_x\,\|_{L^1(0,1)} \leq 0,$$

and so,

$$\|\,w_x\,\|_{L^1(0,1)} \leq \|\,(u_0)_x\,\|_{L^1(0,1)}, \qquad \forall\,t \in (0,T).$$

When the function u_0 has discontinuities, the same result holds with the total variation of u_0, $|\,u_0\,|_{TV(0,1)}$, replacing the quantity $\|\,(u_0)_x\,\|_{L^1(0,1)}$; these two quantities coincide when $u_0 \in W^{1,1}(0,1)$.

With the two above ingredients, the following error estimate, obtained in 1976 by Kuznetsov, can be proved:

Theorem 4. (L^1-error estimate) *We have*

$$\|\,u(T) - w(T)\,\|_{L^1(0,1)} \leq |\,u_0\,|_{TV(0,1)}\sqrt{8\,T\,\nu},$$

where $\nu = \sup_{s\in[\inf u_0,\sup u_0]}\nu(s)$.

The piecewise-constant case. Let consider the simple case of the DG-method that uses a piecewise-constant approximate solution:

$$\forall \; j = 1, \dots, N :$$
$$\partial_t u_j + \{h(u_j, u_{j+1}) - h(u_{j-1}, u_j)\}/\Delta_j = 0,$$
$$u_j(0) = \frac{1}{\Delta_j} \int_{I_j} u_0(x)\, dx,$$

where we have dropped the superindex '0.' We pick the numerical flux h to be the Engquist-Osher flux.

According to the model provided by the continuous case, we must obtain (i) an entropy inequality and (ii) the uniform boundedness of the total variation of u_h.

To obtain the entropy inequality, we multiply our equation by $U'(u_j - c)$:

$$\partial_t U(u_j - c) + U'(u_j - c)\{h(u_j, u_{j+1}) - h(u_{j-1}, u_j)\}/\Delta_j = 0.$$

The second term in the above equation needs to be carefully treated. First, we rewrite the Engquist-Osher flux in the following form:

$$h^{EO}(a, b) = f^+(a) + f^-(b),$$

and, accordingly, rewrite the second term of the equality above as follows:

$$ST_j = U'(u_j - c)\{f^+(u_j) - f^+(u_{j-1})\}$$
$$+ U'(u_j - c)\{f^-(u_{j+1}) - f^-(u_j)\}.$$

Using the simple identity

$$U'(a - c)(g(a) - g(b)) = G(a, c) - G(b, c) + \int_a^b (g(b) - g(\rho))\, U''(\rho - c)\, d\rho.$$

where $G(a, c) = \int_c^a U'(\rho - c)\, g'(\rho)\, d\rho$, we get

$$ST_j = F^+(u_j, c) - F^+(u_{j-1}, c)$$
$$+ \int_{u_j}^{u_{j-1}} (f^+(u_{j-1}) - f^+(\rho))\, U''(\rho - x)\, d\rho$$
$$+ F^-(u_{j+1}, c) - F^-(u_j, c)$$
$$- \int_{u_j}^{u_{j+1}} (f^-(u_{j+1}) - f^-(\rho))\, U''(\rho - x)\, d\rho$$
$$= F(u_j, u_{j+1}; c) - F(u_{j-1}, u_j; c) + \Theta_{diss,j}$$

where

$$F(a, b; c) = F^+(a, c) + F^-(b, c),$$
$$\Theta_{diss,j} = + \int_{u_j}^{u_{j-1}} (f^+(u_{j-1}) - f^+(\rho))\, U''(\rho - x)\, d\rho$$
$$- \int_{u_j}^{u_{j+1}} (f^-(u_{j+1}) - f^-(\rho))\, U''(\rho - x)\, d\rho.$$

We thus get

$$\partial_t U(u_j - c) + \{F(u_j, u_{j+1}; c) - F(u_{j-1}, u_j; c)\}/\Delta_j + \Theta_{diss,j}/\Delta_j = 0.$$

Since, f^+ and $-f^-$ are nondecreasing functions, we easily see that

$$\Theta_{diss,j} \geq 0,$$

and we obtain our entropy inequality:

$$\partial_t U(u_j - c) + \{F(u_j, u_{j+1}; c) - F(u_{j-1}, u_j; c)\}/\Delta_j \leq 0.$$

Next, we obtain the uniform boundedness on the total variation. To do that, we follow our model and multiply our equation by a discrete version of $-(U'(w_x))_x$, namely,

$$v_j^0 = -\frac{1}{\Delta_j}\left\{U'\left(\frac{u_{j+1} - u_j}{\Delta_{j+1/2}}\right) - U'\left(\frac{u_j - u_{j-1}}{\Delta_{j-1/2}}\right)\right\},$$

where $\Delta_{j+1/2} = (\Delta_j + \Delta_{j+1})/2$, multiply it by Δ_j and sum over j from 1 to N. We easily obtain

$$\frac{d}{dt}| u_h |_{TV(0,1)} + \sum_{1 \leq j \leq N} v_j^0 \{h(u_j, u_{j+1}) - h(u_{j-1}, u_j)\} = 0,$$

where

$$| u_h |_{TV(0,1)} \equiv \sum_{1 \leq j \leq N} | u_{j+1} - u_j |.$$

According to our continuous model, the second term in the above equality should be positive. This is indeed the case since the expression

$$v_j^0 \{h(u_j, u_{j+1}) - h(u_{j-1}, u_j)\}$$

is equal to the quantity

$$v_j^0 \{f^+(u_j) - f^+(u_{j-1})\} + v_j^0 \{f^-(u_{j+1}) - f^-(u_j)\},$$

which is nonnegative by the definition of v_j^0, f^+, and f^-. This implies that

$$| u_h(t) |_{TV(0,1)} \leq | u_h(0) |_{TV(0,1)} \leq | u_0 |_{TV(0,1)}. \tag{2.14}$$

With the two above ingredients, the following error estimate, obtained in 1976 by Kuznetsov, can be proved:

Theorem 5. (L^1-error estimate) *We have*

$$\| u(T) - u_h(T) \|_{L^1(0,1)} \leq \| u_0 - u_h(0) \|_{L^1(0,1)} + C | u_0 |_{TV(0,1)} \sqrt{T \, \Delta x}.$$

The general case. Error estimates for the case of arbitrary k have not been obtained, yet. However, Jiang and Shu [51] found a very interesting result in the case in which the nonlinear flux f is strictly convex or concave. In such a situation, the existence of a discrete, local entropy inequality for the scheme for only a *single* entropy is enough to guarantee that the limit of the scheme, if it exists, is the entropy solution. Jiang and Shu [51] found such a discrete, local entropy inequality for the DG-method.

To describe the main idea of their result, let us first consider the model equation

$$u_t + f(u)_x = (\nu \, u_x)_x.$$

If we multiply the equation by u we obtain, after very simple manipulations,

$$\frac{1}{2} (u)_t^2 + (F(u) - \frac{\nu}{2} (u)_x^2)_x + \Theta = 0,$$

where

$$F(u) = u \, f(u) - \int^u f(s) \, ds,$$

and

$$\Theta = \nu \, (u_x)^2.$$

Since $\Theta \geq 0$, we immediately obtain the following entropy inequality:

$$\frac{1}{2} (u)_t^2 + (F(u) - \frac{\nu}{2} (u)_x^2)_x \leq 0,$$

Now, we only need to mimic the above procedure using the numerical scheme (2.7) instead of the above parabolic equation and obtain a discrete version of the above entropy inequality. To do that, we simply take $v_h = u_h$ in (2.7) and rearrange terms in a suitable way. If we use the following notation:

$$\bar{u}_{j+1/2} = (u_{j+1/2}^+ + u_{j+1/2}^-)/2,$$
$$[u]_{j+1/2} = (u_{j+1/2}^+ - u_{j+1/2}^-),$$

the result can be expressed as follows.

Proposition 6. *We have, for $j = 1, \ldots, N$,*

$$\frac{1}{2} \frac{d}{dt} \int_{I_j} u_h^2(x, \cdot) \, dx + \overset{..}{F}_{j+1/2} - \overset{..}{F}_{j-1/2} + \Theta_j = 0,$$

where

$$\hat{F}_{j+1/2} = \bar{u}_{j+1/2} \, h(u_h)_{j+1/2} - \int^{\bar{u}_{j+1/2}} f(s) \, ds,$$

and

$$\Theta_j = \int_{u_{j+1/2}^-}^{\bar{u}_{j+1/2}} (f(s) - h(u_h)_{j+1/2}) \, ds + \int_{\bar{u}_{j-1/2}}^{u_{j-1/2}^+} (f(s) - h(u_h)_{j-1/2}) \, ds.$$

Since the quantity Θ_j is nonnegative (because the numerical flux in nondecreasing in its first argument and nonincreasing in its second argument), we immediately obtain the following discrete, local entropy inequality:

$$\frac{1}{2}\frac{d}{dt}\int_{I_j} u_h^2(x,\cdot)\,dx + \hat{F}_{j+1/2} - \hat{F}_{j-1/2} \leq 0.$$

As a consequence, we have the following result.

Theorem 7. *Let f be a strictly convex or concave function. Then, for any $k \geq 0$, if the numerical solution given by the DG method converges, it converges to the entropy solution.*

There is no other formally high-order accurate numerical scheme that has the above property. See Jiang and Shu [51] for further developments of the above result.

2.3   The TVD-Runge-Kutta time discretization

To discretize our ODE system in time, we use the TVD Runge Kutta time discretization introduced in [83]; see also [80] and [81].

The discretization Thus, if $\{t^n\}_{n=0}^N$ is a partition of $[0,T]$ and $\Delta t^n = t^{n+1} - t^n$, $n = 0, ..., N-1$, our time-marching algorithm reads as follows:

- Set $u_h^0 = u_{0h}$;
- For $n = 0, ..., N-1$ compute u_h^{n+1} from u_h^n as follows:
  1. set $u_h^{(0)} = u_h^n$;
  2. for $i = 1, ..., k+1$ compute the intermediate functions:

  $$u_h^{(i)} = \left\{\sum_{l=0}^{i-1}\alpha_{il}u_h^{(l)} + \beta_{il}\Delta t^n L_h(u_h^{(l)})\right\};$$

  3. set $u_h^{n+1} = u_h^{(k+1)}$.

Note that this method is very easy to code since *only a single subroutine defining* $L_h(u_h)$ *is needed.* Some Runge-Kutta time discretization parameters are displayed on the table below.

Table 1

Runge-Kutta discretization parameters			
order	α_{il}	β_{il}	$\max\{\beta_{il}/\alpha_{il}\}$
2	1 $\frac{1}{2}\ \frac{1}{2}$	1 $0\ \frac{1}{2}$	1
3	1 $\frac{3}{4}\ \frac{1}{4}$ $\frac{1}{3}\ 0\ \frac{2}{3}$	1 $0\ \frac{1}{4}$ $0\ 0\ \frac{2}{3}$	1

The stability property Note that all the values of the parameters α_{il} displayed in the table below are nonnegative; this is not an accident. Indeed, this is a condition on the parameters α_{il} that ensures the stability property

$$|u_h^{n+1}| \leq |u_h^n|,$$

provided that the 'local' stability property

$$|w| \leq |v|, \tag{2.15}$$

where w is obtained from v by the following 'Euler forward' step,

$$w = v + \delta\, L_h(v), \tag{2.16}$$

holds for values of $|\delta|$ smaller than a given number δ_0.

For example, the second-order Runke-Kutta method displayed in the table above can be rewritten as follows:

$$u_h^{(1)} = u_h^n + \Delta t\, L_h(u_h^n),$$
$$w_h = u_h^{(1)} + \Delta t\, L_h(u_h^{(1)}),$$
$$u_h^{n+1} = \frac{1}{2}(u_h^n + w_h).$$

Now, assuming that the stability property (2.15), (2.16) is satisfied for

$$\delta_0 = |\Delta t\, \max\{\beta_{il}/\alpha_{il}\}| = \Delta t,$$

we have

$$|u_h^{(1)}| \leq |u_h^n|, \qquad |w_h| \leq |u_h^{(1)}|,$$

and so,

$$|u_h^{n+1}| \leq \frac{1}{2}(|u_h^n| + |w_h|) \leq |u_h^n|.$$

Note that we can obtain this result because the coefficients α_{il} are positive! Runge-Kutta methods of this type of order up to order 5 can be found in [81].

The above example shows how to prove the following more general result.

Theorem 8. *(Stability of the Runge-Kutta discretization) Assume that the stability property for the single 'Euler forward' step (2.15), (2.16) is satisfied for*

$$\delta_0 = \max_{0 \le n \le N} |\Delta t^n \max\{\beta_{il}/\alpha_{il}\}|.$$

Assume also that all the coefficients α_{il} are nonnegative and satisfy the following condition:

$$\sum_{l=0}^{i-1} \alpha_{il} = 1, \qquad i = 1, \ldots, k+1.$$

Then

$$|u_h^n| \le |u_h^0|, \qquad \forall n \ge 0.$$

This stability property of the TVD-Runge-Kutta methods is crucial since it allows us to obtain the stability of the method from the stability of a single 'Euler forward' step.

Proof of Theorem 8. We start by rewriting our time discretization as follows:

- Set $u_h^0 = u_{0h}$;
- For $n = 0, \ldots, N-1$ compute u_h^{n+1} from u_h^n as follows:
  1. set $u_h^{(0)} = u_h^n$;
  2. for $i = 1, \ldots, k+1$ compute the intermediate functions:

$$u_h^{(i)} = \sum_{l=0}^{i-1} \alpha_{il} \, w_h^{(il)},$$

where

$$w_h^{(il)} = u_h^{(l)} + \frac{\beta_{il}}{\alpha_{il}} \Delta t^n \, L_h(u_h^{(l)});$$

  3. set $u_h^{n+1} = u_h^{(k+1)}$.

We then have

$$|u_h^{(i)}| \le \sum_{l=0}^{i-1} \alpha_{il} \, |w_h^{(il)}|, \quad \text{since } \alpha_{il} \ge 0,$$

$$\le \sum_{l=0}^{i-1} \alpha_{il} \, |u_h^{(l)}|,$$

by the stability property (2.15), (2.16), and finally,

$$|u_h^{(i)}| \le \max_{0 \le l \le i-1} |u_h^{(l)}|,$$

since

$$\sum_{l=0}^{i-1} \alpha_{il} = 1.$$

It is clear now that that Theorem 8 follows from the above inequality by a simple induction argument. This concludes the proof.

Remarks about the stability in the linear case For the linear case $f(u) = c\,u$, Chavent and Cockburn [12] proved that for the case $k = 1$, i.e., for piecewise-linear approximate solutions, the single 'Euler forward' step is *unconditionally* $L^\infty(0,T; L^2(0,1))$-unstable for any fixed ratio $\Delta t/\Delta x$. On the other hand, in [26] it was shown that if a Runge-Kutta method of second order is used, the scheme is $L^\infty(0,T; L^2(0,1))$-stable provided that

$$c\frac{\Delta t}{\Delta x} \le \frac{1}{3}.$$

This means that we cannot deduce the stability of the complete Runge-Kutta method from the stability of the single 'Euler forward' step. As a consequence, we cannot apply Theorem 8 and we must consider the complete method at once.

When polynomial of degree k are used, a Runge-Kutta of order $(k+1)$ must be used. If this is the case, for $k = 2$, the $L^\infty(0,T; L^2(0,1))$-stability condition can be proven to be the following:

$$c\frac{\Delta t}{\Delta x} \le \frac{1}{5}.$$

The stability condition for a general value of k is still not known.

At a first glance, this stability condition, also called the Courant-Friedrichs-Levy (CFL) condition, seems to compare unfavorably with that of the well-known finite difference schemes. However, we must remember that in the DG-methods there are $(k+1)$ degrees of freedom in each element of size Δx whereas for finite difference schemes there is a single degree of freedom of each cell of size Δx. Also, if a finite difference scheme is of order $(k+1)$ its so-called stencil must be of at least $(2k+1)$ points, whereas the DG-scheme has a stencil of $(k+1)$ elements only.

Convergence analysis in the nonlinear case Now, we explore what is the impact of the explicit Runge-Kutta time-discretization on the convergence properties of the methods under consideration. We start by considering the piecewise-constant case.

The piecewise-constant case. Let us begin by considering the simplest case, namely,

$$\forall\, j = 1, \dots, N :$$
$$(u_j^{n+1} - u_j^n)/\Delta t + \{h(u_j^n, u_{j+1}^n) - h(u_{j-1}^n, u_j^n)\}/\Delta_j = 0,$$
$$u_j(0) = \frac{1}{\Delta_j} \int_{I_j} u_0(x)\,dx,$$

where we pick the numerical flux h to be the Engquist-Osher flux.

According to the model provided by the continuous case, we must obtain (i) an entropy inequality and (ii) the uniform boundedness of the total variation of u_h.

To obtain the entropy inequality, we proceed as in the semidiscrete case and obtain the following result; see [18] for details.

Theorem 9. (Discrete entropy inequality) *We have*

$$\{U(u_j^{n+1} - c) - U(u_j^n - c)\}/\Delta t + \{F(u_j^n, u_{j+1}^n; c) - F(u_{j-1}^n, u_j^n; c)\}/\Delta_j \\ + \Theta_{diss,j}^n/\Delta t = 0,$$

where

$$\Theta_{diss,j}^n = \int_{u_j^{n+1}}^{u_j^n} (p_j(u_j^n) - p_j(\rho)) \, U''(\rho - x) \, d\rho$$

$$+ \frac{\Delta t}{\Delta_j} \int_{u_j^{n+1}}^{u_{j-1}^n} (f^+(u_{j-1}^n) - f^+(\rho)) \, U''(\rho - x) \, d\rho$$

$$- \frac{\Delta t}{\Delta_j} \int_{u_j^{n+1}}^{u_{j+1}^n} (f^-(u_{j+1}^n) - f^-(\rho)) \, U''(\rho - x) \, d\rho,$$

and

$$p_j(w) = w - \frac{\Delta t}{\Delta_j}(f^+(w) - f^-(w)).$$

Moreover, if the following CFL condition is satisfied

$$\max_{1 \le j \le N} \frac{\Delta t}{\Delta_j} |f'| \le 1,$$

then $\Theta_{diss,j}^n \ge 0$, *and the following entropy inequality holds:*

$$\{U(u_j^{n+1} - c) - U(u_j^n - c)\}/\Delta t + \{F(u_j^n, u_{j+1}; c) - F(u_{j-1}, u_j; c)\}/\Delta_j \le 0.$$

Note that $\Theta_{diss,j}^n \ge 0$ because $f^+, -f^-$, are nondecreasing and because p_j is also nondecreasing under the above CFL condition.

Next, we obtain the uniform boundedness on the total variation. Proceeding as before, we easily obtain the following result.

Theorem 10. (TVD property) *We have*

$$|u_h^{n+1}|_{TV(0,1)} - |u_h^n|_{TV(0,1)} + \Theta_{TV}^n = 0,$$

where

$$\Theta_{TV}^n = \sum_{1 \le j \le N} \left(U'^n_{j+1/2} - U'^{n+1}_{j+1/2} \right) (p_{j+1/2}(u_{j+1}^n) - p_{j+1/2}(u_j^n))$$

$$+ \sum_{1 \le j \le N} \frac{\Delta t}{\Delta_j} \left(U'^n_{j-1/2} - U'^{n+1}_{j+1/2} \right) (f^+(u_j^n) - f^+(u_{j-1}^n))$$

$$- \sum_{1 \le j \le N} \frac{\Delta t}{\Delta_j} \left(U'^n_{j+1/2} - U'^{n+1}_{j-1/2} \right) (f^-(u_{j+1}^n) - f^-(u_j^n))$$

where

$$U'^m_{i+1/2} = U' \left(\frac{u_{i+1}^m - u_i^m}{\Delta_{i+1/2}} \right),$$

and

$$p_{j+1/2}(w) = s - \frac{\Delta t}{\Delta_{j+1}}\, f^+(w) + \frac{\Delta t}{\Delta_j}\, f^-(w).$$

Moreover, if the following CFL condition is satisfied

$$\max_{1 \leq j \leq N} \frac{\Delta t}{\Delta_j}\, |f'| \leq 1,$$

then $\Theta_{TV}^n \geq 0$, *and we have*

$$|u_h^n|_{TV(0,1)} \leq |u_0|_{TV(0,1)}.$$

With the two above ingredients, the following error estimate, obtained in 1976 by Kuznetsov, can be proved:

Theorem 11. (L^1-error estimate for monotone schemes) *We have*

$$\| u(T) - u_h(T) \|_{L^1(0,1)} \leq \| u_0 - u_h(0) \|_{L^1(0,1)} + C\, |u_0|_{TV(0,1)} \sqrt{T\, \Delta x}.$$

The general case. The study of the general case is much more difficult than the study of the monotone schemes. In these notes, we restrict ourselves to the study of the stability of the RKDG schemes. Hence, we restrict ourselves to the task of studying under what conditions the total variation of the *local means* is uniformly bounded.

If we denote by $\bar{u}_j$ the mean of u_h on the interval I_j, by setting $v_h = 1$ in the equation (2.7), we obtain,

$$\forall\, j = 1, \ldots, N :$$
$$(\bar{u}_j)_t + \{h(u_{j+1/2}^-, u_{j+1/2}^+) - h(u_{j-1/2}^-, u_{j-1/2}^+)\}/\Delta_j = 0,$$

where $u_{j+1/2}^-$ denotes the limit from the left and $u_{j+1/2}^+$ the limit from the right. We pick the numerical flux h to be the Engquist-Osher flux.

This shows that if we set w_h equal to the Euler forward step $u_h + \delta\, L_h(u_h)$, we obtain

$$\forall\, j = 1, \ldots, N :$$
$$(\bar{w}_j - \bar{u}_j)/\delta + \{h(u_{j+1/2}^-, u_{j+1/2}^+) - h(u_{j-1/2}^-, u_{j-1/2}^+)\}/\Delta_j = 0.$$

Proceeding exactly as in the piecewise-constant case, we obtain the following result for the total variation of the averages,

$$|\bar{u}_h|_{TV(0,1)} \equiv \sum_{1 \leq j \leq N} |\bar{u}_{j+1} - \bar{u}_j|.$$

Theorem 12. (The TVDM property) *We have*

$$|\bar{w}_h|_{TV(0,1)} - |\bar{u}_h|_{TV(0,1)} + \Theta_{TVM} = 0,$$

where

$$\Theta_{TVM} = \sum_{1 \le j \le N} \left(U'(u)_{j+1/2} - U'(w)_{j+1/2} \right) (p_{j+1/2}(u_h|_{I_{j+1}}) - p_{j+1/2}(u_h|_{I_j})$$

$$+ \sum_{1 \le j \le N} \frac{\delta}{\Delta_j} \left(U'(u)_{j-1/2} - U'(w)_{j+1/2} \right) (f^+(u_{j+1/2}^-) - f^+(u_{j-1/2}^-))$$

$$- \sum_{1 \le j \le N} \frac{\delta}{\Delta_j} \left(U'(u)_{j+1/2} - U'(w)_{j-1/2} \right) (f^-(u_{j+1/2}^+) - f^-(u_{j-1/2}^+)),$$

where

$$U'(v)_{i+1/2} = U'\left(\frac{\overline{v}_{i+1} - \overline{v}_i}{\Delta_{i+1/2}} \right),$$

and

$$p_{j+1/2}(u_h|_{I_m}) = \overline{u}_m - \frac{\delta}{\Delta_{j+1}} f^+(u_{m+1/2}^-) + \frac{\delta}{\Delta_j} f^-(u_{m-1/2}^+).$$

From the above result, we see that the total variation of the means of the Euler forward step is nonincreasing if the following **sign conditions** are satisfied:

$$sgn(\overline{u}_{j+1} - \overline{u}_j) = sgn(p_{j+1/2}(u_h|_{I_{j+1}}) - p_{j+1/2}(u_h|_{I_j})), \qquad (2.17)$$

$$sgn(\overline{u}_j - \overline{u}_{j-1}) = sgn(u_{j+1/2}^{n,-} - u_{j-1/2}^{n,-}), \qquad (2.18)$$

$$sgn(\overline{u}_{j+1} - \overline{u}_j) = sgn(u_{j+1/2}^{n,+} - u_{j-1/2}^{n,+}). \qquad (2.19)$$

Note that if the *sign* conditions (2.17) and (2.18) are satisfied, then the *sign* condition (2.19) can always be satisfied for a small enough values of $|\delta|$.

Of course, the numerical method under consideration does not provide an approximate solution automatically satisfying the above conditions. It is thus necessary to *enforce* them by means of a suitably defined generalized slope limiter, $\Lambda\Pi_h$.

2.4   The generalized slope limiter

High-order accuracy versus the TVDM property: Heuristics The ideal generalized slope limiter $\Lambda\Pi_h$

- Maintains the conservation of *mass* element by element,
- Satisfies the *sign* properties (2.17), (2.18), and (2.19),
- Does not degrade the accuracy of the method.

The first requirement simply states that the slope limiting must not change the total mass contained in each interval, that is, if $u_h = \Lambda\Pi_h(v_h)$,

$$\overline{u}_j = \overline{v}_j, \qquad j = 1, \ldots, N.$$

This is, of course a very sensible requirement because after all we are dealing with conservation laws. It is also a requirement very easy to satisfy.

The second requirement, states that if $u_h = \Lambda \Pi_h(v_h)$ and $w_h = u_h + \delta\, L_h(u_h)$ then

$$|\overline{w}_h|_{TV(0,1)} \leq |\overline{u}_h|_{TV(0,1)},$$

for small enough values of $|\delta|$.

The third requirement deserves a more delicate discussion. Note that if u_h is a very good approximation of a smooth solution u in a neighborhood of the point x_0, it behaves (asymptotically as Δx goes to zero) as a straight line if $u_x(x_0) \neq 0$. If x_0 is an isolated extrema of u, then it behaves like a parabola provided $u_{xx}(x_0) \neq 0$. Now, if u_h is a straight line, it trivially satisfies conditions (2.17) and (2.18). However, if u_h is a parabola, conditions (2.17) and (2.18) are not always satisfied. This shows that *it is impossible* to construct the above ideal generalized 'slope limiter,' or, in other words, that in order to enforce the TVDM property, we must loose high-order accuracy at the local extrema. This is a very well-known phenomenon for TVD finite difference schemes!

Fortunately, it is still possible to construct generalized slope limiters that do preserve high-order accuracy even at local extrema. The resulting scheme will then not be TVDM but total variation bounded in the means (TVBM) as we will show.

In what follows we first consider generalized slope limiters that render the RKDG schemes TVDM. Then we suitably modify them in order to obtain TVBM schemes.

Constructing TVDM generalized slope limiters Next, we look for simple, sufficient conditions on the function u_h that imply the *sign* properties (2.17), (2.18), and (2.19). These conditions will be stated in terms of the *minmod* function m defined as follows:

$$m(a_1, \ldots, a_\nu) = \begin{cases} s \min_{1 \leq n \leq \nu} |a_n| & \text{if } s = sign(a_1) = \cdots = sign(a_\nu), \\ 0 & \text{otherwise.} \end{cases}$$

Proposition 13. Sufficient conditions for the sign properties *Suppose the the following CFL condition is satisfied:*

For all $j = 1, \ldots, N$:

$$|\delta|\,\Big(\frac{|f^+|_{Lip}}{\Delta_{j+1}} + \frac{|f^-|_{Lip}}{\Delta_j}\Big) \leq 1/2. \tag{2.20}$$

Then, conditions (2.17), (2.18), and (2.19) are satisfied if, for all $j = 1, \ldots, N$, we have that

$$u^-_{j+1/2} = \overline{u}_j + m\,(u^-_{j+1/2} - \overline{u}_j,\ \overline{u}_j - \overline{u}_{j-1},\ \overline{u}_{j+1} - \overline{u}_j) \tag{2.21}$$
$$u^+_{j-1/2} = \overline{u}_j - m\,(\overline{u}_j - u^+_{j-1/2},\ \overline{u}_j - \overline{u}_{j-1},\ \overline{u}_{j+1} - \overline{u}_j).$$

Proof. Let us start by showing that the property (2.18) is satisfied. We have:

$$u^-_{j+1/2} - u^-_{j-1/2} = (u^-_{j+1/2} - \overline{u}_j) + (\overline{u}_j - \overline{u}_{j-1}) + (\overline{u}_{j-1} - u^-_{j-1/2})$$
$$= \Theta\,(\overline{u}_j - \overline{u}_{j-1}),$$

where

$$\Theta = 1 + \frac{u_{j+1/2}^- - \overline{u}_j}{\overline{u}_j - \overline{u}_{j-1}} - \frac{u_{j-1/2}^- - \overline{u}_{j-1}}{\overline{u}_j - \overline{u}_{j-1}} \in [0, 2],$$

by conditions (2.21) and (2.22). This implies that the property (2.18) is satisfied. Properties (2.19) and (2.17) are proven in a similar way. This completes the proof.

Examples of TVDM generalized slope limiters

a. The MUSCL limiter. In the case of piecewise linear approximate solutions, that is,

$$v_h|_{I_j} = \overline{v}_j + (x - x_j)\, v_{x,j}, \qquad j = 1, \ldots, N,$$

the following generalized slope limiter does satisfy the conditions (2.21) and (2.22):

$$u_h|_{I_j} = \overline{v}_j + (x - x_j)\, m\left(v_{x,j}, \frac{\overline{v}_{j+1} - \overline{v}_j}{\Delta_j}, \frac{\overline{v}_j - \overline{v}_{j-1}}{\Delta_j}\right).$$

This is the well-known slope limiter of the MUSCL schemes of van Leer [88,89].

b. The less restrictive limiter $\Lambda\Pi_h^1$. The following less restrictive slope limiter also satisfies the conditions (2.21) and (2.22):

$$u_h|_{I_j} = \overline{v}_j + (x - x_j)\, m\left(v_{x,j}, \frac{\overline{v}_{j+1} - \overline{v}_j}{\Delta_j/2}, \frac{\overline{v}_j - \overline{v}_{j-1}}{\Delta_j/2}\right).$$

Moreover, it can be rewritten as follows:

$$u_{j+1/2}^- = \overline{v}_j + m\left(v_{j+1/2}^- - \overline{v}_j, \overline{v}_j - \overline{v}_{j-1}, \overline{v}_{j+1} - \overline{v}_j\right) \tag{2.22}$$
$$u_{j-1/2}^+ = \overline{v}_j - m\left(\overline{v}_j - v_{j-1/2}^+, \overline{v}_j - \overline{v}_{j-1}, \overline{v}_{j+1} - \overline{v}_j\right).$$

We denote this limiter by $\Lambda\Pi_h^1$.

Note that we have that

$$\|\overline{v}_h - \Lambda\Pi_h^1(v_h)\|_{L^1(0,1)} \le \frac{\Delta x}{2} |\overline{v}_h|_{TV(0,1)}.$$

See Theorem 16 below.

c. The limiter $\Lambda\Pi_h^k$. In the case in which the approximate solution is piecewise a polynomial of degree k, that is, when

$$v_h(x, t) = \sum_{\ell=0}^{k} v_j^\ell\, \varphi_\ell(x),$$

where

$$\varphi_\ell(x) = P_\ell(2\,(x - x_j)/\Delta_j),$$

and P_ℓ are the Legendre polynomials, we can define a generalized slope limiter in a very simple way. To do that, we need the define what could be called the P^1-part of v_h:

$$v_h^1(x, t) = \sum_{\ell=0}^{1} v_j^\ell\, \varphi_\ell(x),$$

We define $u_h = \Lambda\Pi_h(v_h)$ as follows:

– For $j = 1, ..., N$ compute $u_h|_{I_j}$ as follows:
  1. Compute $u^-_{j+1/2}$ and $u^+_{j-1/2}$ by using (2.22) and (2.23),
  2. If $u^-_{j+1/2} = v^-_{j+1/2}$ and $u^+_{j-1/2} = v^+_{j-1/2}$ set $u_h|_{I_j} = v_h|_{I_j}$,
  3. If not, take $u_h|_{I_j}$ equal to $\Lambda\Pi^1_h(v^1_h)$.

d. The limiter $\Lambda\Pi^k_{h,\alpha}$. When instead of (2.22) and (2.23), we use

$$u^-_{j+1/2} = \overline{v}_j + m\left(v^-_{j+1/2} - \overline{v}_j, \overline{v}_j - \overline{v}_{j-1}, \overline{v}_{j+1} - \overline{v}_j, C(\Delta x)^\alpha\right) \qquad (2.23)$$
$$u^+_{j-1/2} = \overline{v}_j - m\left(\overline{v}_j - v^+_{j-1/2}, \overline{v}_j - \overline{v}_{j-1}, \overline{v}_{j+1} - \overline{v}_j, C(\Delta x)^\alpha\right),$$

for some fixed constant C and $\alpha \in (0,1)$, we obtain a generalized slope limiter we denote by $\Lambda\Pi^k_{h,\alpha}$.

This generalized slope limiter is never used in practice, but we consider it here because it is used for theoretical purposes; see Theorem 16 below.

The complete RKDG method Now that we have our generalized slope limiters, we can display the complete RKDG method. It is contained in the following algorithm:

– Set $u^0_h = \Lambda\Pi_h \, P_{V_h}(u_0)$;
– For $n = 0, ..., N - 1$ compute u^{n+1}_h as follows:
  1. set $u^{(0)}_h = u^n_h$;
  2. for $i = 1, ..., k + 1$ compute the intermediate functions:

$$u^{(i)}_h = \Lambda\Pi_h \left\{ \sum_{l=0}^{i-1} \alpha_{il} \, u^{(l)}_h + \beta_{il} \Delta t^n L_h(u^{(l)}_h) \right\};$$

  3. set $u^{n+1}_h = u^{(k+1)}_h$.

This algorithm describes the complete RKDG method. Note how the generalized slope limiter has to be applied at each intermediate computation of the Runge-Kutta method. This way of applying the generalized slope limiter in the time-marching algorithm ensures that the scheme is TVDM, as we next show.

The TVDM property of the RKDG method To do that, we start by noting that if we set

$$u_h = \Lambda\Pi_h(v_h), \qquad w_h = u_h + \delta \, L_h(u_h),$$

then we have that

$$|\overline{u}_h|_{TV(0,1)} \leq |\overline{v}_h|_{TV(0,1)}, \qquad (2.24)$$
$$|\overline{w}_h|_{TV(0,1)} \leq |\overline{u}_h|_{TV(0,1)}, \quad \forall |\delta| \leq \delta_0, \qquad (2.25)$$

where

$$\delta_0^{-1} = \max_j \left(2\frac{|f^+|_{Lip}}{\Delta_{j+1}} + \frac{|f^-|_{Lip}}{\Delta_j}\right) \qquad j = 1, \ldots, N,$$

by Proposition 13. By using the above two properties of the generalized slope limiter,' it is possible to show that the RKDG method is TVDM.

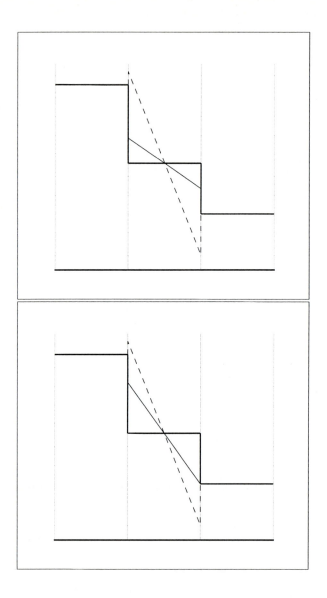

Fig. 2.1. Example of slope limiters: The MUSCL limiter (top) and the less restrictive $\Lambda\Pi_h^1$ limiter (bottom). Displayed are the local means of u_h (thick line), the linear function u_h in the element of the middle before limiting (dotted line) and the resulting function after limiting (solid line).

Theorem 14. (Stability induced by the generalized slope limiter) *Assume that the generalized slope limiter $\Lambda\Pi_h$ satisfies the properties (2.24) and (2.25). Assume also that all the coefficients α_{il} are nonnegative and satisfy the following condition:*

$$\sum_{l=0}^{i-1} \alpha_{il} = 1, \qquad i = 1, \ldots, k+1.$$

Then

$$|\overline{u}_h^n|_{TV(0,1)} \leq |u_0|_{TV(0,1)}, \qquad \forall\, n \geq 0.$$

Proof. The proof of this result is very similar to that of Theorem 8. Thus, we start by rewriting our time discretization as follows:

- Set $u_h^0 = u_{0h}$;
- For $n = 0, \ldots, N-1$ compute u_h^{n+1} from u_h^n as follows:
  1. set $u_h^{(0)} = u_h^n$;
  2. for $i = 1, \ldots, k+1$ compute the intermediate functions:

$$u_h^{(i)} = \Lambda\Pi_h \left\{ \sum_{l=0}^{i-1} \alpha_{il}\, w_h^{(il)} \right\},$$

  where

$$w_h^{(il)} = u_h^{(l)} + \frac{\beta_{il}}{\alpha_{il}}\, \Delta t^n\, L_h(u_h^{(l)});$$

  3. set $u_h^{n+1} = u_h^{(k+1)}$.

Then have,

$$|\overline{u}_h^{(i)}|_{TV(0,1)} \leq |\sum_{l=0}^{i-1} \alpha_{il}\, \overline{w}_h^{(il)}|_{TV(0,1)}, \quad \text{by (2.24)},$$

$$\leq \sum_{l=0}^{i-1} \alpha_{il}\, |\overline{w}_h^{(il)}|_{TV(0,1)}, \quad \text{since } \alpha_{il} \geq 0,$$

$$\leq |\sum_{l=0}^{i-1} \alpha_{il}\, \overline{u}_h^{(l)}|_{TV(0,1)}, \quad \text{by (2.25)},$$

$$\leq \max_{0 \leq l \leq i-1} |\overline{u}_h^{(l)}|_{TV(0,1)},$$

since

$$\sum_{l=0}^{i-1} \alpha_{il} = 1.$$

It is clear now that that the inequality

$$|\overline{u}_h^n|_{TV(0,1)} \leq |\overline{u}_h^0|_{TV(0,1)}, \qquad \forall\, n \geq 0.$$

follows from the above inequality by a simple induction argument. To obtain the result of the theorem, it is enough to note that we have

$$|\overline{u}_h^0|_{TV(0,1)} \leq |u_0|_{TV(0,1)},$$

by the definition of the initial condition u_h^0. This completes the proof.

TVBM generalized slope limiters As was pointed out before, it is possible to modify the generalized slope limiters displayed in the examples above in such a way that the degradation of the accuracy at local extrema is avoided. To achieve this, we follow Shu [82] and modify the definition of the generalized slope limiters by simply replacing the *minmod* function m by the TVB corrected *minmod* function $\bar{m}$ defined as follows:

$$\bar{m}(a_1,...,a_m) = \begin{cases} a_1 & if \ |a_1| \le M(\Delta x)^2, \\ m(a_1,...,a_m) & otherwise, \end{cases} \qquad (2.26)$$

where M is a given constant. We call the generalized slope limiters thus constructed, TVBM slope limiters.

The constant M is, of course, an upper bound of the absolute value of the second-order derivative of the solution at local extrema. In the case of the nonlinear conservation laws under consideration, it is easy to see that, if the initial data is piecewise C^2, we can take

$$M = \sup\{\,|\,(u_0)_{xx}(y)\,|, y: (u_0)_x(y) = 0\}.$$

See [24] for other choices of M.

Thus, if the constant M is is taken as above, there is no degeneracy of accuracy at the extrema and the resulting RKDG scheme retains its optimal accuracy. Moreover, we have the following stability result.

Theorem 15. (The TVBM property) *Assume that the generalized slope limiter $\Lambda\Pi_h$ is a TVBM slope limiter. Assume also that all the coefficients α_{il} are nonnegative and satisfy the following condition:*

$$\sum_{l=0}^{i-1} \alpha_{il} = 1, \qquad i = 1,\dots,k+1.$$

Then

$$|\,\overline{u}_h^n\,|_{TV(0,1)} \le |\,\overline{u}_0\,|_{TV(0,1)} + C\,M, \qquad \forall\, n \ge 0,$$

where C depends on k only.

Convergence in the nonlinear case By using the stability above stability results, we can use the Ascoli-Arzelá theorem to prove the following convergence result.

Theorem 16. (Convergence to the entropy solution) *Assume that the generalized slope limiter $\Lambda\Pi_h$ is a TVDM or a TVBM slope limiter. Assume also that all the coefficients α_{il} are nonnegative and satisfy the following condition:*

$$\sum_{l=0}^{i-1} \alpha_{il} = 1, \qquad i = 1,\dots,k+1.$$

Then there is a subsequence $\{\overline{u}_{h'}\}_{h'>0}$ of the sequence $\{\overline{u}_h\}_{h>0}$ generate by the RKDG scheme that converges in $L^\infty(0,T;L^1(0,1))$ to a weak solution of the problem (2.1), (2.2).

Moreover, if the TVBM version of the slope limiter $\Lambda\Pi^k_{h,\alpha}$ is used, the weak solution is the entropy solution and the whole sequence converges.

Finally, if the generalized slope limiter $\Lambda\Pi_h$ is such that

$$\| \bar{v}_h - \Lambda\Pi_h(v_h) \|_{L^1(0,1)} \leq C \, \Delta x \, | \bar{v}_h |_{TV(0,1)},$$

then the above results hold not only to the sequence of the means $\{ \bar{u}_h \}_{h>0}$ but to the sequence of the functions $\{ u_h \}_{h>0}$.

Error estimates for an *implicit* version of the discontinuous Galerkin method (with the so-called shock-capturing terms) have been obtained by Cockburn and Gremaud [19].

2.5   Computational results

In this section, we display the performance of the RKDG schemes in two simple but typical test problems. We use piecewise linear ($k = 1$) and piecewise quadratic ($k = 2$) elements; the $\Lambda\Pi^k_h$ generalized slope limiter is used.

The first test problem. We consider the simple transport equation with periodic boundary conditions:

$$u_t + u_x = 0,$$

$$u(x,0) = \begin{cases} 1 & .4 < x < .6, \\ 0 & \text{otherwise.} \end{cases}$$

We use this test problem to show that the use of high-order polynomial approximation does improve the approximation of the discontinuities (or, in this case, 'contacts'). To amplify the effect of the dissipation of the method, we take $T = 100$, that is, we let the solution travel 100 times across the domain. We run the scheme with $CFL = 0.9 * 1 = 0.9$ for $k = 0$, $CFL = 0.9 * 1/3 = 0.3$ for $k = 1$, and $CFL = 0.9 * 1/5 = 0.18$ for $k = 2$. In Figure 2.2, we can see that the dissipation effect decreases as the degree of the polynomial k increases; we also see that the dissipation effect for a given k decreases as the Δx decreases, as expected. Other experiments in this direction have been performed by Atkins and Shu [1]. For example, they show that when polynomials of degree $k = 11$ are used, there is no detectable decay of the approximate solution.

To assess if the use of high degree polynomials is advantageous, we must compare the efficiencies of the schemes; we only compare the efficiencies of the method for $k = 1$ and $k = 2$. We define the inverse of the efficiency of the method as the product of the error times the number of operations. Since the RKDG method that uses quadratic elements has $0.3/0.2$ times more time steps, $3/2$ times more inner iterations per time step, and $3 \times 3/2 \times 2$ times more operations per element, its number of operations is $81/16$ times bigger than the one of the RKDG method using linear elements. Hence, the ratio of the efficiency of the RKDG method with quadratic elements to that of the RKDG method with linear elements is

$$eff.ratio = \frac{16}{81} \frac{error(RKDG(k=1))}{error(RKDG(k=2))}.$$

In Table 2, we see that the use of a higher degree does result in a more efficient resolution of the contact discontinuities. This fact remains true for systems as we can see from the numerical experiments for the double Mach reflection problem in the next chapter.

The second test problem. We consider the standard Burgers equation with periodic boundary conditions:

$$u_t + (\frac{u^2}{2})_x = 0,$$

$$u(x,0) = u_0(x) = \frac{1}{4} + \frac{1}{2} \sin(\pi(2x - 1)).$$

Our purpose is to show that (i) when the constant M is properly chosen, the RKDG method using polynomials of degree k is is order $k+1$ in the uniform norm away from the discontinuities, that (ii) it is computationally more efficient to use high-degree polynomial approximations, and that (iii) shocks are captured in a few elements without production of spurious oscillations

The exact solution is smooth at $T = .05$ and has a well developed shock at $T = 0.4$; notice that there is a sonic point. In Tables 3,4, and 5, the history of convergence of the RKDG method using piecewise linear elements is displayed and in Tables 6,7, and 8, the history of convergence of the RKDG method using piecewise quadratic elements. It can be seen that when the TVDM generalized slope limiter is used, i.e., when we take $M = 0$, there is degradation of the accuracy of the scheme, whereas when the TVBM generalized slope limiter is used with a properly chosen constant M, i.e., when $M = 20 \geq 2\pi^2$, the scheme is uniformly high order in regions of smoothness that include critical and sonic points.

Next, we compare the efficiency of the RKDG schemes for $k = 1$ and $k = 2$ for the case $M = 20$ and $T = 0.05$. The results are displayed in Table 9. We can see that the efficiency of the RKDG scheme with quadratic polynomials is several times that of the RKDG scheme with linear polynomials even for very small values of Δx. We can also see that the efficiency ratio is proportional to $(\Delta x)^{-1}$, which is expected for smooth solutions. This indicates that it is indeed more efficient to work with RKDG methods using polynomials of higher degree.

That this is also true when the solution displays shocks can be seen in Figures 2.3, 2.4, and 2.5. In the Figure 2.3, it can be seen that the shock is captured in essentially two elements. Details of these figures are shown in Figures 2.4 and 2.5, where the approximations right in front of the shock are shown. It is clear that the approximation using quadratic elements is superior to the approximation using linear elements. Finally, we illustrate in Figure 2.6 how the schemes follow a shock when it goes through a single element.

2.6   Concluding remarks

In this section, which is the core of these notes, we have devised the general RKDG method for nonlinear scalar conservation laws with periodic boundary conditions.

We have seen that the RKDG are constructed in three steps. First, the Discontinuous Galerkin method is used to discretize in space the conservation law. Then, an explicit TVB-Runge-Kutta time discretization is used to discretize the resulting ODE system. Finally, a generalized slope limiter is introduced that enforces nonlinear stability without degrading the accuracy of the scheme.

We have seen that the numerical results show that the RKDG methods using polynomials of degree $k, k = 1, 2$ are uniformly $(k + 1)$-th order accurate away from discontinuities and that the use of high degree polynomials render the RKDG method more efficient, even close to discontinuities.

All these results can be extended to the initial boundary value problem in a very simple way, see [24]. In what follows, we extend the RKDG methods to multidimensional systems.

Table 2

Comparison of the efficiencies of RKDG schemes for $k = 1$ and $k = 2$
Transport equation with $M = 0$, and $T = 100$.

Δx	L^1-norm	
	$eff.ratio$	order
1/10	0.88	-
1/20	0.93	-0.08
1/40	1.81	-0.96
1/80	2.57	-0.50
1/160	3.24	-0.33

Table 3
P^1, $M = 0$, CFL= 0.3, $T = 0.05$.

	$L^1(0,1) - error$		$L^\infty(0,1) - error$	
Δx	$10^5 \cdot error$	order	$10^5 \cdot error$	order
1/10	1286.23	-	3491.79	-
1/20	334.93	1.85	1129.21	1.63
1/40	85.32	1.97	449.29	1.33
1/80	21.64	1.98	137.30	1.71
1/160	5.49	1.98	45.10	1.61
1/320	1.37	2.00	14.79	1.61
1/640	0.34	2.01	4.85	1.60
1/1280	0.08	2.02	1.60	1.61

Table 4
P^1, $M = 20$, CFL= 0.3, $T = 0.05$.

	$L^1(0,1) - error$		$L^\infty(0,1) - error$	
Δx	$10^5 \cdot error$	order	$10^5 \cdot error$	order
1/10	1073.58	-	2406.38	-
1/20	277.38	1.95	628.12	1.94
1/40	71.92	1.95	161.65	1.96
1/80	18.77	1.94	42.30	1.93
1/160	4.79	1.97	10.71	1.98
1/320	1.21	1.99	2.82	1.93
1/640	0.30	2.00	0.78	1.86
1/1280	0.08	2.00	0.21	1.90

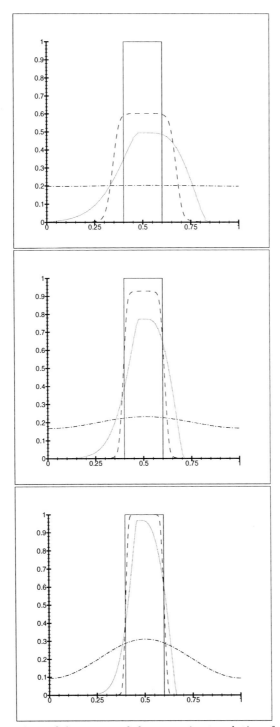

Fig. 2.2. Comparison of the exact and the approximate solutions for the linear case $f(u) = u$. Top: $\Delta x = 1/40$, middle: $\Delta x = 1/80$, bottom: $\Delta x = 1/160$. Exact solution (solid line), piecewise-constant elements (dash/dotted line), piecewise-linear elements (dotted line) and piecewise-quadratic elements (dashed line).

Table 5

Errors in smooth region $\Omega = \{x : |x - shock| \geq 0.1\}$.
P^1, $M = 20$, CFL= 0.3, $T = 0.4$.

Δx	$L^1(\Omega)$ − error		$L^\infty(\Omega)$ − error	
	$10^5 \cdot error$	order	$10^5 \cdot error$	order
1/10	1477.16	-	17027.32	-
1/20	155.67	3.25	1088.55	3.97
1/40	38.35	2.02	247.35	2.14
1/80	9.70	1.98	65.30	1.92
1/160	2.44	1.99	17.35	1.91
1/320	0.61	1.99	4.48	1.95
1/640	0.15	2.00	1.14	1.98
1/1280	0.04	2.00	0.29	1.99

Table 6

P^2, $M = 0$, CFL= 0.2, $T = 0.05$.

Δx	$L^1(0,1)$ − error		$L^\infty(0,1)$ − error	
	$10^5 \cdot error$	order	$10^5 \cdot error$	order
1/10	2066.13	-	16910.05	-
1/20	251.79	3.03	3014.64	2.49
1/40	42.52	2.57	1032.53	1.55
1/80	7.56	2.49	336.62	1.61

Table 7

P^2, $M = 20$, CFL$= 0.2$, $T = 0.05$.

Δx	$L^1(0,1) - error$		$L^\infty(0,1) - error$	
	$10^5 \cdot error$	order	$10^5 \cdot error$	order
1/10	37.31	-	101.44	-
1/20	4.58	3.02	13.50	2.91
1/40	0.55	3.05	1.52	3.15
1/80	0.07	3.08	0.19	3.01

Table 8

Errors in smooth region $\Omega = \{x : |x - shock| \geq 0.1\}$.
P^2, $M = 20$, CFL$= 0.2$, $T = 0.4$.

Δx	$L^1(\Omega) - error$		$L^\infty(\Omega) - error$	
	$10^5 \cdot error$	order	$10^5 \cdot error$	order
1/10	786.36	-	16413.79	-
1/20	5.52	7.16	86.01	7.58
1/40	0.36	3.94	15.49	2.47
1/80	0.06	2.48	0.54	4.84

Table 9

Comparison of the efficiencies of RKDG schemes for $k = 1$ and $k = 2$
Burgers equation with $M = 20$, and $T = 0.05$.

	L^1-norm		L^∞-norm	
Δx	$eff.ratio$	order	$eff.ratio$	order
1/10	5.68	-	4.69	-
1/20	11.96	-1.07	31.02	-2.73
1/40	25.83	-1.11	70.90	-1.19
1/80	52.97	-1.04	148.42	-1.07

2.7   Appendix: Proof of the L^2-error estimates

Proof of the L^2-stability In this section, we prove the the stability result of
Proposition 1. To do that, we first show how to obtain the corresponding stability
result for the exact solution and then mimic the argument to obtain Proposition 1.

The continuous case as a model. We start by rewriting the equations (2.4)
in *compact form*. If in the equations (2.4) we replace $v(x)$ by $v(x,t)$, sum on j from
1 to N, and integrate in time from 0 to T, we obtain

$$\forall \, v : v(t) \text{ is smooth} \quad \forall \, t \in (0,T) :$$
$$B(u, v) = 0, \tag{2.27}$$

where

$$B(u, v) = \int_0^T \int_0^1 \left\{ \partial_t u(x, t)\, v(x, t) - c\, u(x, t)\, \partial_x v(x, t) \right\} dx\, dt.$$

Taking $v = u$, we easily see that we see that

$$B(u, u) = \frac{1}{2} \| u(T) \|^2_{L^2(0,1)} - \frac{1}{2} \| u_0 \|^2_{L^2(0,1)},$$

and since

$$B(u, u) = 0,$$

by (2.27), we immediately obtain the following L^2-stability result:

$$\frac{1}{2} \| u(T) \|^2_{L^2(0,1)} = \frac{1}{2} \| u_0 \|^2_{L^2(0,1)}.$$

This is the argument we have to mimic in order to prove Proposition 1.

The discrete case. Thus, we start by finding the discrete version of the form $B(\cdot,\cdot)$. If we replace $v(x)$ by $v_h(x,t)$ in the equation (2.7), sum on j from 1 to N, and integrate in time from 0 to T, we obtain

$$\forall\, v_h : v_h(t) \in V_h^k \quad \forall\, t \in (0,T):$$
$$B_h(u_h, v_h) = 0, \tag{2.28}$$

where

$$B_h(u_h, v_h) = \int_0^T \int_0^1 \partial_t u_h(x,t)\, v_h(x,t)\, dx\, dt \tag{2.29}$$
$$- \int_0^T \sum_{1 \le j \le N} \int_{I_j} c\, u_h(x,t)\, \partial_x v_h(x,t)\, dx\, dt$$
$$- \int_0^T \sum_{1 \le j \le N} h(u_h)_{j+1/2}(t)\, [\,v_h(t)\,]_{j+1/2}\, dt.$$

Following the model provided by the continuous case, we next obtain an expression for $B_h(w_h, w_h)$. It is contained in the following result which will proved later.

Lemma 17. *We have*

$$B_h(w_h, w_h) = \frac{1}{2}\| w_h(T) \|_{L^2(0,1)}^2 + \Theta_T(w_h) - \frac{1}{2}\| w_h(0) \|_{L^2(0,1)}^2,$$

where

$$\Theta_T(w_h) = \frac{|c|}{2} \int_0^T \sum_{1 \le j \le N} [\, w_h(t)\,]_{j+1/2}^2\, dt.$$

Taking $w_h = u_h$ in the above result and noting that by (2.28),

$$B_h(u_h, u_h) = 0,$$

we get the equality

$$\tfrac{1}{2}\| u_h(T) \|_{L^2(0,1)}^2 + \Theta_T(u_h) = \tfrac{1}{2}\| u_h(0) \|_{L^2(0,1)}^2,$$

from which Proposition 1 easily follows, since

$$\frac{1}{2}\| u_h(T) \|_{L^2(0,1)}^2 \le \frac{1}{2}\| u_0 \|_{L^2(0,1)}^2,$$

by (2.8). It only remains to prove Lemma 17.

Proof of Lemma 17. After setting $u_h = v_h = w_h$ in the definition of B_h, (2.29), we get

$$B_h(w_h, w_h) = \frac{1}{2}\| w_h(T) \|_{L^2(0,1)}^2 + \int_0^T \Theta_{diss}(t)\, dt - \frac{1}{2}\| w_h(0) \|_{L^2(0,1)}^2,$$

where

$$\Theta_{diss}(t) = -\sum_{1 \le j \le N} \left\{ h(w_h)_{j+1/2}(t)\, [\,w_h(t)\,]_{j+1/2} + \int_{I_j} c\, w_h(x,t)\, \partial_x w_h(x,t)\, dx \right\}.$$

We only have to show that $\int_0^T \Theta_{diss}(t)\, dt = \Theta_T(w_h)$. To do that, we proceed as follows. Dropping the dependence on the variable t and setting

$$\overline{w}_h(x_{j+1/2}) = \frac{1}{2}\left(w_h(x_{j+1/2}^-) + w_h(x_{j+1/2}^+)\right),$$

we have, by the definition of the flux h, (2.11),

$$-\sum_{1 \le j \le N}\int_{I_j} h(w_h)_{j+1/2}\,[w_h]_{j+1/2} = -\sum_{1 \le j \le N}\{c\,\overline{w}_h\,[w_h] - \frac{|c|}{2}\,[w_h]^2\,\}_{j+1/2},$$

and

$$-\sum_{1 \le j \le N}\int_{I_j} c\,w_h(x)\,\partial_x\,w_h(x)\,dx = \frac{c}{2}\sum_{1 \le j \le N}[w_h^2]_{j+1/2}$$

$$= c\sum_{1 \le j \le N}\{\overline{w}_h\,[w_h]\}_{j+1/2}.$$

Hence

$$\Theta_{diss}(t) = \frac{|c|}{2}\sum_{1 \le j \le N}[u_h(t)]_{j+1/2}^2,$$

and the result follows. This completes the proof of Lemma 17.
    This completes the proof of Proposition 1.

Proof of Theorem 2 In this section, we prove the error estimate of Theorem 2 which holds for the linear case $f(u) = c\,u$. To do that, we first show how to estimate the error between the solutions $w_\nu = (u_\nu, q_\nu)^t$, $\nu = 1, 2$, of

$$\partial_t\,u_\nu + \partial_x\,f(u_\nu) = 0 \quad \text{in } (0, T) \times (0, 1),$$
$$u_\nu(t = 0) = u_{0,\nu}, \quad \text{on } (0, 1).$$

Then, we mimic the argument in order to prove Theorem 2.
    The continuous case as a model. By the definition of the form $B(\cdot, \cdot)$, (2.7), we have, for $\nu = 1, 2$,

$$B(w_\nu, v) = 0, \qquad \forall\, v : v(t) \text{ is smooth} \quad \forall\, t \in (0, T).$$

Since the form $B(\cdot, \cdot)$ is bilinear, from the above equation we obtain the so-called *error equation*:

$$\forall\, v : v(t) \text{ is smooth} \quad \forall\, t \in (0, T) :$$
$$B(e, v) = 0, \tag{2.30}$$

where $e = w_1 - w_2$. Now, since

$$B(e, e) = \frac{1}{2}\|e(T)\|_{L^2(0,1)}^2 - \frac{1}{2}\|e(0)\|_{L^2(0,1)}^2,$$

and

$$B(e, e) = 0,$$

by the error equation (2.30), we immediately obtain the error estimate we sought:

$$\frac{1}{2}\| e(T)\|^2_{L^2(0,1)} = \frac{1}{2}\| u_{0,1} - u_{0,2}\|^2_{L^2(0,1)}.$$

To prove Theorem 2, we only need to obtain a discrete version of this argument.
   The discrete case. Since,

$$B_h(u_h, v_h) = 0, \qquad \forall\, v_h : v(t) \in V_h \quad \forall\, t \in (0,T),$$
$$B_h(u, v_h) = 0, \qquad \forall\, v_h : v_h(t) \in V_h \quad \forall\, t \in (0,T),$$

by (2.7) and by equations (2.4), respectively, we easily obtain our *error equation*:

$$\forall\, v_h : v_h(t) \in V_h \quad \forall\, t \in (0,T):$$
$$B_h(e, v_h) = 0, \tag{2.31}$$

where $e = w - w_h$.

   Now, according to the continuous case argument, we should consider next the quantity $B_h(e, e)$; however, since $e(t)$ is not in the finite element space V_h, it is more convenient to consider $B_h(P_h(e), P_h(e))$, where $P_h(e(t))$ is the L^2-projection of the error $e(t)$ into the finite element space V_h^k.

   The L^2-projection of the function $p \in L^2(0,1)$ into V_h, $P_h(p)$, is defined as the only element of the finite element space V_h such that

$$\forall\, v_h \in V_h :$$
$$\int_0^1 \big(P_h(p)(x) - p(x)\big) v_h(x)\, dx = 0. \tag{2.32}$$

Note that in fact $u_h(t = 0) = P_h(u_0)$, by (2.8).
   Thus, by Lemma 17, we have

$$B_h(P_h(e), P_h(e)) = \frac{1}{2}\| P_h(e(T))\|^2_{L^2(0,1)} + \Theta_T(P_h(e)) - \frac{1}{2}\| P_h(e(0))\|^2_{L^2(0,1)},$$

and since

$$P_h(e(0)) = P_h(u_0 - u_h(0)) = P_h(u_0) - u_h(0) = 0,$$

and

$$B_h(P_h(e), P_h(e)) = B_h(P_h(e) - e, P_h(e)) = B_h(P_h(u) - u, P_h(e)),$$

by the *error equation* (2.31), we get

$$\frac{1}{2}\| P_h(e(T))\|^2_{L^2(0,1)} + \Theta_T(P_h(e)) = B_h(P_h(u) - u, P_h(e)). \tag{2.33}$$

It only remains to estimate the right-hand side

$$B(P_h(u) - u, P_h(e)),$$

which, according to our continuous model, should be small.

   Estimating the right-hand side. To show that this is so, we must suitably treat the term $B(P_h(w) - w, P_h(e))$. We start with the following remarkable result.

Lemma 18. *We have*

$$B_h(P_h(u) - u, P_h(e)) = -\int_0^T \sum_{1 \le j \le N} h(P_h(u) - u)_{j+1/2}(t) \, [\, P_h(e)(t)\,]_{j+1/2} \, dt.$$

Proof Setting $p = P_h(u) - u$ and $v_h = P_h(e)$ and recalling the definition of $B_h(\cdot, \cdot)$, (2.29), we have

$$\begin{aligned}
B_h(p, v_h) &= \int_0^T \int_0^1 \partial_t p(x,t) \, v_h(x,t) \, dx \, dt \\
&\quad - \int_0^T \sum_{1 \le j \le N} \int_{I_j} c \, p(x,t) \, \partial_x v_h(x,t) \, dx \, dt \\
&\quad - \int_0^T \sum_{1 \le j \le N} h(p)_{j+1/2}(t) \, [\, v_h(t)\,]_{j+1/2} \, dt \\
&= - \int_0^T \sum_{1 \le j \le N} h(p)_{j+1/2}(t) \, [\, v_h(t)\,]_{j+1/2} \, dt,
\end{aligned}$$

by the definition of the L^2-projection (2.32). This completes the proof.

Now, we can see that a simple application of Young's inequality and a standard approximation result should give us the estimate we were looking for. The approximation result we need is the following.

Lemma 19. *If $w \in H^{k+1}(I_j \cup I_{j+1})$, then*

$$|\, h(P_h(w) - w)(x_{j+1/2})\,| \le c_k \, (\Delta x)^{k+1/2} \frac{|c|}{2} \, |\, w \,|_{H^{k+1}(I_j \cup I_{j+1})},$$

where the constant c_k depends solely on k.

Proof. Dropping the argument $x_{j+1/2}$ we have, by the definition (2.11) of the flux h,

$$\begin{aligned}
|\, h(P(w) - w)\,| &= \Big|\, \frac{c}{2}(P_h(w)^+ + P_h(w)^-) - \frac{|c|}{2}(P_h(w)^+ - P_h(w)^-) - c\,w \,\Big| \\
&= \Big|\, \frac{c - |c|}{2}(P_h(w)^+ - w) + \frac{c + |c|}{2}(P_h(w)^- - w) \,\Big| \\
&\le |c| \max\{\,|\, P_h(w)^+ - w\,|, |\, P_h(w)^- - w\,|\,\}
\end{aligned}$$

and the result follows from the properties of P_h after a simple application of the Bramble-Hilbert lemma; see [16]. This completes the proof.

An immediate consequence of this result is the estimate we wanted.

Lemma 20. *We have*

$$B_h(P_h(u) - u, P_h(e)) \le c_k^2 \, (\Delta x)^{2k+1} \frac{|c|}{2} \, T \, |\, u_0 \,|_{H^{k+1}(0,1)}^2 + \frac{1}{2} \Theta_T(P_h(e)),$$

where the constant c_k depends solely on k.

Proof. After using Young's inequality in the right-hand side of Lemma 18, we get

$$B_h(P_h(u) - u, P_h(e)) \leq \int_0^T \sum_{1 \leq j \leq N} \frac{1}{|c|} \, | \, h(P_h(u) - u)_{j+1/2}(t) \, |^2$$
$$+ \int_0^T \sum_{1 \leq j \leq N} \frac{|c|}{4} [\, P_h(e)(t) \,]_{j+1/2}^2 \, dt.$$

By Lemma 19 and the definition of the form Θ_T, we get

$$B_h(P_h(u) - u, P_h(e)) \leq c_k^2 \, (\Delta x)^{2k+1} \frac{|c|}{4} \int_0^T \sum_{1 \leq j \leq N} |u|_{H^{k+1}(I_j \cup I_{j+1})}^2 + \frac{1}{2} \Theta_T(P_h(e))$$
$$\leq c_k^2 \, (\Delta x)^{2k+1} \frac{|c|}{2} T \, | \, u_0 \, |_{H^{k+1}(0,1)}^2 \frac{1}{2} \Theta_T(P_h(e)).$$

This completes the proof.

Conclusion. Finally, inserting in the equation (2.33) the estimate of its right hand side obtained in Lemma 20, we get

$$\| P_h(e(T)) \|_{L^2(0,1)}^2 + \Theta_T(P_h(e)) \leq c_k \, (\Delta x)^{2k+1} \, | \, c \, | \, T \, | \, u_0 \, |_{H^{k+1}(0,1)}^2,$$

Theorem 2 now follows from the above estimate and from the following inequality:

$$\| e(T) \|_{L^2(0,1)} \leq \| u(T) - P_h(u(T)) \|_{L^2(0,1)} + \| P_h(e(T)) \|_{L^2(0,1)}$$
$$\leq c_k' \, (\Delta x)^{k+1} \, | \, u_0 \, |_{H^{k+1}(0,1)} + \| P_h(e(T)) \|_{L^2(0,1)}.$$

Proof of Theorem 3 To prove Theorem 3, we only have to suitably modify the proof of Theorem 2. The modification consists in *replacing* the L^2-projection of the error, $P_h(e)$, by *another* projection that we denote by $R_h(e)$.

Given a function $p \in L^\infty(0,1)$ that is continuous on each element I_j, we define $R_h(p)$ as the only element of the finite element space V_h such that

$$\forall j = 1, \ldots, N :$$
$$R_h(p)(x_{j,\ell}) - p(x_{j,\ell}) = 0, \qquad \ell = 0, \ldots, k, \tag{2.34}$$

where the points $x_{j,\ell}$ are the Gauss-Radau quadrature points of the interval I_j. We take

$$x_{j,k} = \begin{cases} x_{j+1/2} & \text{if } c > 0, \\ x_{j-1/2} & \text{if } c < 0. \end{cases} \tag{2.35}$$

The special nature of the Gauss-Radau quadrature points is captured in the following property:

$$\forall \varphi \in P^\ell(I_j), \quad \ell \leq k, \quad \forall p \in P^{2k-\ell}(I_j) :$$
$$\int_{I_j} (R_h(p)(x) - p(x)) \, \varphi(x) \, dx = 0. \tag{2.36}$$

Compare this equality with (2.32).

The quantity $B_h(R_h(e), R_h(e))$. To prove our error estimate, we start by considering the quantity $B_h(R_h(e), R_h(e))$. By Lemma 17, we have

$$B_h(R_h(e), R_h(e)) = \frac{1}{2}\| R_h(e(T)) \|_{L^2(0,1)}^2 + \Theta_T(R_h(e)) - \frac{1}{2}\| R_h(e(0)) \|_{L^2(0,1)}^2,$$

and since

$$B_h(R_h(e), R_h(e)) = B_h(R_h(e) - e, R_h(e)) = B_h(R_h(u) - u, R_h(e)),$$

by the *error equation* (2.31), we get

$$\frac{1}{2}\| R_h(e(T)) \|_{L^2(0,1)}^2 + \Theta_T(R_h(e)) = \frac{1}{2}\| R_h(e(0)) \|_{L^2(0,1)}^2 + B_h(R_h(u) - u, R_h(e)).$$

Next, we estimate the term $B(R_h(u) - u, R_h(e))$.

Estimating $B(R_h(u) - u, R_h(e))$. The following result corresponds to Lemma 18.

Lemma 21. *We have*

$$B_h(R_h(u) - u, v_h) = \int_0^T \int_0^1 (R_h(\partial_t u)(x,t) - \partial_t u(x,t))\, v_h(x,t)\, dx\, dt$$
$$- \int_0^T \sum_{1 \le j \le N} \int_{I_j} c\, (R_h(u)(x,t) - u(x,t))\, \partial_x v_h(x,t)\, dx\, dt.$$

Proof Setting $p = R_h(u) - u$ and $v_h = R_h(e)$ and recalling the definition of $B_h(\cdot, \cdot)$, (2.29), we have

$$B_h(p, v_h) = \int_0^T \int_0^1 \partial_t p(x,t)\, v_h(x,t)\, dx\, dt$$
$$- \int_0^T \sum_{1 \le j \le N} \int_{I_j} c\, p(x,t)\, \partial_x v_h(x,t)\, dx\, dt$$
$$- \int_0^T \sum_{1 \le j \le N} h(p)_{j+1/2}(t)\, [\, v_h(t)\,]_{j+1/2}\, dt.$$

But, from the definition (2.11) of the flux h, we have

$$h(R(u) - u) = \frac{c}{2}(R_h(u)^+ + R_h(u)^-) - \frac{|c|}{2}(R_h(u)^+ - R_h(u)^-) - c\, u$$
$$= \frac{c - |c|}{2}(R_h(u)^+ - u) + \frac{c + |c|}{2}(R_h(u)^- - u)$$
$$= 0,$$

by (2.35) and the result follows.

Next, we need some approximation results.

Lemma 22. *If $w \in H^{k+2}(I_j)$, and $v_h \in P^k(I_j)$, then*

$$\left| \int_{I_j} (R_h(w) - w)(x)\, v_h(x)\, dx \right| \le c_k \,(\Delta x)^{k+1}\, | w \,|_{H^{k+1}(I_j)}\, \| v_h \|_{L^2(I_j)},$$

and

$$\left| \int_{I_j} (R_h(w) - w)(x)\, \partial_x v_h(x)\, dx \right| \le c_k \,(\Delta x)^{k+1}\, | w \,|_{H^{k+2}(I_j)}\, \| v_h \|_{L^2(I_j)},$$

where the constant c_k depends solely on k.

Proof. The first inequality follows from the property (2.36) with $\ell = k$ and from standard approximation results. The second follows in a similar way from the property (2.36) with $\ell = k - 1$ and a standard scaling argument. This completes the proof.

An immediate consequence of this result is the estimate we wanted.

Lemma 23. *We have*

$$B_h(R_h(u) - u, R_h(e)) \le c_k \,(\Delta x)^{k+1}\, | u_0 \,|_{H^{k+2}(0,1)} \int_0^T \| R_h(e(t)) \|_{L^2(0,1)}\, dt,$$

where the constant c_k depends solely on k and $|c|$.

Conclusion. Finally, inserting in the equation (2.33) the estimate of its right hand side obtained in Lemma 23, we get

$$\| R_h(e(T)) \|^2_{L^2(0,1)} + \Theta_T(R_h(e)) \le \| R_h(e(0)) \|^2_{L^2(0,1)}$$
$$+ c_k \,(\Delta x)^{k+1}\, | u_0 \,|_{H^{k+2}(0,1)} \int_0^T \| R_h(e(t)) \|_{L^2(0,1)}\, dt.$$

After applying a simple variation of the Gronwall lemma, we obtain

$$\| R_h(e(T)) \|_{L^2(0,1)} \le \| R_h(e(0))(x) \|_{L^2(0,1)} + c_k \,(\Delta x)^{k+1}\, T\, | u_0 \,|_{H^{k+2}(0,1)}$$
$$\le c'_k (\Delta x)^{k+1}\, | u_0 \,|_{H^{k+2}(0,1)}.$$

Theorem 3 now follows from the above estimate and from the following inequality:

$$\| e(T) \|_{L^2(0,1)} \le \| u(T) - R_h(u(T)) \|_{L^2(0,1)} + \| R_h(e(T)) \|_{L^2(0,1)}$$
$$\le c'_k \,(\Delta x)^{k+1}\, | u_0 \,|_{H^{k+1}(0,1)} + \| R_h(e(T)) \|_{L^2(0,1)}.$$

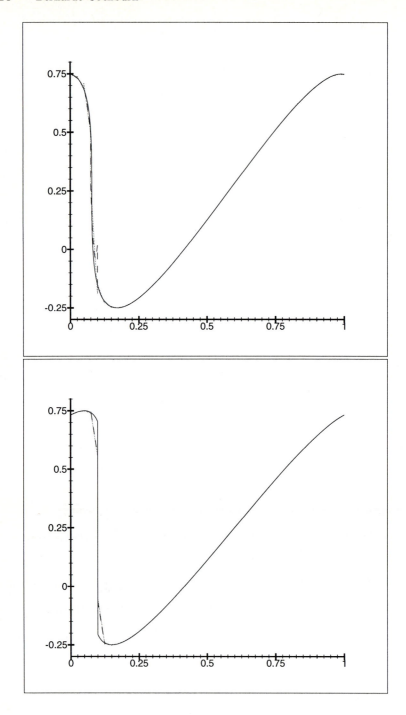

Fig. 2.3. Comparison of the exact and the approximate solutions obtained with $M = 20$, $\Delta x = 1/40$ at $T = 1/\pi$ (top) and at $T = 0.40$ (bottom): Exact solution (solid line), piecewise linear solution (dotted line), and piecewise quadratic solution (dashed line).

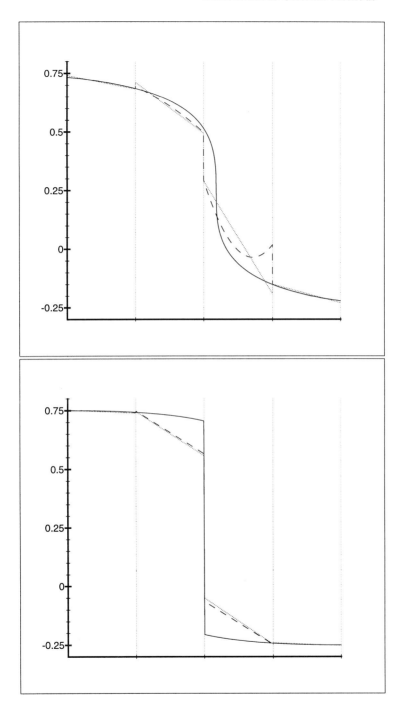

Fig. 2.4. Detail of previous figures. Behavior of the approximate solutions four elements around the shock at $T = 1/\pi$ (top) and at $T = 0.40$ (bottom): Exact solution (solid line), piecewise linear solution (dotted line), and piecewise quadratic solution (dashed line).

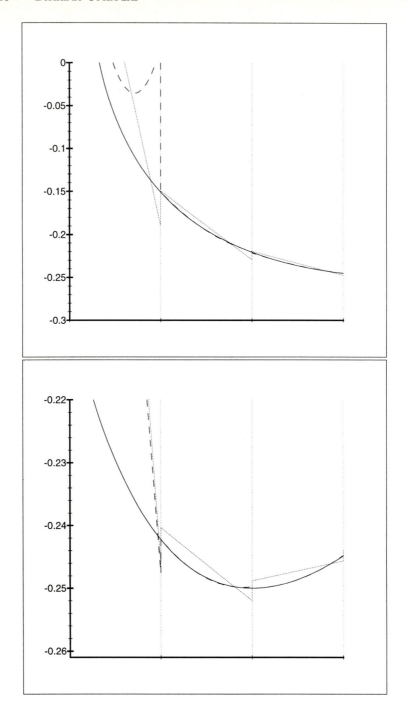

Fig. 2.5. Detail of previous figures. Behavior of the approximate solutions two elements in front of the shock at $T = 1/\pi$ (top) and at $T = 0.40$ (bottom): Exact solution (solid line), piecewise linear solution (dotted line), and piecewise quadratic solution (dashed line).

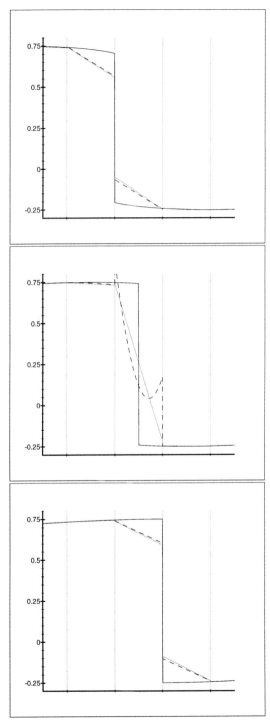

Fig. 2.6. Comparison of the exact and the approximate solutions obtained with $M = 20$, $\Delta x = 1/40$ as the shock passes through one element. Exact solution (solid line), piecewise linear elements (dotted line) and piecewise quadratic elements (dashed line). Top: $T = 0.40$, middle: $T = 0.45$, and bottom: $T = 0.50$.

3   The Hamilton-Jacobi equations in one space dimension

3.1   Introduction

In this chapter, we extend the RKDG method to the following simple problem for the Hamilton-Jacobi equation

$$\varphi_t + H(\varphi_x) = 0, \qquad \text{in } (0,1) \times (0,T), \qquad (3.1)$$
$$\varphi(x,0) = \varphi_0(x), \qquad \forall\, x \in (0,1), \qquad (3.2)$$

where we take periodic boundary conditions. The material in this section is based in the work of Hu and Shu [43].

3.2   The RKDG method

The main idea to extend the RKDG method to this case, is to realize that $u = \varphi_x$ satisfies the following problem:

$$u_t + H(u)_x = 0, \qquad \text{in } (0,1) \times (0,T), \qquad (3.3)$$
$$u(x,0) = (\varphi_0)_x(x), \qquad \forall\, x \in (0,1), \qquad (3.4)$$

and that φ can be computed from u by solving the following problem:

$$\varphi_t = -H(u), \qquad \text{in } (0,1) \times (0,T), \qquad (3.5)$$
$$\varphi(x,0) = \varphi_0(x), \qquad \forall\, x \in (0,1). \qquad (3.6)$$

A straightforward application of the RKDG method to the equations (3.3), (3.4) produces a piecewise polynomial approximation u_h to $u = \varphi_x$. If the approximating polynomials are taken to be of degree $(k-1)$, it is reasonable to seek an approximation φ_h to φ that is piecewise a polynomial of degree k. To obtain it, we can discretize (3.5), (3.6) in one of the following ways:

(i) Take $\varphi_h(\cdot, t)$ in V_h^k such that

$$\forall j = 1, \ldots, N, \quad v_h \in P^k(I_j):$$
$$\int_{I_j} \partial_t \varphi_h(x,t)\, v_h(x)\, dx = -\int_{I_j} H(u_h(x,t))\, v_h(x)\, dx,$$
$$\int_{I_j} \varphi(x,0)\, v_h(x)\, dx = \int_{I_j} \varphi_0(x)\, v_h(x)\, dx.$$

(ii) Take $\varphi_h(\cdot, t)$ in V_h^k such that

$$\forall j = 1, \ldots, N: \quad \partial_x \varphi_h(x,t) = u(x,t) \qquad \forall x \in I_j.$$

This determines φ_h up to a constant. To find this constant, we impose the following condition:

$$\forall j = 1, \ldots, N:$$
$$\frac{d}{dt} \int_{I_j} \varphi_h(x,t)\, dx = -\int_{I_j} H(u_h(x,t))\, dx,$$
$$\int_{I_j} \varphi(x,0)\, dx = \int_{I_j} \varphi_0(x)\, dx.$$

(iii) Pick one element, say I_J, and determine the values of φ_h on it by requiring the following conditions:

$$\partial_x \varphi_h(x,t) = u(x,t) \qquad \forall x \in I_J,$$

$$\frac{d}{dt} \int_{I_J} \varphi_h(x,t)\,dx = -\int_{I_J} H(u_h(x,t))\,dx,$$

$$\int_{I_J} \varphi(x,0)\,dx = \int_{I_J} \varphi_0(x)\,dx.$$

Then, compute φ_h as follows:

$$\varphi_h(x,t) = \varphi_h(x_J,t) + \int_{x_J}^{x} u(s,t)\,ds.$$

Note that, unlike the previous approaches, the approximate solution φ_h is now continuous.

The advantage of the first two approaches is that they can be carried out in parallel. On the other hand, in the third approach, only a single ODE has to be solved; moreover, the integration in space takes place just at the very end of the whole computation. This approach is much more efficient.

It could be argued that in the third approach, the recovered values of φ depend upon the choice of the starting point x_1. However, this difference is on the level of truncation errors and does not affect the order of accuracy. Hu and Shu [43] used both the second and third approaches; they report that their numerical experience is that, when there are singularities in the derivatives, the second approach will often produce dents and bumps when the integral path in time passes through the singularities. This can be avoided in the third approach. Indeed, the main idea of using the third approach is to choose the element I_J so that the time integral paths *do not cross* derivative singularities. This cannot be always be done with a single element I_J, but it is always possible to switch to *another* element *before* the singularity in the derivative hits the current element I_J. If the number of discontinuities in the derivative is finite, this needs to be done only a finite number of times. This maintains the efficiency of the method.

Note that all the properties fo the RKDG method obtained in the previous section apply to the approximate solution u_h. In particular, a consequence of the work of Jiang and Shu [51], is the following result for the approximation to the derivative φ_x, u_h; see also Proposition 6 and Theorem 7.

Theorem 24. *For any of the above methods and any polynomial degree $k \geq 0$, we have*

$$\frac{d}{dt} \int_0^1 u_h^2(x,t)\,dx \leq 0. \tag{3.7}$$

Moreover, if the Hamiltonian H is a strictly convex or concave function, for any $k \geq 0$, if the numerical solution given by the DG method converges, it converges to the viscosity solution.

Note that the above result trivially implies the TVB (total variation bounded) property for the numerical solution φ_h. Indeed

$$TV(\varphi_h(t)) = \int_a^b |u_h(x,t)|\, dx \leq \sqrt{b-a}\, \|\varphi_0'\|_{L^2(a,b)}.$$

This is a rather strong stability result, considering that it holds independently of the degree of the polynomial approximation even when the derivative of the solution φ_x develops discontinuities and without the application of the generalized slope limiter!

3.3   Computational results

In this section, we present the numerical experiments of Hu and Shu [43] showing the performance of the RKDG method. Our main purpose is to asses the accuracy of the method and see if the generalized slope limiter needs to be used. We display the results obtained with the third approach to compute φ_h.

The first test problem. One dimensional Burgers' equation:

$$\varphi_t + \frac{(\varphi_x + 1)^2}{2} = 0, \qquad \text{in } (-1,1) \times (0,T),$$
$$\varphi(x,0) = -\cos(\pi x), \qquad \forall\, x \in (-1,1),$$

with periodic boundary conditions.

The local Lax-Friedrichs flux is used. At $T = 0.5/\pi^2$, the solution is still smooth. We list the errors and the numerical orders of accuracy in Table 3.1. We observe that, except for the P^1 case which seems to be only first order, P^k for $k > 1$ seems to provide close to $(k+1)$-th order accuracy. The meshes used are all uniform, and errors are computed at the middle point of each interval.

To investigate the accuracy problem further, we use non-uniform meshes obtained by randomly shifting the cell boundaries in a uniform mesh in the range $[-0.1h, 0.1h]$. In order to avoid possible superconvergence at cell centers, we also give the "real" L^2 error (computed by a 6-point Gaussian quadrature in each cell). The results are shown in Table 3.2.

At $T = 3.5/\pi^2$, the solution has developed a discontinuous derivative. In Fig. 3.1, we show the sharp corner-like numerical solution with 41 elements obtained with P^k for $k = 1,2,3,4$ with a uniform mesh. Here and below, the solid line is the exact solution, the circles are numerical solutions (only one point per element is drawn).

The second test problem. One dimensional equation with a non-convex flux:

$$\varphi_t - \cos(\varphi_x + 1) = 0, \qquad \text{in } (-1,1) \times (0,T),$$
$$\varphi(x,0) = -\cos(\pi x), \qquad \forall\, x \in (-1,1),$$

with periodic boundary conditions.

The local Lax-Friedrichs flux and uniform meshes are used. At $T = 0.5/\pi^2$, the solution is still smooth. The accuracy of the numerical solution is listed in Table 3.3. We observe similar accuracy as in the previous example.

Table 3.1. Accuracy for 1D Burgers equation (uniform mesh), $T = 0.5/\pi^2$.

N	P^1 L^1 error	order	P^2 L^1 error	order	P^3 L^1 error	order	P^4 L^1 error	order
10	0.17E+00	—	0.14E-02	—	0.21E-03	—	0.57E-05	—
20	0.78E-01	1.12	0.18E-03	2.92	0.13E-04	3.94	0.73E-06	2.97
40	0.35E-01	1.16	0.24E-04	2.97	0.75E-06	4.17	0.32E-07	4.52
80	0.16E-01	1.12	0.28E-05	3.08	0.43E-07	4.12	0.12E-08	4.79
160	0.76E-02	1.02	0.31E-06	3.19	0.25E-08	4.10	0.48E-10	4.59

N	P^1 L^∞ error	order	P^2 L^∞ error	order	P^3 L^∞ error	order	P^4 L^∞ error	order
10	0.29E+00	—	0.24E-02	—	0.69E-03	—	0.13E-04	—
20	0.13E+00	1.13	0.33E-03	2.88	0.61E-04	3.51	0.16E-05	2.99
40	0.58E-01	1.15	0.37E-04	3.15	0.58E-05	3.39	0.13E-06	3.64
80	0.27E-01	1.11	0.48E-05	2.97	0.38E-06	3.93	0.59E-08	4.44
160	0.13E-01	1.07	0.59E-06	3.00	0.23E-07	4.07	0.25E-09	4.57

Table 3.2. Accuracy for 1D Burgers equation (non-uniform mesh), $T = 0.5/\pi^2$.

N	P^1 L^2 error	order	P^2 L^2 error	order	P^3 L^2 error	order	P^4 L^2 error	order
10	0.74E+00	—	0.34E-02	—	0.32E-03	—	0.53E-04	—
20	0.34E+00	1.11	0.51E-03	2.76	0.24E-04	3.72	0.20E-05	4.71
40	0.15E+00	1.19	0.65E-04	2.96	0.17E-05	3.82	0.71E-07	4.84
80	0.67E-01	1.17	0.90E-05	2.86	0.13E-06	3.72	0.20E-08	5.15
160	0.31E-01	1.13	0.11E-05	3.02	0.81E-08	4.02	0.74E-10	4.76

N	P^1 L^1 error	order	P^2 L^1 error	order	P^3 L^1 error	order	P^4 L^1 error	order
10	0.53E+00	—	0.17E-02	—	0.23E-03	—	0.30E-05	—
20	0.24E+00	1.13	0.21E-03	3.05	0.14E-04	4.01	0.40E-06	2.89
40	0.11E+00	1.19	0.26E-04	2.99	0.78E-06	4.20	0.16E-07	4.65
80	0.47E-01	1.17	0.37E-05	2.82	0.47E-07	4.05	0.61E-09	4.70
160	0.21E-01	1.13	0.41E-06	3.16	0.27E-08	4.15	0.26E-10	4.56

N	P^1 L^∞ error	order	P^2 L^∞ error	order	P^3 L^∞ error	order	P^4 L^∞ error	order
10	0.62E+00	—	0.36E-02	—	0.69E-03	—	0.11E-04	—
20	0.29E+00	1.11	0.47E-03	2.94	0.61E-04	3.52	0.16E-05	2.81
40	0.13E+00	1.16	0.67E-04	2.80	0.47E-05	3.70	0.13E-06	3.64
80	0.58E-01	1.14	0.17E-04	2.01	0.62E-06	2.91	0.59E-08	4.45
160	0.27E-01	1.11	0.19E-05	3.11	0.31E-07	4.32	0.33E-09	4.17

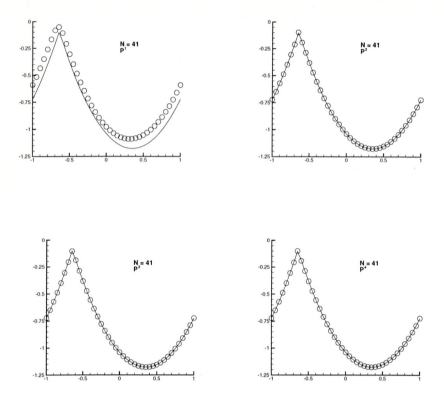

Fig. 3.1. One-dimensional Burgers' equation, $T = 3.5/\pi^2$.

At $T = 1.5/\pi^2$, the solution has developed corner-like discontinuity in the derivative. The numerical result with 41 elements is shown in Fig. 3.2.

The third test problem. Riemann problem for the one dimensional equation with a non-convex flux:

$$\varphi_t + \frac{1}{4}(\varphi_x^2 - 1)(\varphi_x^2 - 4) = 0, \qquad \text{in } (-1, 1) \times (0, T),$$

$$\varphi(x, 0) = -2|x|, \qquad \qquad \forall\, x \in (-1, 1),$$

For this test problem, the use of the generalized slope limiter proved to be *essential* since otherwise the approximate solution does not converge to the viscosity solution; this is the only example in which we use the nonlinear limiting. We remark that for the finite difference schemes, such nonlinear limiting or the adaptive stencil in ENO is needed in most cases in order to enforce stability and to obtain non-oscillatory results.

Numerical results at $T = 1$ with 81 elements, using the local Lax-Friedrichs flux, is shown in Fig. 3.3. The results of using the Godunov flux is shown in Fig. 3.4. We can see that while for P^1, the results of using two different monotone fluxes

Table 3.3. Accuracy for 1D non-convex, $H(u) = -\cos(u+1)$, $T = 0.5/\pi^2$.

	P^1		P^2		P^3		P^4	
N	L^1 error	order	L^1 error	order	L^1 error	order	L^1 error	order
10	0.84E-01	—	0.10E-02	—	0.34E-03	—	0.24E-04	—
20	0.36E-01	1.23	0.15E-03	2.75	0.30E-04	3.49	0.13E-05	4.28
40	0.15E-01	1.26	0.21E-04	2.84	0.15E-05	4.33	0.59E-07	4.42
80	0.68E-02	1.14	0.27E-05	2.97	0.94E-07	4.00	0.21E-08	4.78

	P^1		P^2		P^3		P^4	
N	L^∞ error	order	L^∞ error	order	L^∞ error	order	L^∞ error	order
10	0.18E+00	—	0.15E-02	—	0.11E-02	—	0.99E-04	—
20	0.73E-01	1.31	0.27E-03	2.43	0.22E-03	2.35	0.13E-04	2.95
40	0.31E-01	1.24	0.47E-04	2.54	0.18E-04	3.63	0.59E-06	4.44
80	0.14E-01	1.16	0.85E-05	2.47	0.14E-05	3.75	0.26E-07	4.49

are significantly different in resolution, this difference is greatly reduced for higher order of accuracy. In most of the high order cases, the simple local Lax-Friedrichs flux is a very good choice.

3.4 Concluding Remarks.

In this section, we have extended the RKDG method, originally devised for nonlinear conservation laws, to the Hamilton-Jacobi equations. The extension was carried out by exploiting the fact that the derivative of the solution of the Hamilton-Jacobi equation satisfies a nonlinear conservation law.

The numerical experiments show that when polynomials of degree k are used, the method is of order $(k+1)$ in L^2, except when $k = 1$; this phenomenon remains to be explained. Also, we have seen that the use of slope limiters was only needed in the third test problem- otherwise the convergence to the viscosity solution did not take place.

The scheme can be extended to the case of a bounded domain in a very simple way. The extension of the scheme to the multidimensional case is not quite straightforward and will be carried out after we study how to define the RKDG method for multidimensional conservation laws.

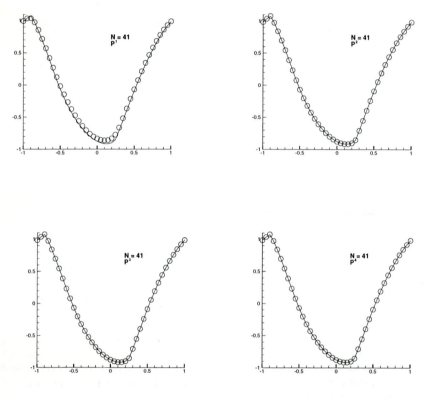

Fig. 3.2. One dimension non-convex, $H(u) = -\cos(u + 1)$, $T = 1.5/\pi^2$.

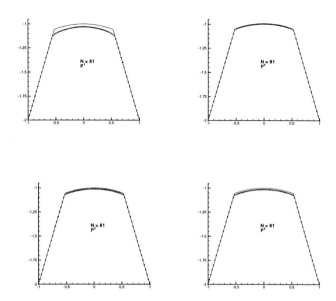

Fig. 3.3. One dimension Riemann problem, local Lax-Friedrichs flux, $H(u) = \frac{1}{4}(u^2 - 1)(u^2 - 4), T = 1$.

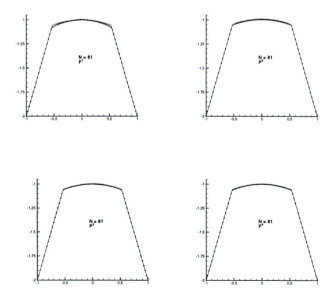

Fig. 3.4. One dimension Riemann problem, Godunov flux, $H(u) = \frac{1}{4}(u^2 - 1)(u^2 - 4), T = 1$.

4   The RKDG method for multi-dimensional systems

4.1   Introduction

In this section, we extend the RKDG methods to multidimensional systems:

$$u_t + \nabla f(u) = 0, \qquad \text{in } \Omega \times (0, T), \tag{4.1}$$

$$u(x, 0) = u_0(x), \qquad \forall\, x \in \Omega, \tag{4.2}$$

and periodic boundary conditions. For simplicity, we assume that Ω is the unit cube.

This section is essentially devoted to the description of the algorithms and their implementation details. The practitioner should be able to find here all the necessary information to completely code the RKDG methods.

This section also contains two sets of numerical results for the Euler equations of gas dynamics in two space dimensions. The first set is devoted to transient computations and domains that have corners; the effect of using triangles or rectangles and the effect of using polynomials of degree one or two are explored. The main conclusions from these computations are that (i) the RKDG method works as well with triangles as it does with rectangles and that (ii) the use of high-order polynomials does not deteriorate the approximation of strong shocks and is advantageous in the approximation of contact discontinuities.

The second set concerns steady state computations with smooth solutions. For these computations, no generalized slope limiter is needed. The effect of (i) the quality of the approximation of curved boundaries and of (ii) the degree of the polynomials on the quality of the approximate solution is explored. The main conclusions from these computations are that (i) a high-order approximation of the curve boundaries introduces a dramatic improvement on the quality of the solution and that (ii) the use of high-degree polynomials is advantageous when smooth solutions are sought.

This section contains material from the papers [21], [20], and [28]. It also contains numerical results from the paper by Bassi and Rebay [4] in two dimensions and from the paper by Warburton, Lomtev, Kirby and Karniadakis [91] in three dimensions.

4.2   The general RKDG method

The RKDG method for multidimensional systems has the same structure it has for one-dimensional scalar conservation laws, that is,

- Set $u_h^0 = \Lambda\Pi_h\, P_{V_h}(u_0)$;
- For $n = 0, ..., N - 1$ compute u_h^{n+1} as follows:
    1. set $u_h^{(0)} = u_h^n$;
    2. for $i = 1, ..., k + 1$ compute the intermediate functions:

$$u_h^{(i)} = \Lambda\Pi_h \left\{ \sum_{l=0}^{i-1} \alpha_{il} u_h^{(l)} + \beta_{il} \Delta t^n L_h(u_h^{(l)}) \right\};$$

    3. set $u_h^{n+1} = u_h^{(k+1)}$.

In what follows, we describe the operator L_h that results form the DG-space discretization, and the generalized slope limiter $\Lambda\Pi_h$.

The Discontinuous Galerkin space discretization To show how to discretize in space by the DG method, it is enough to consider the case in which u is a scalar quantity since to deal with the general case in which u, we apply the same procedure component by component.

Once a triangulation $\mathcal{T}_h$ of Ω has been obtained, we determine $L_h(\cdot)$ as follows. First, we multiply (4.1) by v_h in the finite element space V_h, integrate over the element K of the triangulation $\mathcal{T}_h$ and replace the exact solution u by its approximation $u_h \in V_h$:

$$\frac{d}{dt}\int_K u_h(t,x)\,v_h(x)\,dx + \int_K div\, f(u_h(t,x))\,v_h(x)\,dx = 0, \ \forall v_h \in V_h. \qquad (4.3)$$

Integrating by parts formally we obtain

$$\frac{d}{dt}\int_K u_h(t,x)\,v_h(x)\,dx + \sum_{e \in \partial K}\int_e f(u_h(t,x))\cdot n_{e,K}\ v_h(x)\,d\Gamma$$

$$-\int_K f(u_h(t,x))\cdot \nabla v_h(x)\,dx = 0, \quad \forall v_h \in V_h,$$

where $n_{e,K}$ is the outward unit normal to the edge e. Notice that $f(u_h(t,x))\cdot n_{e,K}$ does not have a precise meaning, for u_h is discontinuous at $x \in e \in \partial K$. Thus, as in the one dimensional case, we replace $f(u_h(t,x))\cdot n_{e,K}$ by the function $h_{e,K}(u_h(t,x^{int(K)}), u_h(t,x^{ext(K)}))$. The function $h_{e,K}(\cdot,\cdot)$ is any consistent two–point monotone Lipschitz flux, consistent with $f(u)\cdot n_{e,K}$.

In this way we obtain

$$\frac{d}{dt}\int_K u_h(t,x)v_h(x)\,dx + \sum_{e \in \partial K}\int_e h_{e,K}(t,x)\ v_h(x)\,d\Gamma$$

$$-\int_K f(u_h(t,x))\cdot \nabla v_h(x)\,dx = 0, \quad \forall\, v_h \in V_h.$$

Finally, we replace the integrals by quadrature rules that we shall choose as follows:

$$\int_e h_{e,K}(t,x)\,v_h(x)\,d\Gamma \approx \sum_{l=1}^{L}\omega_l\, h_{e,K}(t,x_{el})\,v(x_{el})|e|, \qquad (4.4)$$

$$\int_K f(u_h(t,x))\cdot \nabla v_h(x)\,dx \approx \sum_{j=1}^{M}\omega_j\, f(u_h(t,x_{Kj}))\cdot \nabla v_h(x_{Kj})|K|. \qquad (4.5)$$

Thus, we finally obtain, for each element $K \in \mathcal{T}_h$, the weak formulation:

$$\frac{d}{dt}\int_K u_h(t,x)v_h(x)dx + \sum_{e \in \partial K}\sum_{l=1}^{L}\omega_l\, h_{e,K}(t,x_{el})\,v(x_{el})|e|$$

$$-\sum_{j=1}^{M}\omega_j\, f(u_h(t,x_{Kj}))\cdot \nabla v_h(x_{Kj})|K| = 0, \quad \forall v_h \in V_h.$$

These equations can be rewritten in ODE form as $\frac{d}{dt}u_h = L_h(u_h, \gamma_h)$. This defines the operator $L_h(u_h)$, which is a discrete approximation of $-div\, f(u)$. The following result gives an indication of the quality of this approximation.

Proposition 25. *Let $f(u) \in W^{k+2,\infty}(\Omega)$, and set $\gamma = trace(u)$. Let the quadrature rule over the edges be exact for polynomials of degree $(2k+1)$, and let the one over the element be exact for polynomials of degree $(2k)$. Assume that the family of triangulations $\mathcal{F} = \{\mathcal{T}_h\}_{h>0}$ is regular, i.e., that there is a constant σ such that:*

$$\frac{h_K}{\rho_K} \geq \sigma, \quad \forall K \in \mathcal{T}_h, \quad \forall \mathcal{T}_h \in \mathcal{F}, \tag{4.6}$$

where h_K is the diameter of K, and ρ_K is the diameter of the biggest ball included in K. Then, if $V(K) \supset P^k(K)$, $\forall K \in \mathcal{T}_h$:

$$\|L_h(u,\gamma) + div\, f(u)\|_{L^\infty(\Omega)} \leq C\, h^{k+1}|f(u)|_{W^{k+2,\infty}(\Omega)}.$$

For a proof, see [20].

The form of the generalized slope limiter $\Lambda\Pi_h$. The construction of generalized slope limiters $\Lambda\Pi_h$ for several space dimensions is not a trivial matter and will not be discussed in these notes; we refer the interested reader to the paper by Cockburn, Hou, and Shu [20].

In these notes, we restrict ourselves to displaying very simple, practical, and effective generalized slope limiters $\Lambda\Pi_h$ which are closely related to the generalized slope limiters $\Lambda\Pi_h^k$ of the previous section.

To compute $\Lambda\Pi_h u_h$, we rely on the *assumption* that spurious oscillations are present in u_h only if they are present in its P^1 part u_h^1, which is its L^2-projection into the space of piecewise linear functions V_h^1. Thus, if they are not present in u_h^1, i.e., if

$$u_h^1 = \Lambda\Pi_h\, u_h^1,$$

then we assume that they are not present in u_h and hence do not do any limiting:

$$\Lambda\Pi_h\, u_h = u_h\, .$$

On the other hand, if spurious oscillations are present in the P^1 part of the solution u_h^1, i.e., if

$$u_h^1 \neq \Lambda\Pi_h\, u_h^1,$$

then we chop off the higher order part of the numerical solution, and limit the remaining P^1 part:

$$\Lambda\Pi_h\, u_h = \Lambda\Pi_h\, u_h^1.$$

In this way, in order to define $\Lambda\Pi_h$ for arbitrary space V_h, we only need to actually define it for piecewise linear functions V_h^1. The exact way to do that, both for the triangular elements and for the rectangular elements, will be discussed in the next section.

4.3   Algorithm and implementation details

In this section we give the algorithm and implementation details, including numerical fluxes, quadrature rules, degrees of freedom, fluxes, and limiters of the RKDG method for both piecewise-linear and piecewise-quadratic approximations in both triangular and rectangular elements.

Fluxes The numerical flux we use is the simple Lax-Friedrichs flux:

$$h_{e,K}(a,b) = \frac{1}{2}\left[\mathbf{f}(a)\cdot n_{e,K} + \mathbf{f}(b)\cdot n_{e,K} - \alpha_{e,K}(b-a)\right].$$

The numerical viscosity constant $\alpha_{e,K}$ should be an estimate of the biggest eigenvalue of the Jacobian $\frac{\partial}{\partial u}\mathbf{f}(u_h(x,t))\cdot n_{e,K}$ for (x,t) in a neighborhood of the edge e.

For the triangular elements, we use the local Lax-Friedrichs recipe:

- Take $\alpha_{e,K}$ to be the larger one of the largest eigenvalue (in absolute value) of $\frac{\partial}{\partial u}\mathbf{f}(\bar{u}_K)\cdot n_{e,K}$ and that of $\frac{\partial}{\partial u}\mathbf{f}(\bar{u}_{K'})\cdot n_{e,K}$, where $\bar{u}_K$ and $\bar{u}_{K'}$ are the means of the numerical solution in the elements K and K' sharing the edge e.

For the rectangular elements, we use the local Lax-Friedrichs recipe :

- Take $\alpha_{e,K}$ to be the largest of the largest eigenvalue (in absolute value) of $\frac{\partial}{\partial u}\mathbf{f}(\bar{u}_{K''})\cdot n_{e,K}$, where $\bar{u}_{K''}$ is the mean of the numerical solution in the element K'', which runs over all elements on the same line (horizontally or vertically, depending on the direction of $n_{e,K}$) with K and K' sharing the edge e.

Quadrature rules According to the analysis done in [20], the quadrature rules for the edges of the elements, (4.4), must be exact for polynomials of degree $2k+1$, and the quadrature rules for the interior of the elements, (4.5), must be exact for polynomials of degree $2k$, if P^k methods are used. Here we discuss the quadrature points used for P^1 and P^2 in the triangular and rectangular element cases.

The rectangular elements For the edge integral, we use the following two point Gaussian rule

$$\int_{-1}^{1} g(x)dx \approx g\left(-\frac{1}{\sqrt{3}}\right) + g\left(\frac{1}{\sqrt{3}}\right), \tag{4.1}$$

for the P^1 case, and the following three point Gaussian rule

$$\int_{-1}^{1} g(x)dx \approx \frac{5}{9}\left[g\left(-\frac{3}{5}\right) + g\left(\frac{3}{5}\right)\right] + \frac{8}{9}g(0), \tag{4.2}$$

for the P^2 case, suitably scaled to the relevant intervals.

For the interior of the elements, we could use a tensor product of (4.1), with four quadrature points, for the P^1 case. But to save cost, we "recycle" the values of the fluxes at the element boundaries, and only add one new quadrature point in the middle of the element. Thus, to approximate the integral $\int_{-1}^{1}\int_{-1}^{1} g(x,y)dxdy$, we use the following quadrature rule:

$$\approx \frac{1}{4}\left[g\left(-1,\frac{1}{\sqrt{3}}\right) + g\left(-1,-\frac{1}{\sqrt{3}}\right)\right.$$
$$+ g\left(-\frac{1}{\sqrt{3}},-1\right) + g\left(\frac{1}{\sqrt{3}},-1\right)$$
$$+ g\left(1,-\frac{1}{\sqrt{3}}\right) + g\left(1,\frac{1}{\sqrt{3}}\right)$$
$$\left. + g\left(\frac{1}{\sqrt{3}},1\right) + g\left(-\frac{1}{\sqrt{3}},1\right)\right]$$
$$+ 2\,g(0,0). \tag{4.3}$$

For the P^2 case, we use a tensor product of (4.2), with 9 quadrature points.

The triangular elements For the edge integral, we use the same two point or three point Gaussian quadratures as in the rectangular case, (4.1) and (4.2), for the P^1 and P^2 cases, respectively.

For the interior integrals (4.5), we use the three mid-point rule

$$\int_K g(x,y)dxdy \approx \frac{|K|}{3} \sum_{i=1}^{3} g(m_i),$$

where m_i are the mid-points of the edges, for the P^1 case. For the P^2 case, we use a seven-point quadrature rule which is exact for polynomials of degree 5 over triangles.

Basis and degrees of freedom We emphasize that the choice of basis and degrees of freedom does not affect the algorithm, as it is completely determined by the choice of function space $V(h)$, the numerical fluxes, the quadrature rules, the slope limiting, and the time discretization. However, a suitable choice of basis and degrees of freedom may simplify the implementation and calculation.

The rectangular elements For the P^1 case, we use the following expression for the approximate solution $u_h(x,y,t)$ inside the rectangular element $[x_{i-\frac{1}{2}}, x_{i+\frac{1}{2}}] \times [y_{j-\frac{1}{2}}, y_{j+\frac{1}{2}}]$:

$$u_h(x,y,t) = \bar{u}(t) + u_x(t)\phi_i(x) + u_y(t)\psi_j(y) \tag{4.4}$$

where

$$\phi_i(x) = \frac{x - x_i}{\Delta x_i/2}, \qquad \psi_j(y) = \frac{y - y_j}{\Delta y_j/2}, \tag{4.5}$$

and

$$\Delta x_i = x_{i+\frac{1}{2}} - x_{i-\frac{1}{2}}, \qquad \Delta y_j = y_{j+\frac{1}{2}} - y_{j-\frac{1}{2}}.$$

The degrees of freedoms, to be evolved in time, are then

$$\bar{u}(t), \quad u_x(t), \quad u_y(t).$$

Here we have omitted the subscripts ij these degrees of freedom should have, to indicate that they belong to the element ij which is $[x_{i-\frac{1}{2}}, x_{i+\frac{1}{2}}] \times [y_{j-\frac{1}{2}}, y_{j+\frac{1}{2}}]$.

Notice that the basis functions

$$1, \quad \phi_i(x), \quad \psi_j(y),$$

are orthogonal, hence the local mass matrix is diagonal:

$$M = \Delta x_i \Delta y_j \, diag\left(1, \frac{1}{3}, \frac{1}{3}\right).$$

For the P^2 case, the expression for the approximate solution $u_h(x, y, t)$ inside the rectangular element $[x_{i-\frac{1}{2}}, x_{i+\frac{1}{2}}] \times [y_{j-\frac{1}{2}}, y_{j+\frac{1}{2}}]$ is:

$$u_h(x, y, t) = \bar{u}(t) + u_x(t)\phi_i(x) + u_y(t)\psi_j(y)$$
$$+ u_{xy}(t)\phi_i(x)\psi_j(y)$$
$$+ u_{xx}(t)\left(\phi_i^2(x) - \frac{1}{3}\right)$$
$$+ u_{yy}(t)\left(\psi_j^2(y) - \frac{1}{3}\right), \tag{4.6}$$

where $\phi_i(x)$ and $\psi_j(y)$ are defined by (4.5). The degrees of freedoms, to be evolved in time, are

$$\bar{u}(t), \quad u_x(t), \quad u_y(t), \quad u_{xy}(t), \quad u_{xx}(t), \quad u_{yy}(t).$$

Again the basis functions

$$1, \quad \phi_i(x), \quad \psi_j(y), \quad \phi_i(x)\psi_j(y), \quad \phi_i^2(x) - \frac{1}{3}, \quad \psi_j^2(y) - \frac{1}{3},$$

are orthogonal, hence the local mass matrix is diagonal:

$$M = \Delta x_i \Delta y_j \, diag\left(1, \frac{1}{3}, \frac{1}{3}, \frac{1}{9}, \frac{4}{45}, \frac{4}{45}\right).$$

The triangular elements For the P^1 case, we use the following expression for the approximate solution $u_h(x, y, t)$ inside the triangle K:

$$u_h(x, y, t) = \sum_{i=1}^{3} u_i(t)\varphi_i(x, y)$$

where the degrees of freedom $u_i(t)$ are values of the numerical solution at the midpoints of edges, and the basis function $\varphi_i(x, y)$ is the linear function which takes the value 1 at the mid-point m_i of the i-th edge, and the value 0 at the mid-points of the two other edges. The mass matrix is diagonal

$$M = |K| diag\left(\frac{1}{3}, \frac{1}{3}, \frac{1}{3}\right).$$

For the P^2 case, we use the following expression for the approximate solution $u_h(x, y, t)$ inside the triangle K:

$$u_h(x, y, t) = \sum_{i=1}^{6} u_i(t)\xi_i(x, y)$$

where the degrees of freedom, $u_i(t)$, are values of the numerical solution at the three midpoints of edges and the three vertices. The basis function $\xi_i(x, y)$, is the quadratic function which takes the value 1 at the point i of the six points mentioned above (the three midpoints of edges and the three vertices), and the value 0 at the remaining five points. The mass matrix this time is not diagonal.

Limiting We construct slope limiting operators $\Lambda\Pi_h$ on piecewise linear functions u_h in such a way that the following properties are satisfied:

1. Accuracy: if u_h is linear then $\Lambda\Pi_h u_h = u_h$.
2. Conservation of mass: for every element K of the triangulation $\mathcal{T}_h$, we have:

$$\int_K \Lambda\Pi_h u_h = \int_K u_h.$$

3. Slope limiting: on each element K of $\mathcal{T}_h$, the gradient of $\Lambda\Pi_h u_h$ is not bigger than that of u_h.

The actual form of the slope limiting operators is closely related to that of the slope limiting operators studied in [24] and [20].

The rectangular elements The limiting is performed on u_x and u_y in (4.4), using the differences of the means. For a scalar equation, u_x would be limited (replaced) by

$$\bar{m}\left(u_x, \bar{u}_{i+1,j} - \bar{u}_{ij}, \bar{u}_{ij} - \bar{u}_{i-1,j}\right) \tag{4.7}$$

where the function $\bar{m}$ is the TVB corrected *minmod* function defined in the previous section.

The TVB correction is needed to avoid unnecessary limiting near smooth extrema, where the quantity u_x or u_y is on the order of $O(\Delta x^2)$ or $O(\Delta y^2)$. For an estimate of the TVB constant M in terms of the second derivatives of the function, see [24]. Usually, the numerical results are not sensitive to the choice of M in a large range. In all the calculations in this paper we take M to be 50.

Similarly, u_y is limited (replaced) by

$$\bar{m}\left(u_y, \bar{u}_{i,j+1} - \bar{u}_{ij}, \bar{u}_{ij} - \bar{u}_{i,j-1}\right).$$

with a change of Δx to Δy in (4.7).

For systems, we perform the limiting in the local characteristic variables. To limit the vector u_x in the element ij, we proceed as follows:

- Find the matrix R and its inverse R^{-1}, which diagonalize the Jacobian evaluated at the mean in the element ij in the x-direction:

$$R^{-1} \frac{\partial f_1(\bar{u}_{ij})}{\partial u} R = \Lambda,$$

where Λ is a diagonal matrix containing the eigenvalues of the Jacobian. Notice that the columns of R are the right eigenvectors of $\frac{\partial f_1(\bar{u}_{ij})}{\partial u}$ and the rows of R^{-1} are the left eigenvectors.
- Transform all quantities needed for limiting, i.e., the three vectors $u_{x\,ij}$, $\bar{u}_{i+1,j} - \bar{u}_{ij}$ and $\bar{u}_{ij} - \bar{u}_{i-1,j}$, to the characteristic fields. This is achieved by left multiplying these three vectors by R^{-1}.
- Apply the scalar limiter (4.7) to each of the components of the transformed vectors.
- The result is transformed back to the original space by left multiplying R on the left.

The triangular elements To construct the slope limiting operators for triangular elements, we proceed as follows. We start by making a simple observation. Consider the triangles in Figure 4.1, where m_1 is the mid-point of the edge on the boundary of K_0 and b_i denotes the barycenter of the triangle K_i for $i = 0, 1, 2, 3$.

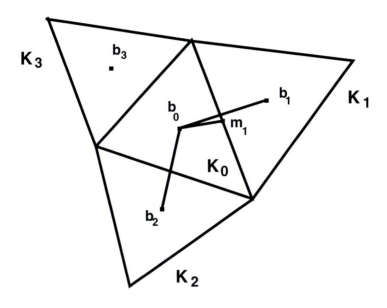

Fig. 4.1. Illustration of limiting.

Since we have that

$$m_1 - b_0 = \alpha_1 (b_1 - b_0) + \alpha_2 (b_2 - b_0),$$

for some nonnegative coefficients α_1, α_2 which depend only on m_1 and the geometry, we can write, for any linear function u_h,

$$u_h(m_1) - u_h(b_0) = \alpha_1 (u_h(b_1) - u_h(b_0)) + \alpha_2 (u_h(b_2) - u_h(b_0)),$$

and since

$$\bar{u}_{K_i} = \frac{1}{|K_i|} \int_{K_i} u_h = u_h(b_i), \qquad i = 0, 1, 2, 3,$$

we have that

$$\begin{aligned}
\tilde{u}_h(m_1, K_0) &\equiv u_h(m_1) - \bar{u}_{K_0} \\
&= \alpha_1 (\bar{u}_{K_1} - \bar{u}_{K_0}) + \alpha_2 (\bar{u}_{K_2} - \bar{u}_{K_0}) \\
&\equiv \Delta \bar{u}(m_1, K_0).
\end{aligned}$$

Now, we are ready to describe the slope limiting. Let us consider a piecewise linear function u_h, and let $m_i, i = 1, 2, 3$ be the three mid-points of the edges of the triangle K_0. We then can write, for $(x, y) \in K_0$,

$$u_h(x, y) = \sum_{i=1}^{3} u_h(m_i)\varphi_i(x, y)$$

$$= \bar{u}_{K_0} + \sum_{i=1}^{3} \tilde{u}_h(m_i, K_0)\varphi_i(x, y).$$

To compute $\Lambda\Pi_h u_h$, we first compute the quantities

$$\Delta_i = \bar{m}(\tilde{u}_h(m_i, K_0), \nu \, \Delta\bar{u}(m_i, K_0)),$$

where $\bar{m}$ is the TVB modified *minmod* function and $\nu > 1$. We take $\nu = 1.5$ in our numerical runs. Then, if $\sum_{i=1}^{3} \Delta_i = 0$, we simply set

$$\Lambda\Pi_h u_h(x, y) = \bar{u}_{K_0} + \sum_{i=1}^{3} \Delta_i \, \varphi_i(x, y).$$

If $\sum_{i=1}^{3} \Delta_i \neq 0$, we compute

$$pos = \sum_{i=1}^{3} \max(0, \Delta_i), \qquad neg = \sum_{i=1}^{3} \max(0, -\Delta_i),$$

and set

$$\theta^+ = \min\left(1, \frac{neg}{pos}\right), \qquad \theta^- = \min\left(1, \frac{pos}{neg}\right).$$

Then, we define

$$\Lambda\Pi_h u_h(x, y) = \bar{u}_{K_0} + \sum_{i=1}^{3} \hat{\Delta}_i \, \varphi_i(x, y),$$

where

$$\hat{\Delta}_i = \theta^+ \max(0, \Delta_i) - \theta^- \max(0, -\Delta_i).$$

It is very easy to see that this slope limiting operator satisfies the three properties listed above.

For systems, we perform the limiting in the local characteristic variables. To limit Δ_i, we proceed as in the rectangular case, the only difference being that we work with the following Jacobian

$$\frac{\partial}{\partial u} f(\bar{u}_{K_0}) \cdot \frac{m_i - b_0}{|m_i - b_0|}.$$

4.4 Computational results: Transient, nonsmooth solutions

In this section we present several numerical results obtained with the P^1 and P^2 (second and third order accurate) RKDG methods with either rectangles or triangles in the triangulation. These are standard test problems for Euler equations of compressible gas dynamics.

The double-Mach reflection problem Double Mach reflection of a strong shock. This problem was studied extensively in Woodward and Colella [92] and later by many others. We use exactly the same setup as in [92], namely a Mach 10 shock initially makes a 60° angle with a reflecting wall. The undisturbed air ahead of the shock has a density of 1.4 and a pressure of 1.

For the rectangle based triangulation, we use a rectangular computational domain $[0, 4] \times [0, 1]$, as in [92]. The reflecting wall lies at the bottom of the computational domain for $\frac{1}{6} \leq x \leq 4$. Initially a right-moving Mach 10 shock is positioned at $x = \frac{1}{6}, y = 0$ and makes a 60° angle with the x-axis. For the bottom boundary, the exact post-shock condition is imposed for the part from $x = 0$ to $x = \frac{1}{6}$, to mimic an angled wedge. Reflective boundary condition is used for the rest. At the top boundary of our computational domain, the flow values are set to describe the exact motion of the Mach 10 shock. Inflow/outflow boundary conditions are used for the left and right boundaries. As in [92], only the results in $[0, 3] \times [0, 1]$ are displayed.

For the triangle based triangulation, we have the freedom to treat irregular domains and thus use a true wedged computational domain. Reflective boundary conditions are then used for all the bottom boundary, including the sloped portion. Other boundary conditions are the same as in the rectangle case.

Uniform rectangles are used in the rectangle based triangulations. Four different meshes are used: 240×60 rectangles ($\Delta x = \Delta y = \frac{1}{60}$); 480×120 rectangles ($\Delta x = \Delta y = \frac{1}{120}$); 960×240 rectangles ($\Delta x = \Delta y = \frac{1}{240}$); and 1920×480 rectangles ($\Delta x = \Delta y = \frac{1}{480}$). The density is plotted in Figure 4.2 for the P^1 case and in 4.3 for the P^2 case.

To better appreciate the difference between the P^1 and P^2 results in these pictures, we show a "blowed up" portion around the double Mach region in Figure 4.4 and show one-dimensional cuts along the line $y = 0.4$ in Figures 4.5 and 4.6. In Figure 4.4, w can see that P^2 with $\Delta x = \Delta y = \frac{1}{240}$ has qualitatively the same resolution as P^1 with $\Delta x = \Delta y = \frac{1}{480}$, for the fine details of the complicated structure in this region. P^2 with $\Delta x = \Delta y = \frac{1}{480}$ gives a much better resolution for these structures than P^1 with the same number of rectangles.

Moreover, from Figure 4.5, we clearly see that the difference between the results obtained by using P^1 and P^2, on the same mesh, increases dramatically as the mesh size decreases. This indicates that the use of polynomials of high degree might be beneficial for capturing the above mentioned structures. From Figure 4.6, we see that the results obtained with P^1 are qualitatively similar to those obtained with P^2 in a coarser mesh; the similarity increases as the meshsize decreases. The conclusion here is that, if one is interested in the above mentioned fine structures, then one can use the third order scheme P^2 with only half of the mesh points in each direction as in P^1. This translates into a reduction of a factor of 8 in space-time grid points for 2D time dependent problems, and will more than off-set the increase of cost per mesh point and the smaller CFL number by using the higher order P^2 method. This saving will be even more significant for 3D.

The optimal strategy, of course, is to use adaptivity and concentrate triangles around the interesting region, and/or change the order of the scheme in different regions.

The forward-facing step problem Flow past a forward facing step. This problem was again studied extensively in Woodward and Colella [92] and later by many others. The set up of the problem is the following: A right going Mach 3 uniform flow enters a wind tunnel of 1 unit wide and 3 units long. The step is 0.2 units high and is located 0.6 units from the left-hand end of the tunnel. The problem is initialized by a uniform, right-going Mach 3 flow. Reflective boundary conditions are applied along the walls of the tunnel and in-flow and out-flow boundary conditions are applied at the entrance (left-hand end) and the exit (right-hand end), respectively.

The corner of the step is a singularity, which we study carefully in our numerical experiments. Unlike in [92] and many other papers, we do not modify our scheme near the corner in any way. It is well known that this leads to an errorneous entropy layer at the downstream bottom wall, as well as a spurious Mach stem at the bottom wall. However, these artifacts decrease when the mesh is refined. In Figure 4.7, second order P^1 results using rectangle triangulations are shown, for a grid refinement study using $\Delta x = \Delta y = \frac{1}{40}$, $\Delta x = \Delta y = \frac{1}{80}$, $\Delta x = \Delta y = \frac{1}{160}$, and $\Delta x = \Delta y = \frac{1}{320}$ as mesh sizes. We can clearly see the improved resolution (especially at the upper slip line from the triple point) and decreased artifacts caused by the corner, with increased mesh points. In Figure 4.8, third order P^2 results using the same meshes are shown.

To have a better idea of the nature of the singularity at the corner, we display the values of the density and the entropy along the line $y = 0.2$; note that the corner is located on this line at $x = 0.6$. In Figure 4.9, we show the results obtained with P^1 and in Figure 4.10, the results obtained with P^2. At the corner ($x = 0.6$), we can see that there is a jump both in the entropy and in the density. As the meshsize decreases, the jump in the entropy does not vary significantly; however, the jump in the density does. The sharp decrease in the density right after the corner can be interpreted as a cavitation effect that the scheme seems to be able to better approximate as the meshsize decreases.

In order to verify that the erroneous entropy layer at the downstream bottom wall and the spurious Mach stem at the bottom wall are both artifacts caused by the corner singularity, we use our triangle code to locally refine near the corner progressively; we use the meshes displayed in Figure 4.11. In Figure 4.12, we plot the density obtained by the P^1 triangle code, with triangles (roughly the resolution of $\Delta x = \Delta y = \frac{1}{40}$, except around the corner). In Figure 4.13, we plot the entropy around the corner for the same runs. We can see that, with more triangles concentrated near the corner, the artifacts gradually decrease. Results with P^2 codes in Figures 4.14 and 4.15 show a similar trend.

4.5   Computational results: Steady state, smooth solutions

In this section, we present some of the numerical results of Bassi and Rebay [4] in two dimensions and Warburton, Lomtev, Kirby and Karniadakis [91] in three dimensions.

The purpose of the numerical results of Bassi and Rebay [4] we are presenting is to assess (i) the effect of the quality of the approximation of curved boundaries and of (ii) the effect of the degree of the polynomials on the quality of the approximate solution. The test problem we consider here is the two-dimensional steady-state, subsonic flow around a disk at Mach number $M_\infty = 0.38$. Since the solution is smooth and can be computed analytically, the quality of the approximation can be easily assessed.

In the figures 4.16, 4.17, 4.18, and 4.19, details of the meshes around the disk are shown together with the approximate solution given by the RKDG method using piecewise linear elements. These meshes approximate the circle with a polygonal. It can be seen that the approximate solution are of very low quality even for the most refined grid. This is an effect caused by the kinks of the polygonal approximating the circle.

This statement can be easily verified by taking a look to the figures 4.20, 4.21, 4.22, and 4.23. In these pictures the approximate solutions with piecewise linear, quadratic, and cubic elements are shown; the meshes have been modified to render *exactly* the circle. It is clear that the improvement in the quality of the approximation is enormous. Thus, a high-quality approximation of the boundaries has a dramatic improvement on the quality of the approximations.

Also, it can be seen that the higher the degree of the polynomials, the better the quality of the approximations, in particular from figures 4.20 and 4.21. In [4], Bassi and Rebay show that the RKDG method using polynomilas of degree k are $(k + 1)$-th order accurate for $k = 1, 2, 3$. As a consequence, a RKDG method using polynomials of a higher degree is more efficient than a RKDG method using polynomials of lower degree.

In [91], Warburton, Lomtev, Kirby and Karniadakis present the same test problem in a three dimensional setting. In Figure 4.24, we can see the three-dimensional mesh and the density isosurfaces. We can also see how, while the mesh is being kept fixed and the degree of the polynomials k is increased from 1 to 9, the maximum error on the entropy goes exponentialy to zero. (In the picture, a so-called 'mode' is equal to $k + 1$).

4.6   Concluding remarks

In this section, we have extended the RKDG methods to multidimensional systems. We have described in full detail the algorithms and displayed numerical results showing the performance of the methods for the Euler equations of gas dynamics.

The flexibility of the RKDG method to handle nontrivial geometries and to work with different elements has been displayed. Moreover, it has been shown that the use of polynomials of high degree not only does not degrade the resolution of strong shocks, but enhances the resolution of the contact discontinuities and renders the scheme more efficient on smooth regions.

Next, we extend the RKDG methods to convection-dominated problems.

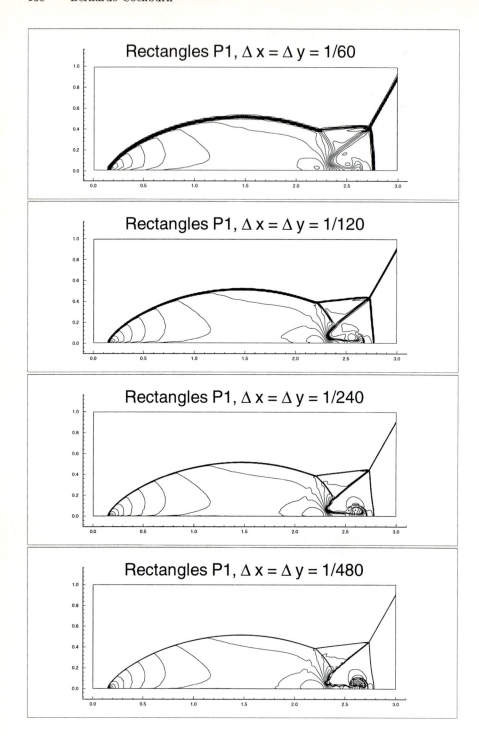

Fig. 4.2. Double Mach reflection problem. Second order P^1 results. Density ρ. 30 equally spaced contour lines from $\rho = 1.3965$ to $\rho = 22.682$. Mesh refinement study. From top to bottom: $\Delta x = \Delta y = \frac{1}{60}, \frac{1}{120}, \frac{1}{240}$, and $\frac{1}{480}$.

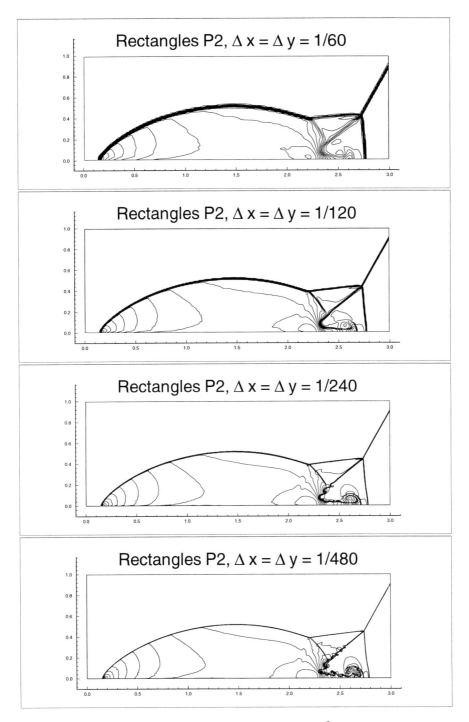

Fig. 4.3. Double Mach reflection problem. Third order P^2 results. Density ρ. 30 equally spaced contour lines from $\rho = 1.3965$ to $\rho = 22.682$. Mesh refinement study. From top to bottom: $\Delta x = \Delta y = \frac{1}{60}, \frac{1}{120}, \frac{1}{240}$, and $\frac{1}{480}$.

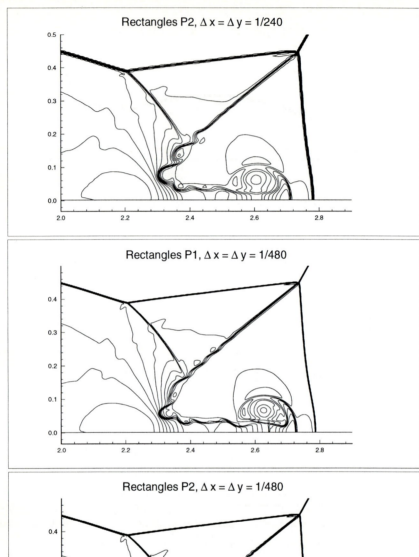

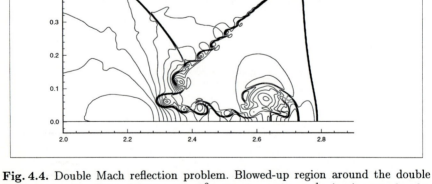

Fig. 4.4. Double Mach reflection problem. Blowed-up region around the double Mach stems. Density ρ. Third order P^2 with $\Delta x = \Delta y = \frac{1}{240}$ (top); second order P^1 with $\Delta x = \Delta y = \frac{1}{480}$ (middle); and third order P^2 with $\Delta x = \Delta y = \frac{1}{480}$ (bottom).

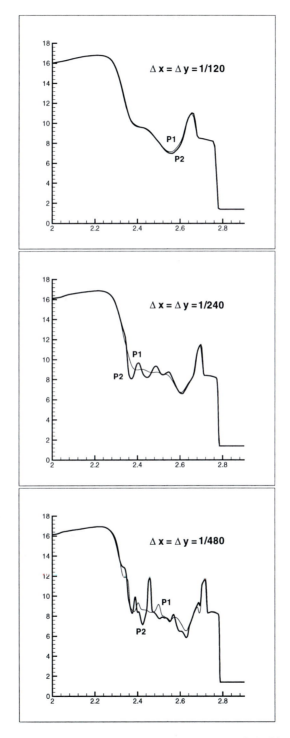

Fig. 4.5. Double Mach reflection problem. Cut at $y = 0.04$ of the blowed-up region. Density ρ. Comparison of second order P^1 with third order P^2 on the same mesh

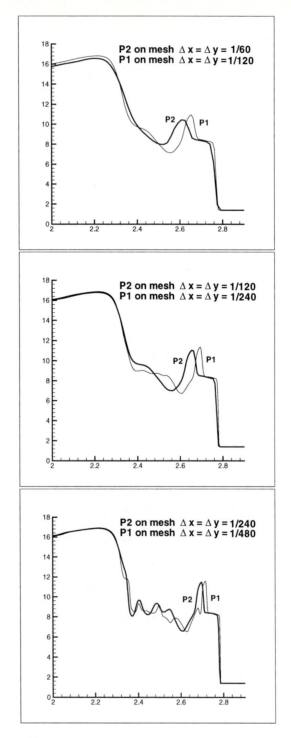

Fig. 4.6. Double Mach reflection problem. Cut at $y = 0.04$ of the blowed-up region. Density ρ. Comparison of second order P^1 with third order P^2 on a coarser mesh

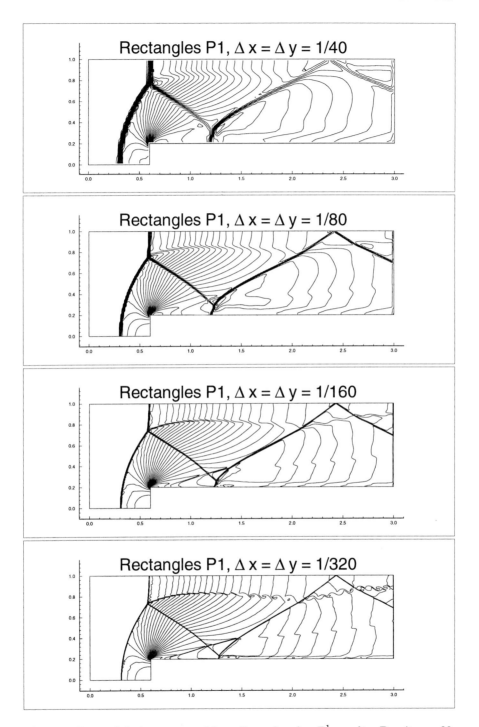

Fig. 4.7. Forward facing step problem. Second order P^1 results. Density ρ. 30 equally spaced contour lines from $\rho = 0.090338$ to $\rho = 6.2365$. Mesh refinement study. From top to bottom: $\Delta x = \Delta y = \frac{1}{40}, \frac{1}{80}, \frac{1}{160}$, and $\frac{1}{320}$.

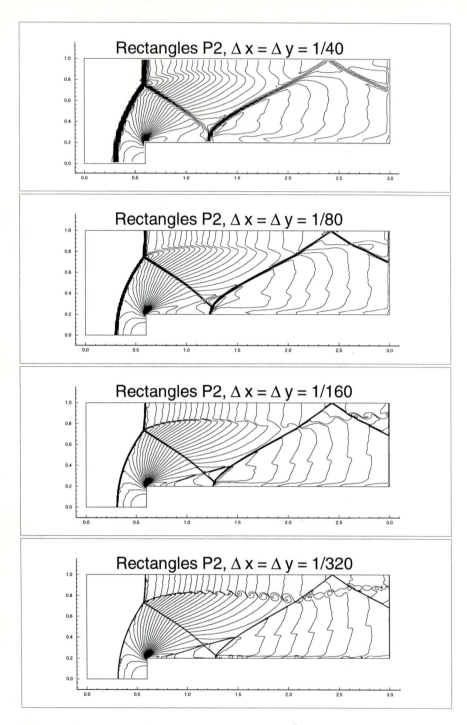

Fig. 4.8. Forward facing step problem. Third order P^2 results. Density ρ. 30 equally spaced contour lines from $\rho = 0.090338$ to $\rho = 6.2365$. Mesh refinement study. From top to bottom: $\Delta x = \Delta y = \frac{1}{40}, \frac{1}{80}, \frac{1}{160},$ and $\frac{1}{320}$.

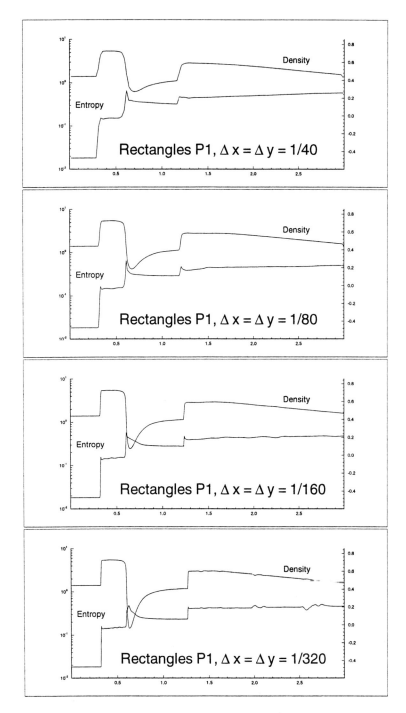

Fig. 4.9. Forward facing step problem. Second order P^1 results. Values of the density and entropy along the line $y = .2$. Mesh refinement study. From top to bottom: $\Delta x = \Delta y = \frac{1}{40}, \frac{1}{80}, \frac{1}{160},$ and $\frac{1}{320}$.

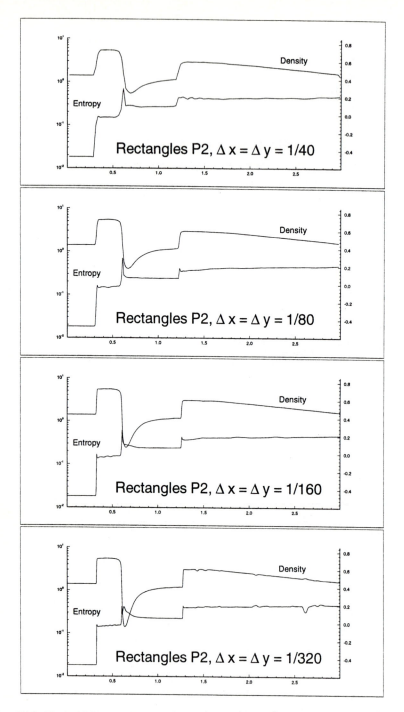

Fig. 4.10. Forward facing step problem. Third order P^2 results. Values of the density and entropy along the line $y = .2$. Mesh refinement study. From top to bottom: $\Delta x = \Delta y = \frac{1}{40}, \frac{1}{80}, \frac{1}{160}$, and $\frac{1}{320}$.

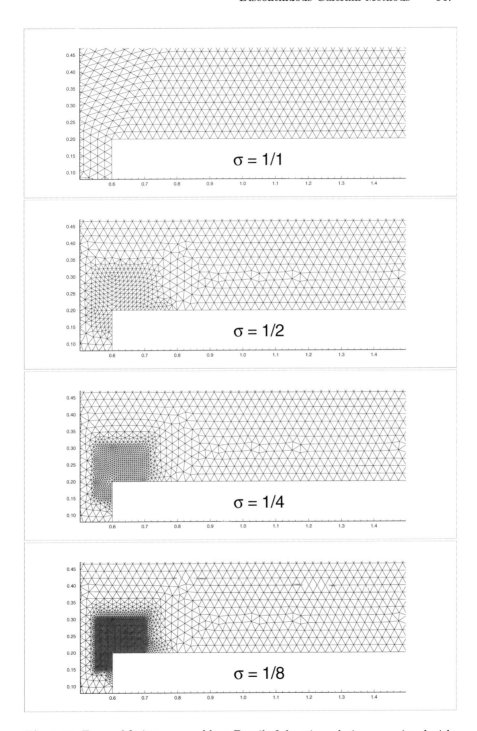

Fig. 4.11. Forward facing step problem. Detail of the triangulations associated with the different values of σ. The parameter σ is the ratio between the typical size of the triangles near the corner and that elsewhere.

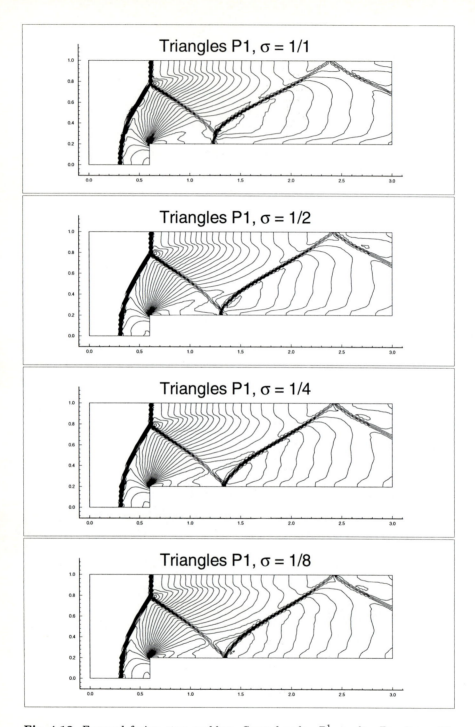

Fig. 4.12. Forward facing step problem. Second order P^1 results. Density ρ. 30 equally spaced contour lines from $\rho = 0.090338$ to $\rho = 6.2365$. Triangle code. Progressive refinement near the corner

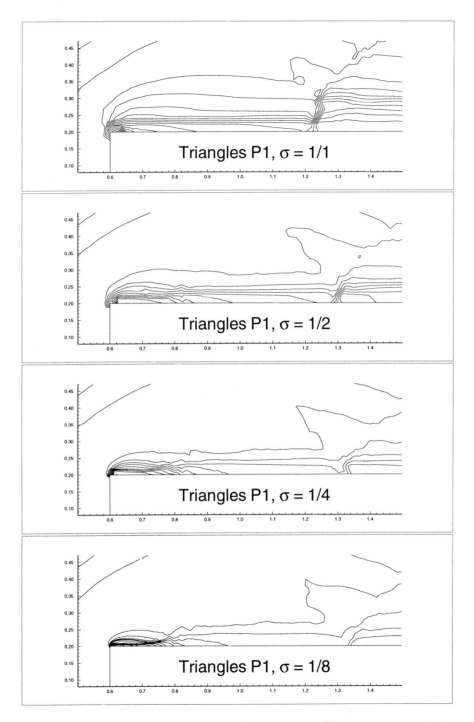

Fig. 4.13. Forward facing step problem. Second order P^1 results. Entropy level curves around the corner. Triangle code. Progressive refinement near the corner

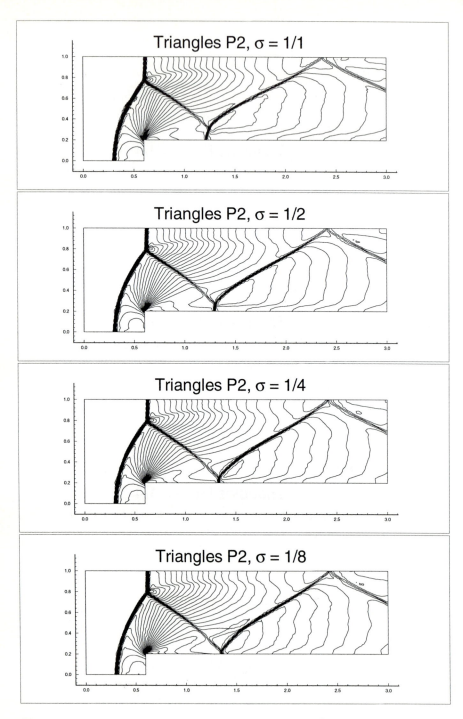

Fig. 4.14. Forward facing step problem. Third order P^2 results. Density ρ. 30 equally spaced contour lines from $\rho = 0.090338$ to $\rho = 6.2365$. Triangle code. Progressive refinement near the corner

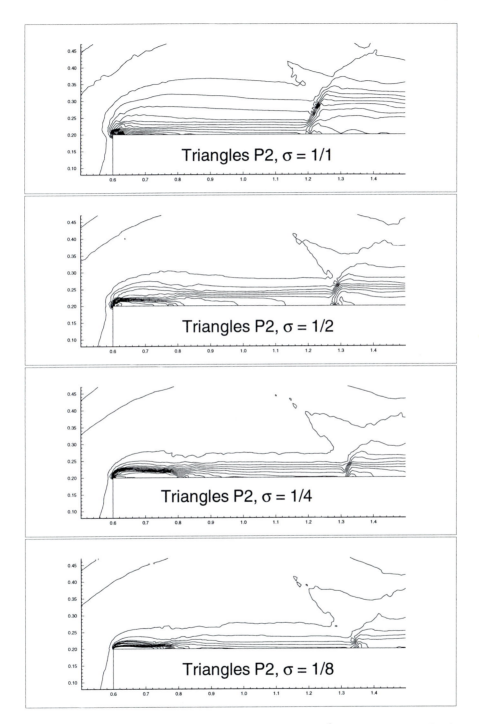

Fig. 4.15. Forward facing step problem. Third order P^1 results. Entropy level curves around the corner. Triangle code. Progressive refinement near the corner

Fig. 4.16. Grid "16 × 8" with a piecewise linear approximation of the circle (top) and the corresponding solution (Mach isolines) using P^1 elements (bottom).

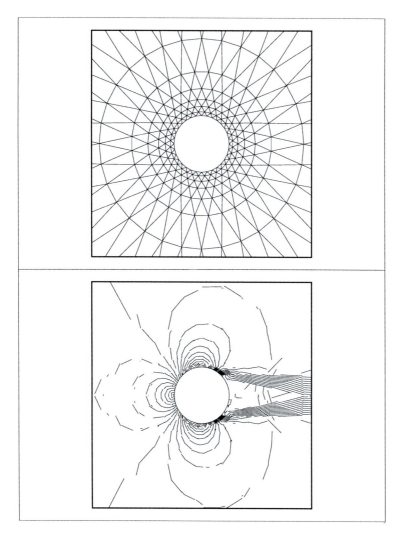

Fig. 4.17. Grid "32 × 8" with a piecewise linear approximation of the circle (top) and the corresponding solution (Mach isolines) using P^1 elements (bottom).

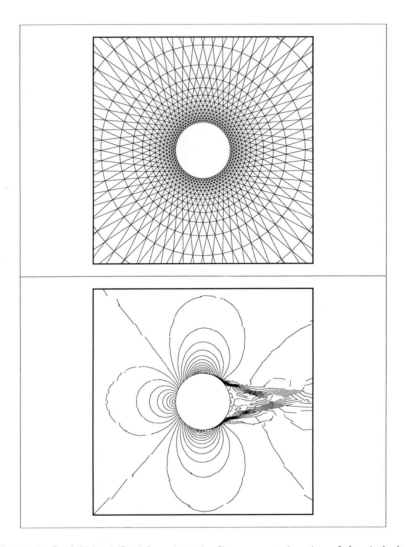

Fig. 4.18. Grid "64 × 16" with a piecewise linear approximation of the circle (top) and the corresponding solution (Mach isolines) using P^1 elements (bottom).

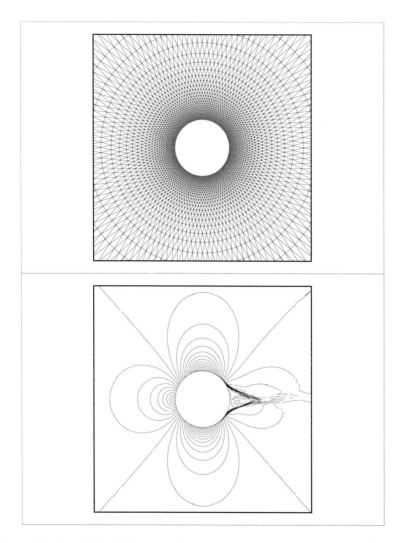

Fig. 4.19. Grid "128 × 32" a piecewise linear approximation of the circle (top) and the corresponding solution (Mach isolines) using P^1 elements (bottom).

Fig. 4.20. Grid "16 × 4" with exact rendering of the circle and the corresponding P^1 (top), P^2(middle), and P^3 (bottom) approximations (Mach isolines).

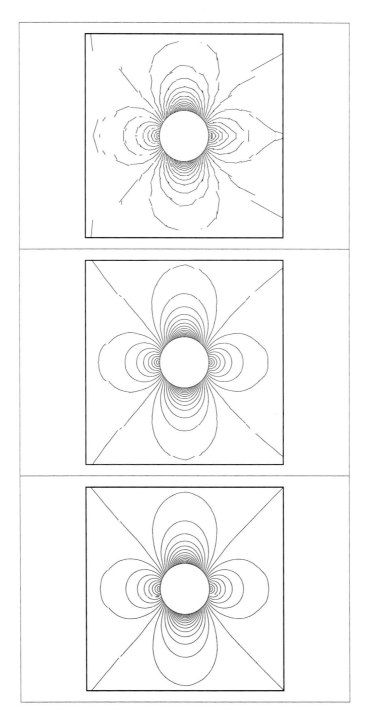

Fig. 4.21. Grid "32 × 8" with exact rendering of the circle and the corresponding P^1 (top), P^2(middle), and P^3 (bottom) approximations (Mach isolines).

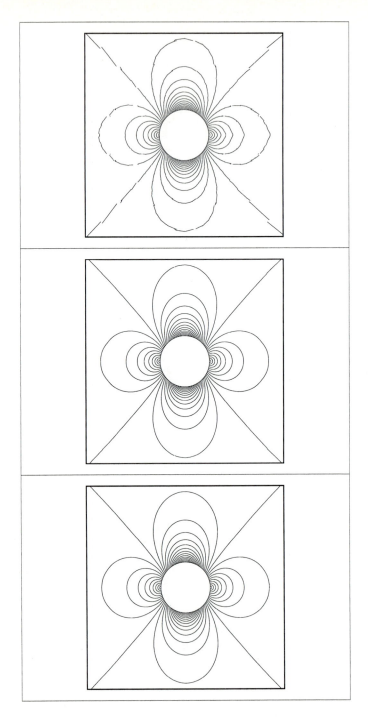

Fig. 4.22. Grid "64 × 16" with exact rendering of the circle and the corresponding P[1] (top), P[2](middle), and P[3] (bottom) approximations (Mach isolines).

Fig. 4.23. Grid "128×32" with exact rendering of the circle and the corresponding P^1 (top), P^2(middle), and P^3 (bottom) approximations (Mach isolines).

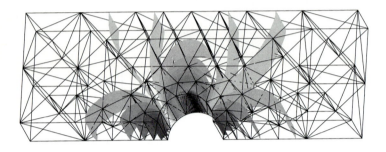

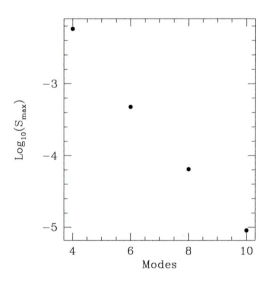

Fig. 4.24. Three-dimensional flow over a semicircular bump. Mesh and density isosurfaces (top) and history of convergence with p-refinement of the maximum entropy generated (bottom). The degree of the polynomial plus one is plotted on the 'modes' axis.

5   The Hamilton-Jacobi equations in several space dimensions

5.1   Introduction

In this chapter, we consider the RKDG method for multidimensional Hamilton-Jacobi equations. The model problem we consider is the following:

$$\varphi_t + H(\varphi_x, \varphi_y) = 0, \qquad \text{in } (0,1)^2 \times (0,T), \tag{5.1}$$
$$\varphi(x,y,0) = \varphi_0(x,y), \qquad \forall \ (x,y) \in (0,1)^2, \tag{5.2}$$

where we take periodic boundary conditions. The material in this section is based in the work of Hu and Shu [43].

5.2   The RKDG method

As in the one-dimensional case, the main idea to extend the RKDG method to this case, is to realize that $u = \varphi_x$ and $v = \varphi_y$ satisfy the following problem:

$$u_t + H(u,v)_x = 0, \qquad \text{in } (0,1)^2 \times (0,T), \tag{5.3}$$
$$v_t + H(u,v)_y = 0, \qquad \text{in } (0,1)^2 \times (0,T), \tag{5.4}$$
$$u(x,y,0) = (\varphi_0)_x(x,y), \qquad \forall \ (x,y) \in (0,1)^2, \tag{5.5}$$
$$v(x,y,0) = (\varphi_0)_y(x,y), \qquad \forall \ (x,y) \in (0,1)^2, \tag{5.6}$$

and that φ can be computed from u and v by solving the following problem:

$$\varphi_t = -H(u,v), \qquad \text{in } (0,1) \times (0,T), \tag{5.7}$$
$$\varphi(x,0) = \varphi_0(x), \qquad \forall \ x \in (0,1). \tag{5.8}$$

Again, a straightforward application of the RKDG method to (5.3), (5.5), produces an approximation u_h to $u = \varphi_x$; and a straightforward application of the RKDG method to (5.4), (5.6), produces an approximation v_h to $v = \varphi_y$. Both u_h and v_h are taken to be piecewise polynomials of degree $(k-1)$. Then, φ_h is computed in one of the following ways:

(i) Take $\varphi_h(\cdot, t)$ in V_h^k such that

$$\forall K \in \mathcal{T}_h, \quad w_h \in P^k(K):$$

$$\int_K \partial_t \varphi_h(x,y,t) \, w_h(x) \, dx \, dy = - \int_K H(u_h(x,t), v_h(x,y)) \, w_h(x,y) \, dx \, dy$$

$$\int_K \varphi(x,y,0) \, w_h(x,y) \, dx \, dy = \int_K \varphi_0(x,y) \, w_h(x,y) \, dx \, dy.$$

(ii) Take $\varphi_h(\cdot, t)$ in V_h^k such that, $\forall K \in \mathcal{T}_h$:

$$\| \nabla \varphi_h - (u_h, v_h) \|_{L^2(K)} = \min_{\psi \in P^k(K)} \| \nabla \psi - (u_h, v_h) \|_{L^2(K)}.$$

This determines φ_h up to a constant. To find this constant, we impose the following condition:

$$\forall K \in \mathcal{T}_h :$$

$$\frac{d}{dt} \int_K \varphi_h(x, y, t) \, dx \, dy = - \int_K H(u_h(x, y, t), v_h(x, y, t)) \, dx \, dy,$$

$$\int_K \varphi(x, y, 0) \, dx \, dy = \int_K \varphi_0(x, y) \, dx \, dy.$$

(iii) Pick the element K_J and determine the values of φ_h on it in such a way that

$$\| \nabla \varphi_h - (u_h, v_h) \|_{L^2(K_J)} = \min_{\psi \in P^k(K)} \| \nabla \psi - (u_h, v_h) \|_{L^2(K_J)}.$$

and that

$$\frac{d}{dt} \int_{K_J} \varphi_h(x, y, t) \, dx \, dy = - \int_{K_J} H(u_h(x, y, t)) \, dx \, dy,$$

$$\int_{K_J} \varphi(x, y, 0) \, dx \, dy = \int_{K_J} \varphi_0(x, y) \, dx \, dy. \tag{5.9}$$

Then, compute φ_h as follows:

$$\varphi(B, t) = \varphi(A, t) + \int_A^B (\varphi_x \, dx + \varphi_y \, dy). \tag{5.10}$$

to determine the missing constant. The path should be taken to avoid crossing a derivative discontinuity, if possible.

We remark again that, in the third approach, the recovered values of φ_h depend on the choice of the starting point A as well as the integration path. However this difference is on the level of truncation errors and does not affect the order of accuracy as is shown in the computational results we show next.

5.3   Computational results

The purpose of the numerical experiments we report in this section is to asses the accuracy of the method, to see if the generalized slope limiter is actually needed, and to evaluate the effect of changing the integration path. The third approach is used.

First test problem. Two dimensional Burgers' equation:

$$\varphi_t + \frac{(\varphi_x + \varphi_y + 1)^2}{2} = 0, \qquad \text{in } (-2, 2)^2 \times (0, T),$$

$$\varphi(x, y, 0) = -\cos\left(\frac{\pi(x + y)}{2}\right), \qquad \forall \, (x, y) \in (-2, 2)^2,$$

with periodic boundary conditions.

We first use uniform rectangular meshes and the local Lax-Friedrichs flux. At $T = 0.5/\pi^2$, the solution is still smooth. The errors (computed at the center of the cells) and orders of accuracy are listed in Table 5.1. It seems that only k-th order of accuracy is achieved when φ is a piecewise polynomial of degree k. Next, as in the one dimensional case, we use non-uniform rectangular meshes obtained from the tensor product of one dimensional nonuniform meshes (the meshes in two directions are independent). Again, we give the "real" L^2-errors computed by a 6×6 point Gaussian quadrature as well. The results are shown in Table 5.2.

The results in Tables 5.1 and 5.1 are obtained by updating the element at the left-lower corner with time, and then taking an integration path consisting of line segments starting from the corner and parallel to the x-axis first, then vertically to the point. To further address the issue of the dependency of the computed values of the solution φ on the integration path and starting point, we use another path which starts vertically, then parallelly with the x-axis to reach the point. In Table 5.3, we list the *difference* of two recovered solutions φ from these two different integration paths, for the non-uniform mesh cases. We can see that these differences are at the levels of local truncation errors and decay in the same order as the errors. Thus the choice of integration path in recovering φ does not affect accuracy.

At $T = 1.5/\pi^2$, the solution has discontinuous derivatives. Fig. 5.1 is the graph of the numerical solution with 40×40 elements (uniform mesh).

Finally we use triangle based triangulation, the mesh with $h = \frac{1}{4}$ is shown in Fig. 5.2. The accuracy at $T = 0.5/\pi^2$ is shown in Table 5.4. Similar accuracy pattern is observed as in the rectangular case. The result at $T = 1.5/\pi^2$, when the derivative is discontinuous, is shown in Fig. 5.3.

Table 5.1. Accuracy for 2D Burgers equation, uniform rectangular mesh, $T = 0.5/\pi^2$.

	P^1		P^2		P^3	
$N \times N$	L^1 error	order	L^1 error	order	L^1 error	order
10×10	8.09E-02	—	8.62E-03	—	3.19E-03	—
20×20	3.36E-02	1.268	1.72E-03	2.325	3.49E-04	3.192
40×40	1.48E-02	1.183	3.93E-04	2.130	6.64E-05	2.394
80×80	6.88E-03	1.105	9.74E-05	2.013	1.14E-05	2.542
160×160	3.31E-03	1.056	2.45E-05	1.991	1.68E-06	2.763

	P^1		P^2		P^3	
$N \times N$	L^∞ error	order	L^∞ error	order	L^∞ error	order
10×10	2.62E-01	—	3.56E-02		8.65E-03	—
20×20	1.14E-01	1.201	8.40E-03	2.083	1.16E-03	2.899
40×40	5.00E-02	1.189	2.02E-03	2.056	1.98E-04	2.551
80×80	2.39E-02	1.065	4.92E-04	2.038	3.13E-05	2.661
160×160	1.16E-02	1.043	1.21E-04	2.024	4.41E-06	2.827

Table 5.2. Accuracy for 2D Burgers equation, non-uniform rectangular mesh, $T = 0.5/\pi^2$.

	P^1		P^2		P^3	
$N \times N$	L^2 error	order	L^2 error	order	L^2 error	order
10×10	4.47E-01	—	6.28E-02	—	1.61E-02	—
20×20	1.83E-01	1.288	1.50E-02	2.066	2.06E-03	2.966
40×40	8.01E-02	1.192	3.63E-03	2.047	3.48E-04	2.565
80×80	3.82E-02	1.068	9.17E-04	1.985	6.03E-05	2.529
160×160	1.87E-02	1.031	2.34E-04	1.970	8.58E-06	2.813

	P^1		P^2		P^3	
$N \times N$	L^1 error	order	L^1 error	order	L^1 error	order
10×10	8.16E-02	—	9.16E-03	—	3.39E-03	—
20×20	3.41E-02	1.259	2.09E-03	2.132	4.12E-04	3.041
40×40	1.50E-02	1.185	5.21E-04	2.004	7.03E-05	2.551
80×80	7.16E-03	1.067	1.42E-04	1.875	1.24E-05	2.503
160×160	3.50E-03	1.033	3.85E-05	1.883	1.76E-06	2.817

	P^1		P^2		P^3	
$N \times N$	L^∞ error	order	L^∞ error	order	L^∞ error	order
10×10	2.83E-01	—	4.68E-02	—	1.00E-02	—
20×20	1.25E-01	1.179	1.23E-02	1.928	1.39E-03	2.847
40×40	5.74E-02	1.123	3.54E-03	1.797	2.29E-04	2.602
80×80	2.78E-02	1.046	1.15E-03	1.622	5.11E-05	2.164
160×160	1.42E-02	0.969	2.72E-04	2.080	7.16E-06	2.835

Table 5.3. Differences of the solution φ recovered by two different integration paths, non-uniform mesh, Burgers equation.

	P^1		P^2		P^3	
$N \times N$	L^1 error	order	L^1 error	order	L^1 error	order
10×10	8.61E-03	—	2.90E-03	—	1.15E-03	—
20×20	4.64E-03	0.892	1.28E-03	1.180	2.44E-04	2.237
40×40	2.54E-03	0.869	4.12E-04	1.635	3.76E-05	2.698
80×80	1.81E-03	0.489	1.39E-04	1.568	6.71E-06	2.486
160×160	1.09E-03	0.732	3.66E-05	1.925	8.79E-07	2.932

Table 5.4. Accuracy for 2D Burgers equation, triangular mesh, $T = 0.5/\pi^2$.

	P^2				P^3			
h	L^1 error	order	L^∞ error	order	L^1 error	order	L^∞ error	order
1	5.48E-02	—	1.52E-01	—	1.17E-02	—	2.25E-02	—
1/2	1.35E-02	2.02	6.26E-02	1.28	1.35E-03	3.12	4.12E-03	2.45
1/4	2.94E-03	2.20	1.55E-02	2.01	1.45E-04	3.22	4.31E-04	3.26
1/8	6.68E-04	2.14	3.44E-03	2.17	1.71E-05	3.08	7.53E-05	2.52

Second test problem. We consider the following problem:

$$\varphi_t - \cos(\varphi_x + \varphi_y + 1) = 0, \qquad \text{in } (-2,2)^2 \times (0,T),$$
$$\varphi(x,y,0) = -\cos\left(\frac{\pi(x+y)}{2}\right), \qquad \forall\, (x,y) \in (-2,2)^2,$$

with periodic boundary conditions.

For this example we use uniform rectangular meshes. The local Lax-Friedrichs flux is used. The solution is smooth at $T = 0.5/\pi^2$. The accuracy of the numerical solution is shown in Table 5.5.

Table 5.5. Accuracy, 2D, $H(u,v) = -\cos(u + v + 1), T = 0.5/\pi^2$.

	P^1		P^2		P^3	
$N \times N$	L^1 error	order	L^1 error	order	L^1 error	order
10 × 10	6.47E-02	—	8.31E-03	—	1.35E-02	—
20 × 20	2.54E-02	1.349	1.93E-03	2.106	1.57E-03	3.104
40 × 40	1.05E-02	1.274	4.58E-04	2.075	2.39E-04	2.716
80 × 80	4.74E-03	1.147	1.13E-04	2.019	2.89E-05	3.048
160 × 160	2.23E-03	1.088	2.83E-05	1.997	4.38E-06	2.722

	P^1		P^2		P^3	
$N \times N$	L^∞ error	order	L^∞ error	order	L^∞ error	order
10 × 10	1.47E-01	—	1.88E-02	—	2.36E-02	—
20 × 20	6.75E-02	1.123	7.34E-03	1.357	3.44E-03	2.778
40 × 40	2.65E-02	1.349	1.83E-03	2.004	4.59E-04	2.906
80 × 80	1.18E-02	1.167	4.55E-04	2.008	5.78E-05	2.989
160 × 160	2.23E-03	1.088	1.13E-04	2.010	8.54E-06	2.759

The solution has developed a discontinuous derivative at $T = 1.5/\pi^2$. Results with 40 × 40 elements are shown in Fig. 5.4.

Third test problem. The level set equation in a domain with a hole:

$$\varphi_t + sign(\varphi_0)(\sqrt{\varphi_x^2 + \varphi_y^2} - 1) = 0, \qquad \text{in } \Omega \times (0, T),$$

$$\varphi(x, y, 0) = -\cos\left(\frac{\pi(x+y)}{2}\right), \qquad \forall \, (x, y) \in \Omega,$$

where $\Omega = \{(x, y) : 1/2 < \sqrt{x^2 + y^2} < 1\}$.

This problem was introduced in [84]. Its exact solution φ has the same zero level set as φ_0, and the steady state solution is the distance function to that zero level curve. We use this problem to test the effect on the accuracy of the approximation of using various integration paths (5.10) when there is a hole in the region. Notice that the exact steady state solution is the distance function to the inner boundary of domain when boundary condition is adequately prescribed. We compute the time dependent problem to reach a steady state solution, using the exact solution for the boundary conditions of φ_x and φ_y. Four symmetric elements near the outer boundary are updated by (5.9), all other elements are recovered from (5.10) by the shortest path to the nearest one of above four elements. The results are shown in Table 5.6. Also shown in Table 5.6 is the error (difference) between the numerical solution φ thus recovered, and the value of φ after another integration along a circular path (starting and ending at the same point in (5.10)). We can see that the difference is small with the correct order of accuracy, further indicating that the dependency of the recovered solution φ on the integration path is on the order of the truncation errors even for such problems with holes. Finally, the mesh with 1432 triangles and the solution with 5608 triangles are shown in Fig. 5.5.

Table 5.6. Errors for the level set equation, triangular mesh with P^2.

	Errors for the Solution				Errors by Integration Path			
N	L^1 error	order	L^∞ error	order	L^1 error	order	L^∞ error	order
403	1.02E-03	—	1.32E-03	—	1.61E-04	—	5.71E-04	—
1432	1.23E-04	3.05	2.73E-04	2.27	5.84E-05	1.46	1.68E-04	1.78
5608	1.71E-05	2.85	3.18E-05	3.10	9.32E-06	2.65	4.36E-05	1.95
22238	2.09E-06	3.03	5.01E-06	2.67	1.43E-06	2.70	6.63E-06	2.72

Fourth test problem. Two dimensional Riemann problem:

$$\varphi_t + \sin(\varphi_x + \varphi_y) = 0, \qquad \text{in } (-1, 1)^1 \times (0, T),$$

$$\varphi(x, y, 0) = \pi(|y| - |x|), \qquad \forall \, (x, y) \in (-1, 1)^2,$$

For this example we use a uniform rectangular mesh with 40×40 elements. The local Lax-Friedrichs flux is used. As was mentioned in Example 4.3, we have found out that a nonlinear limiting is needed, for convergence towards an viscosity solution. We show the numerical solution at $T = 1$ in Fig. 5.6.

Fifth test problem. A problem from optimal control [73]:

$$\varphi_t + (\sin y)\varphi_x + (\sin x + \text{sign}(\varphi_y))\varphi_y = \frac{1}{2}\sin^2 y + (1 - \cos x),$$

$$\varphi(x, y, 0) = 0,$$

where the space domain is $(-\pi, \pi)^2$ and the boundary conditions are periodic. We use a uniform rectangular mesh of 40×40 elements and the local Lax-Friedrichs flux. The solution at $T = 1$ is shown in Fig. 5.7, while the optimal control $w = \text{sign}(\varphi_y)$ is shown in Fig. 5.8.

Notice that our method computes $\nabla\varphi$ as an independent variable. It is very desirable for those problems in which the most interesting features are contained in the first derivatives of φ, as in this optimal control problem.

Sixth test problem. A problem from computer vision [78]:

$$\varphi_t + I(x, y)\sqrt{1 + \varphi_x^2 + \varphi_y^2} - 1 = 0, \qquad \text{in } (-1, 1)^2 \times (0, T),$$

$$\varphi(x, y, 0) = 0, \qquad\qquad \forall\, (x, y) \in (-1, 1)^2,$$

with $\varphi = 0$ as the boundary condition. The steady state solution of this problem is the shape lighted by a source located at infinity with vertical direction. The solution is not unique if there are points at which $I(x, y) = 1$. Conditions must be prescribed at those points where $I(x, y) = 1$. Since our method is a finite element method, we need to prescribe suitable conditions at the correspondent elements. We take

$$I(x, y) = 1/\sqrt{1 + (1 - |x|)^2 + (1 - |y|)^2} \tag{5.1}$$

The exact steady solution is $\varphi(x, y, \infty) = (1 - |x|)(1 - |y|)$. We use a uniform rectangular mesh of 40×40 elements and the local Lax-Friedrichs flux. We impose the exact boundary conditions for $u = \varphi_x, v = \varphi_y$ from the above exact steady solution, and take the exact value at one point (the lower left corner) to recover φ. The results for P^2 and P^3 are presented in Fig. 5.3, while Fig. 5.9 contains the history of iterations to the steady state.

Next we take

$$I(x, y) = 1/\sqrt{1 + 4y^2(1 - x^2)^2 + 4x^2(1 - y^2)^2} \tag{5.2}$$

The exact steady solution is $\varphi(x, y, \infty) = (1 - x^2)(1 - y^2)$. We again use a uniform rectangular mesh of 40×40 elements, the local Lax-Friedrichs flux, impose the exact boundary conditions for $u = \varphi_x, v = \varphi_y$ from the above exact steady solution, and take the exact value at one point (the lower left corner) to recover φ. A continuation method is used, with the steady solution using

$$I_\varepsilon(x, y) = 1/\sqrt{1 + 4y^2(1 - x^2)^2 + 4x^2(1 - y^2)^2 + \varepsilon} \tag{5.3}$$

for bigger ε as the initial condition for smaller ε. The sequence of ε used are $\varepsilon = 0.2, 0.05, 0$. The results for P^2 and P^3 are presented in Fig. 5.10.

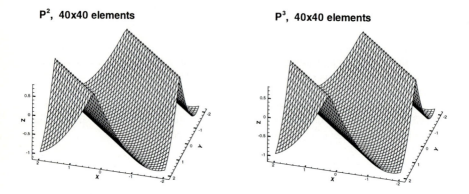

Fig. 5.1. Two dimension Burgers' equation, rectangular mesh, $T=1.5/\pi^2$.

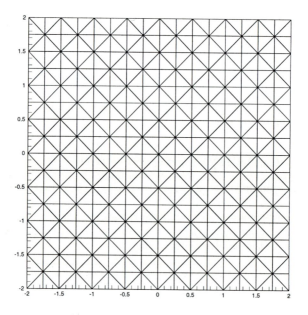

Fig. 5.2. Triangulation for two dimensional Burgers equation, $h = \frac{1}{4}$.

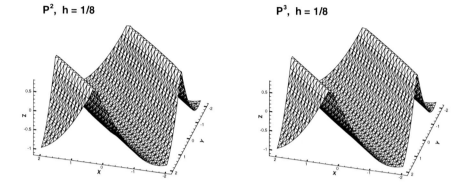

Fig. 5.3. Two dimension Burgers' equation, triangular mesh, $T=1.5/\pi^2$.

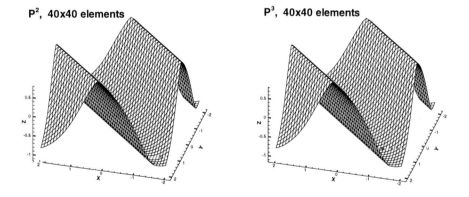

Fig. 5.4. Two dimensional, $H(u,v) = -\cos(u+v+1), T = 1.5/\pi^2$.

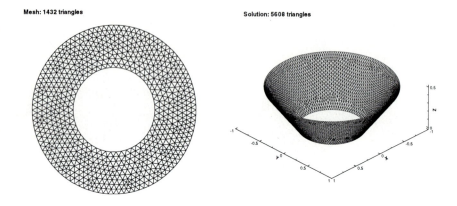

Fig. 5.5. The level set equation, P^2.

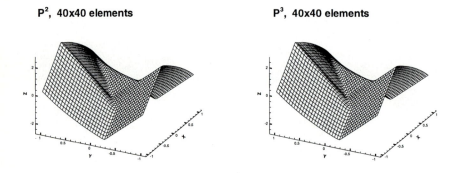

Fig. 5.6. Two dimensional Riemann problem, $H(u,v) = \sin(u+v)$, $T = 1$.

P², 40x40 elements   **P³, 40x40 elements**

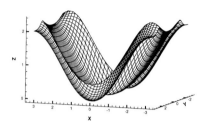 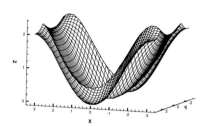

Fig. 5.7. Control problem, $T = 1$.

P², 40x40 elements   **P³, 40x40 elements**

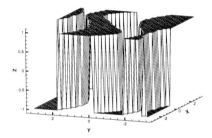 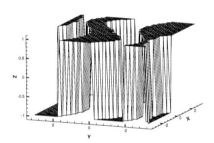

Fig. 5.8. Control problem, $T = 1, w = \operatorname{sign}(\varphi_y)$.

P², 40x40 elements   **P³, 40x40 elements**

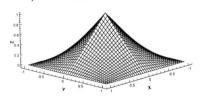 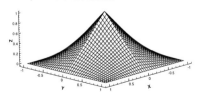

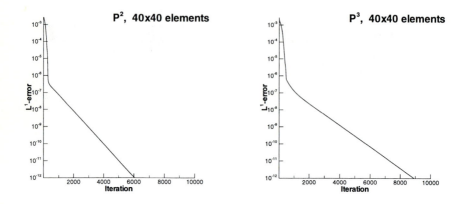

Fig. 5.9. Computer vision problem, history of iterations.

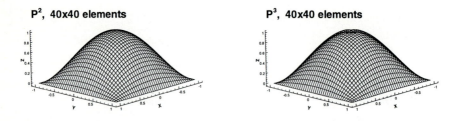

Fig. 5.10. Computer vision problem, $\varphi(x, y, \infty) = (1 - x^2)(1 - y^2)$.

6   Convection diffusion: The LDG method

6.1   Introduction

In this chapter, which follows the work by Cockburn and Shu [27], we restrict ourselves to the semidiscrete LDG methods for convection-diffusion problems with periodic boundary conditions. Our aim is to clearly display the most distinctive features of the LDG methods in a setting as simple as possible; the extension of the method to the fully discrete case is straightforward. In §2, we introduce the LDG methods for the simple one-dimensional case $d = 1$ in which

$$\mathbf{F}(u, Du) = f(u) - a(u)\, \partial_x u,$$

u is a scalar and $a(u) \geq 0$ and show, in §3, some preliminary numerical results displaying the performance of the method. In this simple setting, the main ideas of how to device the method and how to analyze it can be clearly displayed in a simple way. Thus, the L^2-stability of the method is proven in the general nonlinear case and the rate of convergence of $(\Delta x)^k$ in the $L^\infty(0, T; L^2)$-norm for polynomials of degree $k \geq 0$ in the linear case is obtained; this estimate is sharp. In §4, we extend these results to the case in which u is a scalar and

$$\mathbf{F}_i(u, Du) = f_i(u) - \sum_{1 \leq j \leq d} a_{ij}(u)\, \partial_{x_j} u,$$

where a_{ij} defines a positive semidefinite matrix. Again, the L^2-stability of the method is proven for the general nonlinear case and the rate of convergence of $(\Delta x)^k$ in the $L^\infty(0, T; L^2)$-norm for polynomials of degree $k \geq 0$ and arbitrary triangulations is proven in the linear case. In this case, the multidimensionality of the problem and the arbitrariness of the grids increase the technicality of the analysis of the method which, nevertheless, uses the same ideas of the one-dimensional case. In §5, the extension of the LDG method to multidimensional systems is briefly described and in §6, some numerical results for the compressible Navier-Stokes equations from the paper by Bassi and Rebay [3] and from the paper by Lomtev and Karniadakis [63] are presented.

6.2   The LDG methods for the one-dimensional case

In this section, we present and analyze the LDG methods for the following simple model problem:

$$\partial_t u + \partial_x \left(f(u) - a(u)\, \partial_x u \right) = 0 \quad \text{in } Q, \tag{6.1}$$

$$u(t - 0) - u_0 \qquad \text{on } (0, 1), \tag{6.2}$$

where $Q = (0, T) \times (0, 1)$, with periodic boundary conditions.

General formulation and main properties   To define the LDG method, we introduce the new variable $q = \sqrt{a(u)}\, \partial_x u$ and rewrite the problem (6.1), (6.2) as follows:

$$\partial_t u + \partial_x \left(f(u) - \sqrt{a(u)}\, q \right) = 0 \quad \text{in } Q, \tag{6.3}$$

$$q - \partial_x g(u) = 0 \quad \text{in } Q, \tag{6.4}$$

$$u(t = 0) = u_0, \quad \text{on } (0, 1), \tag{6.5}$$

where $g(u) = \int^u \sqrt{a(s)}\, ds$. The LDG method for (6.1), (6.2) is now obtained by simply discretizing the above system with the Discontinuous Galerkin method.

To do that, we follow [24] and [21]. We define the flux $\mathbf{h} = (h_u, h_q)^t$ as follows:

$$\mathbf{h}(u, q) = (f(u) - \sqrt{a(u)}\, q, -g(u))^t. \qquad (6.6)$$

For each partition of the interval $(0, 1)$, $\{x_{j+1/2}\}_{j=0}^N$, we set, for $j = 1, \ldots, N$:

$$I_j = (x_{j-1/2}, x_{j+1/2}), \quad \Delta x_j = x_{j+1/2} - x_{j-1/2}, \qquad (6.7)$$

and

$$\Delta x = \max_{1 \le j \le N} \Delta x_j. \qquad (6.8)$$

We seek an approximation $\mathbf{w}_h = (u_h, q_h)^t$ to $\mathbf{w} = (u, q)^t$ such that for each time $t \in [0, T]$, both $u_h(t)$ and $q_h(t)$ belong to the finite dimensional space

$$V_h = V_h^k = \{v \in L^1(0, 1) : v|_{I_j} \in P^k(I_j), \ j = 1, \ldots, N\}, \qquad (6.9)$$

where $P^k(I)$ denotes the space of polynomials in I of degree at most k. In order to determine the approximate solution (u_h, q_h), we first note that by multiplying (6.3), (6.4), and (6.5) by arbitrary, smooth functions v_u, v_q, and v_i, respectively, and integrating over I_j, we get, after a simple formal integration by parts in (6.3) and (6.4),

$$\int_{I_j} \partial_t u(x, t)\, v_u(x)\, dx - \int_{I_j} h_u(\mathbf{w}(x, t))\, \partial_x v_u(x)\, dx$$
$$+ h_u(\mathbf{w}(x_{j+1/2}, t))\, v_u(x_{j+1/2}^-) - h_u(\mathbf{w}(x_{j-1/2}, t))\, v_u(x_{j-1/2}^+) = 0, \qquad (6.10)$$

$$\int_{I_j} q(x, t)\, v_q(x)\, dx - \int_{I_j} h_q(\mathbf{w}(x, t))\, \partial_x v_q(x)\, dx$$
$$+ h_q(\mathbf{w}(x_{j+1/2}, t))\, v_q(x_{j+1/2}^-) - h_q(\mathbf{w}(x_{j-1/2}, t))\, v_q(x_{j-1/2}^+) = 0, \qquad (6.11)$$

$$\int_{I_j} u(x, 0)\, v_i(x)\, dx = \int_{I_j} u_0(x)\, v_i(x)\, dx. \qquad (6.12)$$

Next, we replace the smooth functions v_u, v_q, and v_i by test functions $v_{h,u}$, $v_{h,q}$, and $v_{h,i}$, respectively, in the finite element space V_h and the exact solution $\mathbf{w} = (u, q)^t$ by the approximate solution $\mathbf{w}_h = (u_h, q_h)^t$. Since this function is discontinuous in each of its components, we must also replace the nonlinear flux $\mathbf{h}(\mathbf{w}(x_{j+1/2}, t))$ by a numerical flux $\hat{\mathbf{h}}(\mathbf{w})_{j+1/2}(t) = (\hat{h}_u(\mathbf{w}_h)_{j+1/2}(t), \hat{h}_q(\mathbf{w}_h)_{j+1/2}(t))$ that will be suitably chosen later. Thus, the approximate solution given by the LDG method is defined as the solution of the following weak formulation:

$\forall \, v_{h,u} \in P^k(I_j):$

$$\int_{I_j} \partial_t \, u_h(x,t) \, v_{h,u}(x) \, dx - \int_{I_j} h_u(\mathbf{w}_h(x,t)) \, \partial_x \, v_{h,u}(x) \, dx$$

$$+\hat{h}_u(\mathbf{w}_h)_{j+1/2}(t) \, v_{h,u}(x_{j+1/2}^-) - \hat{h}_u(\mathbf{w}_h)_{j-1/2}(t) \, v_{h,u}(x_{j-1/2}^+) = 0, \quad (6.13)$$

$\forall \, v_{h,q} \in P^k(I_j):$

$$\int_{I_j} q_h(x,t) \, v_{h,q}(x) \, dx - \int_{I_j} h_q(\mathbf{w}_h(x,t)) \, \partial_x \, v_{h,q}(x) \, dx$$

$$+\hat{h}_q(\mathbf{w}_h)_{j+1/2}(t) \, v_{h,q}(x_{j+1/2}^-) - \hat{h}_q(\mathbf{w}_h)_{j-1/2}(t) \, v_{h,q}(x_{j-1/2}^+) = 0, \quad (6.14)$$

$\forall \, v_{h,i} \in P^k(I_j):$

$$\int_{I_j} (\, u_h(x,0) - u_0(x) \,) \, v_{h,i}(x) \, dx = 0 \qquad (6.15)$$

It only remains to choose the numerical flux $\hat{h}(\mathbf{w}_h)_{j+1/2}(t)$. We use the notation:

$$[p] = p^+ - p^-, \quad \overline{p} = \frac{1}{2}(p^+ + p^-), \quad p_{j+1/2}^{\pm} = p(x_{j+1/2}^{\pm}).$$

To be consistent with the type of numerical fluxes used in the RKDG methods, we consider numerical fluxes of the form

$$\hat{h}(\mathbf{w}_h)_{j+1/2}(t) \equiv \hat{h}(\mathbf{w}_h(x_{j+1/2}^-, t), \mathbf{w}_h(x_{j+1/2}^+, t)),$$

that:

(i)   Are locally Lipschitz and consistent with the flux h,
(ii) Allow for a local resolution of q_h in terms of u_h,
(iii) Reduce to an E-flux (see Osher [71]) when $a(\cdot) \equiv 0$, and that (iv) enforce the L^2-stability of the method.

To reflect the convection-diffusion nature of the problem under consideration, we write our numerical flux as the sum of a convective flux and a diffusive flux:

$$\hat{h}(\mathbf{w}^-, \mathbf{w}^+) = \hat{h}_{conv}(\mathbf{w}^-, \mathbf{w}^+) + \hat{h}_{diff}(\mathbf{w}^-, \mathbf{w}^+). \qquad (6.16)$$

The convective flux is given by

$$\hat{h}_{conv}(\mathbf{w}^-, \mathbf{w}^+) = (\hat{f}(u^-, u^+), 0)^t, \qquad (6.17)$$

where $\hat{f}(u^-, u^+)$ is any locally Lipschitz E-flux consistent with the nonlinearity f, and the diffusive flux is given by

$$\hat{h}_{diff}(\mathbf{w}^-, \mathbf{w}^+) = \left(-\frac{[g(u)]}{[u]} \overline{q}, \, -\overline{g(u)}\right)^t - C_{diff} [\mathbf{w}], \qquad (6.18)$$

where

$$\mathcal{C}_{diff} = \begin{pmatrix} 0 & c_{12} \\ -c_{12} & 0 \end{pmatrix}, \tag{6.19}$$

$$c_{12} = c_{12}(\mathbf{w}^-, \mathbf{w}^+) \quad \text{is locally Lipschitz}, \tag{6.20}$$

$$c_{12} \equiv 0 \quad \text{when } a(\cdot) \equiv 0. \tag{6.21}$$

We claim that this flux satisfies the properties (i) to (iv).

Let us prove our claim. That the flux $\hat{\mathbf{h}}$ is consistent with the flux $\mathbf{h}$ easily follows from their definitions. That $\hat{\mathbf{h}}$ is locally Lipschitz follows from the fact that $\hat{f}(\cdot, \cdot)$ is locally Lipschitz and from (6.19); we assume that $f(\cdot)$ and $a(\cdot)$ are locally Lipschitz functions, of course. Property (i) is hence satisfied.

That the approximate solution q_h can be resolved element by element in terms of u_h by using (6.14) follows from the fact that, by (6.18), the flux

$$\hat{h}_q = -\overline{g(u)} - c_{12}\,[\,u\,]$$

is independent of q_h. Property (ii) is hence satisfied.

Property (iii) is also satisfied by (6.21) and by the construction of the convective flux.

To see that the property (iv) is satisfied, let us first rewrite the flux $\hat{\mathbf{h}}$ in the following way:

$$\hat{\mathbf{h}}(\mathbf{w}^-, \mathbf{w}^+) = \left(\tfrac{[\phi(u)]}{[u]} - \tfrac{[g(u)]}{[u]}\,\overline{q}, -\overline{g(u)} \right)^t - \mathcal{C}\,[\,\mathbf{w}\,],$$

where

$$\mathcal{C} = \begin{pmatrix} c_{11} & c_{12} \\ -c_{12} & 0 \end{pmatrix}, \quad c_{11} = \frac{1}{[u]}\left(\frac{[\phi(u)]}{[u]} - \hat{f}(u^-, u^+) \right). \tag{6.22}$$

with $\phi(u)$ defined by $\phi(u) = \int^u f(s)\,ds$. Since $\hat{f}(\cdot, \cdot)$ is an E-flux,

$$c_{11} = \tfrac{1}{[u]^2} \int_{u^-}^{u^+} \left(f(s) - \hat{f}(u^-, u^+) \right) ds \geq 0,$$

and so, by (6.19), the matrix $\mathcal{C}$ is semipositive definite. The property (iv) follows from this fact and from the following result.

Proposition 26. (Stability) *We have,*

$$\frac{1}{2} \int_0^1 u_h^2(x, T)\,dx + \int_0^T \int_0^1 q_h^2(x, t)\,dx\,dt + \Theta_{T,\mathcal{C}}([\mathbf{w}_h]) \leq \frac{1}{2} \int_0^1 u_0^2(x)\,dx,$$

where $\Theta_{T,\mathcal{C}}([\mathbf{w}_h])$ is the following expression:

$$\int_0^T \sum_{1 \leq j \leq N} \left\{ [\mathbf{w}_h(t)]^t \mathcal{C}\,[\mathbf{w}_h(t)] \right\}_{j+1/2} dt.$$

For a proof, see the appendix. Thus, this shows that the flux $\hat{\mathbf{h}}$ under consideration does satisfy the properties (i) to (iv)- as claimed.

Now, we turn to the question of the quality of the approximate solution defined by the LDG method. In the linear case $f' \equiv c$ and $a(\cdot) \equiv a$, from the above stability result and from the the approximation properties of the finite element space V_h, we can prove the following error estimate. We denote the $L^2(0,1)$-norm of the ℓ-th derivative of u by $|u|_\ell$.

Theorem 27. (Error estimate) *Let* $\mathbf{e}$ *be the approximation error* $\mathbf{w} - \mathbf{w}_h$. *Then we have,*

$$\left\{ \int_0^1 |e_u(x,T)|^2 \, dx + \int_0^T \int_0^1 |e_q(x,t)|^2 \, dx \, dt + \Theta_{T,c}([\mathbf{e}]) \right\}^{1/2} \leq C\,(\varDelta x)^k,$$

where $C = C(k, |u|_{k+1}, |u|_{k+2})$. *In the purely hyperbolic case* $a = 0$, *the constant* C *is of order* $(\varDelta x)^{1/2}$. *In the purely parabolic case* $c = 0$, *the constant* C *is of order* $\varDelta x$ *for even values of* k *for uniform grids and for* C *identically zero.*

For a proof, see the appendix. The above error estimate gives a suboptimal order of convergence, but it is sharp for the LDG methods. Indeed, Bassi *et al* [5] report an order of convergence of order $k+1$ for even values of k and of order k for odd values of k for a steady state, purely elliptic problem for uniform grids and for C identically zero. The numerical results for a purely parabolic problem that will be displayed later lead to the same conclusions; see Table 5 in the section §2.b.

The error estimate is also sharp in that the optimal order of convergence of $k + 1/2$ is recovered in the purely hyperbolic case, as expected. This improvement of the order of convergence is a reflection of the *semipositive definiteness* of the matrix C, which enhances the stability properties of the LDG method. Indeed, in the purely hyperbolic case, the quantity

$$\int_0^T \sum_{1 \leq j \leq N} \left\{ [u_h(t)]^t \, c_{11} \, [u_h(t)] \right\}_{j+1/2} \, dt,$$

is uniformly bounded. This additional control on the jumps of the variable u_h is reflected in the improvement of the order of accuracy from k in the general case to $k + 1/2$ in the purely hyperbolic case.

However, this can only happen in the purely hyperbolic case for the LDG methods. Indeed, since $c_{11} = 0$ for $c = 0$, the control of the jumps of u_h is not enforced in the purely parabolic case. As indicated by the numerical experiments of Bassi *et al.* [5] and those of section §2.b below, this can result in the effective degradation of the order of convergence. To remedy this situation, the control of the jumps of u_h in the purely parabolic case can be easily enforced by letting c_{11} be strictly positive if $|c| + |a| > 0$. Unfortunately, this is not enough to guarantee an improvement of the accuracy: an additional control on the jumps of q_h is required! This can be easily achieved by allowing the matrix C to be *symmetric and positive definite* when $a > 0$. In this case, the order of convergence of $k + 1/2$ can be easily obtained for the general convection-diffusion case. However, this would force the matrix entry c_{22} to be nonzero and the property (ii) of local resolvability of q_h in terms of u_h would not be satisfied anymore. As a consequence, the high parallelizability of the LDG would be lost.

The above result shows how strongly the order of convergence of the LDG methods depend on the choice of the matrix C. In fact, the numerical results of section §2.b in uniform grids indicate that with yet another choice of the matrix C, see (6.23), the LDG method converges with the optimal order of $k + 1$ in the general case. The analysis of this phenomenon constitutes the subject of ongoing work.

6.3   Numerical results in the one-dimensional case

In this section we present some numerical results for the schemes discussed in this paper. We will only provide results for the following one dimensional, linear convection diffusion equation

$$\partial_t u + c\,\partial_x u - a\,\partial_x^2 u = 0 \quad \text{in } (0,T) \times (0, 2\pi),$$
$$u(t = 0, x) = \sin(x), \quad \text{on } (0, 2\pi),$$

where c and $a \geq 0$ are both constants; periodic boundary conditions are used. The exact solution is $u(t, x) = e^{-at}\sin(x - ct)$. We compute the solution up to $T = 2$, and use the LDG method with C defined by

$$C = \begin{pmatrix} \frac{|c|}{2} & -\frac{\sqrt{a}}{2} \\ \frac{\sqrt{a}}{2} & 0 \end{pmatrix}. \tag{6.23}$$

We notice that, for this choice of fluxes, the approximation to the convective term cu_x is the standard upwinding, and that the approximation to the diffusion term $a\,\partial_x^2 u$ is the standard three point central difference, for the P^0 case. On the other hand, if one uses a central flux corresponding to $c_{12} = -c_{21} = 0$, one gets a spread-out five point central difference approximation to the diffusion term $a\,\partial_x^2 u$.

The LDG methods based on P^k, with $k = 1, 2, 3, 4$ are tested. Elements with equal size are used. Time discretization is by the third-order accurate TVD Runge-Kutta method [81], with a sufficiently small time step so that error in time is negligible comparing with spatial errors. We list the L_∞ errors and numerical orders of accuracy, for u_h, as well as for its derivatives suitably scaled $\Delta x^m \partial_x^m u_h$ for $1 \leq m \leq k$, at the center of of each element. This gives the complete description of the error for u_h over the whole domain, as u_h in each element is a polynomial of degree k. We also list the L_∞ errors and numerical orders of accuracy for q_h at the element center.

In all the convection-diffusion runs with $a > 0$, accuracy of at least $(k + 1)$-th order is obtained, for both u_h and q_h, when P^k elements are used. See Tables 1 to 3. The P^4 case for the purely convection equation $a = 0$ seems to be not in the asymptotic regime yet with $N = 40$ elements (further refinement with $N = 80$ suffers from round-off effects due to our choice of non-orthogonal basis functions), Table 4. However, the absolute values of the errors are comparable with the convection dominated case in Table 3.

Finally, to show that the order of accuracy could really degenerate to k for P^k, as was already observed in [5], we rerun the heat equation case $a = 1, c = 0$ with the central flux

$$C = \begin{pmatrix} 0 & 0 \\ 0 & 0 \end{pmatrix}.$$

This time we can see that the global order of accuracy in L_∞ is only k when P^k is used with an odd value of k.

Table 1

The heat equation $a = 1$, $c = 0$. L_∞ errors and numerical order of accuracy, measured at the center of each element, for $\Delta x^m \partial_x^m u_h$ for $0 \le m \le k$, and for q_h.

k	variable	$N = 10$ error	$N = 20$ error	order	$N = 40$ error	order
1	u	4.55E-4	5.79E-5	2.97	7.27E-6	2.99
	$\Delta x\, \partial_x u$	9.01E-3	2.22E-3	2.02	5.56E-4	2.00
	q	4.17E-5	2.48E-6	4.07	1.53E-7	4.02
2	u	1.43E-4	1.76E-5	3.02	2.19E-6	3.01
	$\Delta x\, \partial_x u$	7.87E-4	1.03E-4	2.93	1.31E-5	2.98
	$(\Delta x)^2\, \partial_x^2 u$	1.64E-3	2.09E-4	2.98	2.62E-5	2.99
	q	1.42E-4	1.76E-5	3.01	2.19E-6	3.01
3	u	1.54E-5	9.66E-7	4.00	6.11E-8	3.98
	$\Delta x\, \partial_x u$	3.77E-5	2.36E-6	3.99	1.47E-7	4.00
	$(\Delta x)^2\, \partial_x^2 u$	1.90E-4	1.17E-5	4.02	7.34E-7	3.99
	$(\Delta x)^3\, \partial_x^3 u$	2.51E-4	1.56E-5	4.00	9.80E-7	4.00
	q	1.48E-5	9.66E-7	3.93	6.11E-8	3.98
4	u	2.02E-7	5.51E-9	5.20	1.63E-10	5.07
	$\Delta x\, \partial_x u$	1.65E-6	5.14E-8	5.00	1.61E-9	5.00
	$(\Delta x)^2\, \partial_x^2 u$	6.34E-6	2.04E-7	4.96	6.40E-9	4.99
	$(\Delta x)^3\, \partial_x^3 u$	2.92E-5	9.47E-7	4.95	2.99E-8	4.99
	$(\Delta x)^4\, \partial_x^4 u$	3.03E-5	9.55E-7	4.98	2.99E-8	5.00
	q	2.10E-7	5.51E-9	5.25	1.63E-10	5.07

Table 2

The convection diffusion equation $a = 1$, $c = 1$. L_∞ errors and numerical order of accuracy,   measured at the center of each element,   for $\Delta x^m \partial_x^m u_h$ for $0 \le m \le k$, and for q_h.

k	variable	$N = 10$ error	$N = 20$ error	order	$N = 40$ error	order
1	u	6.47E-4	1.25E-4	2.37	1.59E-5	2.97
	$\Delta x\,\partial_x u$	9.61E-3	2.24E-3	2.10	5.56E-4	2.01
	q	2.96E-3	1.20E-4	4.63	1.47E-5	3.02
2	u	1.42E-4	1.76E-5	3.02	2.18E-6	3.01
	$\Delta x\,\partial_x u$	7.93E-4	1.04E-4	2.93	1.31E-5	2.99
	$(\Delta x)^2\,\partial_x^2 u$	1.61E-3	2.09E-4	2.94	2.62E-5	3.00
	q	1.26E-4	1.63E-5	2.94	2.12E-6	2.95
3	u	1.53E-5	9.75E-7	3.98	6.12E-8	3.99
	$\Delta x\,\partial_x u$	3.84E-5	2.34E-6	4.04	1.47E-7	3.99
	$(\Delta x)^2\,\partial_x^2 u$	1.89E-4	1.18E-5	4.00	7.36E-7	4.00
	$(\Delta x)^3\,\partial_x^3 u$	2.52E-4	1.56E-5	4.01	9.81E-7	3.99
	q	1.57E-5	9.93E-7	3.98	6.17E-8	4.01
4	u	2.04E-7	5.50E-9	5.22	1.64E-10	5.07
	$\Delta x\,\partial_x u$	1.68E-6	5.19E-8	5.01	1.61E-9	5.01
	$(\Delta x)^2\,\partial_x^2 u$	6.36E-6	2.05E-7	4.96	6.42E-8	5.00
	$(\Delta x)^3\,\partial_x^3 u$	2.99E-5	9.57E-7	4.97	2.99E-8	5.00
	$(\Delta x)^4\,\partial_x^4 u$	2.94E-5	9.55E-7	4.95	3.00E-8	4.99
	q	1.96E-7	5.35E-9	5.19	1.61E-10	5.06

Table 3

The convection dominated convection diffusion equation $a = 0.01$, $c = 1$. L_∞ errors and numerical order of accuracy, measured at the center of each element, for $\Delta x^m \partial_x^m u_h$ for $0 \le m \le k$, and for q_h.

k	variable	$N = 10$	$N = 20$		$N = 40$	
		error	error	order	error	order
1	u	7.14E-3	9.30E-4	2.94	1.17E-4	2.98
	$\Delta x\, \partial_x u$	6.04E-2	1.58E-2	1.93	4.02E-3	1.98
	q	8.68E-4	1.09E-4	3.00	1.31E-5	3.05
2	u	9.59E-4	1.25E-4	2.94	1.58E-5	2.99
	$\Delta x\, \partial_x u$	5.88E-3	7.55E-4	2.96	9.47E-5	3.00
	$(\Delta x)^2\, \partial_x^2 u$	1.20E-2	1.50E-3	3.00	1.90E-4	2.98
	q	8.99E-5	1.11E-5	3.01	1.10E-6	3.34
3	u	1.11E-4	7.07E-6	3.97	4.43E-7	4.00
	$\Delta x\, \partial_x u$	2.52E-4	1.71E-5	3.88	1.07E-6	4.00
	$(\Delta x)^2\, \partial_x^2 u$	1.37E-3	8.54E-5	4.00	5.33E-6	4.00
	$(\Delta x)^3\, \partial_x^3 u$	1.75E-3	1.13E-4	3.95	7.11E-6	3.99
	q	1.18E-5	7.28E-7	4.02	4.75E-8	3.94
4	u	1.85E-6	4.02E-8	5.53	1.19E-9	5.08
	$\Delta x\, \partial_x u$	1.29E-5	3.76E-7	5.10	1.16E-8	5.01
	$(\Delta x)^2\, \partial_x^2 u$	5.19E-5	1.48E-6	5.13	4.65E-8	4.99
	$(\Delta x)^3\, \partial_x^3 u$	2.21E-4	6.93E-6	4.99	2.17E-7	5.00
	$(\Delta x)^4\, \partial_x^4 u$	2.25E-4	6.89E-6	5.03	2.17E-7	4.99
	q	3.58E-7	3.06E-9	6.87	5.05E-11	5.92

Table 4

The convection equation $a = 0$, $c = 1$. L_∞ errors and numerical order of accuracy, measured at the center of each element, for $\Delta x^m \partial_x^m u_h$ for $0 \leq m \leq k$.

k	variable	$N = 10$ error	$N = 20$ error	order	$N = 40$ error	order
1	u	7.24E-3	9.46E-4	2.94	1.20E-4	2.98
	$\Delta x\, \partial_x u$	6.09E-2	1.60E-2	1.92	4.09E-3	1.97
2	u	9.96E-4	1.28E-4	2.96	1.61E-5	2.99
	$\Delta x\, \partial_x u$	6.00E-3	7.71E-4	2.96	9.67E-5	3.00
	$(\Delta x)^2\, \partial_x^2 u$	1.23E-2	1.54E-3	3.00	1.94E-4	2.99
3	u	1.26E-4	7.50E-6	4.07	4.54E-7	4.05
	$\Delta x\, \partial_x u$	1.63E-4	2.00E-5	3.03	1.07E-6	4.21
	$(\Delta x)^2\, \partial_x^2 u$	1.52E-3	9.03E-5	4.07	5.45E-6	4.05
	$(\Delta x)^3\, \partial_x^3 u$	1.35E-3	1.24E-4	3.45	7.19E-6	4.10
4	u	3.55E-6	8.59E-8	5.37	3.28E-10	8.03
	$\Delta x\, \partial_x u$	1.89E-5	1.27E-7	7.22	1.54E-8	3.05
	$(\Delta x)^2\, \partial_x^2 u$	8.49E-5	2.28E-6	5.22	2.33E-8	6.61
	$(\Delta x)^3\, \partial_x^3 u$	2.36E-4	5.77E-6	5.36	2.34E-7	4.62
	$(\Delta x)^4\, \partial_x^4 u$	2.80E-4	8.93E-6	4.97	1.70E-7	5.72

Table 5

The heat equation $a = 1$, $c = 0$. L_∞ errors and numerical order of accuracy, measured at the center of each element, for $\Delta x^m \partial_x^m u_h$ for $0 \leq m \leq k$, and for q_h, using the central flux.

k	variable	$N = 10$ error	$N = 20$ error	order	$N = 40$ error	order
1	u	3.59E-3	8.92E-4	2.01	2.25E-4	1.98
	$\Delta x\, \partial_x u$	2.10E-2	1.06E-2	0.98	5.31E-3	1.00
	q	2.39E-3	6.19E-4	1.95	1.56E-4	1.99
2	u	6.91E-5	4.12E-6	4.07	2.57E-7	4.00
	$\Delta x\, \partial_x u$	7.66E-4	1.03E-4	2.90	1.30E-5	2.98
	$(\Delta x)^2\, \partial_x^2 u$	2.98E-4	1.68E-5	4.15	1.03E-6	4.02
	q	6.52E-5	4.11E-6	3.99	2.57E-7	4.00
3	u	1.62E-5	1.01E-6	4.00	6.41E-8	3.98
	$\Delta x\, \partial_x u$	1.06E-4	1.32E-5	3.01	1.64E-6	3.00
	$(\Delta x)^2\, \partial_x^2 u$	1.99E-4	1.22E-5	4.03	7.70E-7	3.99
	$(\Delta x)^3\, \partial_x^3 u$	6.81E-4	8.68E-5	2.97	1.09E-5	2.99
	q	1.54E-5	1.01E-6	3.93	6.41E-8	3.98
4	u	8.25E-8	1.31E-9	5.97	2.11E-11	5.96
	$\Delta x\, \partial_x u$	1.62E-6	5.12E-8	4.98	1.60E-9	5.00
	$(\Delta x)^2\, \partial_x^2 u$	1.61E-6	2.41E-8	6.06	3.78E-10	6.00
	$(\Delta x)^3\, \partial_x^3 u$	2.90E-5	9.46E-7	4.94	2.99E-8	4.99
	$(\Delta x)^4\, \partial_x^4 u$	5.23E-6	7.59E-8	6.11	1.18E-9	6.01
	q	7.85E-8	1.31E-9	5.90	2.11E-11	5.96

6.4   The LDG methods for the multidimensional case

In this section, we consider the LDG methods for the following convection-diffusion model problem

$$\partial_t u + \sum_{1 \le i \le d} \partial_{x_i} \left(f_i(u) - \sum_{1 \le j \le d} a_{ij}(u) \partial_{x_j} u \right) = 0 \quad \text{in } Q, \tag{6.24}$$

$$u(t = 0) = u_0 \qquad \text{on } (0,1)^d, \tag{6.25}$$

where $Q = (0,T) \times (0,1)^d$, with periodic boundary conditions. Essentially, the one-dimensional case and the multidimensional case can be studied in exactly the same way. However, there are two important differences that deserve explicit discussion. The first is the treatment of the matrix of entries $a_{ij}(u)$, which is assumed to be *symmetric, semipositive definite* and the introduction of the variables q_ℓ, and the second is the treatment of arbitrary meshes.

To define the LDG method, we first notice that, since the matrix $a_{ij}(u)$ is assumed to be symmetric and semipositive definite, there exists a symmetric matrix $b_{ij}(u)$ such that

$$a_{ij}(u) = \sum_{1 \le \ell \le d} b_{i\ell}(u)\, b_{\ell j}(u). \tag{6.26}$$

Then we define the new scalar variables $q_\ell = \sum_{1 \le j \le d} b_{\ell j}(u)\, \partial_{x_j} u$ and rewrite the problem (6.24), (6.25) as follows:

$$\partial_t u + \sum_{1 \le i \le d} \partial_{x_i} \left(f_i(u) - \sum_{1 \le \ell \le d} b_{i\ell}(u)\, q_\ell \right) = 0 \quad \text{in } Q, \tag{6.27}$$

$$q_\ell - \sum_{1 \le j \le d} \partial_{x_j} g_{\ell j}(u) = 0 \quad \ell = 1, \dots d, \qquad \text{in } Q, \tag{6.28}$$

$$u(t = 0) = u_0 \qquad \text{on } (0,1)^d, \tag{6.29}$$

where $g_{\ell j}(u) = \int^u b_{\ell j}(s)\, ds$. The LDG method is now obtained by discretizing the above equations by the Discontinuous Galerkin method.

We follow what was done in §2. So, we set $\mathbf{w} = (u, \mathbf{q})^t = (u, q_1, \cdots, q_d)^t$ and, for each $i = 1, \cdots, d$, introduce the flux

$$\mathbf{h}_i(\mathbf{w}) = \left(f_i(u) - \sum_{1 \le \ell \le d} b_{i\ell}(u)\, q_\ell, -g_{1i}(u), \dots, -g_{di}(u) \right)^t. \tag{6.30}$$

We consider triangulations of $(0,1)^d$, $\mathcal{T}_{\Delta x} = \{K\}$, made of non-overlapping polyhedra. We require that for any two elements K and K', $\overline{K} \cap \overline{K}'$ is either a face e of both K and K' with nonzero $(d-1)$-Lebesgue measure $|e|$, or has Hausdorff dimension less than $d-1$. We denote by $\mathcal{E}_{\Delta x}$ the set of all faces e of the border of K for all $K \in \mathcal{T}_{\Delta x}$. The diameter of K is denoted by Δx_K and the maximum Δx_K, for $K \in \mathcal{T}_{\Delta x}$ is denoted by Δx. We require, for the sake of simplicity, that the triangulations $\mathcal{T}_{\Delta x}$ be regular, that is, there is a constant independent of Δx such that

$$\frac{\Delta x_K}{\rho_K} \le \sigma \quad \forall K \in \mathcal{T}_{\Delta x},$$

where ρ_K denotes the diameter of the maximum ball included in K.

We seek an approximation $\mathbf{w}_h = (u_h, \mathbf{q}_h)^t = (u_h, q_{h1}, \cdots, q_{hd})^t$ to $\mathbf{w}$ such that for each time $t \in [0, T]$, each of the components of $\mathbf{w}_h$ belong to the finite element space

$$V_h = V_h^k = \{ v \in L^1((0,1)^d) : v|_K \in P^k(K) \ \forall \, K \in \mathcal{T}_{\Delta x} \}, \qquad (6.31)$$

where $P^k(K)$ denotes the space of polynomials of total degree at most k. In order to determine the approximate solution $\mathbf{w}_h$, we proceed exactly as in the one-dimensional case. This time, however, the integrals are made on each element K of the triangulation $\mathcal{T}_{\Delta x}$. We obtain the following weak formulation on each element K of the triangulation $\mathcal{T}_{\Delta x}$:

$$\forall \, v_{h,u} \in P^k(K) :$$

$$\int_K \partial_t \, u_h(x,t) \, v_{h,u}(x) \, dx - \sum_{1 \le i \le d} \int_K h_{i\,u}(\mathbf{w}_h(x,t)) \, \partial_{x_i} \, v_{h,u}(x) \, dx$$

$$+ \int_{\partial K} \hat{h}_u(\mathbf{w}_h, \mathbf{n}_{\partial K})(x,t) \, v_{h,u}(x) \, d\Gamma(x) = 0, \qquad (6.32)$$

for $\ell = 1, \cdots, d :$

$$\forall \, v_{h,u} \in P^k(K) :$$

$$\int_K q_{h\ell}(x,t) \, v_{h,q_\ell}(x) \, dx - \sum_{1 \le j \le d} \int_K h_{j\,q_\ell}(\mathbf{w}_h(x,t)) \, \partial_{x_j} \, v_{h,q_\ell}(x) \, dx$$

$$+ \int_{\partial K} \hat{h}_{q_\ell}(\mathbf{w}_h, \mathbf{n}_{\partial K})(x,t) \, v_{h,q_\ell}(x) \, d\Gamma(x) = 0, \qquad (6.33)$$

$$\forall \, v_{h,u} \in P^k(K) :$$

$$\int_K u_h(x,0) \, v_{h,i}(x) \, dx = \int_K u_0(x) \, v_{h,i}(x) \, dx, \qquad (6.34)$$

where $\mathbf{n}_{\partial K}$ denotes the outward unit normal to the element K at $x \in \partial K$. It remains to choose the numerical flux $(\hat{h}_u, \hat{h}_{q_1}, \cdots, \hat{h}_{q_d})^t \equiv \hat{\mathbf{h}} \equiv \hat{\mathbf{h}}(\mathbf{w}_h, \mathbf{n}_{\partial K})(x,t)$.

As in the one-dimensional case, we require that the fluxes $\hat{\mathbf{h}}$ be of the form

$$\hat{\mathbf{h}}(\mathbf{w}_h, \mathbf{n}_{\partial K})(x) \equiv \hat{\mathbf{h}}(\mathbf{w}_h(x^{int_K}, t), \mathbf{w}_h(x^{ext_K}, t); \mathbf{n}_{\partial K}),$$

where $\mathbf{w}_h(x^{int_K})$ is the limit at x taken from the interior of K and $\mathbf{w}_h(x^{ext_K})$ the limit at x from the exterior of K, and consider fluxes that:

(i) Are locally Lipschitz, conservative, that is,

$$\hat{\mathbf{h}}(\mathbf{w}_h(x^{int_K}), \mathbf{w}_h(x^{ext_K}); \mathbf{n}_{\partial K}) \hat{\mathbf{h}}(\mathbf{w}_h(x^{ext_K}), \mathbf{w}_h(x^{int_K}); -\mathbf{n}_{\partial K}) = 0,$$

and consistent with the flux

$$\sum_{1 \le i \le d} \mathbf{h}_i \, n_{\partial K, i},$$

(ii) Allow for a local resolution of each component of $\mathbf{q}_h$ in terms of u_h *only*,
(iii) Reduce to an E-flux when $a(\cdot) \equiv 0$,
(iv) Enforce the L^2-stability of the method.

Again, we write our numerical flux as the sum of a convective flux and a diffusive flux:

$$\hat{\mathbf{h}} = \hat{\mathbf{h}}_{conv} + \hat{\mathbf{h}}_{diff},$$

where the convective flux is given by

$$\hat{\mathbf{h}}_{conv}(\mathbf{w}^-, \mathbf{w}^+; \mathbf{n}) = \left(\hat{f}(u^-, u^+; \mathbf{n}), 0\right)^t,$$

where $\hat{f}(u^-, u^+; \mathbf{n})$ is any locally Lipschitz E-flux which is conservative and consistent with the nonlinearity

$$\sum_{1 \le i \le d} f_i(u)\, n_i,$$

and the diffusive flux $\hat{\mathbf{h}}_{diff}(\mathbf{w}^-, \mathbf{w}^+; \mathbf{n})$ is given by

$$\left(-\sum_{1 \le i,\ell \le d} \frac{[g_{i\ell}(u)]}{[u]}\, \overline{q_\ell}\, n_i, \; -\sum_{1 \le i \le d} \overline{g_{i1}(u)}\, n_i, \cdots, \; -\sum_{1 \le i \le d} \overline{g_{id}(u)}\, n_i\,\right)^t - \mathcal{C}_{diff}\,[\mathbf{w}],$$

where

$$\mathcal{C}_{diff} = \begin{pmatrix} 0 & c_{12} & c_{13} & \cdots & c_{1d} \\ -c_{12} & 0 & 0 & \cdots & 0 \\ -c_{13} & 0 & 0 & \cdots & 0 \\ \vdots & \vdots & \vdots & \ddots & \vdots \\ -c_{1d} & 0 & 0 & \cdots & 0 \end{pmatrix},$$

$$c_{1j} = c_{1j}(\mathbf{w}^-, \mathbf{w}^+) \quad \text{is locally Lipschitz for } j = 1, \cdots, d,$$

$$c_{1j} \equiv 0 \quad \text{when } a(\cdot) \equiv 0 \quad \text{for } j = 1, \cdots, d.$$

We claim that this flux satisfies the properties (i) to (iv).

To prove that properties (i) to (iii) are satisfied is now a simple exercise. To see that the property (iv) is satisfied, we first rewrite the flux $\hat{\mathbf{h}}$ in the following way:

$$\left(-\sum_{1 \le i,\ell \le d} \frac{[g_{i\ell}(u)]}{[u]}\, \overline{q_\ell}\, n_i, \; -\sum_{1 \le i \le d} \overline{g_{i1}(u)}\, n_i, \cdots, \; -\sum_{1 \le i \le d} \overline{g_{id}(u)}\, n_i\,\right)^t - \mathcal{C}\,[\mathbf{w}],$$

where

$$\mathcal{C} = \begin{pmatrix} c_{11} & c_{12} & c_{13} & \cdots & c_{1d} \\ -c_{12} & 0 & 0 & \cdots & 0 \\ -c_{13} & 0 & 0 & \cdots & 0 \\ \vdots & \vdots & \vdots & \ddots & \vdots \\ -c_{1d} & 0 & 0 & \cdots & 0 \end{pmatrix},$$

$$c_{11} = \frac{1}{[u]}\left(\sum_{1 \le i \le d} \frac{[\phi_i(u)]}{[u]}\, n_i - \hat{f}(u^-, u^+; \mathbf{n})\right),$$

where $\phi_i(u) = \int^u f_i(s)\,ds$. Since $\hat{f}(\cdot,\cdot;\mathbf{n})$ is an E-flux,

$$c_{11} = \frac{1}{[u]^2} \int_{u^-}^{u^+} \Big(\sum_{1\le i\le d} f_i(s)\, n_i - \hat{f}(u^-, u^+; \mathbf{n}) \Big)\, ds$$

$$\ge 0,$$

and so the matrix $\mathcal{C}$ is semipositive definite. The property (iv) follows from this fact and from the following result.

Proposition 28. (Stability) *We have,*

$$\frac{1}{2}\int_{(0,1)^d} u_h^2(x,T)\,dx + \int_0^T \int_{(0,1)^d} |\mathbf{q}_h(x,t)|^2\,dx\,dt + \Theta_{T,\mathcal{C}}([\mathbf{w}_h])$$

$$\le \frac{1}{2}\int_{(0,1)^d} u_0^2(x)\,dx,$$

where the quantity $\Theta_{T,\mathcal{C}}([\mathbf{w}_h])$ is given by

$$\int_0^T \sum_{e\in\mathcal{E}_{\Delta x}} \int_e [\mathbf{w}_h(x,t)]^t \mathcal{C}\,[\mathbf{w}_h(x,t)]\,d\,\Gamma(x)\,dt.$$

We can also prove the following error estimate. We denote the integral over $(0,1)^d$ of the sum of the squares of all the derivatives of order $(k+1)$ of u by $|u|_{k+1}^2$.

Theorem 29. (Error estimate) *Let e be the approximation error $\mathbf{w} - \mathbf{w}_h$. Then we have, for arbitrary, regular grids,*

$$\left\{ \int_{(0,1)^d} |e_u(x,T)|^2\,dx + \int_0^T \int_{(0,1)^d} |e_q(x,t)|^2\,dx\,dt + \Theta_{T,\mathcal{C}}([e]) \right\}^{1/2}$$

$$\le C\,(\Delta x)^k,$$

where $C = C(k, |u|_{k+1}, |u|_{k+2})$. In the purely hyperbolic case $a_{ij} = 0$, the constant C is of order $(\Delta x)^{1/2}$. In the purely parabolic case $c = 0$, the constant C is of order Δx for even values of k and of order 1 otherwise for Cartesian products of uniform grids and for C identically zero provided that the local spaces Q^k are used instead of the spaces P^k, where Q^k is the space of tensor products of one dimensional polynomials of degree k.

6.5   Extension to multidimensional systems

In this chapter, we have considered the so-called LDG methods for convection-diffusion problems. For scalar problems in multidimensions, we have shown that they are L^2-stable and that in the linear case, they are of order k if polynomials of order k are used. We have also shown that this estimate is sharp and have displayed the strong dependence of the order of convergence of the LDG methods on the choice of the numerical fluxes.

The main advantage of these methods is their extremely high parallelizability and their high-order accuracy which render them suitable for computations of convection-dominated flows. Indeed, although the LDG method have a large amount of degrees of freedom per element, and hence more computations per element are necessary, its extremely local domain of dependency allows a very efficient parallelization that by far compensates for the extra amount of local computations.

The LDG methods for multidimensional systems, like for example the compressible Navier-Stokes equations and the equations of the hydrodynamic model for semiconductor device simulation, can be easily defined by simply applying the procedure described for the multidimensional scalar case to each component of $\mathbf{u}$. In practice, especially for viscous terms which are not symmetric but still semipositive definite, such as for the compressible Navier-Stokes equations, we can use $\mathbf{q} = (\partial_{x_1} u, ..., \partial_{x_d} u)$ as the auxiliary variables. Although with this choice, the L^2-stability result will not be available theoretically, this would not cause any problem in practical implementations.

6.6   Some numerical results

Next, we present some numerical results from the papers by Bassi and Rebay [3] and Lomtev and Karniadakis [63].

• **Smooth, steady state solutions**. We start by displaying the convergence of the method for a p-refinement done by Lomtev and Karniadakis [63]. In Figure 6.1, we can see how the maximum errors in density, momentum, and energy decrease exponentially to zero as the degree k of the approximating polynomials increases while the grid is kept fixed; details about the exact solution can be found in [63].

Now, let us consider the laminar, transonic flow around the NACA0012 airfoil at an angle of attack of ten degrees, free stream Mach number $M = 0.8$, and Reynolds number (based on the free stream velocity and the airfoil chord) equal to 73; the wall temperature is set equal to the free stream total temperature. Bassy and Rebay [3] have computed the solution of this problem with polynomials of degree $1, 2$, and 3 and Lomtev and Karniadakis [63] have tried the same test problem with polynomials of degree $2, 4$, and 6 in a mesh of 592 elements which is about four times less elements than the mesh used by Bassi and Rebay [3]. In Figure 6.3, taken from [63], we display the pressure and drag coefficient distributions computed by Bassi and Rebay [3] with polynomials on degree 3 and the ones computed by Lomtev and Karniadakis [63] computed with polynomials of degree 6. We can see good agreement of both computations. In Figure 6.2, taken from [63], we see the mesh and the Mach isolines obtained with polynomials of degree two and four; note the improvement of the solution.

Next, we show a result from the paper by Bassi and Rebay [3]. We consider the laminar, subsonic flow around the NACA0012 airfoil at an angle of attack of zero degrees, free stream Mach number $M = 0.5$, and Reynolds number equal to 5000. In figure 6.4, we can see the Mach isolines corresponding to linear, quadratic, and cubic elements. In the figures 6.5, 6.6, and 6.7 details of the results with cubic elements are shown. Note how the boundary layer is captured within a few layers of elements and how its separation at the trailing edge of the airfoil has been clearly resolved. Bassi and Rebay [3] report that these results are comparable to common structured and unstructured finite volume methods on much finer grids- a result consistent with the computational results we have displayed in these notes.

Finally, we present a not-yet-published result kindly provided by Lomtev and Karniadakis about the simulation of an expansion pipe flow. The smaller cylinder has a diameter of 1 and the larger cylinder has a diameter of 2. In Figure 6.8, we display the velocity profile and some streamlines for a Reynolds number equal to 50 and Mach number 0.2. The computation was made with polynomials of degree 5 and a mesh of 600 tetrahedra; of course the tetrahedra have curved faces to accommodate the exact boundaries. In Figure 6.9, we display a comparison between computational and experimental results. As a function of the Reynolds number, two quantities are plotted. The first is the distance between the step and the center of the vertex (lower branch) and the second is the distance from the step to the separation point (upper branch). The computational results are obtained by the method under consideration with polynomials of degree 5 for the compressible Navier Stokes equations, and by a standard Galerkin formulation in terms of velocity-pressure (NEKTAR), by Sherwin and Karniadakis [79], or in terms of velocity-vorticity (IVVA), by Trujillo [87], for the *incompressible* Navier Stokes equations; results produced by the code called PRISM are also included, see Newmann [69]. The experimental data was taken from Macagno and Hung [67]. The agreement between computations and experiments is remarkable.

• **Unsteady solutions**. To end this chapter, we present the computation of an unsteady solution by Lomtev and Karniadakis [63]. The test problem is the classical problem of a flow around a cylinder in two space dimensions. The Reynolds number is $10,000$ and the Mach number 0.2.

In Figure 6.10, the streamlines are shown for a computation made on a grid of 680 triangles (with curved sides fitting the cylinder) and polynomials whose degree could vary from element to element; the maximum degree was 5. In Figure 6.11, details of the mesh and the density around the cylinder are shown. Note how the method is able to capture the shear layer instability observed experimentally. For more details, see [63].

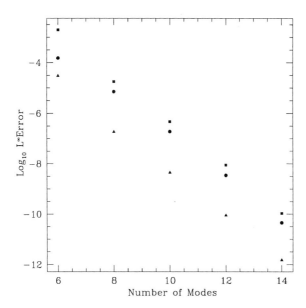

Fig. 6.1. Maximum errors of the density (triangles), momemtum (circles) and energy (squares) as a function of the degree of the approximating polynomial plus one (called "number of modes" in the picture).

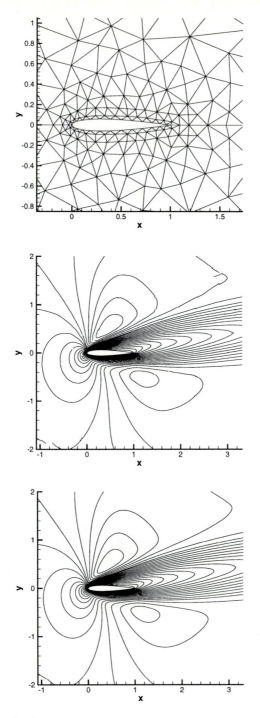

Fig. 6.2. Mesh (top) and Mach isolines around the NACA0012 airfoil, ($Re = 73$, $M = 0.8$, angle of attack of ten degrees) for quadratic (middle) and quartic (bottom) elements.

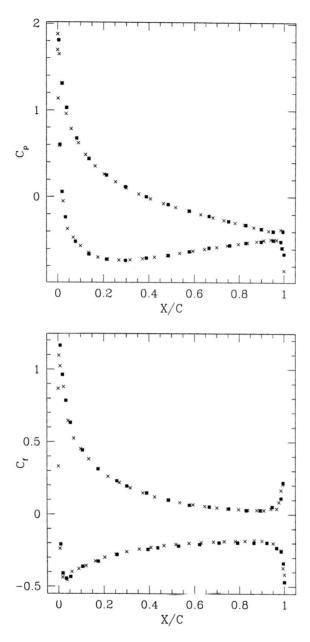

Fig. 6.3. Pressure (top) and drag(bottom) coefficient distributions. The squares were obtained by Bassi and Rebay [3] with cubics and the crosses by Lomtev and Karniadakis [63] with polynomials of degree 6.

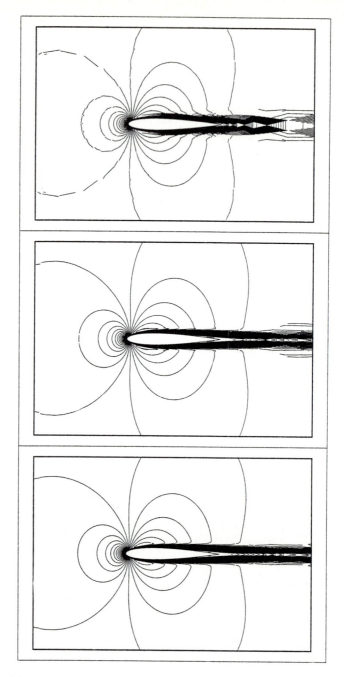

Fig. 6.4. Mach isolines around the NACA0012 airfoil, ($Re = 5000, M = 0.5$, zero angle of attack) for the linear (top), quadratic (middle), and cubic (bottom) elements.

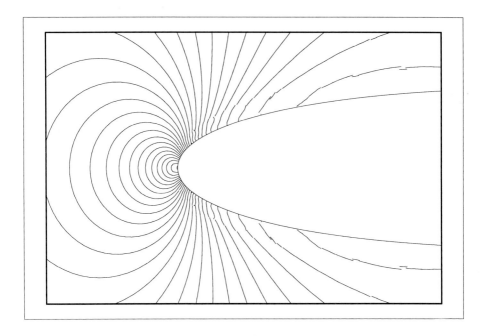

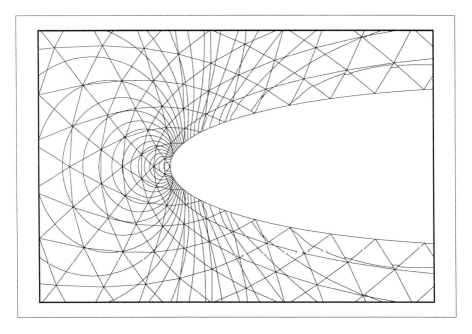

Fig. 6.5. Pressure isolines around the NACA0012 airfoil, ($Re = 5000, M = 0.5$, zero angle of attack) for the for cubic elements without (top) and with (bottom) the corresponding grid.

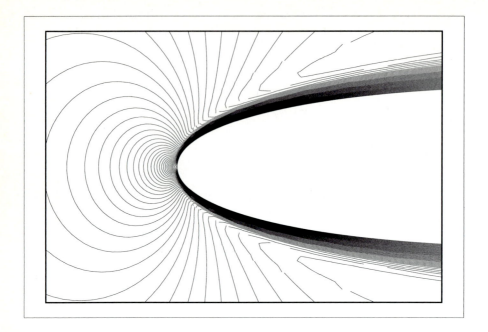

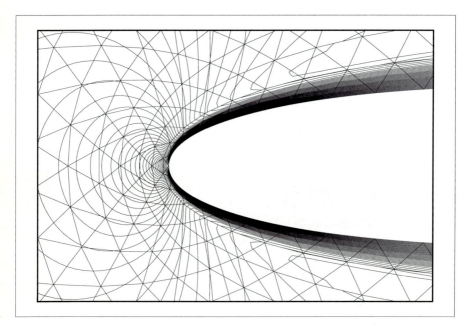

Fig. 6.6. Mach isolines around the leading edge of the NACA0012 airfoil, ($Re = 5000$, $M = 0.5$, zero angle of attack) for the for cubic elements without (top) and with (bottom) the corresponding grid.

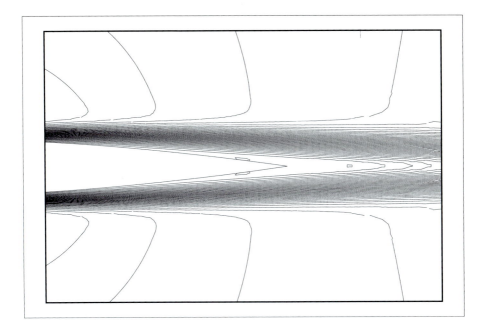

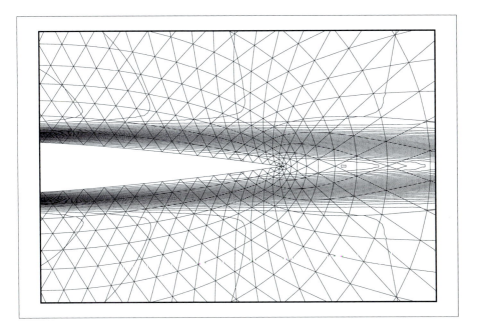

Fig. 6.7. Mach isolines around the trailing edge of the NACA0012 airfoil, ($Re = 5000$, $M = 0.5$, zero angle of attack) for the for cubic elements without (top) and with (bottom) the corresponding grid.

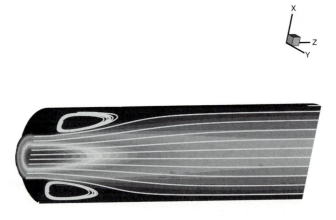

Fig. 6.8. Expansion pipe flow at Reynolds number 50 and Mach number 0.2. Velocity profile and streamlines computed with a mesh of 600 elements and polynomials of degree 5.

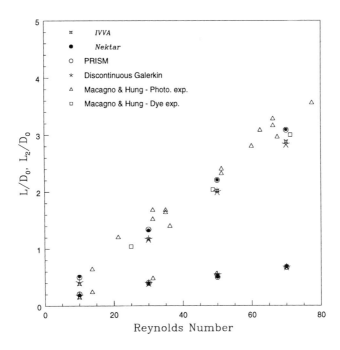

Fig. 6.9. Expansion pipe flow: Comparison between computational and experimental results.

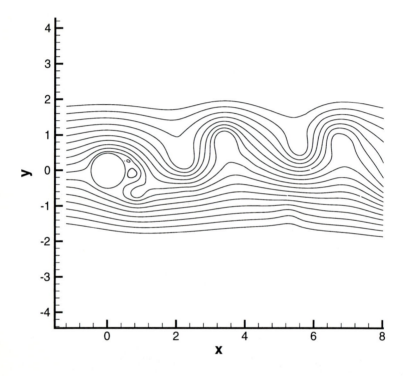

Fig. 6.10. Flow around a cylinder with Reynolds number $10,000$ and Mach number 0.2. Streamlines. A mesh of 680 elements was used with polynomials that could change degree from element to element; the maximum degree was 5.

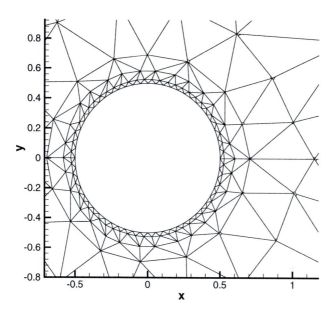

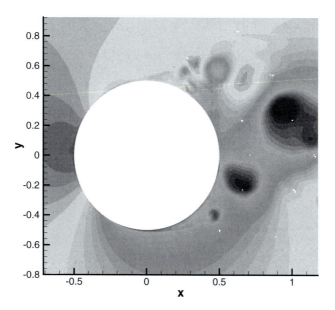

Fig. 6.11. Flow around a cylinder with Reynolds number 10,000 and Mach number 0.2. Detail of the mesh (top) and density (bottom) around the cylinder.

6.7   Appendix: Proof of the L²-error estimates

Proof of Proposition 26 In this section, we prove the the nonlinear stability result of Proposition 26. To do that, we first show how to obtain the corresponding stability result for the exact solution and then mimic the argument to obtain Proposition 26.

The continuous case as a model. We start by rewriting the equations (6.10) and (6.11), in *compact form*. If in equations (6.10) and (6.11) we replace $v_u(x)$ and $v_q(x)$ by $v_u(x,t)$ and $v_q(x,t)$, respectively, add the resulting equations, sum on j from 1 to N, and integrate in time from 0 to T, we obtain that

$$B(\mathbf{w}, \mathbf{v}) = 0, \qquad \forall \text{ smooth } \mathbf{v}, \tag{6.35}$$

where

$$B(\mathbf{w}, \mathbf{v}) = \int_0^T \int_0^1 \partial_t u(x,t)\, v_u(x,t)\, dx\, dt \tag{6.36}$$

$$+ \int_0^T \int_0^1 q(x,t)\, v_q(x,t)\, dx\, dt$$

$$- \int_0^T \int_0^1 \mathbf{h}(\mathbf{w}(x,t))^t\, \partial_x\, \mathbf{v}(x,t)\, dx\, dt.$$

Note that if we use the fact that

$$\mathbf{h}(\mathbf{w}(x,t))^t\, \partial_x \mathbf{w}(x,t) = \partial_x(\, \phi(u) - g(u)\, q\,)$$

is a complete derivative, we see that

$$B(\mathbf{w}, \mathbf{w}) = \frac{1}{2} \int_0^1 u^2(x,T)\, dx + \int_0^T \int_0^1 q^2(x,t)\, dx\, dt$$

$$- \frac{1}{2} \int_0^1 u_0^2(x)\, dx, \tag{6.37}$$

and that $B(\mathbf{w}, \mathbf{w}) = 0$, by (6.35). As a consequence, we immediately obtain the following L²-stability result:

$$\frac{1}{2} \int_0^1 u^2(x,T)\, dx + \int_0^T \int_0^1 q^2(x,t)\, dx\, dt = \tfrac{1}{2} \int_0^1 u_0^2(x)\, dx.$$

This is the argument we have to mimic in order to prove Proposition 26.

The discrete case. Thus, we start by finding a *compact form* of equations (6.13) and (6.14). If we replace $v_{h,u}(x)$ and $v_{h,q}(x)$ by $v_{h,u}(x,t)$ and $v_{h,q}(x,t)$ in the equations (6.13) and (6.14), add them up, sum on j from 1 to N and integrate in time from 0 to T, we obtain

$$B_h(\mathbf{w}_h, \mathbf{v}_h) = 0, \tag{6.38}$$

$$\forall \mathbf{v}_h(t) \in V_h^k \times V_h^k, \quad \forall t \in (0,T).$$

where

$$B_h(\mathbf{w}_h, \mathbf{v}_h) = \int_0^T \int_0^1 \partial_t u_h(x,t)\, v_{h,u}(x,t)\, dx\, dt$$

$$+ \int_0^T \int_0^1 q_h(x,t)\, v_{h,q}(x,t)\, dx\, dt$$

$$- \int_0^T \sum_{1 \le j \le N} \hat{\mathbf{h}}(\mathbf{w}_h)^t_{j+1/2}(t)\, [\,\mathbf{v}_h(t)\,]_{j+1/2}\, dt$$

$$- \int_0^T \sum_{1 \le j \le N} \int_{I_j} \mathbf{h}(\mathbf{w}_h(x,t))^t\, \partial_x \mathbf{v}_h(x,t)\, dx\, dt. \qquad (6.39)$$

Next, we obtain an expression for $B_h(\mathbf{w}_h, \mathbf{w}_h)$. It is contained in the following result.

Lemma 30. *We have*

$$B_h(\mathbf{w}_h, \mathbf{w}_h) = \frac{1}{2} \int_0^1 u_h^2(x,T)\, dx$$

$$+ \int_0^T \int_0^1 q_h^2(x,t)\, dx\, dt + \Theta_{T,C}([\mathbf{w}_h])$$

$$- \frac{1}{2} \int_0^1 u_h^2(x,0)\, dx,$$

where $\Theta_{T,C}([\mathbf{w}_h])$ is defined in Proposition 26.

Next, since $B_h(\mathbf{w}_h, \mathbf{w}_h) = 0$, by (6.38), we get the equality

$$\frac{1}{2} \int_0^1 u_h^2(x,T)\, dx + \int_0^T \int_0^1 q_h^2(x,t)\, dx\, dt + \Theta_{T,C}([\mathbf{w}_h]) = \frac{1}{2} \int_0^1 u_h^2(x,0)\, dx$$

from which Proposition 26 easily follows since

$$\tfrac{1}{2} \int_0^1 u_h^2(x,0)\, dx \le \tfrac{1}{2} \int_0^1 u_0^2(x)\, dx,$$

by (6.12). It remains to prove Lemma 30.

Proof of Lemma (30). After setting $\mathbf{v}_h = \mathbf{w}_h$ in (6.39), we get

$$B(\mathbf{w}_h, \mathbf{w}_h) = \frac{1}{2} \int_0^1 u_h^2(x,T)\, dx + \int_0^T \int_0^1 q_h^2(x,t)\, dx\, dt$$

$$+ \int_0^T \Theta_{diss}(t)\, dt - \frac{1}{2} \int_0^1 u_h^2(x,0)\, dx,$$

where $\Theta_{diss}(t)$ is given by

$$- \sum_{1 \le j \le N} \left\{ \hat{\mathbf{h}}(\mathbf{w}_h)^t_{j+1/2}(t)\, [\,\mathbf{w}_h(t)\,]_{j+1/2} + \int_{I_j} \mathbf{h}(\mathbf{w}_h(x,t))^t\, \partial_x \mathbf{w}_h(x,t)\, dx \right\}.$$

It only remains to show that

$$\int_0^T \Theta_{diss}(t)\, dt = \Theta_{T,C}([\mathbf{w}_h]).$$

To do that, we proceed as follows. Since

$$
\begin{aligned}
\mathbf{h}(\mathbf{w}_h(x,t))^t\, \partial_x\, \mathbf{w}_h(x,t)b &= \big(f(u_h) - \sqrt{a(u_h)}\, q_h\big)\, \partial_x\, u_h - g(u_h)\, \partial_x\, q_h \\
&= \partial_x \left(\int^{u_h} f(s)\, ds - g(u_h)\, q_h \right) \\
&= \partial_x \left(\phi(u_h) - g(u_h)\, q_h \right) \\
&\equiv \partial_x\, H(\mathbf{w}_h(x,t)),
\end{aligned}
$$

we get

$$
\begin{aligned}
\Theta_{diss}(t) &= \sum_{1 \le j \le N} \left\{ [\, H(\mathbf{w}_h(t))\,]_{j+1/2} - \hat{\mathbf{h}}(\mathbf{w}_h)^t_{j+1/2}(t)\, [\, \mathbf{w}_h(t)\,]_{j+1/2} \right\} \\
&\equiv \sum_{1 \le j \le N} \left\{ [\, H(\mathbf{w}_h(t))\,] - \hat{\mathbf{h}}(\mathbf{w}_h)^t(t)\, [\, \mathbf{w}_h(t)\,] \right\}_{j+1/2}
\end{aligned}
$$

Since, by the definition of H,

$$
\begin{aligned}
[\, H(\mathbf{w}_h(t))\,] &= [\, \phi(u_h(t))\,] - [\, g(u_h(t))\, q_h(t)\,] \\
&= [\, \phi(u_h(t))\,] - [\, g(u_h(t))\,]\, \overline{q}_h(t) - [\, q_h(t)\,]\, \overline{g(u_h(t))},
\end{aligned}
$$

and since $(\hat{h}_u, \hat{h}_q)^t = \hat{\mathbf{h}}$, we get

$$
\begin{aligned}
\Theta_{diss}(t) \\
&= \sum_{1 \le j \le N} \left\{ [\, \phi(u_h(t))\,] - [\, g(u_h(t))\,]\, \overline{q}_h(t) - [\, u_h(t)\,]\, \hat{h}_u \right\}_{j+1/2} \\
&+ \sum_{1 \le j \le N} \left\{ -[\, q_h(t)\,]\, \overline{g(u_h)(t)} - [\, q_h(t)\,]\, \hat{h}_q \right\}_{j+1/2},
\end{aligned}
$$

This is the crucial step to obtain the L^2-stability of the LDG methods, since the above expression gives us key information about the form that the flux $\hat{\mathbf{h}}$ should have in order to make $\Theta_{diss}(t)$ a nonnegative quantity and hence enforce the L²-stability of the LDG methods. Thus, by taking $\hat{\mathbf{h}}$ as in (6.16), we get

$$\Theta_{diss}(t) = \sum_{1 \le j \le N} \left\{ [\mathbf{w}_h(t)]^t C\, [\mathbf{w}_h(t)] \right\}_{j+1/2},$$

and the result follows. This completes the proof.

This completes the proof of Proposition 26.

Proof of Theorem 27 In this section, we prove the error estimate of Theorem 27 which holds for the linear case $f'(\cdot) \equiv c$ and $a(\cdot) \equiv a$. To do that, we first show how to estimate the error between the solutions $\mathbf{w}_\nu = (u_\nu, q_\nu)^t$, $\nu = 1, 2$, of

$$\partial_t u_\nu + \partial_x \left(f(u_\nu) - \sqrt{a(u_\nu)}\, q_\nu \right) = 0 \quad \text{in } (0, T) \times (0, 1),$$
$$q_\nu - \partial_x g(u_\nu) = 0 \quad \text{in } (0, T) \times (0, 1),$$
$$u_\nu(t = 0) = u_{0,\nu}, \quad \text{on } (0, 1).$$

Then, we mimic the argument in order to prove Theorem 27.

The continuous case as a model. By the definition of the form $B(\cdot, \cdot)$, (6.36), we have, for $\nu = 1, 2$,

$$B(\mathbf{w}_\nu, \mathbf{v}) = 0, \qquad \forall \text{ smooth } \mathbf{v}(t), \quad \forall t \in (0, T).$$

Since in this case, the form $B(\cdot, \cdot)$ is bilinear, from the above equation we obtain the so-called *error equation*:

$$B(\mathbf{e}, \mathbf{v}) = 0, \qquad \forall \text{ smooth } \mathbf{v}(t), \quad \forall t \in (0, T),$$

where $\mathbf{e} = \mathbf{w}_1 - \mathbf{w}_2$. Now, from (6.37), we get that

$$B(\mathbf{e}, \mathbf{e}) = \frac{1}{2} \int_0^1 e_u^2(x, T)\, dx + \int_0^T \int_0^1 e_q^2(x, t)\, dx\, dt - \frac{1}{2} \int_0^1 e_u^2(x, 0)\, dx,$$

and since $e_u(x, 0) = u_{0,1}(x) - u_{0,2}(x)$ and $B(\mathbf{e}, \mathbf{e}) = 0$, by the *error equation*, we immediately obtain the error estimate we sought:

$$\frac{1}{2} \int_0^1 e_u^2(x, T)\, dx + \int_0^T \int_0^1 e_q^2(x, t)\, dx\, dt = \frac{1}{2} \int_0^1 \left(u_{0,1}(x) - u_{0,2}(x) \right)^2 dx \ . \tag{6.40}$$

To prove Theorem 27, we only need to obtain a discrete version of this argument.

The discrete case. Since,

$$B_h(\mathbf{w}_h, \mathbf{v}_h) = 0, \qquad \forall \, \mathbf{v}_h(t) \in V_h \times V_h, \quad \forall t \in (0, T),$$
$$B_h(\mathbf{w}, \mathbf{v}_h) = 0, \qquad \forall \, \mathbf{v}_h(t) \in V_h \times V_h, \quad \forall t \in (0, T),$$

by (6.38) and by equations (6.10) and (6.11), respectively, we immediately obtain our *error equation*:

$$B_h(\mathbf{e}, \mathbf{v}_h) = 0, \qquad \forall \, \mathbf{v}_h(t) \in V_h \times V_h, \quad \forall t \in (0, T),$$

where $\mathbf{e} = \mathbf{w} - \mathbf{w}_h$. Now, according to the continuous case argument, we should consider next the quantity $B_h(\mathbf{e}, \mathbf{e})$; however, since $\mathbf{e}$ is not in the finite element space, it is more convenient to consider $B_h(P_h(\mathbf{e}), P_h(\mathbf{e}))$, where

$$P_h(\mathbf{e}(t)) = \left(P_h(e_u(t)), P_h(e_q(t)) \right)$$

is the so-called L^2-projection of $\mathbf{e}(t)$ into the finite element space $V_h^k \times V_h^k$. The L^2-projection of the function p into V_h, $P_h(p)$, is defined as the only element of the finite element space V_h such that

$$\forall\, v_h \in V_h : \qquad \int_0^1 \big(\, P_h(p)(x) - p(x)\,\big)\, v_h(x)\, dx = 0. \qquad (6.41)$$

Note that, in fact $u_h(t = 0) = P_h(u_0)$, by (6.15).

Thus, by Lemma 30, we have

$$B_h(P_h(\mathbf{e}), P_h(\mathbf{e})) = \frac{1}{2}\int_0^1 |\,P_h(e_u(T))(x)\,|^2\, dx$$
$$+ \int_0^T \int_0^1 |\,P_h(e_q(t))(x)\,|^2\, dx\, dt$$
$$+ \Theta_{T,C}([P_h(\mathbf{e})])$$
$$- \frac{1}{2}\int_0^1 |\,P_h(e_u(0))(x)\,|^2\, dx,$$

and since

$$P_h(e_u(0)) = P_h(u_0 - u_h(0)) = P_h(u_0) - u_h(0) = 0,$$

by (6.15) and (6.41), and

$$B_h(P_h(\mathbf{e}), P_h(\mathbf{e})) = B_h(P_h(\mathbf{e}) - \mathbf{e}, P_h(\mathbf{e})) = B_h(P_h(\mathbf{w}) - \mathbf{w}, P_h(\mathbf{e})),$$

by the *error equation*, we get

$$\frac{1}{2}\int_0^1 |\,P_h(e_u(T))(x)\,|^2\, dx + \int_0^T \int_0^1 |\,P_h(e_q(t))(x)\,|^2\, dx\, dt + \Theta_{T,C}([P_h(\mathbf{e})])$$
$$= B_h(P_h(\mathbf{w}) - \mathbf{w}, P_h(\mathbf{e})). \qquad (6.42)$$

Note that since in our continuous model, the right-hand side is zero, we expect the term $B(P_h(\mathbf{w}) - \mathbf{w}, P_h(\mathbf{e}))$ to be small.

Estimating the right-hand side. To show that this is so, we must suitably treat the term $B(P_h(\mathbf{w}) - \mathbf{w}, P_h(\mathbf{e}))$.

Lemma 31. *For* $\mathbf{p} = P_h(\mathbf{w}) - \mathbf{w}$, *we have*

$$B_h(\mathbf{p}, P_h(\mathbf{e})) = \frac{1}{2}\Theta_{T,C}(\overline{\mathbf{p}}) + \frac{1}{2}\int_0^T \int_0^1 |\,P_h(e_q(t))(x)\,|^2\, dx\, dt$$
$$+ \frac{1}{2}(\Delta x)^{2k}\int_0^T C_1(t)\, dt$$
$$+ (\Delta x)^k \int_0^T C_2(t)\left\{\int_0^1 |\,P_h(e_u(t))(x)\,|^2\, dx\right\}^{1/2} dt,$$

where

$$C_1(t) = 2\,c_k^2 \left\{ \left(\frac{(|c| + c_{11})^2}{c_{11}} \, \Delta x + 4\,|\,c_{12}\,|^2\, d_k^2 \right) |\,u(t)\,|_{k+1}^2 \right.$$

$$\left. +4\,a\,d_k^2\,(\Delta x)^{2\,(\hat{k}-k)} |\,u(t)\,|_{\hat{k}+1}^2 \right\},$$

$$C_2(t) = \sqrt{8}\,c_k\,d_k \left\{ \sqrt{a}\,|\,c_{12}\,|\,u(t)\,|_{k+2} \right.$$

$$\left. +a\,(\Delta x)^{(\hat{k}-k)} |\,u(t)\,|_{\hat{k}+2} \right\}.$$

where the constants c_k and d_k depend solely on k, and $\hat{k} = k$ except when the grids are uniform and k is even, in which case $\hat{k} = k + 1$.

Note how c_{11} appears in the denominator of $C_1(t)$. However, $C_1(t)$ remains bounded as c_{11} goes to zero since the convective numerical flux is an E-flux.

To prove this result, we will need the following auxiliary lemmas. We denote by $|\,u\,|_{H^{(k+1)}(J)}^2$ the integral over J of the square of the $(k + 1)$-the derivative of u.

Lemma 32. *For* $\mathbf{p} = P_h(\mathbf{w}) - \mathbf{w}$, *we have*

$$|\,\overline{p_u}_{j+1/2}\,| \le c_k\,(\Delta x)^{\hat{k}+1/2}\,|\,u\,|_{H^{(\hat{k}+1)}(J_{j+1/2})},$$

$$|\,[p_u]_{j+1/2}\,| \le c_k\,(\Delta x)^{k+1/2}\,|\,u\,|_{H^{(k+1)}(J_{j+1/2})},$$

$$|\,\overline{p_q}_{j+1/2}\,| \le c_k\,\sqrt{a}\,(\Delta x)^{\hat{k}+1/2}\,|\,u\,|_{H^{(\hat{k}+2)}(J_{j+1/2})},$$

$$|\,[p_q]_{j+1/2}\,| \le c_k\,\sqrt{a}\,(\Delta x)^{k+1/2}\,|\,u\,|_{H^{(k+2)}(J_{j+1/2})},$$

where $J_{j+1/2} = I_j \cup I_{j+1}$, the constant c_k depends solely on k, and $\hat{k} = k$ except when the grids are uniform and k is even, in which case $\hat{k} = k + 1$.

Proof. The two last inequalities follow from the first two and from the fact that $q = \sqrt{a}\,\partial_x u$. The two first inequalities with $\hat{k} = k$ follow from the definitions of $\overline{p_u}$ and $[p_u]$ and from the following estimate:

$$|\,P_h(u)(x_{j+1/2}^{\pm}) - u_{j+1/2}\,| \le \frac{1}{2} c_k\,(\Delta x)^{k+1/2}\,|\,u\,|_{H^{(k+1)}(J_{j+1/2})},$$

where the constant c_k depends solely on k. This inequality follows from the fact that

$$P_h(u)(x_{j+1/2}^{\pm}) - u_{j+1/2} = 0$$

when u is a polynomial of degree k and from a simple application of the Bramble-Hilbert lemma.

To prove the inequalities in the case in which $\hat{k} = k + 1$, we only need to show that if u is a polynomial of degree $k + 1$ for k even, then $\overline{p_u} = 0$. It is clear that it is enough to show this equality for the particular choice

$$u(x) = \left((x - x_{j+1/2})/(\Delta x/2) \right)^{k+1}.$$

To prove this, we recall that if P_ℓ denotes the Legendre polynomials of order ℓ:

(i) $\int_{-1}^{1} P_\ell(s) P_m(s)\, ds = \frac{2}{2\ell+1} \delta_{\ell m}$,

(ii) $P_\ell(\pm 1) = (\pm 1)^\ell$, and

(iii) $P_\ell(s)$ is a linear combination of odd (even) powers of s for odd (even) values of ℓ.

Since we are assuming that the grid is uniform, $\Delta x_j = \Delta x_{j+1} = \Delta x$, we can write, by (i), that

$$P_h(u)(x) = \sum_{0 \le \ell \le k} \frac{2\ell + 1}{2} \left\{ \int_{-1}^{1} P_\ell(s)\, u(x_j + \frac{1}{2}\Delta x\, s)\, ds \right\} P_\ell \left(\frac{x - x_j}{\Delta x/2} \right),$$

for $x \in I_j$. Hence, for our particular choice of u, we have that the value of $\overline{p_u}_{j+1/2}$ is given by

$$\frac{1}{2} \sum_{0 \le \ell \le k} \frac{2\ell + 1}{2} \int_{-1}^{1} P_\ell(s) \cdot \left\{ (s - 1)^{k+1} P_\ell(1) + (s + 1)^{k+1} P_\ell(-1) \right\} ds$$

$$= \frac{1}{2} \sum_{0 \le \ell, i \le k} \frac{2\ell + 1}{2} \binom{k+1}{i} \int_{-1}^{1} P_\ell(s)\, s^i \left\{ (-1)^{k+1-i} P_\ell(1) + P_\ell(-1) \right\} ds$$

$$= \frac{1}{2} \sum_{0 \le \ell, i \le k} \frac{2\ell + 1}{2} \binom{k+1}{i} \int_{-1}^{1} P_\ell(s)\, s^i \left\{ (-1)^{k+1-i} + (-1)^\ell \right\} ds,$$

by (ii). When the factor $\left\{ (-1)^{k+1-i} + (-1)^\ell \right\}$ is different from zero, $|\,k + 1 - i + \ell\,|$ is even and since k is also even, $|\,i - \ell\,|$ is odd. In this case, by (iii),

$$\int_{-1}^{1} P_\ell(s)\, s^i\, ds = 0,$$

and so $\overline{p_u}_{j+1/2} = 0$. This completes the proof.

We will also need the following result that follows from a simple scaling argument.

Lemma 33. *We have*

$$|\,[P_h(p)]_{j+1/2}\,| \le d_k\, (\Delta x)^{-1/2} \,\| P_h(p) \|_{L^2(J_{j+1/2})},$$

where $J_{j+1/2} = I_j \cup I_{j+1}$ and the constant d_k depends solely on k.

We are now ready to prove Lemma 31.

Proof of Lemma 31. To simplify the notation, let us set $\mathbf{v}_h = P_h\mathbf{e}$. By the definition of $B_h(\cdot, \cdot)$, we have

$$
\begin{aligned}
B_h(\mathbf{p}, \mathbf{v}_h) &= \int_0^T \int_0^1 \partial_t p_u(x, t)\, v_{h,u}(x, t)\, dx\, dt \\
&+ \int_0^T \int_0^1 p_q(x, t)\, v_{h,q}(x, t)\, dx\, dt \\
&- \int_0^T \sum_{1 \le j \le N} \hat{\mathbf{h}}(\mathbf{p})^t_{j+1/2}(t)\, [\,\mathbf{v}_h(t)\,]_{j+1/2}\, dt \\
&- \int_0^T \sum_{1 \le j \le N} \int_{I_j} \mathbf{h}(\mathbf{p}(x, t))^t\, \partial_x\, \mathbf{v}_h(x, t)\, dx\, dt \\
&= - \int_0^T \sum_{1 \le j \le N} \hat{\mathbf{h}}(\mathbf{p})^t_{j+1/2}(t)\, [\,\mathbf{v}_h(t)\,]_{j+1/2}\, dt,
\end{aligned}
$$

by the definition of the L^2-projection (6.41).

Now, recalling that $\mathbf{p} = (p_u, p_q)^t$ and that $\mathbf{v}_h = (v_u, v_q)^t$, we have

$$
\begin{aligned}
\hat{\mathbf{h}}(\mathbf{p})^t\, [\,\mathbf{v}_h(t)\,] &= (c\,\overline{p_u} - c_{11}\,[\,p_u\,])\,[\,v_u\,] \\
&+ (-\sqrt{a}\,\overline{p_q} - c_{12}\,[\,p_q\,])\,[\,v_u\,] \\
&+ (-\sqrt{a}\,\overline{p_u} + c_{12}\,[\,p_u\,])\,[\,v_q\,] \\
&\equiv \theta_1 + \theta_2 + \theta_3.
\end{aligned}
$$

By Lemmas 32 and 33, and writing J instead $J_{j+1/2}$, we get

$$
\begin{aligned}
|\,\theta_1\,| &\le c_k\,(\Delta x)^{k+1/2}\,|\,u\,|_{H^{k+1}(J)}\,(|\,c\,| + c_{11})\,|\,[\,v_u\,]\,|, \\
|\,\theta_2\,| &\le c_k\,d_k\,(\Delta x)^k\,\big(a\,|\,u\,|_{H^{k+2}(J)}\,(\Delta x)^{k-k} \\
&\quad + \sqrt{a}\,|\,c_{12}\,|\,|\,u\,|_{H^{k+2}(J)}\big)\,\|\,v_u\,\|_{L^2(J)}, \\
|\,\theta_3\,| &\le c_k\,d_k\,(\Delta x)^k\,\big(\sqrt{a}\,|\,u\,|_{H^{k+1}(J)}\,(\Delta x)^{k-k} \\
&\quad + |\,c_{12}\,|\,|\,u\,|_{H^{k+1}(J)}\big)\,\|\,v_q\,\|_{L^2(J)}.
\end{aligned}
$$

This is the crucial step for obtaining our error estimates. Note that the treatment of θ_1 is very different than the treatment of θ_2 and θ_3. The reason for this difference is that the upper bound for θ_1 can be controlled by the form $\Theta_{T,C}([\,\mathbf{v}_h\,])$- we recall that $\mathbf{v}_h = P_h(\mathbf{e})$. This is not the case for the upper bound for θ_2 because $\Theta_{T,C}[\,\mathbf{v}_h\,] \equiv 0$ if $c = 0$ nor it is the case for the upper bound for θ_3 because $\Theta_{T,C}[\,\mathbf{v}_h\,]$ does not involve the jumps $[\,v_q\,]$!

Thus, after a suitable application of Young's inequality and simple algebraic manipulations, we get

$$\hat{\mathbf{h}}(\mathbf{p})^t\,[\,\mathbf{v}_h(t)\,] \le \frac{1}{2}\,c_{11}\,[\,v_u\,]^2 + \frac{1}{4}\|\,v_q\,\|_{L^2(J)}^2$$
$$+ \frac{1}{4}\,C_{1,J}(t)\,(\Delta x)^{2k} + C_{2,J}(t)\,(\Delta x)^k\,\|\,v_u\,\|_{L^2(J)},$$

where

$$C_{1,J}(t) = c_k^2\,\Big(\frac{(|\,c\,|+c_{11})^2}{c_{11}}\,\Delta x + 4\,|\,c_{12}\,|^2\,d_k^2\,\Big)\,|\,u(t)\,|_{H^{k+1}(J)}^2$$
$$+\,4\,a\,c_k^2\,d_k^2\,(\Delta x)^{2\,(\hat{k}-k)}\,|\,u(t)\,|_{H^{\hat{k}+1}(J)}^2,$$

and

$$C_{2,J}(t) = c_k\,d_k\Big\{\,\sqrt{a}\,|\,c_{12}\,|\,|\,u(t)\,|_{H^{k+2}(J)} + a\,(\Delta x)^{(\hat{k}-k)}\,|\,u(t)\,|_{H^{\hat{k}+2}(J)}\,\Big\}.$$

Since

$$B_h(\mathbf{p},\mathbf{v}_h) \le \int_0^T\sum_{1\le j\le N}\,\big|\,\hat{\mathbf{h}}(\mathbf{p})_{j+1/2}^t(t)\,[\,\mathbf{v}_h(t)\,]_{j+1/2}\,\big|\,dt,$$

and since $J_{j+1/2} = I_j \cup I_{j+1}$, the result follows after simple applications of the Cauchy-Schwartz inequality. This completes the proof.

Conclusion. Combining the equation (6.42) with the estimate of Lemma 31, we easily obtain, after a simple application of Gronwall's lemma,

$$\Big\{\int_0^1|\,P_h(e_u(T))(x)\,|^2\,dx + \int_0^T\!\!\int_0^1|\,P_h(e_q(t))(x)\,|^2\,dx\,dt + \Theta_{T,C}([P_h(\mathbf{e})])\Big\}^{1/2}$$
$$\le (\Delta x)^k\,\Big\{\sqrt{\int_0^T C_1(t)\,dt} + \int_0^T C_2(t)\,dt\Big\}.$$

Theorem 27 follows easily from this inequality, Lemma 33, and from the following simple approximation result:

$$\|\,p - P_h(p)\,\|_{L^2(0,1)} \le g_k\,(\Delta x)^{k+1}\,|\,p\,|_{H^{(k+1)}(0,1)}$$

where g_k depends solely on k.

7   The LDG method for other nonlinear parabolic problems: Propagating surfaces

7.1   Introduction

In this chapter, we briefly show how to extend the LDG method to nonlinear second-order parabolic equations. We consider the following model problem:

$$\varphi_t + F(D\varphi, D^2\varphi) = 0, \quad \text{in } (0,1)^d \times (0,T),$$
$$\varphi(x,0) = \varphi_0(x), \qquad \forall\ (x) \in (0,1)^d,$$

where we take periodic boundary conditions and assume that F is nonincreasing in the second variable. For the definition and properties of the *viscosity solution* of this and more general problems of this type, see the work by Crandall, Ishii, and Lions [29].

For simplicity, we only consider the two-dimensional case, $d = 2$:

$$\varphi_t + F(\varphi_x, \varphi_y, \varphi_{xx}, \varphi_{xy}, \varphi_{yy}) = 0, \quad \text{in } (0,1)^2 \times (0,T), \tag{7.1}$$
$$\varphi(x,0) = \varphi_0(x), \qquad \forall\ (x) \in (0,1)^2, \tag{7.2}$$

with periodic boundary conditions. The material presented in this section is based in the work of Hu and Shu [43].

7.2   The method

To idea to extend the LDG method to this case, is to rewrite the problem (7.1), (7.2) for φ as follows:

$$\varphi_t = -F(u,v,p,q,r), \qquad \text{in } (0,1) \times (0,T), \tag{7.3}$$
$$\varphi(x,0) = \varphi_0(x), \qquad \forall\ x \in (0,1). \tag{7.4}$$

where (u,v,p,qp,r) solves the following problem:

$$u_t + F(u,v,p,q,r)_x = 0, \qquad \text{in } (0,1)^2 \times (0,T), \tag{7.5}$$
$$v_t + H(u,v,p,q,r)_y = 0, \qquad \text{in } (0,1)^2 \times (0,T), \tag{7.6}$$
$$p - u_x = 0, \qquad \text{in } (0,1)^2 \times (0,T), \tag{7.7}$$
$$q - u_y = 0, \qquad \text{in } (0,1)^2 \times (0,T), \tag{7.8}$$
$$r - v_x = 0, \qquad \text{in } (0,1)^2 \times (0,T), \tag{7.9}$$
$$u(x,y,0) = (\varphi_0)_x(x,y), \qquad \forall\ (x,y) \in (0,1)^2, \tag{7.10}$$
$$v(x,y,0) = (\varphi_0)_y(x,y), \qquad \forall\ (x,y) \in (0,1)^2. \tag{7.11}$$

Again, a straightforward application of the LDG method to the above problem produces an approximation $(u_h, v_h, p_h, q_h, r_h)$ to (u,v,p,qp,r). We can take each of the approximate solutions to be piecewise a polynomial of degree $k-1$. Then, we define the approximation φ_h to ϕ by solving the problem (7.3), (7.4) in the manner described in the chapter on RKDG methods for multidimensional Hamilton-Jacobi equations.

7.3 Computational results

We present a couple of numerical results that display the good performance of the method. Our main purpose is to show that the method works well if both quadrangles and triangles are used.

First test problem. We consider the problem of a propagating surface:

$$\begin{cases} \varphi_t - (1 - \varepsilon K)\sqrt{1 + \varphi_x^2 + \varphi_y^2} = 0, & 0 < x < 1, 0 < y < 1 \\ \varphi(x,y,0) = 1 - \frac{1}{4}(\cos(2\pi x - 1))(\cos(2\pi y - 1)) \end{cases} \qquad (7.12)$$

where K is the mean curvature defined by

$$K = -\frac{\varphi_{xx}(1 + \varphi_y^2) - 2\varphi_{xy}\varphi_x\varphi_y + \varphi_{yy}(1 + \varphi_x^2)}{(1 + \varphi_x^2 + \varphi_y^2)^{\frac{3}{2}}}, \qquad (7.13)$$

and ε is a small constant. Periodic boundary condition is used.

This problem was studied in [72] by using the finite difference ENO schemes.

We first use a uniform rectangular mesh of 50×50 elements and the local Lax-Friedrichs flux. The results of $\varepsilon = 0$ (pure convection) and $\varepsilon = 0.1$ are presented in Fig. 7.1 and Fig. 7.2, respectively. Notice that the surface at $T = 0$ is shifted downward by 0.35 in order to show the detail of the solution at $T = 0.3$.

Next we use a triangulation shown in Fig. 7.3. We refine the mesh around the center of domain where the solution develops discontinuous derivatives (for the $\varepsilon = 0$ case). There are 2146 triangles and 1128 nodes in this triangulation. The solutions are displayed in Fig. 7.4 and Fig. 7.5, respectively, for $\varepsilon = 0$ (pure convection) and $\varepsilon = 0.1$. Notice that we again shift the solution at $T = 0.0$ downward by 0.35 to show the detail of the solutions at later time.

Second test problem. The problem of a propagating surface on a unit disk. The equation is the same as (7.12) in the previous example, but it is solved on a unit disk $x^2 + y^2 < 1$ with an initial condition

$$\varphi(x,y,0) = \sin\left(\frac{\pi(x^2 + y^2)}{2}\right)$$

and a Neumann type boundary condition $\nabla\varphi = 0$.

It is difficult to use rectangular meshes for this problem. Instead we use the triangulation shown in Fig. 7.6. Notice that we have again refined the mesh near the center of the domain where the solution develops discontinuous derivatives. There are 1792 triangles and 922 nodes in this triangulation. The solutions with $\varepsilon = 0$ are displayed in Fig. 7.7. Notice that the solution at $t = 0$ is shifted downward by 0.2 to show the detail of the solution at later time.

The solution with $\varepsilon = 0.1$ are displayed in Fig. 7.8. Notice that the solution at $t = 0$ is again shifted downward by 0.2 to show the detail of the solution at later time.

P², 50x50 elements   **P³, 50x50 elements**

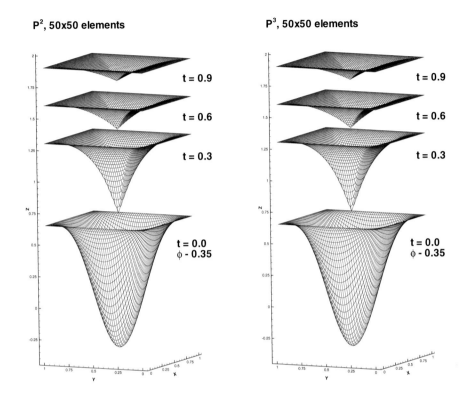

Fig. 7.1. Propagating surfaces, rectangular mesh, $\varepsilon = 0$.

7.4   Concluding remarks

We have shown, briefly, how to extend the LDG method originally devised for nonlinear convection-diffusion equations to second-order parabolic equations that have a viscosity solution. We have shown that the method works well without slope limiting and that it works well in both quadrangles and triangles.

P², 50x50 elements   **P³, 50x50 elements**

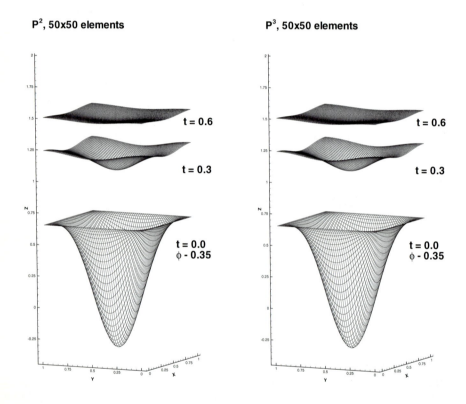

Fig. 7.2. Propagating surfaces, rectangular mesh, $\varepsilon = 0.1$.

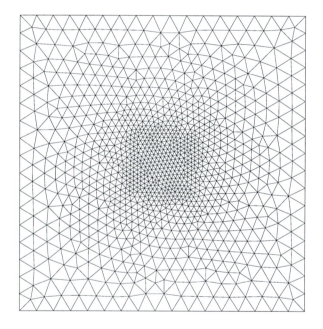

Fig. 7.3. Triangulation used for the propagating surfaces.

P², triangles   **P³, triangles**

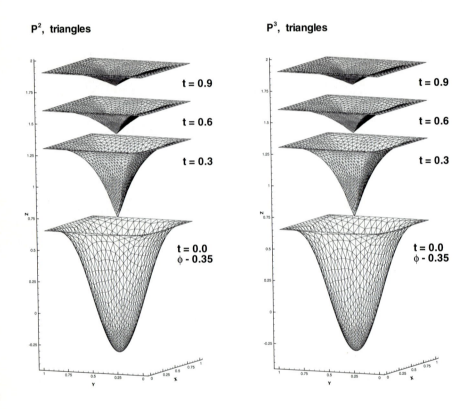

Fig. 7.4. Propagating surfaces, triangular mesh, $\varepsilon = 0$.

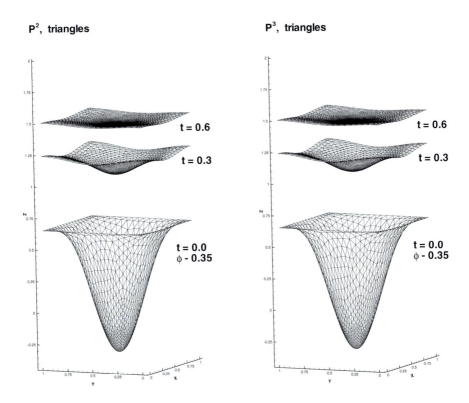

Fig. 7.5. Propagating surfaces, triangular mesh, $\varepsilon = 0.1$.

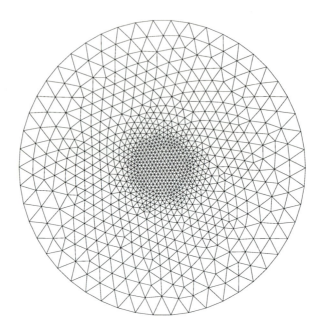

Fig. 7.6. Triangulation for the propagating surfaces on a disk.

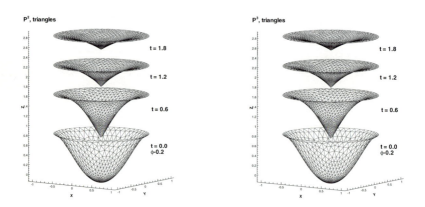

Fig. 7.7. Propagating surfaces on a disk, triangular mesh, $\varepsilon = 0$.

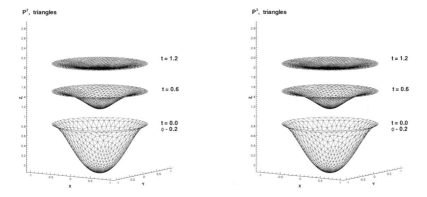

Fig. 7.8. Propagating surfaces on a disk, triangular mesh, $\varepsilon = 0.1$.

References

1. H.L. Atkins and C.-W. Shu. Quadrature-free implementation of discontinuous Galerkin methods for hyperbolic equations. Technical Report 96-51, ICASE, 1996. To appear in AIAA J.
2. I. Babuška, C.E. Baumann, and J.T. Oden. A discontinuous hp finite element method for diffusion problems: 1-D analysis. Technical Report 22, TICAM, 1997.
3. F. Bassi and S. Rebay. A high-order accurate discontinuous finite element method for the numerical solution of the compressible Navier-Stokes equations. *J. Comput. Phys.*, 131:267–279, 1997.
4. F. Bassi and S. Rebay. High-order accurate discontinuous finite element solution of the 2D Euler equations. *J. Comput. Phys.*, 138:251–285, 1997.
5. F. Bassi, S. Rebay, M. Savini, G. Mariotti, and S. Pedinotti. A high-order accurate discontinuous finite element method for inviscid and viscous turbo-machinery flows. In *Proceedings of the Second European Conference on Turbomachinery Fluid Dynamics and Thermodynamics*, 1997. Antwerpen, Belgium.
6. C.E. Baumann. *An hp-adaptive discontinuous Galerkin method for computational fluid dynamics*. PhD thesis, The University of Texas at Austin, 1997.
7. C.E. Baumann and J.T. Oden. A discontinuous hp finite element method for convection-diffusion problems. *Comput. Methods Appl. Mech. Engrg.* To appear.
8. C.E. Baumann and J.T. Oden. A discontinuous hp finite element method for the Navier-Stokes equations. In *10th. International Conference on Finite Element in Fluids*, 1998.
9. C.E. Baumann and J.T. Oden. A discontinuous hp finite element method for the solution of the Euler equation of gas dynamics. In *10th. International Conference on Finite Element in Fluids*, 1998.
10. K.S. Bey and J.T. Oden. A Runge-Kutta discontinuous Galerkin finite element method for high speed flows. *10th. AIAA Computational Fluid Dynamics Conference, Honolulu, Hawaii, June 24-27*, 1991.
11. R. Biswas, K.D. Devine, and J. Flaherty. Parallel, adaptive finite element methods for conservation laws. *Appl. Numer. Math.*, 14:255–283, 1994.
12. G. Chavent and B. Cockburn. The local projection P^0 P^1-discontinuous-Galerkin finite element method for scalar conservation laws. M^2AN, 23:565–592, 1989.
13. G. Chavent and G. Salzano. A finite element method for the 1D water flooding problem with gravity. *J. Comput. Phys.*, 45:307–344, 1982.
14. Z. Chen, B. Cockburn, C. Gardner, and J. Jerome. Quantum hydrodynamic simulation of hysteresis in the resonant tunneling diode. *J. Comput. Phys.*, 117:274–280, 1995.
15. Z. Chen, B. Cockburn, J. Jerome, and C.-W. Shu. Mixed-RKDG finite element method for the drift-diffusion semiconductor device equations. *VLSI Design*, 3:145–158, 1995.
16. P. Ciarlet. *The finite element method for elliptic problems*. North Holland, 1975.
17. B. Cockburn. An introduction to the discontinuous Galerkin method for convection-dominated problems. In *Advanced numerical approximation of nonlinear hyperbolic equations*, A. Quarteroni, editor, Lecture Notes in Mathematics, CIME subseries. Springer Verlag. To appear.

18. B. Cockburn and P.-A. Gremaud. A priori error estimates for numerical methods for scalar conservation laws. part I: The general approach. *Math. Comp.*, 65:533–573, 1996.

19. B. Cockburn and P.A. Gremaud. Error estimates for finite element methods for nonlinear conservation laws. *SIAM J. Numer. Anal.*, 33:522–554, 1996.

20. B. Cockburn, S. Hou, and C.W. Shu. TVB Runge-Kutta local projection discontinuous Galerkin finite element method for conservation laws IV: The multidimensional case. *Math. Comp.*, 54:545–581, 1990.

21. B. Cockburn, S.Y. Lin, and C.W. Shu. TVB Runge-Kutta local projection discontinuous Galerkin finite element method for conservation laws III: One dimensional systems. *J. Comput. Phys.*, 84:90–113, 1989.

22. B. Cockburn, M. Luskin, C.-W. Shu, and E. Süli. A priori error estimates for the discontinuous Galerkin method. in preparation.

23. B. Cockburn and C. Schwab. *hp*-error analysis for the local discontinuous Galerkin method. In preparation.

24. B. Cockburn and C.W. Shu. TVB Runge-Kutta local projection discontinuous Galerkin finite element method for scalar conservation laws II: General framework. *Math. Comp.*, 52:411–435, 1989.

25. B. Cockburn and C.W. Shu. The P^1-RKDG method for two-dimensional Euler equations of gas dynamics. Technical Report 91-32, ICASE, 1991.

26. B. Cockburn and C.W. Shu. The Runge-Kutta local projection P^1-discontinuous Galerkin method for scalar conservation laws. M^2AN, 25:337–361, 1991.

27. B. Cockburn and C.W. Shu. The local discontinuous Galerkin finite element method for convection-diffusion systems. *SIAM J. Numer. Anal.*, 35:2440–2463, 1998.

28. B. Cockburn and C.W. Shu. The Runge-Kutta discontinuous Galerkin finite element method for conservation laws V: Multidimensional systems. *J. Comput. Phys.*, 141:199–224, 1998.

29. M.G. Crandall and H. Ishiiand P.L. Lions. User's guide to viscosity solutions of second-order partial differential equations. *Bull. Amer. Math. Soc.*, 27:1–67, 1992.

30. H.L. deCougny, K.D. Devine, J.E. Flaherty, R.M. Loy, C. Ozturan, and M.S. Shephard. Load balancing for the parallel adaptive solution of partial differential equations. *Appl. Numer. Math.*, 16:157–182, 1994.

31. K.D. Devine and J.E. Flaherty. Parallel adaptive *hp*-refinement techniques for conservation laws. *Appl. Numer. Math.*, 20:367–386, 1996.

32. K.D. Devine, J.E. Flaherty, R.M. Loy, and S.R. Wheat. Parallel partitioning strategies for the adaptive solution of conservation laws. In I Babuška, W.D. Henshaw, J.E. Hopcroft, J.E. Oliger, and T. Tezduyar, editors, *Modeling, mesh generation, and adaptive numerical methods for partial differential equations*, volume 75, pages 215–242, 1995.

33. K.D. Devine, J.E. Flaherty, S.R. Wheat, and A.B. Maccabe. A massively parallel adaptive finite element method with dynamic load balancing. In *Proceedings Supercomputing'93*, pages 2–11, 1993.

34. K. Eriksson and C. Johnson. Adaptive finite element methods for parabolic problems I: A linear model problem. *SIAM J. Numer. Anal.*, 28:43–77, 1991.

35. K. Eriksson and C. Johnson. Adaptive finite element methods for parabolic problems II: Optimal error estimates in $l_\infty l_2$ and $l_\infty l_\infty$. *SIAM J. Numer. Anal.*, 32:706–740, 1995.

36. K. Eriksson and C. Johnson. Adaptive finite element methods for parabolic problems IV: A nonlinear model problem. *SIAM J. Numer. Anal.*, 32:1729–1749, 1995.

37. K. Eriksson and C. Johnson. Adaptive finite element methods for parabolic problems V: Long time integration. *SIAM J. Numer. Anal.*, 32:1750–1762, 1995.

38. K. Eriksson, C. Johnson, and V. Thomée. Time discretization of parabolic problems by the discontinuous Galerkin method. *RAIRO, Anal. Numér.*, 19:611–643, 1985.

39. R.S. Falk and G.R. Richter. Explicit finite element methods for symmetric hyperbolic equations. *SIAM J. Numer. Anal.* To appear.

40. J.E. Flaherty, R.M. Loy, M.S. Shephard, B.K. Szymanski, J.D. Teresco, and L.H. Ziantz. Adaptive refinement with octree load-balancing for the parallel solution of three-dimensional conservation laws. Technical report, IMA Preprint Series # 1483, 1997.

41. J. Goodman and R. LeVeque. On the accuracy of stable schemes for 2D scalar conservation laws. *Math. Comp.*, 45:15–21, 1985.

42. P. Houston, C. Schwab, and E. Süli. Stabilized *hp*-finite element methods for hyperbolic problems. *SIAM J. Numer. Anal.* To appear.

43. C. Hu and C.-W. Shu. A discontinuous Galerkin finite element method for Hamilton-Jacobi equations. *SIAM J. Sci. Comput.* To appear.

44. T. Hughes and A. Brook. Streamline upwind-Petrov-Galerkin formulations for convection dominated flows with particular emphasis on the incompressible Navier-Stokes equations. *Comput. Methods Appl. Mech. Engrg.*, 32:199–259, 1982.

45. T. Hughes, L.P. Franca, M. Mallet, and A. Misukami. A new finite element formulation for computational fluid dynamics, I. *Comput. Methods Appl. Mech. Engrg.*, 54:223–234, 1986.

46. T. Hughes, L.P. Franca, M. Mallet, and A. Misukami. A new finite element formulation for computational fluid dynamics, II. *Comput. Methods Appl. Mech. Engrg.*, 54:341–355, 1986.

47. T. Hughes, L.P. Franca, M. Mallet, and A. Misukami. A new finite element formulation for computational fluid dynamics, III. *Comput. Methods Appl. Mech. Engrg.*, 58:305–328, 1986.

48. T. Hughes, L.P. Franca, M. Mallet, and A. Misukami. A new finite element formulation for computational fluid dynamics, IV. *Comput. Methods Appl. Mech. Engrg.*, 58:329–336, 1986.

49. T. Hughes and M. Mallet. A high-precision finite element method for shock-tube calculations. *Finite Element in Fluids*, 6:339–, 1985.

50. P. Jamet. Galerkin-type approximations which are discontinuous in time for parabolic equations in a variable domain. *SIAM J. Numer. Anal.*, 15:912–928, 1978.

51. G. Jiang and C.-W. Shu. On cell entropy inequality for discontinuous Galerkin methods. *Math. Comp.*, 62:531–538, 1994.

52. C. Johnson and J. Pitkaranta. An analysis of the discontinuous Galerkin method for a scalar hyperbolic equation. *Math. Comp.*, 46:1–26, 1986.

53. C. Johnson and J. Saranen. Streamline diffusion methods for problems in fluid mechanics. *Math. Comp.*, 47:1–18, 1986.

54. C. Johnson and A. Szepessy. On the convergence of a finite element method for a non-linear hyperbolic conservation law. *Math. Comp.*, 49:427–444, 1987.

55. C. Johnson, A. Szepessy, and P. Hansbo. On the convergence of shock capturing streamline diffusion finite element methods for hyperbolic conservation laws. *Math. Comp.*, 54:107–129, 1990.

56. D.S. Kershaw, M.K. Prasad, and M.J. Shawand J.L. Milovich. 3D unstructured mesh ALE hydrodynamics with the upwind discontinuous Galerkin method. *Comput. Methods Appl. Mech. Engrg.*, 158:81–116, 1998.

57. D.A. Kopriva. A staggered-grid multidomain spectral method for the compressible Navier-Stokes equations. Technical Report 97-66, Florida State University-SCRI, 1997.

58. P. LeSaint and P.A. Raviart. On a finite element method for solving the neutron transport equation. In C. de Boor, editor, *Mathematical aspects of finite elements in partial differential equations*, pages 89–145. Academic Press, 1974.

59. Q. Lin, N. Yan, and A.-H. Zhou. An optimal error estimate of the discontinuous Galerkin method. *Journal of Engineering Mathematics*, 13:101–105, 1996.

60. Q. Lin and A.-H. Zhou. Convergence of the discontinuous Galerkin method for a scalar hyperbolic equation. *Acta Math. Sci.*, 13:207–210, 1993.

61. W. B. Lindquist. Construction of solutions for two-dimensional Riemann problems. *Comp. & Maths. with Appls.*, 12:615–630, 1986.

62. W. B. Lindquist. The scalar Riemann problem in two spatial dimensions: Piecewise smoothness of solutions and its breakdown. *SIAM J. Numer. Anal.*, 17:1178–1197, 1986.

63. I. Lomtev and G.E. Karniadakis. A discontinuous Galerkin method for the Navier-Stokes equations. *Int. J. Num. Meth. Fluids.* in press.

64. I. Lomtev and G.E. Karniadakis. A discontinuous spectral/ hp element Galerkin method for the Navier-Stokes equations on unstructured grids. In *Proc. IMACS WC'97*, 1997. Berlin, Germany.

65. I. Lomtev and G.E. Karniadakis. Simulations of viscous supersonic flows on unstructured hp-meshes. *AIAA-97-0754*, 1997. 35th. Aerospace Sciences Meeting, Reno.

66. I. Lomtev, C.W. Quillen, and G.E. Karniadakis. Spectral/hp methods for viscous compressible flows on unstructured 2D meshes. *J. Comput. Phys.*, 144:325–357, 1998.

67. E.O. Macagno and T. Hung. Computational and experimental study of a captive annular eddy. *J.F.M.*, 28:43–XX, 1967.

68. X. Makridakis and I. Babusŝka. On the stability of the discontinuous Galerkin method for the heat equation. *SIAM J. Numer. Anal.*, 34:389–401, 1997.

69. Newmann. *A Computational Study of Fluid/Structure Interactions: Flow-Induced Vibrations of a Flexible Cable.* PhD thesis, Princeton University, 1996.

70. J.T. Oden, Ivo Babuška, and C.E. Baumann. A discontinuous hp finite element method for diffusion problems. *J. Comput. Phys.*, 146:491–519, 1998.

71. S. Osher. Riemann solvers, the entropy condition and difference approximations. *SIAM J. Numer. Anal.*, 21:217–235, 1984.

72. S. Osher and J. Sethian. Fronts propagating with curvature-dependent speed: Algorithms based on hamilton-jacobi formulation. *J. Comput. Phys.*, 79:12–49, 1988.

73. S. Osher and C.-W. Shu. High-order essentially nonoscillatory schemes for Hamilton-Jacobi equations. *SIAM J. Numer. Anal.*, 28:907–922, 1991.

74. C. Ozturan, H.L. deCougny, M.S. Shephard, and J.E. Flaherty. Parallel adaptive mesh refinement and redistribution on distributed memory computers. *Comput. Methods Appl. Mech. Engrg.*, 119:123–137, 1994.

75. T. Peterson. A note on the convergence of the discontinuous Galerkin method for a scalar hyperbolic equation. *SIAM J. Numer. Anal.*, 28:133–140, 1991.

76. W.H. Reed and T.R. Hill. Triangular mesh methods for the neutron transport equation. Technical Report LA-UR-73-479, Los Alamos Scientific Laboratory, 1973.

77. G.R. Richter. An optimal-order error estimate for the discontinuous Galerkin method. *Math. Comp.*, 50:75–88, 1988.

78. E. Rouy and A. Tourin. A viscosity solutions approach to shape-from-shading. *SIAM J. Numer. Anal.*, 29:867–884, 1992.

79. S.J. Sherwin and G. Karniadakis. Thetrahedral *hp*-finite elements: Algorithms and flow simulations. *J. Comput. Phys.*, 124:314–345, 1996.

80. C.-W. Shu and S. Osher. Efficient implementation of essentially non-oscillatory shock-capturing schemes. *J. Comput. Phys.*, 77:439–471, 1988.

81. C.-W. Shu and S. Osher. Efficient implementation of essentially non-oscillatory shock capturing schemes, II. *J. Comput. Phys.*, 83:32–78, 1989.

82. C.W. Shu. TVB uniformly high order schemes for conservation laws. *Math. Comp.*, 49:105–121, 1987.

83. C.W. Shu. TVD time discretizations. *SIAM J. Sci. Stat. Comput.*, 9:1073–1084, 1988.

84. M. Sussman, P. Smereka, and S. Osher. A level set approach for computing solution to incompressible two-phase flow. *J. Comput. Phys.*, 114:146–159, 1994.

85. C. Tong and G.Q. Chen. Some fundamental concepts about systems of two spatial dimensional conservation laws. *Acta Mathematica Scientia (English Ed.)*, 6:463–474, 1986.

86. C. Tong and Y.-X. Zheng. Two dimensional Riemann problems for a single conservation law. *Trans. Amer. Math. Soc.*, 312:589–619, 1989.

87. J.R. Trujillo. *Effective high-order vorticity-velocity formulation*. PhD thesis, Princeton University, 1997.

88. B. van Leer. Towards the ultimate conservation difference scheme, II. *J. Comput. Phys.*, 14:361–376, 1974.

89. B. van Leer. Towards the ultimate conservation difference scheme, V. *J. Comput. Phys.*, 32:1–136, 1979.

90. D. Wagner. The Riemann problem in two space dimensions for a single conservation law. *SIAM J. Math. Anal.*, 14:534–559, 1983.

91. T.C. Warburton, I. Lomtev, R.M. Kirby, and G.E. Karniadakis. A discontinuous Galerkin method for the Navier-Stokes equations in hybrid grids. In M. Hafez and J.C. Heirich, editors, *10th. International Conference on Finite Elements in Fluids, Tucson, Arizona*, 1998.

92. P. Woodward and P. Colella. The numerical simulation of two-dimensional fluid flow with strong shocks. *J. Comput. Phys.*, 54:115–173, 1984.

93. A.-H. Zhou and Q. Lin. Optimal and superconvergence estimates of the finite element method for a scalar hyperbolic equation. *Acta Math. Sci.*, 14:90–94, 1994.

Adaptive Spectral Element Methods for Turbulence and Transition

Ronald D. Henderson

Aeronautics & Applied Mathematics
California Institute of Technology
Pasadena, California 91125
ron@galcit.caltech.edu

Abstract. These notes present an introduction to the spectral element method with applications to fluid dynamics. The method is introduced for one-dimensional problems, followed by the discretization of the advection and diffusion operators in multi-dimensions, and efficient ways of dealing with these operators numerically. We also discuss the mortar element method, a technique for incorporating local mesh refinement using nonconforming elements; this is the foundation for adaptive methods. An adaptive strategy based on analyzing the local polynomial spectrum is presented and shown to give accurate solutions even for problems with weak singularities. Finally we describe techniques for integrating the incompressible Navier-Stokes equations, including methods for performing computational linear and nonlinear stability analysis of non-parallel and time-periodic flows.

Table of Contents

1   Introduction

High-order numerical methods have been used almost exclusively in the direct numerical simulation of turbulent flows in the last two decades. Under the broad heading of "high-order methods" we include expansions based on Fourier series, orthogonal polynomial series, and compact finite difference schemes. These methods have been used in studies of transition and turbulence because they offer fast convergence, have small numerical dissipation and dispersion errors, and can be implemented efficiently on most modern computer architectures, including vector and parallel supercomputers. Although they have a higher computational cost per grid point than low-order finite difference, finite volume, or finite element schemes, they are ultimately more efficient for the long-time integration of unsteady flow problems [55].

For all their advantages, there are two key issues that prevent these methods from being applied to more general problems in fluid dynamics: the ability to simulate flows through geometric complex domains with general boundary conditions, and the ability to incorporate local mesh refinement as part of the convergence process. In these notes we describe a class of discretizations that have the advantages of global spectral methods outlined above, but are not subject to their limitations of simple geometries and uniform grids. These newer techniques go under the name of spectral and *h-p* finite element methods, or simply "spectral elements" as they will frequently be referred to here.

Spectral element methods combine the generality of finite element methods with some basic ideas from approximation theory about what constitutes a "good" interpolant. By subdividing a complex domain into macro-elements, they can provide accurate solutions to many problems with substantially fewer degrees of freedom than low-order discretizations. High accuracy comes from the use of orthogonal polynomial expansions to represent the solution over a single element. Galerkin projection operators relate the differential and algebraic equations and keep the global system "sparse" by imposing the minimal continuity requirement on the approximate solution. However, the ability to simulate more general problems with arbitrarily high-order accuracy does not come for free! A polynomial spectral code with domain decomposition and adaptive mesh refinement capabilities is much more complex that either its Fourier series or finite element counterpart. One purpose of these notes is assure the reader that the benefits of spectral element methods far outweigh the cost of implementation.

Spectral and h-p finite element methods are most commonly based on Chebyshev and Legendre polynomials. These are complete orthogonal sets that can be computed easily from a three-term reccurence formula. However, other polynomials can be useful for special cases. All of the "good" polynomial series for numerical methods are derived from the same class of Jacobi polynomials, $P_n^{\alpha,\beta}(x)$. These are the eigenfunctions of an appropriately defined singular Sturm–Liouville problem. These polynomials form an *expansion basis* for representing square-integrable functions $u(x) \in L_2$. The unknowns of the expansion could be the nodal values of the function on a selected grid or other coefficients that weight the importance of polynomials (modes) of different order. The details depend on exactly how the basis is formed and implemented.

Eigenfunction expansions based on singular Sturm–Liouville problems converge at a rate governed by the regularity (smoothness) of the function being expanded and not by any special boundary conditions. Numerical solutions of differential equations based on these expansions have the same property. This observation is important for fluid dynamics, especially for simulations of incompressible flows since these flows are free of discontinuities and can typically be approximated well by polynomials. If the solution is sufficiently smooth then the discretization error decays exponentially fast to zero, at least asymptotically. Doubling the grid resolution reduces the error by two orders of magnitude, not by a mere factor of four as in typical methods with second-order algebraic convergence. Fast convergence is one key to the computational efficiency of high-order methods: they often require a higher operation count than low-order methods for a given number of degrees of freedom, but they require fewer degrees of freedom for a given level of accuracy.

Exponential convergence of numerical solutions in practical situations depends on a number of factors. Although frequently cited as the primary motivation for using high-order methods, exponential convergence only occurs once all but the exponentially small high-order components of an approximation have been resolved; it is probably the exception rather than the rule in simulations of complex phenomena like turbulent flows.[1] Convergence is tied closely to issues like the non-uniformity of the mesh, the form of geometric singularities (e.g. corners), discontinuities in the boundary conditions, and so forth. Such features degrade convergence because they propagate into the high-order components of the solution. These features must be isolated or resolved before fast convergence is realized. Multidomain spectral

---

[1] There are other advantages, such as low numerical dissipation and dispersion errors, that make high-order methods attractive candidates for simulating turbulence even though a flow may be marginally resolved.

discretizations like the ones considered here offer such a possibility due to their *dual* path of convergence. The accuracy of the numerical model can be increased in two ways: by increasing the number of subdomain elements (*h*-refinement), or by increasing the polynomial order of a fixed number of elements (*p*-refinement); this flexibility makes the methods robust.

The following example demonstrates some of the advantages and limitations of spectral elements. Figure 1.1 shows results from a simulation of flow past a half-cylinder [43]. This simulation could not be performed with any method based on global expansions because the domain cannot be mapped to a simpler form. Domain decomposition is a natural choice for the discretization. However, the sharp corner of the body and the relatively thin shear layer make the flow difficult to resolve. In the lower image there are obvious "wiggles" in the computed vorticity field indicative of insufficient resolution. These are equivalent to the familiar aliasing errors in Fourier spectral methods, but manifest in the high-order components of the polynomial approximation. Increasing the polynomial order in this case is a particularly inefficient way to improve the approximation — the geometric singularity prevents fast convergence. The fix is to perform local mesh refinement of the boundary layer and near-wake as shown in the upper part of the figure. Again, no method based on global expansions is capable of this path to convergence.

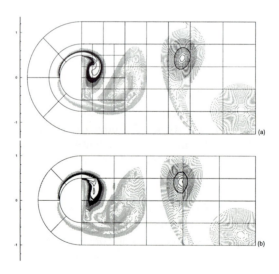

Fig. 1.1. Vorticity in the wake of a half-cylinder at $Re = 250$: (*a*) locally refined mesh using nonconforming spectral elements to resolve the boundary layer and near wake; (*b*) conforming mesh where the solution exhibits "wiggles" due to insufficient resolution. Both simulations are performed with order $p = 7$.

Spectral elements, like finite elements, require that each subdomain in the mesh be *conforming*, that is aligned edge by edge with each neighboring subdomain. This requirement is a natural result of the continuity imposed on the discrete solution. Unlike finite elements, spectral elements represent a coarse discretization of the geometry and achieve high accuracy by using a fine mesh on the interior of each element. Conforming finite elements are not particularly restrictive, but conforming spectral elements make mesh refinement difficult to implement and the improved solution expensive to compute.

Notice that the refined mesh in figure 1.1 contains *nonconforming* elements. These are elements that do not connect to an entire neighboring edge, and as a result special constraints are required to impose the correct continuity conditions on the solution. In spite of the increased complexity, nonconforming elements are key to the efficient implementation of adaptive mesh refinement for spectral element methods. They eliminate the need for refinement boundaries that propagate through the entire domain, allowing refinement to be done locally as dictated by some appropriate error indicator.

Background material for these notes can be found in the monographs by Gottlieb and Orszag [34], Canuto *et al.* [20], and Boyd [17]. These references cover *global* spectral methods extensively, i.e. expansions on a single computational domain. The review article of Maday & Patera [57] also provides background material, concentrating exclusively on conforming discretizations. Early work with spectral elements focused primarily on meshes composed of quadrilateral or hexahedral elements. More recent work has made important advances in the formulation, including meshes of nonconforming elements and triangular and tetrahedral elements. These new tools are the cornerstones of adaptive mesh generation and true h-p refinement. This is the class of algorithms emphasized in these notes. In addition to the basic theory and implementation of spectral element methods, we also discuss a number of applications to the simulation of incompressible flows. Finally we discuss useful methods for studying flow instabilities, transition, and turbulence — all ideal applications of spectral element methods.

2   One-Dimensional Problems

Most of the basic numerical machinery required for spectral element methods can be described in terms of one-dimensional problems. In this section we provide a step-by-step formulation of a spectral element solver for a model advection–diffusion equation to illustrate the procedure before going on to the Navier–Stokes equations. In higher dimensions we have to worry about representing the geometry with more complicated elements, but most of the basic operations are the same.

While reading this section, keep the following point in mind: the procedure used to derive a "spectral" element method is *exactly* the same as that used to derive a finite element method. Any finite element discretization can be extended to higher order using the methods we discuss here. What we emphasize are *efficient* ways to achieve high-order accuracy within the finite element framework, using concepts developed originally for spectral methods. To stress this connection, we try to keep the notation as close as possible to that used in standard finite element textbooks.

2.1   Galerkin formulation

Suppose we want to find u such that

$$u'' + f = 0 \quad \text{on } \Omega, \tag{2.1}$$

where Ω is the unit interval $0 \le x \le 1$ and $f : [0,1] \to \mathcal{R}$ is a given smooth function.[2] At the endpoints we will specify the boundary conditions

$$u(0) = g, \tag{2.2a}$$
$$u'(1) = h. \tag{2.2b}$$

This defines the *strong* form of the problem, the usual starting point for finite difference and spectral collocation schemes.

Consider the following alternative formulation of the same problem. We begin with the equation for the residual,

$$R(u) = \int_\Omega w(u'' + f) \, dx, \tag{2.3}$$

from which we want to find the unique function u that drives the residual to zero. The search will include all functions satisfying the boundary condition $u(0) = g$; each candidate is called a *trial* solution, and we denote the set of all trial solutions by $\mathcal{S}$. The residual is orthogonalized with respect to a second set of functions $w \in \mathcal{V}$ called test functions or *variations*. Each test function should satisfy $w(0) = 0$. To incorporate the Neumann boundary condition we integrate (2.3) once by parts, finding that $R(u) = 0$ if

$$\int_\Omega w'u' \, dx = \int_\Omega wf \, dx + w(1)h. \tag{2.4}$$

For this expression to make sense, both u and w must have square-integrable first derivatives, i.e. $\int_\Omega (u')^2 \, dx < \infty$. Recognizing that such functions belong to the Sobolev space H^1, we can summarize the sets of trial and test functions as:

$$\mathcal{S} = \{u \mid u \in H^1, \ u(0) = g\},$$
$$\mathcal{V} = \{w \mid w \in H^1, \ w(0) = 0\}. \tag{2.5}$$

---

[2] We use the term *smooth* as a qualitative description of a function's higher derivatives. A smooth function $f(x)$ has bounded higher derivatives $f^{(r)}(x)$.

If we identify the symmetric, bilinear forms $a(w, u) = \int_\Omega w'u'\, dx$ and $(w, f) = \int_\Omega wf\, dx$, then we can state the *weak* form as follows: find $u \in S$ such that for every $w \in V$

$$a(w, u) = (w, f) + w(1)h. \qquad (2.6)$$

Equation (2.6) is still an infinite-dimensional problem, because the spaces S and V each contain an infinite number of functions. Galerkin approximation solves (2.6) using a finite collection of functions: find $u^h \in S^h$ such that for every $w^h \in V^h$

$$a(w^h, u^h) = (w^h, f) + w^h(1)h. \qquad (2.7)$$

This method reduces an *infinite*-dimensional problem to an n-dimensional problem by choosing a set of n basis functions $(\phi_1, \phi_2, \dots, \phi_n)$ to represent each member of S^h and V^h. It admits all linear combinations $w^h \in V^h$ as $w^h = c_1\phi_1 + c_2\phi_2 + \dots + c_n\phi_n$, where each $\phi_p(0) = 0$. To generate the trial solutions we need one additional function satisfying $\phi_{n+1}(0) = 1$ so that if $u^h \in S^h$ then

$$u^h = g\phi_{n+1} + \sum_{p=1}^{n} d_p\phi_p. \qquad (2.8)$$

Note that with the exception of ϕ_{n+1}, S^h and V^h are composed of the same functions.

Substituting u^h for u and w^h for w, the weak form becomes

$$\sum_{p=1}^{n} c_p G_p = 0, \qquad (2.9)$$

where

$$G_p = \sum_{q=1}^{n} [a(\phi_p, \phi_q)d_q \\ -(\phi_p, f) - \phi_p(1)h + a(\phi_p, \phi_{n+1})g]. \qquad (2.10)$$

Since this must be true for any choice of the c_p's, we require $G_p \equiv 0$. If we put the coefficients d_p into a vector $\mathbf{d}$, it becomes the matrix problem $\mathbf{Ad} = \mathbf{F}$, where the matrix entries are given by $A_{pq} = a(\phi_p, \phi_q)$ and the components of the vector $\mathbf{F}$ are $F_p = (\phi_p, f) + \phi_p(1)h - a(\phi_p, \phi_{n+1})g$. The solution is $\mathbf{d} = \mathbf{A}^{-1}\mathbf{F}$. Quite literally, this is a best fit of the approximate solution u^h to the true solution u based on the measure of error given in (2.3).

The Galerkin formulation, treated in most standard texts on finite element methods [44, 76], is one example of a general class of techniques called *weighted residual methods* [29]. For certain differential equations it reproduces the underlying *variational principle* if one exists. The idea behind a variational principle is that some physical quantity, such as potential energy, is minimized over the problem domain. For example, the Rayleigh–Ritz principle corresponding to (2.1) minimizes the quadratic form

$$I(u) = \frac{1}{2} \int_{\Omega} (u')^2 \, dx - \int_{\Omega} uf \, dx. \tag{2.11}$$

The Galerkin formulation produces the same solution, but it can be developed even for differential equations that have no corresponding variational form.

2.2   Basis functions

Galerkin approximation is "optimal" in the sense that it gives the best approximation in the restricted space $\mathcal{S}^h$. If the true solution u lies in the intersection of $\mathcal{S}^h$ and $\mathcal{S}$, then $u^h = u$. But the success of the method lies in the selection of the basis functions. If they are too complicated it will be impossible to generate the matrix problem, too simple and they cannot adequately describe the true solution u. The key is to combine computability and accuracy. Spectral elements accomplish this in the following manner.

First, the domain is partitioned into K non-overlapping subintervals, where each subinterval, or *element*, is given by $\Omega^k = [a^k, b^k]$. On element k we want to introduce a set of local functions that provide accuracy of order N for the solution over that piece of the computational domain. For spectral element methods, the basis functions are invariably polynomials.

Often the most convenient approach is to form a set of polynomials from the Lagrangian interpolants through a particular set of *nodes*. Recall that the Lagrangian interpolant takes the value one at some node x_i and is zero at all other nodes. The simplest set of nodes would be the equally spaced points $x_i = a^k + (b^k - a^k) i/N$. Of course, this turns out to be a terrible choice for a high-order method because the basis is almost linearly dependent, resulting in ill-conditioned algebraic systems. It is not the choice of Lagrangian interpolants but the choice of nodes we define them over, so to fix the problem we just need to choose a "good" set of nodes, and this is where spectral methods start to shape the formulation.

To standardize the basis, we introduce a parent domain with the coordinates $-1 \le \xi \le 1$, and a coordinate transformation to the elemental nodes as

$$x_i = a^k + \frac{b^k - a^k}{2}(1 + \xi_i). \tag{2.12}$$

Now we choose the nodes ξ_i to be the solutions of $(1 - \xi^2) L_N'(\xi) = 0$, where $L_N(\xi)$ is the Legendre polynomial of degree N. With this special choice, the Lagrangian interpolants can be written down explicitly as

$$\phi_i(\xi) = -\frac{(1 - \xi^2) L_N'(\xi)}{N(N+1) L_N(\xi_i)(\xi - \xi_i)}. \tag{2.13}$$

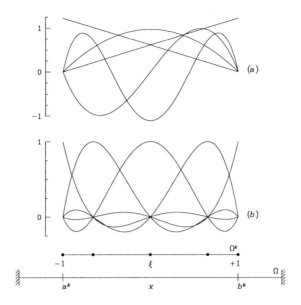

Fig. 2.1. One-dimensional spectral element basis functions for an expansion order of $N = 4$, along with a sketch of the local and global coordinate systems: (a) modal basis constructed from $P_n^{1,1}(\xi)$; (b) Gauss–Lobatto Legendre basis and the set of nodal points that define them as Lagrangian interpolants.

These polynomials are called the Gauss–Lobatto Legendre (GLL) interpolants. Figure 2.1 illustrates the mesh and basis functions for a typical element. We will refer to any basis defined this way as a *nodal* basis.

There are several important reasons for choosing this set of polynomials. First, the expansion of any smooth function using the GLL interpolants, $u \approx u^h = \sum d_i \phi_i(x)$, converges exponentially fast, as can be demonstrated by singular Sturm–Liouville theory [34]. Because these are Lagrangian interpolants, the coefficients d_i are simply the nodal values of the approximate solution: $d_i = u^h(x_i)$. Also, there is a set of integration weights ρ_i associated with the nodes ξ_i so that the integrals appearing in the weak form can be computed via the GLL quadrature

$$\int_{-1}^{1} f \, \mathrm{d}\xi = \sum_{i=0}^{N} \rho_i f(\xi_i) + \epsilon_N, \tag{2.14}$$

where the error $\epsilon_N \sim \mathcal{O}(f^{2N}(\zeta))$ for some point $-1 \leq \zeta \leq 1$; as long as the integrand is a polynomial of degree less than $2N$ this quadrature rule is exact [25]. Finally, and perhaps most importantly, the interpolants, quadrature points, and weights can be generated within a computer program by recursive algorithms that are numerically stable through values of $N \sim 100$, eliminating the need to store static tables of quadrature data.

Legendre polynomials are one example of a broad polynomial class called the *generalized Jacobi polynomials*, which we denote as $P_n^{\alpha,\beta}(\xi)$. Legendre polynomials correspond to the parameter values $\alpha = 0$, $\beta = 0$. Sometimes, especially in higher dimensions and on more complex domains, it is more convenient to work directly with the polynomials rather than an intermediate Lagrangian basis. Jacobi polynomials have the orthogonality property

$$\int_{-1}^{1} (1-\xi)^{\alpha}(1+\xi)^{\beta} P_i^{\alpha,\beta}(\xi) P_j^{\alpha,\beta}(\xi)\,\mathrm{d}\xi = \delta_{ij}. \tag{2.15}$$

We can use Jacobi polynomials directly to represent a function through the expansion $u^h = \sum d_i P_i^{\alpha,\beta}(x)$. The values d_i are the coefficients of the basis functions but they do not correspond to any set of nodal values. In practice, there is a significant advantage if most of the basis functions are orthogonal, so in the one-dimensional case we would use:

$$\begin{aligned}
\phi_0(\xi) &= \tfrac{1}{2}(1+\xi), \\
\phi_1(\xi) &= \tfrac{1}{2}(1-\xi), \\
\phi_i(\xi) &= \tfrac{1}{4}(1+\xi)(1-\xi)P_{i-2}^{1,1}(\xi), \quad i \geq 2.
\end{aligned} \tag{2.16}$$

Figure 2.1 shows the first five basis functions constructed this way. In the nodal basis every function is a polynomial of degree N. In the modal basis there is a *hierarchy* of modes starting with the linear modes, proceeding with the quadratic, the cubic, and so on. Such a basis can accommodate hierarchical p-refinement more readily by increasing the polynomial order. It is also useful to distinguish between hierarchic and non-hierarchic representations. In a hierarchic basis we can easily define a sequence of approximation spaces such that $S^n \subset S^{n+1}$. This ensures that the error decreases monotonically; in non-hierarchic constructions this may or may not be possible [7].

We will refer to spectral elements constructed from a nodal basis as *Lagrange* spectral elements and to those based on a modal basis as *h-p* elements. The latter were first introduced in the early seventies by Szabo [77] who used the integrals of Legendre polynomials as a modal basis, taking $\phi_i(\xi) = \int_{-1}^{\xi} P_{i-1}^{0,0}(s)\,\mathrm{d}s$. However, using the properties of Jacobi polynomials [1] we obtain

$$2n \int_{-1}^{\xi} P_{n-1}^{0,0}(s)\,\mathrm{d}s = (1-\xi)(1+\xi)P_{n-2}^{1,1}(\xi), \tag{2.17}$$

which is the same as the basis in (2.16) except for the normalization.

The choice of which approach to take is somewhat arbitrary since a nodal basis can always be transformed to an equivalent modal basis and vice versa. The Fast Fourier Transform (FFT) is one familiar example of such a transformation onto the basis $\phi_k(\xi) = \exp(ik\xi)$. Unfortunately, there are no "fast transform" methods for Jacobi polynomials and the transforms require matrix multiplication. However, for the values of N used in practice ($N \leq 16$) this is not a serious drawback. Note that for a given polynomial order, the formal accuracy of any basis is the same. Although the modal basis may at first appear to have an advantage for performing local p refinement, the nodal basis can be implemented as a matrix-free method that suffers no penalty for increasing the local polynomial order. The simplicity of working with grid-point values in the nodal basis is an attractive feature. Ultimately, the decision is a matter of personal choice—there is no convincing argument for the exclusive use of one basis type over the other.

For the remainder of this section we will work with the GLL polynomials, but when we introduce the basis on triangular and tetrahedral subdomains we will switch back to the modal point of view.

2.3   Discrete equations

Returning to the problem of solving (2.7), we begin by noting that the integral can be broken into a sum of integrals of each element:

$$a(\phi_p, \phi_q)_\Omega = \sum_{k=1}^{K} a(\phi_p, \phi_q)_{\Omega^k}.$$

Since each basis function is non-zero over a *single* element, the inner product $a(\phi_p, \phi_q)$ is non-zero only if ϕ_p and ϕ_q "belong" to the same element. This makes the global system sparse, and allows us to compute only local matrices. Because of the origin of finite element methods in computational mechanics, these matrices are traditionally called:

"mass" $\mathbf{M}^k_{pq} = \int_{\Omega^k} \phi_p \phi_q \, \mathrm{d}x,$

"stiffness" $\mathbf{A}^k_{pq} = \int_{\Omega^k} \phi'_p \phi'_q \, \mathrm{d}x.$

To construct the right-hand side of the matrix system, $f(x)$ is approximated by collocation at the nodal points to produce $f^h(x)$; the mass matrix provides the coefficients necessary to perform the integration. Now the *elemental* matrix system may be written as

$$\mathbf{A}^k \mathbf{v}^k = \mathbf{F}^k \quad (+ \text{ boundary terms}). \tag{2.18}$$

Just as the integral over the entire domain can be written as a sum of the integral over each element, the global matrices can be computed by summing contributions from the elemental matrices:

$$\mathbf{A} = \sum_{k=1}^{K}{}' \mathbf{A}^k, \quad \mathbf{M} = \sum_{k=1}^{K}{}' \mathbf{M}^k. \tag{2.19}$$

The symbol $\sum'$ represents "direct stiffness summation," the procedure diagrammed for the nodal basis in Fig. 2.2 that maps contributions from the boundary node shared by adjacent elements to the same row of the global matrix $\mathbf{A}$. The global matrix system is

$$\mathbf{A}\,\mathbf{v} = \mathbf{F} \quad (+ \text{ boundary terms}). \tag{2.20}$$

$\mathbf{A}$ is banded as a result of using local basis functions, with all of its non-zero entries located in the N diagonals above and below the main diagonal. It is also symmetric, due to the symmetry of $a(\cdot, \cdot)$, and positive-definite. Thus $\mathbf{A}$ can be computed, stored, and factored economically and efficiently.

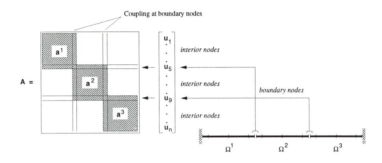

Fig. 2.2. Schematic of the direct stiffness summation of local matrices A^k to form the global matrix $\mathbf{A}$.

Spectral element discretizations encompass both spectral methods and finite elements. Standard approximation error estimates for Galerkin methods applied to elliptic problems on quasi-uniform meshes predict that

$$||u - u^h||_1 \leq \text{const.} \times h^{\mu-1} N^{-(k-1)} ||u||_k, \tag{2.21}$$

where $\mu = \min(k, N + 1)$, N is the polynomial degree appearing in the basis functions, and h is a parameter related to the element size [7]. The constant depends on the degree of mesh quasi-uniformity. There are two ways to improve the approximation: make h smaller ($K \to \infty$), or make N and μ larger ($N \to \infty$). The latter results in *exponential* convergence for smooth solutions. If a solution varies rapidly over a small region, any polynomial fit will oscillate rapidly and the best approach is to reduce the element size until the solution is resolved *locally*. A more effective approach is to combine the two convergence procedures, increasing both K and N simultaneously; this dual path of convergence is known as an *h-p* refinement procedure [77]. The flexibility to adapt the mesh to the solution makes spectral element methods quite robust. The following example clarifies these concepts.

2.4   Example: Burgers equation

Consider the nonlinear differential equation

$$\frac{\partial u}{\partial t} + \frac{1}{2}\frac{\partial u^2}{\partial x} = \nu \frac{\partial^2 u}{\partial x^2}, \tag{2.22}$$

subject to the homogeneous boundary conditions $u(-1) = u(1) = 0$, and smooth initial conditions. Introduced by J. M. Burgers [19], this equation represents a simplified model of the more complicated Navier–Stokes equations that captures the essential features of incompressible fluid dynamics: an unsteady term, a nonlinear advection term, and a viscous diffusion term. Our goal is a numerical method to follow the evolution of a waveform governed by this equation.

Let $u^n(x) \approx u(x, t_n)$ be the approximate solution at time level $t_n = n\Delta t$, where Δt is the time step and n is the time step number. In order to treat the linear and nonlinear terms in the most efficient way possible, we can integrate (2.22) using the two-step splitting scheme

$$\frac{\hat{u} - u^n}{\Delta t} = -\frac{1}{2}\sum_{q=0}^{2} \beta_q \frac{\partial}{\partial x}\left(u^{n-q}\right)^2, \tag{2.23a}$$

$$\frac{u^{n+1} - \hat{u}}{\Delta t} = \frac{\nu}{2}\frac{\partial^2}{\partial x^2}\left(u^{n+1} + u^n\right). \tag{2.23b}$$

The nonlinear term is treated explicitly with a third-order Adams–Bashforth scheme while the linear term is handled with an unconditionally stable, second-order Crank–Nicolson scheme. The values of the β_q's are:

$$\beta_0 = \frac{23}{12}, \quad \beta_1 = -\frac{4}{3}, \quad \beta_2 = \frac{5}{12}. \tag{2.24}$$

Since $\hat{u}$ is just an intermediate solution used to decouple the two steps, boundary conditions will only be applied in the diffusion step to u^{n+1}.

Spectral elements form the spatial discretization, so on element k we have

$$u^n(x) = \sum_{i=0}^{N} u_i \phi_i(\xi) \quad \text{on } \Omega^k, \tag{2.25}$$

where the basis coefficients u_i are to be determined at each new time level. First we take the nonlinear step,

$$\hat{u} = u^n - \frac{\Delta t}{2}\sum_{q=0}^{2} \beta_q \frac{\partial}{\partial x}\left(u^{n-q}\right)^2, \tag{2.26}$$

using explicit collocation:

$$\frac{\partial}{\partial x}(u)^2 = \frac{\partial}{\partial x}\left(\sum_{i=0}^{N} u_i \phi_i(\xi)\right)^2 \quad \text{on } \Omega^k. \tag{2.27}$$

This expression is evaluated at every nodal point. To compute the Galerkin approximation to the diffusion step, (2.23b) is first written in the form

$$(\frac{\partial^2}{\partial x^2} - \frac{2}{\nu \Delta t})v = \frac{2}{\nu \Delta t}(\hat{u} + u^n),$$ (2.28)

where $v = \frac{1}{2}(u^{n+1} + u^n)$. This form, called a Helmholtz equation, is simply (2.1) with an additional term multiplying v. The spectral element approximation of (2.28) results in the algebraic system

$$\left[\mathbf{A} + \frac{2}{\nu \Delta t}\mathbf{M} \right] \mathbf{v} = \left[\frac{2}{\nu \Delta t}\mathbf{M} \right] (\hat{\mathbf{u}} + \mathbf{u}^n),$$ (2.29)

where $\mathbf{A}$ and $\mathbf{M}$ are the global stiffness and mass matrices defined in (2.19), and $\mathbf{v}$, $\hat{\mathbf{u}}$, and $\mathbf{u}^n$ are vectors containing the basis coefficients that determine the approximation $v^h \approx v$, etc. The solution at the new time level is $u^{n+1} = 2v - u^n$.

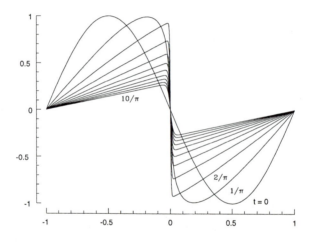

Fig. 2.3. Evolution of a sinusoidal wave governed by the viscous Burgers equation with $\nu = 10^{-2}/\pi$. The structure of the wave is shown at times from $t = 0$ to $t = 10/\pi$.

Burgers' equation can be solved analytically for certain initial conditions. Figure 2.3 shows how an initial sinusoidal wave evolves into a steep sawtooth wave at a time near $t = 1/\pi$. The exact solution is given by

$$u(x,t) = 4\pi\nu \left[\sum_{n=1}^{\infty} n a_n e^{-n^2\pi^2 t\nu} \sin n\pi x / \right.$$
$$\left. (a_0 + 2\sum_{n=1}^{\infty} a_n e^{-n^2\pi^2 t\nu} \cos n\pi x) \right]$$ (2.30)

where $a_n = (-1)^n I_n(1/2\pi\nu)$ and $I_n(z)$ is the modified Bessel function of the first kind [13]. As long as the viscosity ν is finite the profile is continuous but varies rapidly within a narrow region around the origin. The value of the slope at the origin and the time at which it reaches a maximum provide a measure of both spatial and temporal errors in the approximation.

Figure 2.4 shows a sequence of mesh refinements in which the elements near the origin are halved in size while the polynomial order is held fixed at $N = 10$ (h-refinement). On the coarsest mesh the solution begins to oscillate as the wave becomes steeper but eventually recovers as the thin inner layer diffuses outward. Each mesh in Fig. 2.4 contains the same number of points— the only difference is the *size* of the elements, and therefore the distribution of points in the domain. By clustering points near the origin, the final mesh resolves the thin inner layer and improves the solution without increasing the computational cost.

This final mesh, with $(K, N) = (4, 16)$, gives four significant digits for both $\max(|\partial u/\partial x|) = 152.06$ and the corresponding time $\pi t = 1.6033$. Even with a coarser mesh, Fig. 2.5(a) shows that the wave moves at the correct speed towards the origin. Figure 2.5(b) verifies that the approximation to the derivative converges exponentially, and in fact the error $e^h = \partial(u - u^h)/\partial x$ is bounded by

$$\log \|e^h\|_\infty \le \sigma N + \log \|u\|_\infty + \text{const.}, \tag{2.31}$$

where $\sigma \approx -1/4$. The scatter in the convergence data is due in part to the different approximation properties of odd versus even order polynomials. A general comparison of convergence properties and approximation errors for spectral element, finite difference, and global spectral methods applied to the viscous Burgers equation is given in [12].

We have just observed two important properties of spectral element approximations. First, high-order spatial discretizations result in low numerical dissipation, i.e. the correct wave speed was maintained on each mesh. This is an important property for long-time integration of unsteady flows as discussed in the Introduction. Second, spectral accuracy is achieved for rapidly varying solutions as long as the solution is resolved adequately on the scale of a single element. These properties make spectral elements ideally suited for solving the equations governing incompressible fluid dynamics, where similar phenomena appear as boundary layers and shear layers. Local mesh refinement was a simple matter in this one-dimensional example, but for more interesting two- and three-dimensional problems it becomes one of the most important features of the discretization.

3   Multi-Dimensional Problems

3.1   Basis functions in d-dimensions

A key to the efficiency of high-order methods in two- and three-dimensional problems is the formation of a basis from the *tensor product* of one-dimensional

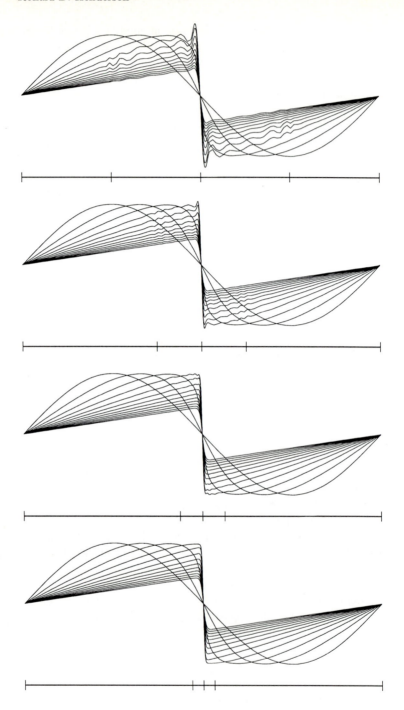

Fig. 2.4. A demonstration of how high-order methods combined with mesh refinement can be used to resolve rapidly varying solutions. The size of the elements used for each calculation is indicated below the corresponding solution.

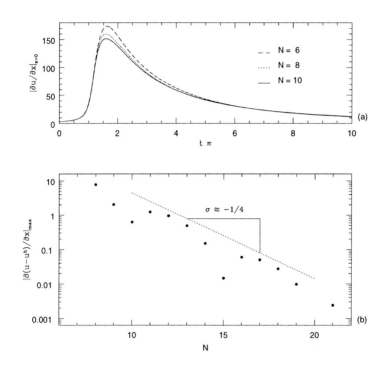

Fig. 2.5. Numerical integration of the viscous Burgers equation: (a) evolution of $|\partial u/\partial x|_{x=0}$ for three different meshes and (b) reduction of the error in $\max(|\partial u/\partial x|)$ with increasing polynomial order N.

functions. Among other things, this allows the computation of integrals and derivatives of the basis functions to be simplified through a procedure called *sum factorization* [65]. It also contributes to the sparse structure of matrix systems for multi-dimensional problems.

In this section we describe the procedure for constructing an efficient, high-order basis on two- and three-dimensional domains. To keep the discussion simple, we only consider the standard domains R^d and T^d, where d is the problem dimension. Figure 3.1 defines the standard rectangle, R^2, and Fig. 3.2 defines the standard triangle, T^2. "Standard" here means that the coordinates are normalized to fall in the range -1 to 1. For $d = 3$, the standard domain is a hexahedral or tetrahedral element. Isoparametric mappings can always be used to transform more general elements to these standard domains, as illustrated in Fig. 3.1. On the standard element, we wish to define a polynomial basis, denoted by $\phi_{ij}(\xi_1, \xi_2)$, so that we can represent a function $u^h(\xi_1, \xi_2)$ by the expansion

$$u^h(\xi_1, \xi_2) = \sum_{i=0}^{N} \sum_{j=0}^{N} u_{ij} \phi_{ij}(\xi_1, \xi_2),$$

where u_{ij} is the coefficient of the basis function ϕ_{ij} and $\boldsymbol{\xi} = (\xi_1, \xi_2)$ is the local coordinate within the element.

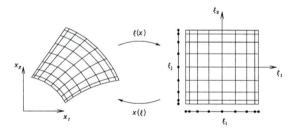

Fig. 3.1. Definition of the standard quadrilateral domain R^2. General curvilinear elements can always be mapped back to the standard element as shown.

For quadrilateral (two-dimensional) and hexahedral (three-dimensional) elements, the procedure is straightforward. For example, on the domain $\Omega^k = R^2$, the basis would be

$$\phi_{ij}(\xi_1, \xi_2) = \phi_i(\xi_1) \, \phi_j(\xi_2),$$

where $\phi_i(\xi)$ is the one-dimensional GLL polynomial defined in § 2. In this case, u_{ij} represents the function value at the node $\boldsymbol{\xi}_{ij}$. The three-dimensional basis on R^3 is exactly analogous to this one.

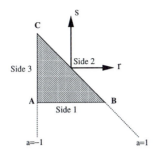

Fig. 3.2. Definition of the standard triangular domain T^2. Here $r \equiv \xi_1$ and $s \equiv \xi_2$.

To introduce the expansion basis for the standard triangle T^2, we first need to define a basic coordinate mapping as illustrated in Fig. 3.3. The rectangular domain R^2 can be mapped into the triangular domain T^2 by the transformation:

$$\eta_1 = \tfrac{1}{2}(1 + \xi_1)(1 - \xi_2) - 1,$$
$$\eta_2 = \xi_2. \tag{3.1}$$

The triangular basis is now partitioned into *interior* modes and *boundary* modes. Interior modes are zero on the boundary of the triangular domain, similar to the *bubble* modes used in p-type finite element methods [6,64]. Boundary modes can be further partitioned into *vertex* and *edge* modes. Vertex modes vary linearly from zero to one along the edge of the triangle. Edge modes are only non-zero along a single edge of the triangle, and are zero along the other edges and at all vertices.

Using the notation shown in Fig. 3.3 and recalling that $P_n^{\alpha,\beta}(\xi)$ refers to the Jacobi polynomial, we can write the triangular basis as follows:

– Vertex modes

$$\phi_{10}^{\text{vertex A}} = \frac{1}{4}(1 - \xi_1)(1 - \xi_2),$$

$$\phi_{10}^{\text{vertex B}} = \frac{1}{4}(1 + \xi_1)(1 - \xi_2),$$

$$\phi_{01}^{\text{vertex C}} = \frac{1}{2}(1 + \xi_2);$$

– Edge modes $(i \geq 2, j \geq 1; i < L, i + j < L)$

$$\phi_{i0}^{\text{edge 1}} = \frac{1}{2^{i+2}}(1 + \xi_1)(1 - \xi_1)(1 - \xi_2)^i P_{i-1}^{1,1}(\xi_1),$$

$$\phi_{1j}^{\text{edge 2}} = \frac{1}{8}(1 + \xi_1)(1 - \xi_2)(1 + \xi_2)^i P_{j-1}^{1,1}(\xi_2),$$

$$\phi_{1j}^{\text{edge 3}} = \frac{1}{8}(1 - \xi_1)(1 - \xi_2)(1 + \xi_2)^i P_{j-1}^{1,1}(\xi_2);$$

– Interior modes $(i \geq 2, j \geq 1; i < L, i+j < L)$

$$\phi_{ij}^{\text{interior}} = \frac{1}{4}(1+\xi_1)(1-\xi_1)P_{i-2}^{1,1}(\xi_1) \times$$
$$\frac{1}{2^{i+1}}(1+\xi_2)(1-\xi_2)^i P_{j-1}^{2i-1,1}(\xi_2)$$

The indices ij refer to the principle polynomial in the ξ_1 and ξ_2 direction. L denotes the total number of modes associated with each direction, i.e. the maximum polynomial order along an edge is $N = L - 1$. For example, if $L = 2$ there are only vertex modes, giving us a linear finite element basis.

This is a polynomial basis in both the η and ξ coordinates. In the ξ coordinate system it forms a tensor product, so basic operations such as integration and differentiation can be performed using equivalent one-dimensional operations just like the tensor product basis on R^d. It also accommodates exact Gauss–Jacobi quadrature and maintains a partial orthogonality between the modes. This partial orthogonality helps keep the matrices formed from inner products of the basis functions sparse. More details about the two-dimensional basis can be found in [27, 75].

The two-dimensional mapping is the foundation for constructing a coordinate system in the tetrahedral domain T^3, starting from the coordinate system for the hexahedral domain R^3. Figure 3.4 shows how R^3 is reduced to T^3 by applying the coordinate transformation given in (3.1) to each pair of coordinates. The inverse mapping is

$$\begin{aligned}
\xi_1 &= -2(1+\eta_1)/(\eta_2+\eta_3) - 1, \\
\xi_2 &= 2(1+\eta_2)/(1-\eta_3) - 1, \\
\xi_3 &= \eta_3.
\end{aligned} \qquad (3.2)$$

For $\eta_3 = -1$, we recover the two-dimensional mapping. The three-dimensional basis for T^3 is then decomposed into vertex modes, edge modes, face modes and interior modes, in analogy with the basis on T^2; details can be found in [74].

In the remainder of this Chapter we will use the following simplified notation. Every index (ijk) in the tensor product basis will be mapped to a single number as $p = i + jN + kN^2$, so there is a one-to-one correspondence between $\phi_p(\boldsymbol{\xi})$ and $\phi_{ijk}(\boldsymbol{\xi})$. This hides the tensor product nature of the basis but makes the discrete equations much easier to write down. When necessary, we can "unroll" the p index to take advantage of the tensor product form. This expression for p is valid for quadrilateral elements only; a modified expression should be used with the triangular domains.

3.2 Data structures

Here we describe the data structures and basic operations required to implement the most common procedures in spectral element methods. We cover:

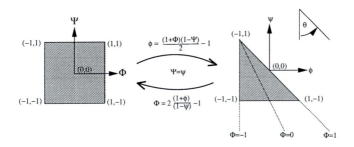

Fig. 3.3. Schematic of the transformation from R^2 to T^2.

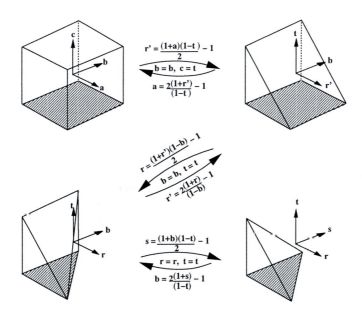

Fig. 3.4. Schematic of the transformation from R^3 to T^3.

representation of the global system, how to transfer global data to local (element) data, direct stiffness summation, and finally the procedures for integration and differentiation of solutions defined on geometrically complex two- and three-dimensional elements.

Implementation First we start with the representation of the solution within a computer program. In this section we give several examples as pseudo-code fragments that follow basic C and $C\text{++}$ syntax. This is not meant to be an in-depth presentation, but simply an illustration of the most important ideas and the basic approach.

In spectral element methods, as in finite element methods, global data is stored as a flat, unstructured array. The basic data structure used to relate the mesh to entries in this array is a table that identifies the global node number of a local node within each element. Since we are interested in both nodal and modal descriptions, we replace "node" with the more general concept of a "degree of freedom" in the global solution. The table of indices can be stored as a two-dimensional array of integers:

$$\texttt{map[k][i]} = \text{global index of local datum } i$$
$$\text{in element } k.$$

Local data can be stored in any convenient, regular format. In our first version, we will assume the number of degrees of freedom in the mesh (`ndof`) and the number of degrees of freedom associated with each element (`edof`) are constant. To perform some global operation, for example to evaluate a function $v = F(u)$, we insert a layer of indirection between the unstructured global data and the structured local data. The following is a template for any such computation:

```
for (i=0; i < ndof; i++)      // Initialize v
   v[i] = 0.;
for (k=0; k < nel; k++) {      // Loop over elements
   for (i=0; i < edof; i++)    // Copy global data
      uk[i] = u[ map[k][i] ];  // -- gather
      compute (uk, vk);        // Compute v=F(u) locally
   for (i=0; i < edof; i++)    // Accumulate the result
      v[ map[k][i] ] += vk[i]; // -- scatter
}
```

Depending on the specific operation, the final result may need to be corrected in some way: rescaled with the global mass matrix, averaged based on the data multiplicity, or some similar global operation. The last loop corresponds to direct stiffness summation, and in our matrix notation we would write this same operation as:

$$\mathbf{v} = \sum_{k=1}^{K}{'}\mathbf{v}^k = \sum_{k=1}^{K}{'}F(\mathbf{u}^k) = F(\mathbf{u}). \tag{3.3}$$

To make this data structure suitable for both hierarchical bases and non-conforming elements (to be developed in § 3.5), we introduce two generalizations. First, we allow the number of degrees of freedom in each element to be different by replacing the constant edof with the array edof[k]. Second, we allow each local degree of freedom to depend on an arbitrary combination of the global degrees of freedom. To implement this we need to introduce two new arrays:

$$idof[k][i] = \text{number of global dependencies for}$$
$$\text{local datum } i \text{ in element } k,$$
$$combine[k][i] = \text{array of coefficients for combining}$$
$$\text{global data to get local data.}$$

And finally, we need to add a new dimension to our index table:

$$map[k][i][j] = \text{global index of the } j\text{th dependency}$$
$$\text{of local datum } i.$$

In effect, we are introducing a set of coefficient matrices Z^k that define a general transformation between global and local degrees of freedom. Using this approach, the global initialization, loop over the elements, and function call for the local computation shown above stay the same, but the procedure for constructing the local data is re-written as follows:

```
for (i=0; i < edof[k]; i++)   // Initialize
   uk[i] = 0.;
for (i=0; i < edof[k]; i++) { // Combine
   real *Z  = combine[k][i];
   for (j=0; j < idof[k][i]; j++)
     uk[i] += Z[j] * u[ map[k][i][j] ];
}
```

Likewise, the accumulation of results uses a similar method for combining local contributions to the global degrees of freedom:

```
for (i=0; i < edof[k]; i++) { // Combine
  real *Z  = combine[k][i];
  for (j=0; j < idof[k][i]; j++)
    v[ map[k][i][j] ] += Z[j] * vk[i];
}
```

We also introduce a new matrix notation for this more general approach. Since the local data is $Z^k\mathbf{u}$, and the local contribution to the global system is

$[Z^k]^T\mathbf{v}^k$, the equivalent procedure for assembling the global system is written as:

$$\mathbf{v} = \sum_{k=1}^{K}{}'[Z^k]^T\mathbf{v}^k = \sum_{k=1}^{K}{}'[Z^k]^T F(Z^k\mathbf{u}) = F(\mathbf{u}) \qquad (3.4)$$

Compare this to (3.3) above, and note that the only change is how we transform *between* the local and global systems. The actual computations at both the local and global level are the same.

In the remaining sections we will describe computations in terms of either the local or global system, omitting the actual "assembly" required to go between them. Equation (3.4) is always implied as the method for recovering local solutions and assembling global ones. This simplifies what would otherwise become a confusing barrage of notation. Along the way we will give more specific information about how the coefficients for the mapping matrix Z^k are chosen. This is a very flexible scheme for storing the global solution and reconstructing the local one. The additional storage and computational overhead is simply the price we pay for new capabilities: variable order of the local basis functions and arbitrary connectivity in the mesh. However, these are the key ingredients for adaptive *h-p* refinement techniques!

Improvements Although the scheme outlined above is complete, it is not an efficient way to implement *h-p* methods: too much of the addressing is done by indirection. One of the computational advantages of high-order elements is the natural partitioning of data into sets that can be operated on as a group. For example, local degrees of freedom are normally partitioned into several groups: vertices, edges, faces, and interior data. Data associated with any of these groups can be operated on as a single entity. For example, all the points on the interior of an element can be identified with the element number and moved around or computed on as a single unit. High-order elements provide better data locality than low-order elements because computations always involve large amounts of data that can be grouped together in memory.

The type of full indirection outlined above is only necessary for the degrees of freedom associated with the surface of an element. These data make up the loosely-coupled components of the global system. This sparse global system forms the "skeleton" of the discretization and shares many characteristics with low-order finite elements. For example, the numbering system stored in the index table can be optimized to reduce its algebraic bandwidth using the same techniques applied in finite element methods (see Sect. 3.6). Unfortunately, more sophisticated data structures than can be described here are required to incorporate these simplifications; we leave this to the reader as an important step in the efficient implementation of spectral element methods.

3.3   Basic operations

Integration The general form for the evaluation of an integral by Gaussian quadrature with weights $(1 - \xi)^\alpha (1 + \xi)^\beta$ can be written as

$$\int_{-1}^{1} (1 - \xi)^\alpha (1 + \xi)^\beta u(\xi) \, d\xi = \sum_{i=0}^{N} \rho_i^{\alpha,\beta} u(\xi_i^{\alpha,\beta}),$$

where $\xi_i^{\alpha,\beta}$ and $\rho_i^{\alpha,\beta}$ are the quadrature points and weights associated with the Jacobi polynomial $P_N^{\alpha,\beta}(\xi)$. The quadrature rule is exact if $u(\xi)$ is a polynomial of degree $2N + 1$ for the Gauss points, $2N$ for the Gauss–Radau points, and $2N - 1$ for the Gauss–Lobatto points.

To integrate a function defined over the standard domain R^2, we simply use the tensor product form to reduce the integral to two one-dimensional quadratures. The integral of a general function is written as

$$\int_{R^2} u(\boldsymbol{\xi}) \, d\xi_1 d\xi_2 = \sum_{i=0}^{N} \sum_{j=0}^{N} \rho_i \rho_j u(\boldsymbol{\xi}_{ij}).$$

The extension to integrals over R^3 is straightforward.

On the triangular domain, we use a coordinate transformation to simplify the integral. The integral of a function over T^2 becomes

$$\int_{T^2} u(\boldsymbol{\eta}) \, d\eta_1 d\eta_2 = \int_{R^2} u(\boldsymbol{\xi}) |J| \, d\xi_1 d\xi_2,$$

where $|J| = (1 - \xi_2)/2$ is the Jacobian determinant of the transformation $\boldsymbol{\eta} \to \boldsymbol{\xi}$. The integral in $\boldsymbol{\xi}$-space can now be evaluated just as the integral over R^2. To include the Jacobian, we use a quadrature rule with $\alpha = 0$, $\beta = 0$ in the ξ_1-direction, and a quadrature rule with $\alpha = 1$, $\beta = 0$ in the ξ_2-direction. Integration over T^3 is performed in a similar way.

Projection To apply the integration rules described above, we need to evaluate a function at a given set of quadrature points. For the nodal basis this is trivial because the basis coefficients *are* the function values at the quadrature points. For a modal basis we need an efficient way to evaluate the full solution at the quadrature points. This, and the related problem of determining the modal expansion coefficients from a set of nodal values, are both called *projections*.

A projection is the procedure for determining the coefficients u_{ijk} so that $u^h \approx u$ for some given function u. First, recall the general form of the expansion:

$$u(\boldsymbol{\xi}) \approx u^h(\boldsymbol{\xi}) = \sum_p u_p \, \phi_p(\boldsymbol{\xi}).$$

The expansion coefficients are determined by taking the inner-product with the basis functions on both sides of this equation:

$$(u, \phi_p)_{\Omega^k} = (u^h, \phi_p)_{\Omega^k} \quad \forall \phi_p \in \{\phi_{ijk}\}. \tag{3.5}$$

Solving this system of equations to determine the approximation u^h is straightforward if the basis $\{\phi_{ijk}\}$ is orthogonal. Otherwise, we have to compute u^h by inverting a matrix.

To describe this for the modal basis, we introduce the following notation:

$\mathbf{u}_p$ = vector of $P \sim N^3$ expansion coefficients, $\mathbf{u}_p \leftarrow u_{ijk}$;

$\tilde{\mathbf{u}}_q$ = vector of Q function values at the quadrature points, $\tilde{\mathbf{u}}_q \leftarrow u(\boldsymbol{\xi}_q)$;

$\mathbf{W}_{qq}$ = diagonal matrix of $Q \times Q$ quadrature weights required to integrate a function over Ω^k;

$\mathbf{B}_{qp}$ = rectangular matrix containing the value of the basis functions at the quadrature points (Q quadrature points $\times$ P basis functions).

Now we can write down the algebraic form of the inner-products given in (3.5). First, the inner product of u with the basis functions:

$$(u, \phi_p)_{\Omega^k} \rightarrow \mathbf{B}^T \mathbf{W} \, \tilde{\mathbf{u}}.$$

Second, the inner product of u^h with the basis functions:

$$(u^h, \phi_p)_{\Omega^k} \rightarrow \mathbf{B}^T \mathbf{W} \mathbf{B} \, \mathbf{u}.$$

The approximation $u^h \approx u$ is determined by matching these two inner products for every basis function:

$$\mathbf{B}^T \mathbf{W} \, \tilde{\mathbf{u}} = \mathbf{B}^T \mathbf{W} \mathbf{B} \, \mathbf{u}. \tag{3.6}$$

This is the fully discrete form of (3.5). Note that the epression on the right-hand-side defines the mass matrix $(\phi_i, \phi_j)_{\Omega^k} \rightarrow \mathbf{B}^T \mathbf{W} \mathbf{B}$, or simply $\mathbf{M} = \mathbf{B}^T \mathbf{W} \mathbf{B}$.

Now we can define the discrete projection operator as

$$\mathbf{u} = \mathcal{P}(\tilde{\mathbf{u}}) \equiv [\mathbf{B}^T \mathbf{W} \mathbf{B}]^{-1} \mathbf{B}^T \mathbf{W} \tilde{\mathbf{u}}.$$

This is also called the *forward transform* of a function from physical space (nodal values) to transform space (modal coefficients). The discrete *inverse transform* is simply the evaluation of the modal basis at a given set of points:

$$\tilde{\mathbf{u}} = \mathcal{P}^{-1}(\mathbf{u}) \equiv \mathbf{B} \mathbf{u}.$$

Finally we note that in the GLL nodal basis, $\mathbf{M}$ is a *diagonal* matrix. This follows directly from the discrete orthogonality of the basis functions and the fact that $\phi_p(\xi_q) = \delta_{pq}$, where ξ_q are the GLL quadrature points. A diagonal mass matrix is a tremendous simplification since multiplication by $\mathbf{M}^{-1}$ is trivial.

Differentiation Since the basis is formed from continuous functions, in principle derivatives can be evaluated by simply differentiating the basis functions:

$$\frac{\partial u^h}{\partial \xi_1} = \sum_{ijk} u_{ijk} \frac{\partial \phi_i}{\partial \xi_1}(\xi_1) \phi_j(\xi_2) \phi_k(\xi_3).$$

In practice we only need the derivatives at certain points, namely the quadrature points. Therefore, the solution is first transformed onto an equivalent Lagrangian interpolant basis defined over the quadrature points. We introduce the one-dimensional Lagrangian derivative matrix

$$\mathbf{D}_{ip} \equiv \left. \frac{d\phi_p}{d\xi} \right|_{\xi_i}.$$

Rather than $\mathcal{O}(N^3)$ terms, the Lagrangian interpolant basis reduces the summation to an equivalent one-dimensional operation. The coefficient of the derivative, u'_{ijk}, is then given by

$$u'_{ijk} = \sum_{p=0}^{N} \mathbf{D}_{ip} u_{pjk}.$$

Since only $\mathcal{O}(N)$ operations are required per point, it takes $\mathcal{O}(N^3)$ operations to compute all derivatives in R^2 or T^2, and $\mathcal{O}(N^4)$ operations to compute all derivatives in R^3 or T^3. In the modal basis, calculation of derivatives is preceded by an inverse transform (to nodal values) and followed by a forward transform (to modal coefficients), therefore increasing the computational cost.

3.4   Spaces and norms

Throughout the rest of these notes we will be concerned primarily with two function spaces $L_2(\Omega)$ and $H^1(\Omega)$. We define the inner-product of two function u and v as:

$$(u, v) = \int_\Omega uv \, d\Omega. \tag{3.7}$$

For reference we define the L_2 norm as:

$$||u|| = (u, u)^{1/2} \quad \forall u \in L_2(\Omega), \tag{3.8}$$

the H^1 norm as:

$$||u||_1 = [(u, u) + (u_{,i}, u_{,i})]^{1/2} \quad \forall u \in H^1(\Omega), \tag{3.9}$$

and the infinity norm as:

$$||u||_\infty = \sup_{x \in \Omega} |u(x)| \quad \forall u \in L_\infty(\Omega). \tag{3.10}$$

For the discrete solution, (3.8) and (3.9) can be evaluated approximately by numerical quadrature; the infinity norm can be estimated from the basis coefficients.

3.5   Global matrix operations

Conforming One of the basic principles for maintaining the sparse structure in the global matrix systems is to enforce only the minimum continuity between elements. For all of the problems we consider here, the global basis is required to be C^0 continuous, i.e. only function values and not derivatives are required to be globally continuous. For discretizations with both Lagrangian and h-p basis functions, this is accomplished by choosing a unique set of global "degrees of freedom" that define the approximation space.

Global continuity in the Lagrangian basis is straightforward. Since the basis functions are defined as the Lagrangian interpolant through the elemental nodes, we only have to use the same set of nodes along the edge of adjacent elements. As long as the elements are conforming (each edge matches up exactly to one other edge) and of equal order (same number of nodes along each edge), C^0 continuity is guaranteed. Figure 3.5 shows a possible global numbering scheme for a simple quadrilateral mesh.

Continuity in the modal basis is more involved because we have to match up all modes. Depending on the orientation chosen for the triangular elements, local modes may be a positive or negative image of the corresponding global mode. This extra bit of information must be tracked as part of the implementation, and we describe it as one use of the mapping matrix $\mathbf{Z}^k$.

Consider a domain made up of two triangular elements as shown in Fig. 3.6. The expansion order is $N = 3$, meaning there are six modes on each triangle: three vertex modes (1, 3, 5) and three edge modes (2, 4, 6), but no interior mode. The number of local degrees of freedom for each element is $n_{eof} = 6$, and the number of global degrees of freedom for the mesh shown is $n_{dof} = 9$. The mapping from global to local degrees of freedom for element Ω^1 is:

$$\mathbf{u}^k = \begin{bmatrix} u_1 \\ u_2 \\ u_3 \\ u_4 \\ u_5 \\ u_6 \end{bmatrix}^k = \begin{bmatrix} 1 & & & & & \\ & 1 & & & & \\ & & 1 & & & \\ & & & 1 & & \\ & & & & 1 & \\ & & & & & 1 \end{bmatrix}^k \begin{bmatrix} u_1 \\ u_2 \\ u_3 \\ u_4 \\ u_5 \\ u_6 \end{bmatrix},$$

or in short form $\mathbf{u}^k = \mathbf{Z}^k \mathbf{u}$. Notice that data for each local mode maps to one and only one global mode, but data for a global mode can be shared by any number of local modes. The number of local modes that contribute data to one global mode is called the *multiplicity* of the global mode.

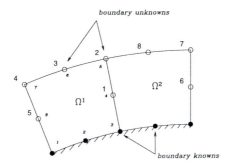

Fig. 3.5. Local and global numbering for a simple domain composed of two quadrilateral elements of order $N = 2$. Points along the boundary do not constitute global "degrees of freedom" and are not assigned indices in the global index set.

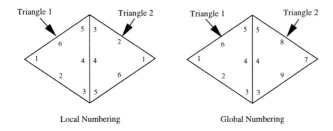

Fig. 3.6. Local and global numbering for a domain containing two triangular elements. Here the expansion order is $N = 3$ so there are $N(N+1)/2 = 6$ modes in each element: three vertex modes (1, 3, 5), and three edge modes (2, 4, 6).

Nonconforming An important extension to the original spectral element method was the introduction of nonconforming elements by Bernardi *et al.* [14]. Here we give only a sketch of the how the method is used to patch together a nonconforming mesh; for a full description of the method, including efficient solution techniques and numerous examples, see the references [2, 14, 39, 40, 59].

The main idea is to use a *constrained approximation*. For a geometrically and functionally nonconforming set of elements, we cannot guarantee global C^0 continuity of the basis. Therefore, we make the basis as continuous as possible by minimizing the difference in function values across each nonconforming interface. We do this by enforcing the following weighted residual equation:

$$\int_\Gamma (u - v)\psi \, ds = 0 \quad \forall \psi \in P_{N-2}(\Gamma). \tag{3.11}$$

The residual is the difference in two functions u and v that we would like to be continuous, and ψ is the weight used to perform the minimization. The algebraic form of this equation is

$$\mathbf{u} = \mathbf{Z}\,\mathbf{v},$$

where $\mathbf{u}$ and $\mathbf{v}$ are the coefficients of whatever basis we choose to represent u and v, and the entries of $\mathbf{Z}$ are determined by evaluating the residual equation using numerical quadrature. We say the values of $\mathbf{v}$ are free and the values of $\mathbf{u}$ are constrained to match them such that (3.11) is satisfied.

To use this as a computational tool, we choose v to be the solution along the edge of some element, and u to be the solution along the edge of an adjacent nonconforming element. Equation (3.11) is used to construct u from v, thereby eliminating u as an "unknown" in the mesh. Since v contributes to the global degrees of freedom in the problem, this is one type of the "combining" described in § 3.2. There is an additional consistency error associated with the nonconforming discretization because the approximation space is no longer a proper subset of the solution space—it admits discontinuous solutions. As bad as this sounds, the consistency error is of the same order as other components of the approximation error, and if implemented properly the method always converges to a continuous solution if one exists.

Nonconforming elements allow quadrilateral meshes to be refined locally, without the conforming restriction propagating refinement across the mesh. It is not as important for triangular and tetrahedral elements where algorithms such as Rivara refinement [66] can be used to perform local refinement and maintain consistency in the mesh. We will give several examples that make use of nonconforming quadrilateral elements in the following sections.

3.6   Solution techniques

In this section we will describe efficient iterative and direct methods for inverting the large algebraic systems that result from nonconforming spectral element discretizations. Iterative methods are more appropriate for steady-state calculations or calculations involving variable properties, such as a changing time step or a Helmholtz equation with a variable coefficient. For direct methods the issue is one of memory management — storing $\mathbf{A}$ as efficiently as possible without sacrificing the performance needed for fast back-substitution.

The development of fast direct and well-preconditioned iterative solvers represents a major advance towards the application of nonconforming spectral element methods to the simulation of turbulent flows on unstructured meshes.

Conjugate gradient iteration Conjugate gradient methods [11] have been particularly successful with spectral elements because the tensor-product form and local structure allows the global Helmholtz inner product to be evaluated using only elemental matrices. To solve the system $\mathbf{Au} = \mathbf{F}$ by the method of conjugate gradients we use the following algorithm:

$$k = 0; \; u_0 = 0; \; r_0 = \mathbf{F};$$
$$\textbf{while } r_k \neq 0$$
$$\qquad \text{Solve } \mathbf{M}q_k = r_k \; ; \; k = k + 1$$
$$\qquad \textbf{if } k = 1$$
$$\qquad\qquad p_1 = q_0$$
$$\qquad \textbf{else}$$
$$\qquad\qquad \beta_k = r_{k-1}^T q_{k-1} / r_{k-2}^T q_{k-2}$$
$$\qquad\qquad p_k = q_{k-1} + \beta_k p_{k-1}$$
$$\qquad \textbf{end}$$
$$\qquad \alpha_k = r_{k-1}^T q_{k-1} / p_k^T \mathbf{A} p_k$$
$$\qquad r_k = r_{k-1} - \alpha_k \mathbf{A} p_k$$
$$\qquad u_k = u_{k-1} + \alpha_k p_k$$
$$\textbf{end}$$
$$\mathbf{u} = u_k$$

where k is the iteration number, r_k is the residual, and p_k is the current search direction. The matrix $\mathbf{M}$ is a preconditioner used to improve the convergence rate of the method and is discussed in detail next.

Selection of a good preconditioner is critical for rapid convergence; the preconditioner must be spectrally close to the full stiffness matrix yet easy to invert. Popular preconditioners for spectral methods include incomplete Cholesky factorization and low-order (finite element, finite difference) approximations [26, 65]. Unfortunately, these preconditioners can be as complicated to construct for an unstructured mesh as the full stiffness matrix $\mathbf{A}$. Next we present three preconditioners which are simple to build and apply even when the mesh is unstructured.

In conjugate gradient methods the number of iterations required to reach a given error level scales as $\sqrt{\kappa_A}$. This is only an estimate, since the actual convergence rate is determined by the *distribution* of eigenvalues — if all of $\mathbf{A}$'s eigenvalues are clustered together, convergence is much faster. To assess the effectiveness of a given preconditioner we begin by looking at the condition number of $\mathbf{M}^{-1}\mathbf{A}$.

Each of the following methods is based on selecting a subset of entries from the full stiffness matrix. The first two preconditioners are diagonal matrices

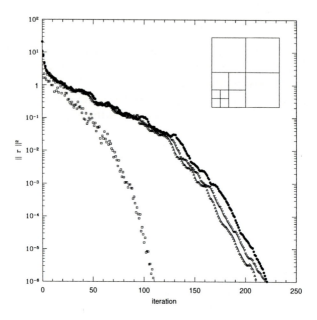

Fig. 3.7. Conjugate gradient iteration convergence history for a Helmholtz equation with $\lambda^2 = 1$: $\bullet$ = none, $\triangle$ = diagonal, $\triangledown$ = row-sum, and $\square$ = block-diagonal preconditioner.

Table 3.1. Condition numbers of $M^{-1}A$ for a Helmholtz equation with $\lambda^2 = 1$.

N	None	Diagonal	Row-Sum	Block-Diagonal
5	177	70	46	34
6	278	108	70	52
7	404	155	99	75
8	558	211	135	104
9	743	277	177	139
10	963	354	226	180
15	3042	961	677	517

given by

$$M_{ii} = A_{ii} \qquad \text{"diagonal"}, \qquad (3.12)$$

$$M_{ii} = \sum_{j=0}^{n_{\text{dof}}} |A_{ij}| \qquad \text{"row-sum"}, \qquad (3.13)$$

where $n_{\text{dof}} = \text{rank}(\mathbf{A})$; the diagonal (3.12) is sometimes called a point Jacobi preconditioner. Both are direct estimates of the spectrum of $\mathbf{A}$, and have the advantage of minimal storage and work. They can be quite effective for diagonally dominant systems such as the viscous correction step of the splitting scheme described in § 5. The third preconditioner is a block-diagonal matrix:

$$M_{ij} = \begin{cases} |A_{ij}| & \text{if } i \le n_{\text{bof}}, \, j = i \\ 0 & \text{if } i \le n_{\text{bof}}, \, j \ne i \\ A_{ij} & \text{otherwise} \end{cases} \qquad (3.14)$$

where n_{bof} is the number of mortar nodes in the mesh. The structure of this matrix assumes that $\mathbf{A}$ is arranged in the static condensation format described in Sect. 3.6. Applying this preconditioner amounts to storing and inverting the isolated blocks of $\mathbf{A}$ associated with the degrees of freedom on the interior of each element, while applying a simple diagonal matrix to the mortar nodes.

The following test examines the iterative solution to a Helmholtz equation for the two extreme cases $\lambda^2 = 1$ and $\lambda^2 = 10\,000$. Convergence is measured with respect to the solution $u(x, y) = \sin \pi x \sin \pi y$. The mesh has $K = 10$ elements generated by recursively subdividing a square domain, with $N = 15$ in each element. Figures 3.7 and 3.8 show the convergence history for the weakly and strongly diagonally dominant systems. The difference in convergence rates is explained in part by the condition numbers of $\mathbf{M}^{-1}\mathbf{A}$, given in Tab. 3.1 and Tab. 3.2. In spite of yielding a lower κ_A, the row-sum preconditioner converges slower and therefore offers no particular advantage over the simpler diagonal preconditioner. The block-diagonal matrix performs significantly better than the other two, effectively doubling the convergence rate in both cases. This preconditioner is fully parallelizable, and offers the most promise in distributed computing environments where the cost per iteration can include significant time performing interprocessor communication; its main drawbacks are the higher operation count and storage requirement. The methods described in the next section for implementing fully direct solvers can also be used to reduce the storage requirement for the block-diagonal preconditioner.

We conclude this section by giving the memory requirements and computational complexity for a preconditioned conjugate gradient (PCG) solver. Since the elemental Helmholtz operator can be evaluated using only the one-dimensional Lagrangian derivative matrix, the required memory is simply storage for the nodal values and geometric factors:

$$S_I = s_1 K N^2. \qquad (3.15)$$

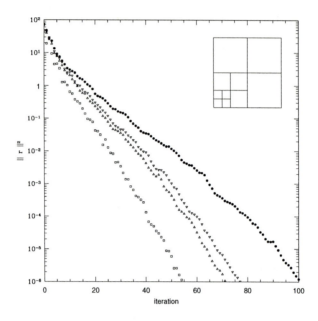

Fig. 3.8. Conjugate gradient iteration convergence history for a Helmholtz equation with $\lambda^2 = 10\,000$: $\bullet$ = none, $\triangle$ = diagonal, $\triangledown$ = row-sum, and $\square$ = block-diagonal preconditioner.

Table 3.2. Condition numbers of $M^{-1}A$ for a Helmholtz equation with $\lambda^2 = 10\,000$.

N	None	Diagonal	Row-Sum	Block-Diagonal
5	325	18.1	17.9	7.37
6	283	20.1	19.6	8.20
7	273	22.1	21.4	8.71
8	247	23.4	22.4	9.44
9	237	25.1	23.7	10.43
10	229	27.1	25.2	11.82
15	243	44.3	36.1	24.40

As mentioned above, the dominant numerical operations are vector-vector and matrix-vector products, although derivative calculations are folded into a more efficient matrix-matrix multiplication. The operation count for the entire solver is

$$C_I = J^\epsilon \left[c_1 K N^3 + c_2 K N^2 + c_3 K N \right],\tag{3.16}$$

where $J^\epsilon \propto \sqrt{K N^3}$ is the number of iterations required to reach a given error level ϵ. Our numerical results (Tables 3.1 and 3.2) show that with these preconditioners J^ϵ is still proportional to $K N^3$, but the constant is reduced. The block matrix operations required to compute the elemental inner products provide good data locality and can be coded efficiently on both vector processors and RISC microprocessors.

Static condensation The static condensation algorithm is a method for reducing the complexity of the stiffness matrices arising in finite element and spectral element methods. Static condensation is particularly attractive for unstructured spectral element methods because of the natural division of equations into those for boundaries (mortars) and element interiors. To apply this method to the discrete Helmholtz equation, we begin by writing partitioning the stiffness matrix into boundary and interior points:

$$\begin{bmatrix} A_{11} & A_{12} \\ A_{21} & A_{22} \end{bmatrix}^k \begin{bmatrix} \mathbf{u}_b \\ \mathbf{u}_i \end{bmatrix}^k = \begin{bmatrix} \mathbf{F}_b \\ \mathbf{F}_i \end{bmatrix}^k,\tag{3.17}$$

where A_{11} is the boundary matrix, $A_{12} = [A_{21}]^T$ is the coupling matrix, and A_{22} is the interior matrix. This system can be factored into one for the boundary (mortar) nodes and one for the interior nodes, so that on Ω^k:

$$[A_{11} - A_{21} A_{22}^{-1} A_{12}] \mathbf{u}_b = \mathbf{F}_b - [A_{21} A_{22}^{-1}] \mathbf{F}_i,\tag{3.18a}$$

$$A_{22} \mathbf{u}_i = \mathbf{F}_i - A_{21} \mathbf{u}_b.\tag{3.18b}$$

During a pre-processing phase, the global boundary matrix is assembled by summing the elemental matrices,

$$\mathbf{A}_{11} = \sum_{k=1}^{K}{}' [A_{11} - A_{21} A_{22}^{-1} A_{12}],\tag{3.19}$$

and prepared for the solution phase by computing its LU factorization. Equation (3.19) may also be recognized as the Schur complement of A_{22} in A. As part of this phase we also compute and store for each element the inverse of the interior matrix $[A_{22}^{-1}]$ and its product with the coupling matrix $[A_{21} A_{22}^{-1}]$. The system is solved by setting up the modified right-hand side of the global boundary equations, solving the boundary equations using back-substitution, and then computing the solution on the interior of each element using direct matrix multiplication. Because the coupling between elements is only C^0,

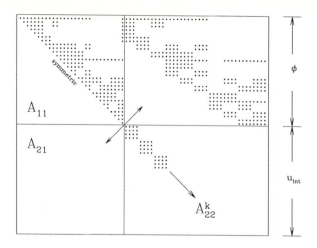

Fig. 3.9. Static condensation form of the spectral element stiffness matrix. The vector $\phi = \mathbf{u}_b$ represents the boundary (mortar) solution, while $\mathbf{u}_i$ represents the interior solution.

the element interiors are independent of each other and on a multiprocessor system this final stage can be solved concurrently.

Figure 3.9 illustrates the structure of a typical spectral element stiffness matrix factored using this approach. To reduce computational time and memory requirements for the boundary phase of the direct solver, we wish to find an optimal form of the discrete system corresponding to a minimum bandwidth for the matrix $\mathbf{A}_{11}$. This is complicated by the irregular connectivity generated by the using of nonconforming elements. One approach to bandwidth optimization is to think of the problem in terms of finding an optimal path through the mesh that visits "nearest neighbors." During each of the K stages of the optimization, an estimate is made of the new bandwidth that results from adding one of the unnumbered elements to the current path. The element corresponding to the largest increase is chosen for numbering, resulting in what is essentially a Greedy algorithm. This basic concept is illustrated in Fig. 3.10. The reduction in bandwidth translates to direct savings in memory and quadratic savings in computational cost. Note that standard methods of bandwidth reduction used for finite elements, e.g. the Reverse Cuthill-McKee algorithm, can also be used, although they only need be applied to the boundary system.

The search for an optimal numbering system can be accomplished during preprocessing, so the extra work has no impact on the simulation cost and can result in significant savings. Table 3.3 shows the results of bandwidth optimization for each of the computational domains pictured in Fig. 3.11. For computers where memory is a limitation, this procedure can determine whether an in-core solution is even possible. Other simple memory optimiza-

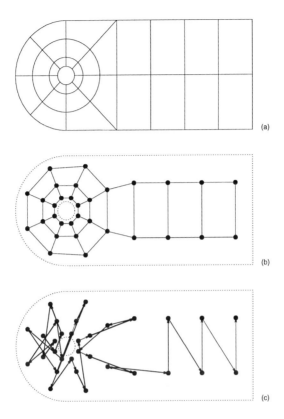

Fig. 3.10. Bandwidth optimization for a spectral element mesh: (a) computational domain, (b) connectivity graph and (c) an optimal path for numbering the boundary nodes in the mesh. Line thickness demonstrates the change in global bandwidth with each step.

Table 3.3. Matrix rank and optimized bandwidth of three complex-geometry domains representative of internal and external flow problems.

Mesh	K	N	rank	original	optimized	savings
riblets	91	9	1484	1483	250	83%
cylinder	114	11	2416	2406	402	83%
half-cylinder	176	7	2177	2156	399	81%

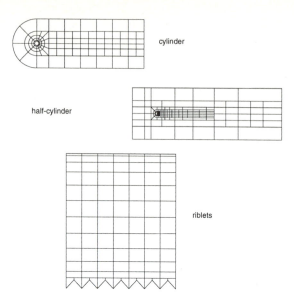

Fig. 3.11. Nonconforming meshes used to test the bandwidth optimization.

tions include storage of only a single copy of the interior and coupling matrices for each element with the same geometry, and evaluation of the force vector **F** using tensor product summation instead of matrix operations. By carefully organizing matrix usage, the overall memory requirement scales as

$$S_D = \frac{1}{2}s_1 K^2 N^2 + s_2 K N^3 + s_3 K N^4. \tag{3.20}$$

As mentioned in the introduction to this section, the direct solver is advantageous only when the cost of factoring this stiffness matrix can be spread over a large number of solutions. Therefore, we consider only the cost of a back-substitution using the factored stiffness matrix, for which the operation count scales as

$$C_D = c_1 K^{3/2} N^2 + c_2 K N^4 + c_3 K N. \tag{3.21}$$

For a well-conditioned, diagonally-dominant system this method usually results in at least a factor of two savings versus an iterative solver. For a system that is not diagonally-dominant, like the Navier–Stokes pressure equation, it can be faster by a full order of magnitude.

3.7   Examples

Advection As a model for the nonlinear term in the Navier–Stokes equations, we now look at a linear advection equation. It can be written as

$$\frac{\partial u}{\partial t} - \boldsymbol{a} \cdot \nabla u = 0 \quad \text{on } \Omega, \tag{3.22}$$

where u is a scalar and $\boldsymbol{a}$ is a given velocity vector field defined on Ω. For simplicity we assume $\boldsymbol{a}$ is constant, divergence free, and normalized so that $|\boldsymbol{a}| = 1$ pointwise. To complete the statement of the problem we must also supply boundary and initial conditions for u, but we leave these open for now. Equation (3.22) represents the transport of u by the velocity field $\boldsymbol{a}$, and it plays an important role in many areas of physics. Here we will be concerned primarily with developing stable time integration schemes to go along with high-order spatial discretizations.

The weak form of (3.22) is: Find $u^h \in \mathcal{S}^h$ such that for all $w^h \in \mathcal{V}^h$

$$\int_\Omega w^h (\dot{u} - \boldsymbol{a} \cdot \nabla u^h) \, d\Omega = 0, \tag{3.23}$$

where $\dot{u} = \partial u^h / \partial t$. The discrete form of the elemental system is

$$\mathbf{M}^k \dot{\mathbf{u}}^k - \mathbf{D}^k \mathbf{u}^k = 0, \tag{3.24}$$

where the elemental mass and advection matrices are

$$\mathbf{M}^k_{pq} = (\phi_p, \phi_q)_{\Omega^k}, \quad \mathbf{D}^k_{pq} = (\boldsymbol{a} \cdot \nabla \phi_p, \phi_q)_{\Omega^k}. \tag{3.25}$$

We interpret the vector $\mathbf{u}^k$ as containing either the nodal values of the solution or the expansion coefficients of the modal basis functions.

Although external boundary conditions are part of the physical statement of the problem, to form the global system and complete the discretization we have to choose "internal" boundary conditions for the subdomain interfaces. One possibility is to use the method of characteristics, which reduces to simple upwinding for the scalar equation. Alternatively, C^0 continuity can be imposed by forming a weighted average of the flux $\boldsymbol{a} \cdot \nabla u$ at element boundaries, using the mass matrix $\mathbf{M}^k$ to provide the weights. This procedure is also stable for smooth solutions, and numerical experiments indicate that for well-resolved problems there is little difference between the accuracy or stability of the two methods. The averaging method is much easier to program since it corresponds to the "direct stiffness summation" described earlier; in this case the global system matrices are formed as

$$\mathbf{M} = \sum_{k=1}^{K}{}' \mathbf{M}^k, \quad \mathbf{D} = \sum_{k=1}^{K}{}' \mathbf{D}^k, \tag{3.26}$$

and the solution is $\dot{\mathbf{u}} = \mathbf{M}^{-1} \mathbf{D} \mathbf{u}$.

Since the GLL nodal basis functions are discretely orthogonal, the associated mass matrix is diagonal and the inversion is trivial. The modal basis is only semi-orthogonal and the corresponding mass matrix is sparse but not diagonal. Since the modal mass matrix is symmetric and positive-definite, we can use iterative methods to invert it like preconditioned conjugate gradient iteration that work well with the discrete Laplacian [21, 24].

To propagate the solution u we discretize time and apply a numerical time integration scheme with some step size Δt. The central question is whether the method and time step we choose result in a stable scheme. For nonlinear equations like Navier–Stokes, explicit methods are generally used for the convective terms and the stability is determined by a CFL-type condition of the form

$$|a|\frac{\Delta t}{\Delta x} \leq \text{const.} \tag{3.27}$$

However, there is no direct analog of the CFL condition for high-order methods we have to make a heuristic estimate for the value of Δt that will keep the scheme stable, and to do this we need to determine the growth rate of the eigenvalues of the discrete system.

Eigenvalues of the linear advection operator are determined by the non-trivial solutions (λ, u) of

$$(a \cdot \nabla - \lambda)u = 0. \tag{3.28}$$

Eigenvalues of the discrete problem are determined by the system

$$(\mathbf{G} - \lambda \mathbf{I})\,\mathbf{u} = 0, \tag{3.29}$$

where $\mathbf{G} = \mathbf{M}^{-1}\mathbf{D}$. This yields the spectrum associated with the spatial discretization, and for stability the eigenvalues of the related matrix $(\mathbf{I}+\Delta t\mathbf{G})$ must lie within the stability region of the time stepping scheme. To state this another way, the time step Δt must balance the largest eigenvalues of $\mathbf{G}$.

First we consider the modal basis on triangular elements, using a periodic domain discretized as shown in Fig. 3.12. We can determine the maximum eigenvalue for wavevectors $a = (\cos\theta, \sin\theta)$ corresponding to various directions of propagation across the domain. The worst case $(\theta = \pi/4)$ corresponds to a wave propagating through the tip of the triangle where the mesh spacing is the smallest. Figure 3.13 shows the maximum eigenvalue versus expansion order N, indicating that $\max(|\lambda|) \sim O(N^2)$. The same result applies to quadrilateral elements using the nodal basis. Figure 3.14 shows the maximum eigenvalues for a simple rectangular domain, again demonstrating $O(N^2)$ growth.

From this, we can form the following heuristic stability criteria:

$$\Delta t \leq \text{const.}/|a|N^2, \tag{3.30}$$

where the constant depends on the particular time stepping method and the uniformity of the mesh. Generally, this criterion should be checked on each element in the mesh and the smallest stable value of Δt chosen for the integration, possibly adapting with each time step. Although the examples we showed were for two-dimensional problems, the same criteria apply to one- and three-dimensional problems as well.

The explanation for the stability limit given by (3.30) is that the polynomial basis clusters the mesh points near the ends of the element, so that near the element boundaries $\Delta x \sim N^{-2}$. This estimate is standard in polynomial spectral methods [34]. On an equispaced grid that might be used with a

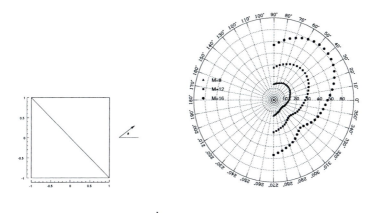

Fig. 3.12. For the periodic domain shown on the left we consider a wave propagating with velocity $\boldsymbol{a} = (\cos\theta, \sin\theta)$. The polar plot on the right shows the maximum eigenvalue of the discrete advection operator for several wave orientations and different number of modes $M \equiv N$.

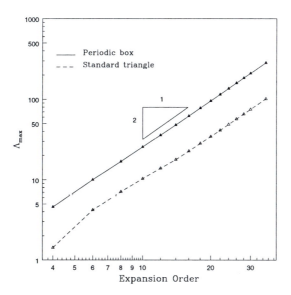

Fig. 3.13. Growth rate of the maximum eigenvalue with respect to polynomial order N for the modal basis [75].

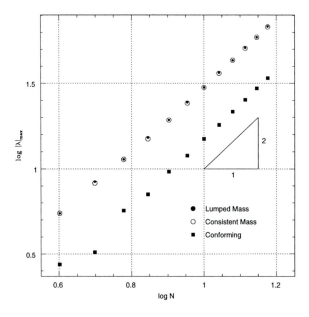

Fig. 3.14. Growth rate of the maximum eigenvalue of the discrete advection operator **G** on conforming and nonconforming meshes versus polynomial order N for the nodal basis.

Fourier spectral method or a finite difference discretization, the mesh spacing is $\Delta x \sim N^{-1}$ and this limit is less strict:

$$\Delta t \leq \text{const.}/|a|N. \tag{3.31}$$

In order to weaken the limit on Δt in (3.30) for polynomial spectral methods, we can redistribute the collocation points to achieve a more even distribution. Although an arbitrary mapping may lead to unstable approximations, stable transformations have been developed that can result in a CFL limit for spectral methods quite close to the finite difference method on uniform grids [53]. In practice, a typical polynomial order is $N \leq 20$ and the difference between (3.30) and (3.31) is not a serious disadvantage to the more straightforward approach.

Diffusion The diffusion of a scalar u with diffusivity ν is described by the equation

$$\frac{\partial u}{\partial t} - \nu \nabla^2 u = 0. \tag{3.32}$$

It represents a type of "averaging" of u that might describe the spreading of heat, momentum, or vorticity in a fluid. It is an important equation that shows up in many branches of physics, but we will put it aside for the moment in favor of another model problem for the approximation of elliptic equations; at the end of this section we show how the two are related.

The Helmholtz problem is: given $\kappa \in \mathcal{R}$ and smooth functions $f : \Omega \to \mathcal{R}$, $g : \Gamma_g \to \mathcal{R}$, and $h : \Gamma_h \to \mathcal{R}$, find u such that

$$\nabla^2 u - \kappa^2 u + f = 0 \quad \text{in } \Omega, \tag{3.33}$$

subject to the boundary conditions

$$u = g \quad \text{on } \Gamma_g, \tag{3.34}$$

$$\boldsymbol{n} \cdot \nabla u = h \quad \text{on } \Gamma_h. \tag{3.35}$$

There are some special cases of equation (3.33): if κ is zero it is called Poisson's equation, and if κ and f are both zero it is called Laplace's equation.

The Galerkin approximation to (3.33) is developed in much the same way as already shown for the one-dimensional problem in Section 2. We only need to extend the ideas to two- and three-dimensions. The variational form of our boundary value problem is: Find $u \in \mathcal{S}$ such that for all $w \in \mathcal{V}$:

$$a(u, w) = (f, w) + (h, w)_{\Gamma_h}, \tag{3.36}$$

where the symmetric, bilinear form $a(\cdot, \cdot)$ is defined as

$$a(u, w) = \int_\Omega (\nabla u \, \nabla w + \kappa^2 \, uw) \, \mathrm{d}\Omega. \tag{3.37}$$

Let $\mathcal{S}^h \subset \mathcal{S}$ be the space of C^0 piecewise polynomial interpolants of degree N that satisfy the essential boundary condition, and $\mathcal{V}^h \subset \mathcal{V}$ a similar space of functions that have value zero on Γ_g; these are our basis functions $\phi_i(\boldsymbol{x})$. To complete the Galerkin approximation to (3.36), we separate the solution into $u^h = g^h + v^h$, where $g^h \in \mathcal{S}^h$ is a polynomial approximation to g and $v^h \in \mathcal{V}^h$ is the unknown part of u^h. Usually, g^h will be an initial guess for u^h that satisfies the boundary conditions but not the weak form, so v^h is simply a correction to make it exact. Evaluation of the integral form (3.36) by numerical quadrature gives the elemental matrices:

$$\mathbf{A}^k_{pq} = a(\phi_p, \phi_q)_{\Omega^k}, \tag{3.38}$$

$$\mathbf{F}^k_p = (f, \phi_p)_{\Omega^k} + (h, \phi_p)_{\Gamma_h} - a(g^h, \phi_p)_{\Omega^k}, \tag{3.39}$$

and the discrete Galerkin equation for the kth element as

$$\mathbf{A}^k \mathbf{v}^k = \mathbf{F}^k. \tag{3.40}$$

To form the global system there is only one choice for the "internal" boundary conditions, and that is to apply direct stiffness summation to get

$$\mathbf{A} = \sum_{k=1}^{K} {}' \mathbf{A}^k, \quad \mathbf{F} = \sum_{k=1}^{K} {}' \mathbf{F}^k. \tag{3.41}$$

The final algebraic system,

$$\mathbf{A}_{pq} \mathbf{v}_q = \mathbf{F}_p, \quad p, q = 1, \dots, n_{\mathrm{dof}} \tag{3.42}$$

requires the inversion of a symmetric, positive-definite matrix $\mathbf{A}$ whose bandwidth is determined by the index set we use to map between the local and global systems.

Now we return to the problem of integrating the diffusion equation. We could follow the same approach used for the advection equation, writing the semi-discrete form as

$$\mathbf{M}\dot{\mathbf{u}} = \nu \mathbf{A}\mathbf{u}. \tag{3.43}$$

However, the discrete Laplace matrix $\mathbf{A} \approx \mathbf{D}^2$ and for stability the time step would scale like $\Delta t \sim N^{-4}$; this has been demonstrated more rigorously for both the nodal as well as the modal basis in [40, 75]. For this reason, the diffusion equation is usually integrated with *implicit* rather than explicit methods. For example, we can approximate the time derivative with an unconditionally stable backward Euler approximation:

$$\frac{u^{n+1} - u^n}{\Delta t} = \nu \nabla^2 u^{n+1}. \tag{3.44}$$

Rearranging this, we get

$$(\nabla^2 - \frac{1}{\nu \Delta t}) u^{n+1} + \frac{1}{\nu \Delta t} u^n = 0, \tag{3.45}$$

which is immediately recognized as the Helmholtz equation with $\kappa = 1/\sqrt{\nu \Delta t}$ and $f = u^n/\nu \Delta t$. After developing appropriate methods for equation (3.33), we can solve any implicit approximation to the diffusion equation.

4   Adaptive Mesh Refinement

Thus far we have looked the development of high-order methods that incorporate the essential features needed to adaptively refine the discrete model of a flow during a simulation. We refer to this as *dynamic refinement*. In spectral or *h-p* finite element methods refinement takes place by decreasing the size of the grid elements (*h*-refinement) or increasing the order of the solution (*p*-refinement). As simple as this sounds, the algorithms for driving adaptive refinement at a "high level" are quite complicated. Adaptive mesh refinement is often as much of an art as a science: it depends on the experienced selection of tolerances and refinement criteria that are highly problem-dependent. The implementation of adaptive methods is equally complex and usually involves the development of irregular, dynamically changing data structures that reflect the complexity of the discrete models.

The basis functions that we have looked at so far have sufficiently flexibility to support the necessary flavors of mesh refinement. High-order expansions on triangular and tetrahedral elements [75] are probably the most straightforward because refinement can be implemented without any fundamental changes in the topology of the mesh. Quadrilateral and hexahedral grids do require a different topology to be efficient, namely nonconforming element boundaries between regions with different spatial resolution. Several choices for high-order expansions on quadrilateral grids have appeared in the recent literature, including Chebyshev polynomials combined with a multipole expansion for the Poisson problem [35], high-order B-spline expansions on locally refined grids [72], and staggered-grid Chebyshev spectral collocation methods for simulating compressible fluid flows [51, 52]. These methods share a common thread in that the grids used to discretize space look similar, but they differ in both the way an approximate solution is represented and how nonconforming elements of the mesh are pieced together. All of these techniques may be classified as *spectral element methods* because of the general combination of domain decomposition and high-order polynomial expansions.

In this section we look at the implementation of a high-order adaptive code based on the nonconforming spectral element method developed in Sect. 3. In practice this method is used with high-order polynomials ($p \approx 4$ to 16) and a mesh of elements that is generated adaptively by *h*-refinement. We will not attempt to refine both the elements and the basis functions simultaneously as the author's experience indicates that uniformly high p and adaptive mesh refinement leads to an efficient solution for a wide variety of problems.

The formulation based on mortar elements [14] allows completely arbitrary assembly of nonconforming elements. However, our goal is to develop automatic procedures for generating an appropriate mesh and this calls for some compromises. To simplify the encoding of the mesh we will require the refinement to propagate down a quadtree (two-dimensional geometries) or octtree (three-dimensional geometries). A basic description of the mesh generation procedure is provided in Sect. 4.2. This is found to be a suitable restriction for problems with smooth solutions and leads to a significant reduction in the complexity of the data structure needed to represent the many levels in the refined grid. For complex geometries the mesh may incorporate multiple trees at the coarse level.

To give a more specific introduction to the goals of developing an adaptive spectral element method, Fig. 4.15 shows a sample calculation for the impulsively started flow past a bluff plate. In this simulation the solution field is generated by integrating the incompressible Navier–Stokes equations from an initial state of zero motion. The characteristic scales in the problem are the free-stream speed u_∞, the plate diameter d, and the kinematic viscosity of the fluid ν. The Reynolds number, defined as $Re \equiv u_\infty d/\nu$, is set to the value $Re = 1000$. The lower part of the figure shows the global domain used to represent the flow around the plate. A symmetry condition is imposed along the centerline so that only one half of the flow field needs to be computed. The upper part of the figure is an enlargement of the near wake region. It shows both the vorticity of the developing flow at an early time and the adaptively generated mesh. Each element is an 8×8 point subdomain ($p = 7$) of the global solution. A large number of separate 'trees' are needed at the coarse level to correctly model the beveled geometry of the finite-thickness plate. The initial stage of mesh generation is done by hand to provide the correct starting geometry. Once the problem is handed to the flow solver the additional adaptivity in the mesh is based on a maximum allowable approximation error in the vorticity field. Because the algorithms for time integration in problems like the one illustrated in Fig. 4.15 are generally semi-implicit, the computational issues that arise are somewhat different when compared to other methods that incorporate adaptive meshes. We are interested primarily in studying *incompressible* flows governed by the Navier–Stokes and Euler equations. Because of the elliptic nature of the governing equations (due in part to the incompressibility constraint), local time-stepping is not usually an option. Therefore, solving the elliptic boundary-value problems that arise in these systems is a particular challenge. Even for two-dimensional flows the resolution needed to maintain sufficiently high accuracy can lead to very large systems of equations, and computational efficiency is an important issue. In the past this meant algorithms that could be *vectorized*, while today it means algorithms that can be *parallelized*. There is a close relationship between spectral elements and finite elements, so when it comes to parallel computing many of the same problems (e.g. load balancing) arise, and similar solutions apply. Section 4.4 addresses the implementation of this method for parallel computers with a programming model based on a weakly coherent shared memory which is synchronized via message passing.

Just as important as overall computational performance are the algorithms used for driving adaptive refinement. Ideally such an algorithm would take as input an error estimate and produce as output a new discrete model or mesh that reduces the error. The basic problems are the lack of an error estimate for nonlinear systems and the unlimited ways in which such

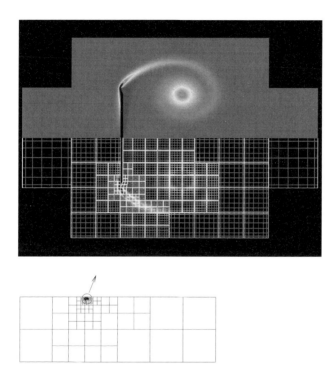

Fig. 4.15. Simulation of the impulsively started flow past a bluff plate at $Re = 1000$ using an adaptive spectral element method: (*top*) close-up of the mesh and vorticity of the flow a short time after the impulsive start; (*bottom*) global computational domain.

an algorithm could improve the discrete model. The latter problem is addressed by restricting 'improvements' to propagating refinement down the tree as described in Sect. 4.2. The former problem is addressed with a pseudoheuristic error estimate based on the local polynomial spectrum as described in Sect. 4.3. Depending on the nonlinearity in the partial differential equations being solved, parts of the spectrum will give an accurate approximation to the true solution and parts will be polluted. We estimate the order of magnitude of the local error by examining the decay along the tail of the local polynomial spectrum. In a general sense, this heuristic flags locations in the mesh where the polynomial basis fails to provide a good description of the solution. For simple problems (linear, one-dimensional) this can be formally related to the true difference between the exact solution and the approximate solution, i.e. the approximation error. For more interesting problems it is shown to be a robust guide for driving adaptivity. The heuristic is easy to compute but is only accurate as an error estimate in computations with sufficiently high p, meaning that the local polynomial coefficients should decay like $|a_n| \sim \exp(-\sigma n)$ for $p = n \gg 1$. This is generally not true near singular points (e.g. corners) and these locations are automatically flagged for refinement. The method based on local spectra is compared to simpler heuristics such as refining in regions with strong gradients and the two are shown to lead to quite different results. In general the local spectrum works well and is a good match to the overall computational strategy.

The effectiveness of this approach is first demonstrated in Sect. 4.5 for scalar problems where the convergence and behavior of the refinement criteria can be checked carefully. More complicated examples are provided in Sect. 5 with several incompressible flow problem. An attempt is made throughout to illustrate both the benefits and difficulties of using this kind of high-order adaptive method, and to point out applications where it may have some advantage over other numerical methods.

4.1   Framework

In this section we restrict our attention to two-dimensional problems. Most of the difficulties arise in two dimensions and there are no fundamental barriers (other than computing power) in extending the method to three dimensions. To begin, let D be some region of space that has been partitioned into K subdomains which we denote $D^{(k)}$. We consider two related problems:

1. Given a discretization tolerance ϵ, generate a spatial discretization $D = \{D^{(k)}\}$ that allows the tolerance to be met;
2. Given a spatial discretization $D = \{D^{(k)}\}$, generate a finite-dimensional approximation $u^h \approx u$. The function u may be given explicitly or implicitly, i.e. as the solution of a boundary-value problem.

Our approach to problem (1) is to create a hierarchy of grids by forming a quadtree partition of D. This provides the computational domain for problem

(2) where we apply a nonconforming spectral element method to approximate u^h.

4.2   Mesh generation

The mesh generation problem is somewhat simpler, so we describe that first. A quadtree is a partition of two-dimensional space into squares. Each square is a *node* of the tree. It has up to four daughters, obtained by bisecting the square along each dimension. Each node in a quadtree has geometrical properties (spatial coordinates, size) and topological properties (parents, daughters, siblings). Geometrical properties of daughter nodes are inherited from parents, and thus the geometrical properties of the entire tree are determined by the root node.

To represent the topological aspects of the tree we use an idea originally developed for gravitational N-body problems [70]. Every possible square $S^{(i)}$ is assigned a unique integer *key*. The root of the tree is $S^{(1)}$ with key 1. The daughters of any node are obtained by a left-shift of two bits of the parent's key, followed by a binary *or* in the range 00–11 (binary) to distinguish each sibling. A node's parent is obtained by a two bit right-shift of its own key. Since the set of keys installed in the tree at any time is obviously much smaller than the set of all possible keys, a hash table is used for storage and lookup. From the complete set of nodes in the tree we choose a certain subset $D^{(k)} \subseteq S^{(i)}$ to form the *active* elements of the computational domain. Figure 4.16 shows a four-level quadtree with thirteen nodes and $K = 10$ active elements. Active elements in the figure are shown with a solid outline while inactive elements are shown with a dashed outline. Inactive elements are retained so that they are available for coarsening the mesh, if necessary. The only requirement enforced on the topology of the mesh is that active elements that share a boundary segment live at most one refinement level apart, limiting adjacent elements to a two-to-one refinement ratio. This imposes a certain smoothness on the change in resolution in the mesh that is appropriate for the class of smooth functions we wish to represent.

4.3   Refinement criteria

The adaptive mesh generation described above and high-order domain decomposition methods described in §3 are coupled through the refinement criteria used to drive adaptivity. Here we consider three types of refinement criteria.

The first is by far the simplest: refine everywhere that solution gradients are large. We can enforce this idea by requiring

$$\| \nabla u^{(k)} \| \leq \epsilon \, \| u^h \|_1 \qquad (4.46)$$

everywhere in the mesh, where $\| \cdot \|$ is the L_2 norm, $\| \cdot \|_1$ is the H^1 norm, and ϵ is the discretization tolerance. This is a common refinement criteria in cases where there is simply no alternative measure of solution errors.

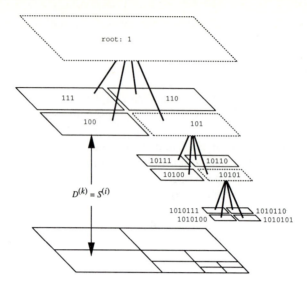

Fig. 4.16. A four-level quadtree mesh, expanded to show the elements that make up each level. Each leaf node $S^{(i)}$ has a unique integer key shown in binary. Daughter keys are generated from a parent's key by a two-bit left shift, followed by a binary *or* in the range 00–11. The active elements $D^{(k)}$ that make up the current discretization are shown with a solid outline.

The second type takes direct advantage of the high-order polynomial basis. Consider the expansion of a given smooth function u over the domain $D = [-1, 1]^2$ in terms of Legendre polynomials:

$$u(x,y) = \sum_{n=0}^{\infty} \sum_{m=0}^{\infty} a_{n,m} P_n(x) P_m(y). \tag{4.47}$$

The expansion coefficients are given by

$$a_{n,m} = \frac{1}{c_n c_m} \int_{-1}^{1} \int_{-1}^{1} u P_n P_m \, |J| \, \mathrm{d}x \, \mathrm{d}y, \tag{4.48}$$

where the normalization constant is $c_i = (i + 1/2)^{-1}$. We have included the Jacobian $|J|$ to include the effects of element size and other geometric transformations, e.g. curvilinear boundaries. There is nothing magical about Legendre polynomials—they are simply a convenient orthogonal basis for projecting the approximation onto. Since our approximate solution $u^h \approx u$ is formed essentially by truncating this expansion at some finite order p, we can form an estimate of the approximation error $\| u - u^h \|$ by examining the tail of the spectrum.

To do so we first average over polynomials in x and y to produce an equivalent one-dimensional spectrum:

$$\bar{a}_p = |a_{p,p}| + \sum_{i=0}^{p-1} |a_{i,p}| + |a_{p,i}|. \tag{4.49}$$

Next we replace the discrete spectrum $\bar{a}_p$ with an approximation to a decaying exponential:

$$\tilde{a}(n) = \text{const.} \times \exp(-\alpha n). \tag{4.50}$$

The function $\tilde{a}(n)$ is a least squares best fit to the last four points in the spectrum $\bar{a}_p$. Our refinement criteria becomes

$$\left(\tilde{a}(p)^2 + \int_{p+1}^{\infty} \tilde{a}(n)^2 \, dn \right)^{1/2} \leq \epsilon \, \| \, u^h \, \| \, . \tag{4.51}$$

The only practical complication here is making sure the decay rate $\alpha > 0$ so that the integral converges. Otherwise, the estimate is ignored and the element is flagged for immediate refinement. This method is analyzed in [60] where it is shown to be an effective refinement criteria for driving h-p refinement.

The third refinement criteria is similar. Since the main contribution to (4.51) comes from the coefficients of order p, we can simply sum along the tail of the spectrum. For an accurate representation of u we require the spectrum to satisfy the discretization tolerance:

$$|a_{p,p}| + \sum_{i=0}^{p-1} |a_{i,p}| + |a_{p,i}| \leq \epsilon \, \| \, u^h \, \| \, . \tag{4.52}$$

This method is somewhat simpler to apply and, as we will see, produces almost identical results.

To use these polynomial spectrum criteria with our spectral element method (based on GLL polynomials) we first perform a Legendre transform of the local solution $u^{(k)} \to a_{n,m}$ and then use (4.51) or (4.52) to decide if the element should be refined. Although we keep p fixed, the error is reduced because we approximate u over a smaller region $D^{(k)}$. This basic idea is illustrated in Fig. 4.17. Here a smooth function $f(x, y)$ has been projected onto the Legendre polynomials of order $p \leq 64$. For a given order p the true approximation error would be given by the sum over all coefficients not contained in the box $m, n \leq p$. We estimate the magnitude of that error by simply summing coefficients along the solid lines.

4.4 Implementation notes

We end this section with a few additional notes on implementation. The algorithms described above have been implemented using a combination of C for

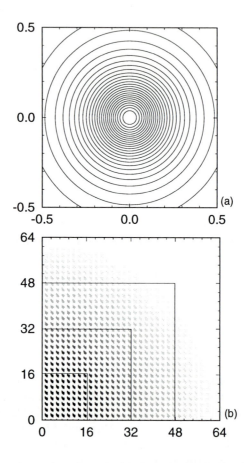

Fig. 4.17. Error estimate based on the local polynomial spectrum: (a) contours of the function $f = 1/(1 + 25r^2)$; (b) scatter plot showing the polynomial coefficients $|a_{n,m}|$. An estimate of the basis accuracy for a given polynomial order is formed by summing $|a_{n,m}|$ along the solid lines: $\epsilon = 0.0962$ ($p = 16$); $\epsilon = 0.000314$ ($p = 32$); $\epsilon = 8.25 \times 10^{-7}$ ($p = 48$).

the computational modules and G+ for high-level data types like `Element` $\equiv D^{(k)}$ and `Field` $\equiv u^h$ that make up the discretization. The logic and control structure needed for most of the code are the same as in any algorithm for finite element methods. The most complex problem is maintaining the connectivity of the mesh dynamically, and the approach taken here is worth mentioning. The geometry and topology of the mesh are closely connected.

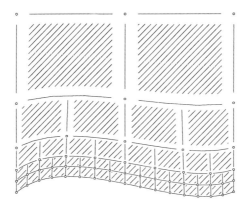

Fig. 4.18. The logical structure of a spectral element mesh can be divided into three geometric parts: (o) vertices, (—) edges, and (*shaded*) interiors. Edges and vertices define the connectivity in the mesh.

Figure 4.18 shows the three geometric elements of the discretization: vertices, edges, and interiors. Obviously interior points are completely local to an element and play no role in the global system. All connectivity in the mesh is through the edges and vertices. Because of the method used to construct the grid these geometric elements are interlocking. The midpoint of each nonconforming edge aligns with the shared vertex of its two adjacent elements. As discussed below, this feature is used to simplify the procedure for setting up the mesh topology.

Figure 4.18 shows one other side effect of the mesh generation. Internal curvilinear boundaries are automatically propagated down the various levels of the refinement tree because of the isoparametric representation of the geometry. In the same way that a solution field is projected onto a new set of elements, the polynomial representation of the geometry can also be projected to a finer grid. On the other hand, external boundaries like the B-spline segment shown as the lower boundary in the figure are explicitly re-evaluated to keep the representation as accurate as possible.

How does one represent the topology of this kind of mesh? One solution is to use pointers. This immediately runs into the problem of interpreting pointers to objects on remote processors if the computation is running in parallel. Instead we use the concept of a *voxel database* (VDB) of geometric positions in the mesh [85]. A VDB may be thought of as register of position–subscript pairs. To each position stored in the VDB we assign a unique integer subscript so that data may be associated with points in space by using the subscript as an index into an array.

The basic idea is illustrated in Fig. 4.19. The number of times a position is registered is its *multiplicity*. Data objects that share positions also share memory by virtue of a common subscript. In essence the VDB provides a natural map of the mesh geometry onto the computer's memory. This basic paradigm can be used to implement many types of finite element or finite volume methods [85].

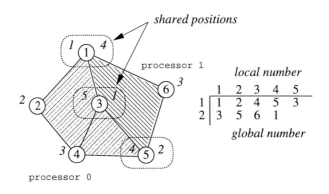

Fig. 4.19. Connectivity and communications are established by building a *voxel database* (VDB) of positions. A VDB maps each position to a unique index or subscript. It also tracks points shared by multiple processors to provide a loosely synchronous shared memory. Points that share memory are those at the same geometric position.

To establish the connectivity of a mesh like the one depicted in Fig. 4.18 we build two separate VDBs: one for the vertices and one for the midpoints of the edges. Every vertex with multiplicity one that does not lie along an external boundary is *virtual* and not part of the true mesh degrees of freedom. Every edge with multiplicity one that does not lie along an external boundary is *nonconforming*. For each nonconforming edge we make a second query to the VDB using the endpoints. If there is a match then the edge is also *virtual* and we store the subscript of the adjacent edge. Otherwise it is simply flagged as an internal nonconforming boundary segment.

The shared memory represented by a VDB is extended across processor boundaries by passing around a list of local positions and comparing against those registered remotely. A communications link is established for each common position. The shared memory at each point is weakly coherent and must be synchronized by explicit message passing. For example, elements on separate processors with a common boundary segment share data along an edge. Each processor may update its edge values independently and then call a synchronization routine that combines local and remote values to produce a globally consistent data set. For further details see [85].

There is very little overhead for the adaptive versus non-adaptive data structure: just one integer (the node key) per element. Likewise, an iterative solver for sparse systems incurs no performance penalty just because the underlying mesh is adaptive. When approached in the right way the conversion to a solution adaptive code is almost trivial. To a large degree this is because of the unstructured nature of the spectral element method we built upon.

4.5   Examples

Next we illustrate the performance of the method with a few simple test problems. First we consider the solution of the Poisson equation $\nabla^2 u = f$ with the right-hand-side given by

$$f(x,y) = (400^2 r^2 - 800)e^{-400r^2/2}, \tag{4.53}$$

where $r^2 = x^2 + y^2$. The exact solution is given by

$$u(x,y) = e^{-400r^2/2}. \tag{4.54}$$

We take the computational domain $D = [-0.5, 0.5]^2$ and impose homogeneous boundary conditions $u = 0$ along the perimeter ∂D. This same test case is studied in [35] to check the performance of a fast multipole method using a similar type of spatial discretization.

Table 4.4. Solution times and relative errors for solving the Poisson equation on a uniform grid with order $p = 7$ elements. Columns (I) and (II) show the estimated error using the exponential fit and summing the trace of the Legendre polynomial spectrum, respectively.

No. levels	No. points	(I)	(II)	$\| u - u^h \|$
0	64			0.534
1	256	0.0127	0.0117	0.0113
2	1024	0.000702	0.000735	0.000389
3	4096	2.625×10^{-6}	2.575×10^{-6}	1.318×10^{-6}
4	16384	1.189×10^{-7}	1.187×10^{-7}	8.212×10^{-8}

Since this problem has a well-defined exact solution, we begin by comparing the error estimates and the true error $\| u - u^h \|$ on a uniformly refined grid (table 4.4). This table shows that the error estimates are actually quite sharp, differing from the L_2 error by only a small multiplicative factor. Also note that the spectrum-based estimate are nearly equivalent. This is true in general. Because the trace is easier and faster to compute, this is the method that will be used from this point forward unless noted otherwise.

Adaptive mesh generation based on the different refinement criteria is illustrated in Fig. 4.20. In this case both methods produce roughly equivalent discretizations. The grid is refined in approximately the same location and to the same depth for a given discretization tolerance. Both methods generate a six-level quadtree with roughly the same number of active elements ($K \approx 300$) using a uniform basis of order $p = 7$.

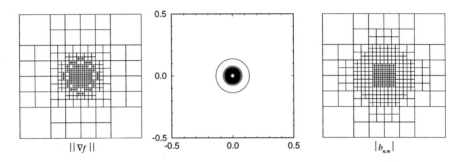

Fig. 4.20. Adaptive mesh generation: (*center*) contours of the function $f(x, y)$ inside the computational domain; (*left*) adaption based on function gradients with a tolerance of $\epsilon = 0.0863$; (*right*) adaption based on the local Legendre polynomial spectrum with a tolerance of $\epsilon = 9.01 \times 10^{-7}$.

For the second example we consider the solution of the Poisson equation

$$\nabla^2 u + 1 = 0, \tag{4.55}$$

on the same domain D with homogeneous boundary conditions $u = 0$ on ∂D. The structure of the solution is quite different, as shown in Fig. 4.21. There is a weak singularity in the corners of the domain where the solution must simultaneously match the curvature and the boundary conditions. However, the solution gradients are largest along the edges of the domain where the structure of u is rather simple. In this case our two refinement criteria lead to nearly complementary grids. Clearly the local polynomial spectrum indicates the correct location for refinement while the magnitude of solution gradients can be misleading. In this case mesh refinement based on solution gradients completely misses the location (e.g. the corners) where the errors are largest.

4.6   Summary

We have outlined the basic features and implementation of an adaptive spectral element method. Perhaps the most interesting part of the method is the

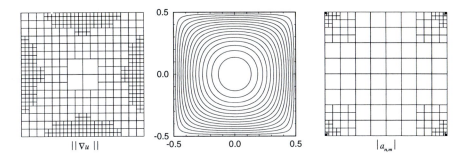

Fig. 4.21. Adaptive solution of the Poisson equation with corner singularities: (*center*) contours of the solution computed on a uniform fine grid; (*left*) adaption based on the solution gradients with a tolerance of $\epsilon = 0.3$; (*right*) adaption based on the local Legendre polynomial spectrum with a tolerance of $\epsilon = 10^{-8}$.

'built-in' refinement criteria provided by the local polynomial spectrum. This provides a heuristic error estimate that is independent of the system being solved. The local spectral properties were shown to be a useful and relatively robust criteria for both simple linear problems (the Poisson equation). In the following section we will look at more complex nonlinear problems.

One area of potentially great improvement is in the algorithms used to implement the sparse matrix solver. For example, recent work on fast multipole methods for spectral elements shows great promise for solving Poisson and Helmholtz equations [35]. There are a host of other possibilities that take better advantage of the multilayer structure of the grid than the more straightforward CG iterations considered here. Also note that direct solvers are still feasible in adaptive calculations as long as a relatively large number of elliptic solves take place between adaption steps.

5 Fluid Dynamics

Advances in both computer technology and numerical methods have opened new possibilities for the study of fluid dynamics through large-scale simulations. Building on the spatial discretizations presented so far, we now turn to the solution of the Navier–Stokes equations for unsteady two and three-dimensional problems and show that high-order splitting methods reduce the computational burden to solving a series of Helmholtz problems.

5.1   Incompressible flows

We consider here Newtonian fluids with constant density ρ and kinematic viscosity ν, the motion of which is governed by the the incompressible Navier–Stokes equations:

$$\nabla \cdot \boldsymbol{u} = 0 \quad \text{in } \Omega, \tag{5.1a}$$

$$\partial_t \boldsymbol{u} = \mathbf{N}(\boldsymbol{u}) - \frac{1}{\rho}\nabla p + \frac{1}{Re}\nabla^2 \boldsymbol{u} \quad \text{in } \Omega, \tag{5.1b}$$

where $\boldsymbol{u} = (u_1, u_2, u_3)$ is the velocity field, p is the static pressure, $Re \equiv UL/\nu$ is the Reynolds number, and Ω is the computational domain. These equations are written in non-dimensional form where velocities are scaled by U and lengths by L. Without loss of generality we take the numerical value of $\rho = 1$ since this simply sets the scale for p. $\mathbf{N}(\boldsymbol{u})$ represents the nonlinear advection term:

$$\mathbf{N}(\boldsymbol{u}) = -(\boldsymbol{u} \cdot \nabla)\boldsymbol{u}, \tag{5.2a}$$

$$= -\frac{1}{2}\left[(\boldsymbol{u} \cdot \nabla)\boldsymbol{u} + \nabla \cdot (\boldsymbol{u}\boldsymbol{u})\right], \tag{5.2b}$$

$$= -\frac{1}{2}\nabla(\boldsymbol{u} \cdot \boldsymbol{u}) - \boldsymbol{u} \times \nabla \times \boldsymbol{u}. \tag{5.2c}$$

We refer to these as the *convective* form, *skew-symmetric* form, and *rotational* form, respectively. These three forms for $\mathbf{N}(\boldsymbol{u})$ are mathematically equivalent but behave differently when implemented for a discrete system. As shown by Zang [91], the skew-symmetric form is the most robust; this form is used in all calculations unless noted otherwise.

The Navier–Stokes equations are coupled through the incompressibility constraint $\nabla \cdot \boldsymbol{u} = 0$ and the nonlinear term $\mathbf{N}(\boldsymbol{u})$. However, the biggest challenge for time-integration comes from the *linear* term:

$$\mathbf{L}(\boldsymbol{u}) \equiv \frac{1}{Re}\nabla^2 \boldsymbol{u}. \tag{5.3}$$

This term is responsible for the fastest time scales in the system and thus poses the most severe constraint on the maximum allowable time step for numerical integration of the fluid equations. Problems associated with the stiffness of the linear operator are handled by treating this term implicitly, while the nonlinear term can be integrated with an easier explicit method.

Semi-discrete formulation To solve the Navier–Stokes equations, (5.1b) is integrated over a single time step to obtain:

$$\boldsymbol{u}(t + \Delta t) = \boldsymbol{u}(t) + \int_t^{t+\Delta t} \left[\mathbf{N}(\boldsymbol{u}) - \frac{1}{\rho}\nabla p + \mathbf{L}(\boldsymbol{u})\right] \mathrm{d}t. \tag{5.4}$$

Next we introduce a discrete set of times $t_n \equiv n\Delta t$ where the solution is to be evaluated, and define $\boldsymbol{u}^n \equiv \boldsymbol{u}(\boldsymbol{x}, t_n)$ as the semi-discrete approximation to the velocity (discrete in time, continuous in space). For reasons that will be explained in a moment, the pressure integral is replaced with:

$$\nabla \tilde{p} \equiv \frac{1}{\Delta t} \int_{t_n}^{t_{n+1}} \frac{1}{\rho} \nabla p \, \mathrm{d}t. \tag{5.5}$$

Next we introduce appropriate integration schemes for the linear and nonlinear terms. The simplest implicit/explicit scheme would be first-order Euler time integration:

$$\int_{t_n}^{t_{n+1}} \mathbf{L}(\boldsymbol{u}) \, \mathrm{d}t \approx \Delta t \, \mathbf{L}(\boldsymbol{u}^{n+1}); \tag{5.6}$$

$$\int_{t_n}^{t_{n+1}} \mathbf{N}(\boldsymbol{u}) \, \mathrm{d}t \approx \Delta t \, \mathbf{N}(\boldsymbol{u}^n). \tag{5.7}$$

Combining (5.5)–(5.7) we get a semi-discrete approximation to the momentum equation:

$$u^{n+1} = u^n + [\mathbf{N}(\boldsymbol{u}^n) - \nabla \tilde{p} + \mathbf{L}(\boldsymbol{u}^{n+1})] \, \Delta t. \tag{5.8}$$

This system of equations can be solved by further splitting (5.8) into three substeps as follows:

$$\boldsymbol{u}^{(1)} - \boldsymbol{u}^n = \Delta t \, \mathbf{N}(\boldsymbol{u}^n), \tag{5.9a}$$
$$\boldsymbol{u}^{(2)} - \boldsymbol{u}^{(1)} = -\Delta t \, \nabla \tilde{p}, \tag{5.9b}$$
$$\boldsymbol{u}^{n+1} - \boldsymbol{u}^{(2)} = \Delta t \, \mathbf{L}(\boldsymbol{u}^{n+1}). \tag{5.9c}$$

Here $\boldsymbol{u}^{(1)}$ and $\boldsymbol{u}^{(2)}$ are intermediate velocity fields that progressively incorporate the nonlinear terms and the incompressibility constraint. The motivation for the splitting is to decouple the pressure term from the advection and diffusion terms.

The classical splitting scheme proceeds by introducing two assumptions: that $\boldsymbol{u}^{(2)}$ satisfies the divergence free condition ($\nabla \cdot \boldsymbol{u}^{(2)} = 0$), and that $\boldsymbol{u}^{(2)}$ satisfies the correct Dirichlet boundary conditions in the direction normal to the boundary ($\boldsymbol{n} \cdot \boldsymbol{u}^{(2)} = \boldsymbol{n} \cdot \boldsymbol{u}^{n+1}$). Incorporating these assumptions, we can derive a separately solvable elliptic problem for the pressure in the form:

$$\nabla^2 \tilde{p} = \frac{1}{\Delta t} (\nabla \cdot \boldsymbol{u}^{(1)}). \tag{5.10}$$

The field $\tilde{p}$ becomes a dynamic variable that couples the divergence-free condition and the momentum equation. The correct Neumann boundary conditions for $\tilde{p}$ come from (5.8), which can be simplified to the form:

$$\frac{\partial \tilde{p}}{\partial \boldsymbol{n}} = \boldsymbol{n} \cdot [\mathbf{N}(\boldsymbol{u}^n) - \frac{1}{Re} \nabla \times \nabla \times \boldsymbol{u}^n]. \tag{5.11}$$

This boundary condition prevents the propagation and accumulation of time differencing errors and ensures that $\tilde{p}$ satisfies the important pressure compatibility condition [48]. Note that the linear term in (5.11) is derived from $\mathbf{L}(\boldsymbol{u}^n)$ rather than $\mathbf{L}(\boldsymbol{u}^{n+1})$. This type of first-order extrapolation is necessary to keep the pressure equation decoupled from the other substeps. The order of the extrapolation should be consistent with the overall time accuracy.

Higher-order schemes It is relatively easy to make the integration scheme outlined above more accurate in time, i.e. to increase the time accuracy to $O(\Delta t^J)$. The basic idea is to use higher-order multi-step schemes for the time integration. Time derivatives can be approximated with a backward difference of the form:

$$\partial_t \boldsymbol{u} \approx \Delta t \left(\gamma_0 \boldsymbol{u}^{n+1} - \sum_{q=0}^{J-1} \alpha_q \boldsymbol{u}^{n-q} \right), \qquad (5.12)$$

where $\gamma_0 = \sum \alpha_q$ for consistency. The nonlinear term can be integrated using an Adams–Bashforth method:

$$\int_{t_n}^{t^{n+1}} \mathbf{N}(\boldsymbol{u}) \, \mathrm{d}t \approx \Delta t \sum_{q=0}^{J-1} \beta_q \, \mathbf{N}(\boldsymbol{u}^{n-q}), \qquad (5.13)$$

where $\sum \beta_q = 1$. The pressure boundary conditions should be integrated with a scheme of the same order to ensure consistent time accuracy:

$$\frac{\partial \tilde{p}}{\partial \boldsymbol{n}} = \boldsymbol{n} \cdot \sum_{q=0}^{J-1} \beta_q \left[\mathbf{N}(\boldsymbol{u}^{n-q}) - \frac{1}{Re} \nabla \times \nabla \times \boldsymbol{u}^{n-q} \right]. \qquad (5.14)$$

Combining these various integration schemes produces the following semi-discrete equations:

$$\boldsymbol{u}^{(1)} - \sum_{q=0}^{J-1} \alpha_q \boldsymbol{u}^{n-q} = \Delta t \sum_{q=0}^{J-1} \beta_q \, \mathbf{N}(\boldsymbol{u}^{n-q}), \qquad (5.15a)$$

$$\boldsymbol{u}^{(2)} - \boldsymbol{u}^{(1)} = -\Delta t \, \nabla \tilde{p} \qquad (5.15b)$$

$$\gamma_0 \boldsymbol{u}^{n+1} - \boldsymbol{u}^{(2)} = \Delta t \, \mathbf{L}(\boldsymbol{u}^{n+1}). \qquad (5.15c)$$

This method would typically be used with $J = 2$ or 3 and an integration rule like one of the schemes given in table 5.1. Overall, (15) provides an very efficient way to integrate the Navier–Stokes equations.

Two-dimensional simulations A single time step using the skew-symmetric form of the nonlinear terms requires ten spatial derivatives plus the solution to one Poisson equation for the pressure and two Helmholtz equations for

Table 5.1. Integration coefficients for multi-step schemes of order J: (*top*) classic Adams–Bashforth schemes; (*bottom*) stiffly-stable schemes with coefficients derived for a model advection–diffusion equation [48].

J	γ_0	α_0	α_1	α_2	β_0	β_1	β_2
1	1	1			1		
2	1	1			3/2	-1/2	
3	1	1			23/12	-4/3	5/12
1	1	1			1		
2	3/2	2	-1/2		2	-1	
3	11/6	3	-3/2	1/3	5/2	-2	1/2

the diffusion in each direction. Most of the computational work is associated with solving these linear systems; collocation is used to integrate the nonlinear terms and makes only a minor contribution. The techniques outlined in Sect. 3 can be applied directly to the solution of the various elliptic subproblems.

Note that the pressure is indeterminant to within a constant in a two-dimensional calculation with all Neumann boundary conditions. This is because for any field $p(x,y)$ that satisfies (5.10), the field $p(x,y) + c$ is also a solution, for any constant c. This ambiguity can be removed in a direct method by setting exactly one pressure degree-of-freedom to zero, typically the last element in the array of pressure boundary unknowns. For iterative methods it is sufficient to set the mean of the initial residual to zero.

Three-dimensional simulations We can simulate three-dimensional flows in one of several ways. If the geometry is fully three-dimensional we have to use hexahedral or tetrahedral spectral elements [73]. If the problem has one of several symmetries — axisymmetric, spherically symmetric, or periodic in one direction — then Fourier expansion in one direction becomes a much more efficient way to represent the flow.

Consider the case of a flow that is periodic in the z-direction and satisfies the symmetry

$$u(x, y, z, t) = u(z, y, z + L, t).$$

Under these conditions u can be projected exactly onto a set of two-dimensional Fourier modes $\hat{u}_q$ as

$$\hat{u}_q(x, y, t) = L^{-1} \int_0^L u(x, y, z, t) e^{-i(2\pi/L)qz} \, dz.$$

Likewise, the Fourier modes $\hat{u}_q$ given the expansion of the velocity field in a Fourier series:

$$u(x, y, z, t) = \sum_{q=-\infty}^{\infty} \hat{u}_q(x, y, t)\, e^{i(2\pi/L)qz}.$$

Substituting the Fourier expansion of the velocity field into the Navier–Stokes equations, we obtain a coupled set of equations for the Fourier modes. To simplify the notation, we define the scaled wavenumber $\beta_q \equiv (2\pi/L)q$ and the q-dependent operators

$$\tilde{\nabla} \equiv (\partial_x, \partial_y, i\beta_q), \quad \tilde{\nabla}^2 \equiv (\partial_x^2, \partial_y^2, -\beta_q^2).$$

The evolution equation for the Fourier modes can then be written as

$$\tilde{\nabla} \cdot \hat{u}_q = 0 \quad \text{in } \Omega, \tag{5.16a}$$

$$\partial_t \hat{u}_q = N_q(u) - \frac{1}{\rho}\tilde{\nabla}\hat{p}_q + \frac{1}{Re}\tilde{\nabla}^2 \hat{u}_q \quad \text{in } \Omega. \tag{5.16b}$$

The nonlinear advection term provides the coupling between all modes. We can denote this term by

$$N_q(u) = L^{-1} \int_0^L N(u)e^{-i(2\pi/L)qz}\, dz. \tag{5.16c}$$

Dissipation becomes important at wavenumbers $\beta_D \sim Re^{1/2}$; at wavenumbers $\beta > \beta_D$ the equations are dominated by viscosity. These high-wavenumber modes contribute little to the dynamics of the flow at large scales because their energy is rapidly dissipated by viscosity. For an adequate description of the dynamics in a system with a given spanwise dimension L we only need a finite set of M Fourier modes to cover the range of scales from $\beta = 0$ (the mean flow) to $\beta_D = (2\pi/L)M \sim Re^{1/2}$, or $M = O(LRe^{1/2})$. We take as our final representation of the velocity field the truncated expansion

$$u(x, y, z, t) = \sum_{q=-M}^{M} \hat{u}_q(x, y, t)e^{i(2\pi/L)qz}.$$

Writing the equations in Fourier space reduces the problem for a three-dimensional flow to a sequence of coupled two-dimensional problems. The only coupling is through the nonlinear term which is again evaluated explicitly. Computationally it is more convenient to follow the evolution of the two-dimensional Fourier modes $\hat{u}_q(x, y, t)$ than the full three-dimensional field $u(x, t)$. Because u is real, the Fourier modes satisfy the symmetry $\hat{u}_{-q} = \hat{u}_q^*$. Therefore, only half of the spectrum ($q \geq 0$) is needed. In addition to convenience, the Fourier representation of the velocity field has other intrinsic advantages. It provides a direct way of linking particular modes of the system with specific three-dimensional spatial patterns. Linear stability theory can predict which modes will have the strongest interaction with the two-dimensional flow to produce these patterns. The time-averaged amplitude of the Fourier modes gives a direct indication of how well-resolved the calculations are. And finally, the time-dependent amplitude of the Fourier modes provides a convenient way of explaining the transfer of energy to different scales in a three-dimensional flow.

The comments in the previous section about solving the pressure equation apply only to the mean flow ($\beta = 0$) of a periodic three-dimensional flow. All other wave numbers determine fluctuations about the mean and are uniquely defined. The same techniques described for solving the two-dimensional problem can be applied to the pressure system for $\beta = 0$.

5.2   Examples

Next we present a variety of examples that illustrate the versatility of spectral element methods for two and three-dimensional flow problems. For most cases we show results using both quadrilateral and triangular spectral elements. We also examine problems where nonconforming elements and adaptive mesh refinement are used to automatically generate an appropriate discretization that achieves a prescribed error tolerance.

Wannier flow The first example is an exact solution to the Stokes equations ($\mathbf{N}(\mathbf{u}) \equiv 0$, $Re = 1$), but for a relatively complicated flow with curvilinear boundaries. It is an exact solution derived by Wannier [82] for the creeping flow past a rotating circular cylinder next to a moving wall. The solution depends only on the cylinder radius, r, its rate of rotation, ω, the distance from the center of the cylinder to the moving wall, d, and the velocity of the wall, U. For convenience we define $s^2 = d^2 - r^2$ and $\Gamma = (d+s)/(d-s)$, and the constants:

$$
\begin{aligned}
a_0 &= U/\ln \Gamma, \\
a_1 &= -d(a_0 + \tfrac{1}{2}r^2\omega/s), \\
a_2 &= (d+s)(a_0 + \tfrac{1}{2}r^2\omega/s), \\
a_3 &= (d-s)(a_0 + \tfrac{1}{2}r^2\omega/s).
\end{aligned}
$$

Next we define the following functions that depend on position (x, y):

$$
\begin{aligned}
Y_1(y) &= y + d, \\
Y_2(y) &= 2Y_1(y), \\
K_1(x, y) &= x^2 + (s + Y_1(y))^2, \\
K_2(x, y) &= x^2 + (s - Y_1(y))^2.
\end{aligned}
$$

In terms of these quantities, the solution can be written as:

$$u_1(x,y) = U - 2(a_1 + a_0 Y_1)\left[\frac{s+Y_1}{K_1} + \frac{s-Y_1}{K_2}\right]$$
$$-a_0 \ln(K_1/K_2)$$
$$-\frac{a_2}{K_1}\left[s+Y_2 - \frac{(s+Y_1)^2 Y_2}{K_1}\right]$$
$$-\frac{a_3}{K_2}\left[s-Y_2 + \frac{(s-Y_1)^2 Y_2}{K_2}\right],$$

$$u_2(x,y) = \frac{2x}{K_1 K_2}(a_1 + a_0 Y_1)(K_2 - K_1)$$
$$-\frac{xa_2(s+Y_1)Y_2}{K_1^2} - \frac{xa_3(s-Y_1)Y_2}{K_2^2}.$$

This problem was solved using nonconforming quadrilaterals and triangular spectral elements. Figure 5.1 shows the corresponding computational domains along with streamlines of the steady-state solution. Since the exact solution is known, Dirichlet boundary conditions for the velocity can be applied along the perimeter of the domain. The nonconforming mesh of quadrilateral elements incorporates some local refinement near the cylinder and uses a total of $K = 40$ elements. The triangular mesh uses $K = 65$ elements to discretize the same region of space but with higher resolution near the cylinder.

We note a few items about the calculation using triangular elements. Since this mesh uses curvilinear elements around the cylinder, it serves to test the convergence of the method on distorted grids. All elements are mapped to the standard triangle when performing integration. Because of their deformed nature, the Jacobian is not constant within a curvilinear element. A non-constant Jacobian destroys the sparsity of the interior–interior coupling submatrices in the global mass matrix and global stiffness matrix. Nevertheless, since there are only a few of these elements performance is not noticeably affected [75].

Figure 5.2 shows the results from a p-convergence study for this flow. The figure shows the H^1 error in the computed velocity field. As expected for a smooth solution, the simulations converge exponentially to the exact velocity field. Although the quadrilateral and triangular elements converge at approximately the same rate, the actual value of the error for a given order p depends on how elements are distributed in the domain. This results in parallel convergence curves with different prefactors that depend on the specifics of the grid.

Because the exact solution is known, this problem also makes a good test case for adaptive mesh refinement. Figure 5.3 shows a convergence plot for the Wannier flow solved by adaptively refining an initial coarse grid. This

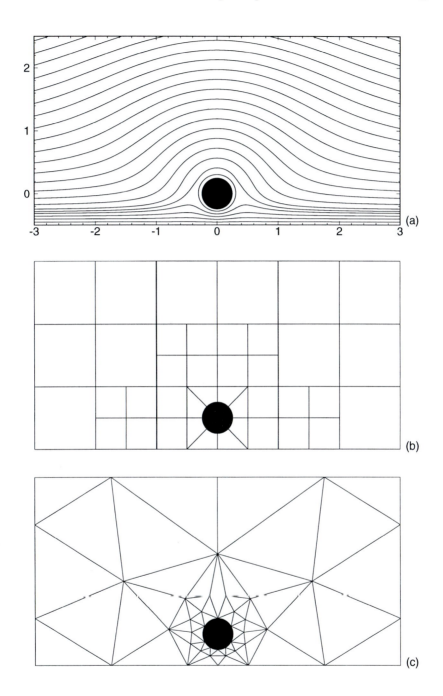

Fig. 5.1. Wannier flow, an exact solution for creeping flow past a rotating circular cylinder near a moving wall: (*a*) streamlines of the exact solution corresponding to the parameters $r = 0.25$, $d = 0.5$, $U = 1$, and $\omega = 2$; computational domain discretized using (*b*) $K = 40$ quadrilateral spectral elements and (*c*) $K = 65$ triangular spectral elements.

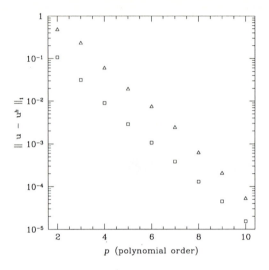

Fig. 5.2. Convergence of the velocity field in the H^1 norm to the exact solution for Wannier flow shown in the previous figure: ($\square$) quadrilateral and ($\triangle$) triangular spectral elements. Note that errors have been normalized by the domain size.

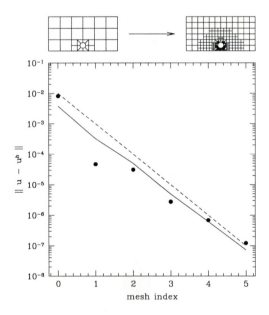

Fig. 5.3. Adaptive solution to the Wannier flow problem: ($\bullet$), computed L_2 error in the velocity field; (—), $10 \times \epsilon$ where ϵ is the error estimated from the trace of the polynomial spectrum; the dashed line is the prescribed error tolerance. Meshes M_0 and M_5 are shown above the plot.

results in convergence via adaptive h-refinements of the initial mesh. We use the trace of the polynomial spectrum to drive adaptivity, and apply the refinement criteria to each component of the velocity vector.

The calculations are performed as follows. We start by solving the Stokes equations on the initial coarse mesh, designated M_0. This mesh is then refined to meet a prescribed value of the refinement parameter ϵ. A new solution is computed and the actual L_2 error is compared to the new estimate. The process is iterated by lowering ϵ by a factor of 10 each time. In pseudo-code the procedure looks like this:

```
do n = 1, 5
    set eps = 1/10^$n          # set tolerance
    refine if trace(u1) > $eps # update grid
    refine if trace(u2) > $eps #
    solve(u1,u2)               # update solution
end
```

This produces a sequence of grids $M_1, M_2, \ldots, M_5$. Only the first and last grids are shown in Fig. 5.3.

There are two curves related to the error estimates shown in Fig. 5.3. Look at the results for grid M_2. The dashed line indicates the precribed error tolerance used to generate that grid. This is an *a priori* error estimate in the sense that the adaptive procedure refines grid $M_1 \rightarrow M_2$ until the new tolerance has been met, but before a new solution is available. The solid line indicates the estimated error on the new grid after the solution has been regenerated. This is an *a posteriori* estimate. Finally, the symbol indicates the true L_2 error on the new grid. The error estimate is sharp in the sense that it follows the true error to within a constant factor, although for this problem that constant is ≈ 10.

Kovasznay flow In 1948, Kovasznay solved the problem of steady, laminar flow behind a two-dimensional grid [54]. This exact solution to the Navier–Stokes equations is given by:

$$u_1(x,y) = 1 - e^{\lambda x} \cos 2\pi y,$$

$$u_2(x,y) = \frac{\lambda}{2\pi} e^{\lambda x} \sin 2\pi y,$$

$$p(x,y) = \frac{1}{2}(1 - e^{\lambda x}) + c,$$

where $\lambda = Re/2 - (Re^2/4 + 4\pi^2)^{\frac{1}{2}}$ and c is an arbitrary constant. We we look at the solution for $Re = 40$.

The Kovasznay flow pattern is similar to the low-speed flow of a viscous fluid past an array of cylinders. Figure 5.4 shows streamlines of the steady solution and computational domains using quadrilateral and triangular spectral elements. The exact solution was used to apply Dirichlet boundary conditions, and the Navier–Stokes equations were integrated to obtain a steady-state solution on the interior of the domain. Figure 5.5 shows the results of a p-convergence study for this problem. Again we observe exponential convergence of the solution in both methods, and at roughly the same rate.

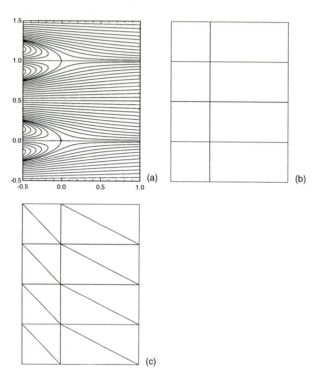

Fig. 5.4. Kovasznay flow, an exact solution to the Navier–Stokes equations: (a) streamlines of the exact solution corresponding to $Re = 40$; computational domain discretized using (b) $K = 8$ quadrilateral spectral elements and (c) $K = 16$ triangular spectral elements.

Next we consider solving this problem using adaptive mesh refinement and nonconforming elements. Figure 5.6 shows the results from the same type of convergence study that was presented in Sect. 5.2 for Wannier flow. In this problem the error estimate differs from the true L_2 error by about a factor of 4. Note that the initial grid is so coarse that the estimate is completely unreliable — the estimated error on the refined mesh M_1 is higher! This emphasizes that fact that the adaptive procedure should really only be applied to a solution that is well-resolved in the sense that $||u^h|| \approx ||u||$. That assumption is, after all, at the heart of the error estimate. Also note that the refinement $M_1 \to M_2$ produces such a large drop in the error that the next

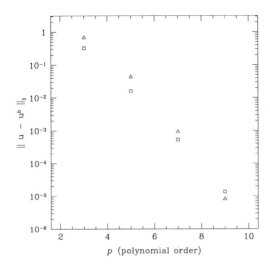

Fig. 5.5. Convergence of the velocity field in the H^1 norm to the exact solution for Kovasznay flow as shown in the previous figure: ($\square$) quadrilateral and ($\triangle$) triangular spectral elements.

iteration $M_2 \to M_3$ does not change it at all.[3] Although this problem is relatively simple, it gives some additional confidence that the error estimate provides a meaningful measure of the approximation error even for a fully nonlinear problem.

Lid-driven cavity The first two examples provide good benchmarks because they are exact solutions to the fluid equations. However, the solutions are really too simple to warrant the use of adaptive mesh refinement. The next example clearly *does* benefit from the use of an automatic procedure to construct appropriate grids.

Consider the case of a lid-driven cavity (LDC). The flow within the cavity is driven from above by a lid moving with unit velocity, and the problem is non-dimensionalized so that the cavity has unit length on each side. Boundary conditions for the velocity are $(u_1, u_2) = (1, 0)$ along the top boundary and $(u_1, u_2) = (0, 0)$ on the three remaining sides. Note that the velocity boundary conditions are discontinuous at the corners, making this an extremely difficult problem to resolve with a high-order method. It is one of the situations where p-refinement degenerates, ruling that out as a practical way to resolve the flow.

---

[3] If you look at Fig. 5.6 and think the actual L_2 error increases from M_2 to M_3, then you are seeing an optical illusion caused by the fact that the error *estimate* decreases slightly.

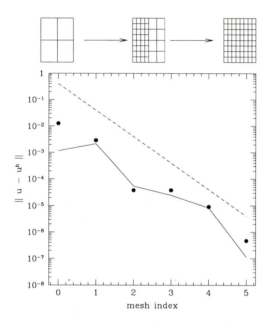

Fig. 5.6. Adaptive solution to the Kovasznay flow problem: (•), computed L_2 error in the velocity field; (—), $4 \times \epsilon$ where ϵ is the error estimated from the trace of the polynomial spectrum; the dashed line is the prescribed error tolerance. Meshes M_0, M_4, and M_5 are shown above the plot.

We will look at two types of parameter variation for this problem: numerical convergence as $\epsilon \to 0$ for fixed $Re = 1000$, and evolution of the grid with increasing Re for fixed $\epsilon = 10^{-6}$. All of these calculations will use a fixed polynomial order of $p = 7$ and apply the trace of the polynomial spectrum as a refinement criteria. Note that temporal refinement is necessary as well — a suitable time step is chosen for each new domain so that the time integration remains stable. The time step in these calculations varies from $\Delta t = 0.01$ to $\Delta t = 0.000625$. Because we are using an implicit method to solve the Navier–Stokes equations, local time stepping is not an option.

First consider the problem of computing the steady-state LDC flow at a fixed value of $Re = 1000$. Figure 5.7 shows the adaptively generated grid and corresponding vorticity field for different values of the refinement parameter ϵ. The initial coarse grid for this calculation was simply the unit square. Several intermediate grids were generated prior to the one shown in Fig. 5.7(a). The refinement procedure proceeds in a similar manner to that described previously: the solution is integrated for a specified amount of time, then the refinement criteria is applied to the components of the velocity field to produce a new grid. The old solution is projected onto the new grid and the next iteration begins.

The solution shown in Fig. 5.7(a) with $\epsilon = 10^{-3}$ is quite coarse and clearly a poor approximation to anything resembling the vorticity of a real flow. The next adaption ($\epsilon = 10^{-4}$) refines the entire domain one level and attempts to resolve the shear layers along the upper and right walls. At $\epsilon = 10^{-5}$ it picks out high vorticity regions along the left and bottom walls and continues to refine the shear layers that emerge from each upper corner. Finally, at $\epsilon = 10^{-6}$ the interior of the cavity is refined uniformly and a fine grid is generated near each corner and in the direction just downstream. This process could be continued to achieve an arbitrarily high degree of accuracy but the solution in Fig. 5.7(d) is certainly a good approximation to the flow at this Reynolds number. Keep in mind that the vorticity is a derived quantity obtained by differentiating the velocity field. It is not even continuous in this approximation, although continuity of higher derivatives is obtained as part of the convergence process. For example, compare Figs. 5.7(a) and 5.7(d). It is comforting that the refinement criteria applied to the velocity field automatically picks out the physically important features of the flow.

We can use the same ideas to study how the flow evolves with changes in Re. At a given value of Re we use the adaptive procedure to generate a steady-state solution with a prescribed tolerance ϵ. That solution serves as an initial guess for the next value of Re. The adaptive procedure keeps the solution well-resolved as the flow develops more complex structure with increasing Re.

Figure 5.8 shows the evolution of the flow and adaptively generated grids for this kind of parameter study. At low Reynolds number the cavity contains a diffuse vorticity field. Vorticity becomes more concentrated along the walls with increasing Reynolds number. The adaptive procedure tracks these

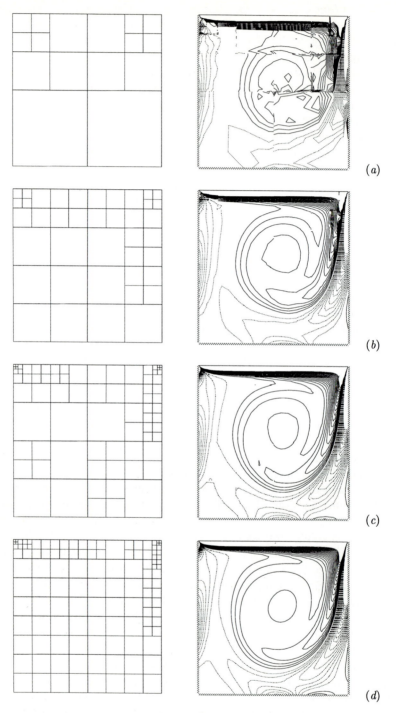

Fig. 5.7. Adaptive calculation for the LDC flow at $Re = 1000$ showing the grid and vorticity field at different values of the refinement parameter: (a) $\epsilon = 10^{-3}$; (b) $\epsilon = 10^{-4}$; (c) $\epsilon = 10^{-5}$; (d) $\epsilon = 10^{-6}$. All calculations used a fixed polynomial order of $p = 7$.

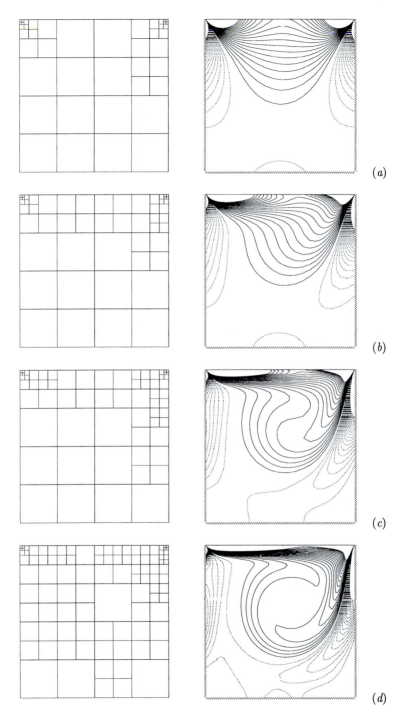

Fig. 5.8. Parameter study in Re using adaptive grids generated to a tolerance of $\epsilon = 10^{-6}$: (a) $Re = 10$; (b) $Re = 100$; (c) $Re = 250$; (d) $Re = 500$. All calculations used a fixed polynomial order of $p = 7$.

changes and refines the grid to an appropriate level at each value of Re. Note that the corner region is refined to the same level at $Re = 10$ and $Re = 1000$. This is because the nature of the boundary condition-induced singularity is independent of Re, as opposed to the physically important behavior that emerges as dissipation is removed from the system.

Although there is obviously no exact solution for this problem, we can compare with high-resolution numerical simulations of the same flow to demonstrate that the adaptive procedure produces an accurate approximation. In a recent study, Botella and Peyret [16] compute solutions to the LDC flow at $Re = 1000$ using a Chebyshev collocation method. To improve the accuracy of the calculations they use an analytic approximation to subtract off the singular part of the solution near the corners and compute the remaining smooth part numerically. By explicitly removing the singular part of the solution they can recover spectral accuracy and exponential convergence. In contrast, the calculations presented here attempt to "resolve" the singularity directly through mesh refinement near the corners.

Figure 5.9 compares profiles of the u- and v-components of velocity along the centerline of the cavity. Data for the comparison is taken from tables 9 and 10 of Botella and Peyret [16]. These values correspond to calculations with $N = 160$ Chebyshev modes in each direction, or 25 600 grid points. The spectral element data corresponds to figure 5.7(d); this mesh has $K \times N^2 \approx$ 7500 grid points. The comparison shows that the two calculations are in extremely close agreement, and demonstrates that the adaptive procedure results in a highly accurate solution for small ϵ with an intelligent distribution of element size.

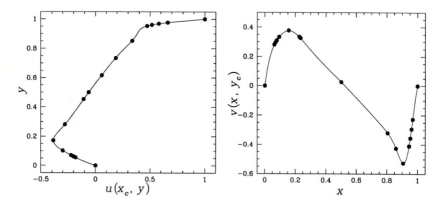

Fig. 5.9. Comparison of velocity profiles through the center of the cavity ($x_c = y_c = \frac{1}{2}$) at $Re = 1000$: $\bullet$, spectral results from Botella & Peyret (1998); —, adaptive spectral element calculation with tolerance $\epsilon = 10^{-6}$.

Additional information on adaptive spectral element calculations of the lid-driven cavity problem can be found in [60], including the use of directional splitting as an efficient way to refine elements in regions where the flow may be resolved in one direction but under-resolved in another.

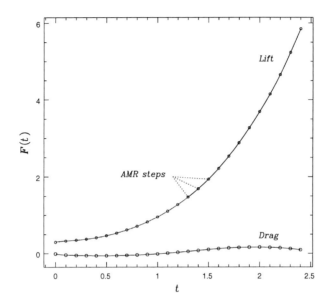

Fig. 5.10. Unsteady forces on an impulsively accelerated NACA 0012 airfoil at $\alpha = 22.5$ degrees and $Re = 852$. Points (∘) indicate refinement steps during the simulation.

Impulsively accelerated airfoil Next we look at an application of this technique to an *unsteady* flow problem. Consider the motion of an airfoil that is set at an angle of attack α and impulsively accelerated into a still fluid. Dimensional parameters are the chord length c, the airfoil acceleration a, and the kinematic viscosity of the fluid ν. From these parameters we need to choose a length scale L and a velocity scale U. The fluid motion satisfies the incompressible Navier–Stokes equations which we will solve in a non-inertial reference frame attached to the accelerating airfoil. In non-dimensional form the governing equations are:

$$\nabla \cdot \boldsymbol{u} = 0, \tag{5.17}$$

$$\partial_t \boldsymbol{u} = \mathbf{N}(\boldsymbol{u}) - \frac{1}{\rho}\nabla p + \frac{\nu}{UL}\nabla^2 \boldsymbol{u} - \frac{L}{U^2}\boldsymbol{a}. \tag{5.18}$$

Note that $\boldsymbol{a} = -a\,(\cos\alpha, \sin\alpha)$ is the frame acceleration. A natural and obvious choice for the reference scales is to normalize for unit acceleration by

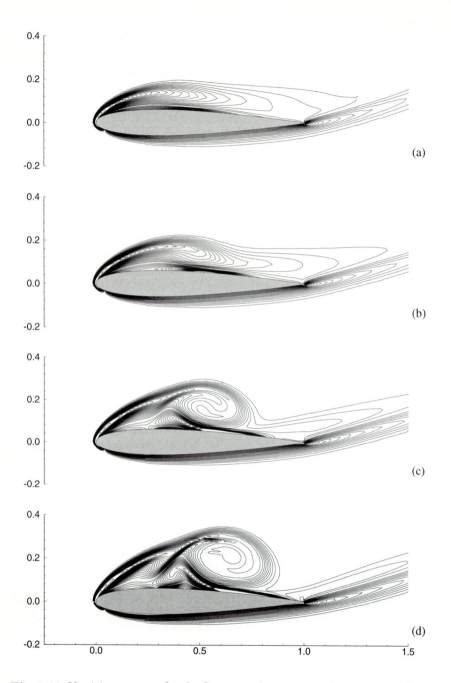

Fig. 5.11. Vorticity contours for the flow around an impulsively accelerated airfoil at $\alpha = 22.5$ degrees and $Re = 852$ (based on chord length and acceleration): (*a*) $t = 1.7$; (*b*) $t = 1.9$; (*c*) $t = 2.2$; (*d*) $t = 2.4$.

taking $L = c$ and $U = \sqrt{ac}$. From these we can also form a time scale $T = L/U = \sqrt{c/a}$. The similarity variable or Reynolds number is then $Re \equiv \sqrt{ac^3}/\nu$, and the problem is completely specified by prescribing α and Re. These equations can be integrated using the same technique described in Sect. 5.1 by including the frame acceleration in the integration of the nonlinear terms and the pressure boundary conditions.

The parameters for this calculation correspond to a companion set of experiments conducted at GALCIT for a NACA 0012 airfoil in water [33]. The airfoil sits at an angle of attack $\alpha = 22.5$ degrees and the Reynolds number is set to $Re = 852$ to match the experimental setup. Note that the free-stream velocity increases with time as $U_\infty = at$ owing to the constant acceleration. The angle of attack and Reynolds number are set to large values so that the flow over the airfoil separates almost immediately and produces a complex vorticity field just above the upper surface.

The problem was solved on a large computational domain with order $p = 7$ elements. The initial grid of $K \approx 140$ elements was built 'by hand' to provide a sufficiently accurate discretization for starting the adaptive procedure. As a metric for adaption we required the vorticity field $\xi \equiv \nabla \times u$ to be represented on the computational grid with a discretization tolerance of $\epsilon = 0.01$ for the local polynomial spectrum. In this case we are applying the refinement criteria to a physically important derived quantity rather than one of the primitive variables. Adaption steps were carried out at constant time intervals of $\Delta T = 0.1$ during the integration from $t = 0$ to $t = 2.4$.

Figure 5.10 shows the unsteady loading on the airfoil as it accelerates. The solid line in this figure connects the force computed at each time step in the simulation. The points indicate the discrete times when the grid is adapted to maintain resolution. This figure is shown primarily to document that the adaptive procedure evolves smoothly and does not produce discontinuous jumps in the loading on the airfoil.

The developing vorticity field is shown in Fig. 5.11. The airfoil leaves a weak starting vortex in its wake and rapidly develops a strong region of separated vorticity along the upper (opposite to the direction of acceleration) surface. The refinement criteria maintains a sharp resolution of the vorticity field at all times. During the course of the calculations the number of active elements in the mesh increases from $K \approx 180$ to $K \approx 480$, giving a total of ≈ 30720 grid points in the final mesh. The most aggressive mesh refinement takes place early in response to the strong vorticity layer near the leading edge of the airfoil and the singularity produced by the sharp trailing edge. Similar to the LDC flow described in Sect. 5.2, the singularity along the boundary requires the most attention from the adaptive procedure. Once these parts of the flow are resolved there are relatively few additional refinements to maintain resolution of the separated vorticity field. To keep the integration stable the time step is reduced by about two orders of magnitude to a final value of $\Delta t \approx 7.5 \times 10^{-5}$. This maintains a relatively constant CFL number

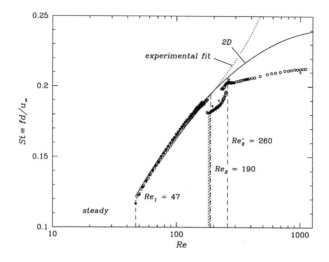

Fig. 5.12. Plot of St versus Re for the flow past a circular cylinder. Experiments: o, Williamson [87]; •, Hammache & Gharib [38]; 3D simulations: +, Henderson [41]. The solid line is a curve fit to two-dimensional simulation data for Re up to 1000 [41].

during the simulation. This necessary reduction in Δt is due to a combination of the mesh refinement and the increasing free-stream velocity U_∞.

A detailed comparison of the computational and experimental results for this problem are the subject of current work.

Cylinder wake Understanding the fluid flow around a straight circular cylinder is one of the most fundamental problems in fluid mechanics. It's a model for flow around bridges, buildings, and many other non-aerodynamic objects. Recent work, both experimental and computational, has revealed some exciting new information about the nature of this flow including intricate three-dimensional structures that emerge just prior to the onset of turbulence in the wake. In this section we describe spectral element calculations of the two-dimensional flow and then pick it back up in Sect. 6.3 to look at methods for studying the subsequent transition to turbulence.

The system considered is an infinitely long cylinder placed perpendicular to an otherwise uniform open flow. The sole parameter for this system in then the Reynolds number: $Re \equiv U_\infty d/\nu$, where U_∞ is the free-stream velocity and d is the cylinder diameter. First we describe some of the physically important behavior in this flow, and then come back to details of how it can be simulated. It helps to begin with a 'road-map' for the sequence of bifurcations that take the flow from simple to more complex states. There are two useful quantities to form such a guide to understanding: the non-dimensional shedding frequency and the mean drag coefficient C_D. Both shedding frequency and drag show distinct changes at the various bifurcation points of the wake

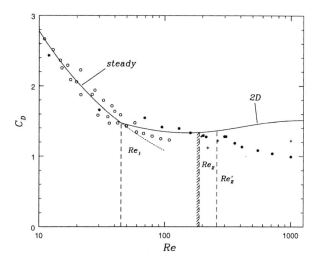

Fig. 5.13. Drag coefficient as a function of Reynolds number for the flow past a circular cylinder. Experiments: (∘,•), Wieselsberger [83]; 3D simulations: +, Henderson [41]. The solid line is a curve fit to two-dimensional simulation data for Re up to 1000 [40].

and can be used as a guide to interpreting changes in the wake structure and dynamics as a function of Reynolds number.

In non-dimensional form the shedding frequency is referred to as the Strouhal number. It is defined as $St \equiv f\,d/u_\infty$, where f is the peak oscillation frequency of the wake. The Strouhal–Reynolds number relationship is shown in Fig. 5.12. At low Reynolds number the flow is steady ($St = 0$) and symmetric about the centerline of the wake. At $Re_1 \simeq 47$ the steady flow becomes unstable and bifurcates to a two-dimensional, time-periodic flow. The shedding frequency of the two-dimensional flow increases smoothly with Reynolds number along the curve shown in Fig. 5.12. Note that each point along the two-dimensional curve represents a perfectly time-periodic flow and there is no evidence of further two-dimensional instabilities for Reynolds numbers up to $Re \approx 1000$. At $Re_2 \simeq 190$ the two-dimensional wake becomes absolutely unstable to long-wavelength spanwise perturbations and bifurcates to a three-dimensional flow (mode A). Experiments and computations indicate a further instability at $Re_2' \simeq 260$ marked by the appearance of fine scale streamwise vortices. We will return to these instabilities in Sect. 6.3.

Figure 5.13 shows the drag curve for flow past a circular cylinder for Reynolds number up to 1000. In the computations the spanwise-averaged fluid force $\mathbf{F}(t)$ is computed by integrating the shear stress and pressure over the surface of the cylinder. The x-component of $\mathbf{F}$ is the drag, the y-component is the lift. Because C_D is determined from an average over the surface of the cylinder, it is much less sensitive to changes in the character

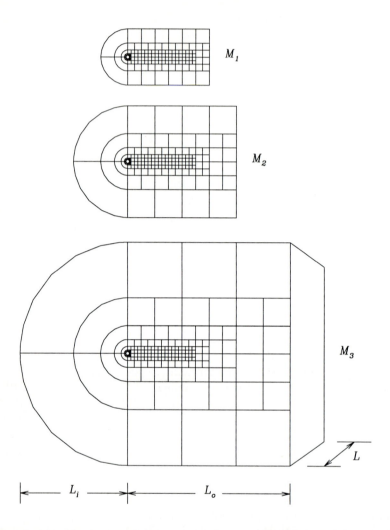

Fig. 5.14. Computational domains used for simulating the flow past a circular cylinder. Each domain is a subset of the largest. The parameters L_o and L_i determine the cross-sectional size, and L determines the spanwise dimension.

of the wake at low Reynolds number than single-point measurements like the shedding frequency. The 'textbook' version of the drag curve is generally plotted on a log-log scale where the only discernible feature is the drag crisis at $Re = O(10^5)$. The flat response of C_D to changes in Reynolds number is compounded by the fact that experimental drag measurements are extremely difficult to make at low Reynolds number, and subtle details of the drag curve are lost in the experimental scatter. The decrease in magnitude of C_D in the steady regime can be fitted to a power-law curve and also makes a sharp but continuous transition at Re_1. Henderson [40] gives the form and coefficients for the steady and unsteady drag curves.

This problem is extremely challenging because it combines several features that are difficult to handle numerically: unsteady separation, thin boundary layers, outflow boundary conditions, and the need for a large computational domain to simulate an open flow. If the computational domain is too small the simulation suffers from blockage. This can have a significant impact on quantities like the shedding frequency, generally producing higher frequencies in the the simulations than are observed in experiments [49]. If resolution near the cylinder is sacrificed for the sake of a larger computational domain then the physically important flow dynamics may not be computed accurately.

Figure 5.14 shows a sequence of computational domains used to simulate both 2D and 3D wakes using nonconforming quadrilateral elements [41]. Boundary conditions are imposed as follows. Along the left, upper, and lower boundaries we use free-stream conditions: $(u_1, u_2, u_3) = (1, 0, 0)$. At the surface of the cylinder the velocity is equal to zero (no-slip). Along the right boundary we use a standard outflow boundary condition for velocity and pressure:

$$p = 0, \quad \partial_x u_i = 0.$$

Along all other boundaries the pressure satisfies (5.11).

These domains use large elements away from the cylinder and outside the wake where the flow is smooth. Local mesh refinement is used to resolve the boundary layer, near wake, and wake regions downstream of the cylinder. In this case the refinement is done beforehand and the mesh is static. Clearly from Figs. 5.12 and 5.13 the simulations predict values of the shedding frequency and drag that agree extremely well with experimental studies up to the point of 3D transition. Just as important as good agreement with experiments, the simulation results are independent of the grid as shown by a detailed h- and p-refinement study [9].

6   Instability, Transition, and Turbulence

In the examples presented thus far we have been building towards more and more complex flows. In this final section we consider methods for studying one of the most complex phenomenon in fluid dynamics: transition to turbulence. Applications in this area are particularly demanding and a good match to the

low numerical dissipation and dispersion errors offered by high-order methods. For example, a physical instability may be suppressed in a numerical
method with excessive artificial viscosity, or it may be triggered prematurely
by numerical dispersion errors. Spectral element methods applied to problems
in transition and turbulence offer the additional ability so simulate geometrically complex domains. This opens a wide range of possibilities for studying
interesting problems in this area.

First we outline some basic tools for computing linear and nonlinear instabilities of a system efficiently. These tools build on the high-order integration
schemes outlined in Sect. 5.1. The discussion here is based on the framework
for bifurcation analysis presented by Tuckerman & Barkley [81]; they discuss additional analysis tools such as efficient methods for computing steady
states and performing continuation.

6.1   Linear stability analysis

For the sake of the following discussion, we can write the Navier–Stokes
equations in the 'schematic' form:

$$\partial_t U = N(U) + LU, \tag{6.19}$$

where $N(U)$ and LU are the operators defined previously. The velocity field
U represents the discretized solution vector whose dimension we denote by
M. We assume this number is quite large, typically $O(10^4)$.

Exponential power method Now consider the problem of determining
the linear stability of steady states. The stability of U is governed by the
eigenvalues λ of the Jacobian $A \equiv N_U + L$:

$$(N_U + L)u = \lambda u. \tag{6.20}$$

This follows from the fact that small perturbations to U evolve according to
the linearized stability equations:

$$\partial_t u = (N_U + L)u \tag{6.21}$$

To determine the stability of U it is sufficient to know whether any eigenvalues have positive real part. Additional information about the leading parts
of the spectrum can also be useful, as well as the structure of the corresponding eigenvectors. In other words, we would like to know complete information
about a *few* of the leading eigenpairs. We assume that the interesting systems
are all too large to construct the Jacobian directly and compute all eigenvalues and eigenvectors via the QR algorithm (operation count $O(M^3)$), so
iterative methods are the key.

The basic iterative technique to compute selected eigenpairs is the power
method. In this method one acts repeatedly with the matrix A on an arbitrary
initial vector u_0 to produce the sequence of vectors $u_n = A^n u_0$. This sequence

approaches the dominant eigenvector, and the sequence of Rayleigh quotients $\lambda_n = u_n^T A u_n / u_n^T u_n$ converges to the corresponding eigenvalue.

Two modifications are needed to make the power method useful for stability analysis. As stated above, we need a few eigenpairs, not just the dominant one. The calculation of several eigenpairs is accomplished by the Arnoldi method or one of its variations [4, 69]. Initially we form the sequence u_0, $A u_0$, ..., $A^{K-1} u_0$, whose span defines the *Krylov space*. K is the number of eigenpairs sought. These vectors are orthonormalized to form a basis v_1, v_2, ..., v_K for the Krylov space. We define the $M \times K$ matrix $V(i, k) \equiv v_k(i)$ and the $K \times K$ Hessenberg matrix $H \equiv V^T A V$. When H is diagonalized, its eigenvalues approximate K of the eigenvalues of A, and V times its eigenvectors approximate K of the eigenvectors of A.

The second modification is to change the region of the complex plane where eigenvalues are sought. The dominant eigenvalues (those largest in magnitude) are not of interest. These correspond to the same exponentially decaying modes that motivated the use of a semi-implicit integration scheme in Sect. 5.1. We want the leading eigenvalues, i.e. those with largest real part.

The solution to the linearized stability problem (6.21) is:

$$u(t + \Delta t) = e^{\Delta t (N_U + L)} u(t). \tag{6.22}$$

The leading eigenvalues of any matrix A are the dominant ones of $\exp(\Delta t A)$ for any positive Δt. The time integration scheme developed for the full Navier–Stokes equations is readily available as an approximation to (6.22), and this is the connection to the power or Arnoldi method: acting with the operator $\exp(\Delta t A)$ is equivalent to integrating the linearized equations over one time step.

A single change is required in the time stepping code: replace the function that computes the nonlinear term $N(U)$ with an equivalent function to compute:

$$N_U u = (U \cdot \nabla) u + (u \cdot \nabla) U.$$

Therefore it is a simple matter to adapt the time stepping algorithm (5.8) to integrate the linearized equations.

Floquet stability analysis The exponential power method can be easily adapted to compute the stability of periodic orbits rather than steady states. Consider a T-periodic solution $U(t \bmod T)$. The operator N_U appearing in the linearized equations (6.21) will also be T-periodic, and it is no longer sufficient to look at the eigenvalues of the constant Jacobian matrix. Instead, stability is determined by the eigenvalues of the operator

$$B \equiv \exp \left(\int_{t_0}^{t_0 + T} (N_U(t') + L) \, dt' \right). \tag{6.23}$$

This operator takes a small perturbation $u(t_0)$ and evolves it once around the orbit to give the perturbation at time $t_0 + T$. In practice the action of B is computed by integrating (6.21) over $T/\Delta t$ time steps.

The eigenvalues μ of B are known as Floquet multipliers. For an initial condition $u(t_0)$ that is an eigenmode of B, the solution to (6.21) is of the form

$$u(t) = \tilde{u}(t \bmod T)e^{\lambda t}, \tag{6.24}$$

where $\lambda = \log(\mu/T)$ is called a Floquet exponent and $\tilde{u}(t \bmod T)$ is called a Floquet mode. The dominant Floquet multipliers (leading Floquet modes) can be computed by applying the exponential power method to the operator B.

Acting with B on a vector u means integrating the linear stability equations over one full period, which in turn means knowing the base flow U at each time step. Because the solutions are time-periodic, a natural simplification is to represent U with a Fourier series in time and only keep enough modes to maintain a level of accuracy consistent with the rest of the computations.

6.2   Nonlinear stability analysis

The final tool we need is a means of distinguishing whether a bifurcation is subcritical or supercritical. Consider the normal form for a pitchfork bifurcation:

$$\partial_t A = \sigma(R - R_c)A - \alpha A^3, \tag{6.25}$$

where A is the amplitude of the bifurcating mode, R is the control parameter, R_c is the bifurcation point, σ is a positive constant relating changes in R to changes in the leading eigenvalue, and α (the Landau coefficient) determines the nonlinear character of the bifurcation. If $\alpha > 0$ the bifurcation is *supercritical* and nonlinearity saturates the growth of A, resulting in a continuous transition. If $\alpha < 0$ the instability is *subcritical* and a sufficiently strong perturbation can trigger a nonlinear instability even for $R < R_c$; the transition is discontinuous and hysteretic.

The critical task is to determine the sign of α. First we compute the steady flow U and the leading eigenmode u for R slightly above R_c. We then start a nonlinear simulation using the initial condition $U + \epsilon u$ for some small ϵ. Choosing some parameter to represent the amplitude A of the bifurcation, we follow the growth of A in time. Initially the simulation shows linear growth consistent with a small positive eigenvalue $\sigma(R - R_c) > 0$. As the flow becomes more nonlinear the time series will begin to deviate from linear growth, in which case it is simple to estimate the value of α directly from the time series. For a supercritical bifurcation the amplitude begins to grow slower than the linear rate, while for a subcritical bifurcation it begins to grow faster. Therefore, the sign of α can be determined quite reliably.

6.3   Examples

We close with two detailed examples showing the application of spectral element methods to complex transition problems: flow over a backward-facing step [8, 46, 47], and flow over a circular cylinder [9, 41, 42]. Numerous other applications can be found in the literature, including: perturbed plane Coutte flow [10], perturbed channel flow [71], turbulent flow past a sphere [80], and turbulent flow over riblets [22, 23].

Backward-facing step The separated flow generated as fluid passes over a backward-facing step is of interest for a variety of reasons. Firstly, separated flows produced by an abrupt change in geometry are of great importance in many engineering applications. This has driven numerous studies of the flow over a backward-facing step during the past 30 years, e.g. [3, 28]. Secondly, from a fundamental perspective, there is a strong interest in understanding instability and transition to turbulence in non-parallel open flows. In this context the flow over a backward-facing step has emerged as a prototype of a nontrivial yet simple geometry in which to examine the onset of turbulence [5, 45–47, 50]. Finally, from a strictly computational perspective, the steady two-dimensional flow over a backward-facing step is an established benchmark in computational fluid dynamics. New computational studies such as the highly accurate stability computations considered help expand the database for this benchmark problem.

The two-dimensional linear stability of this flow has been examined extensively and is discussed in several publications [30, 31, 36]. However, additional computational evidence supports the existence of a local convective instability (again to two-dimensional disturbances) for a sizable portion of the domain at $Re > 525$ [47]. In spite of the numerous investigations of flow over a backward-facing step available in the literature, two of the most basic questions for this flow remain open: in the ideal problem with no sidewalls, at what Reynolds number does the two-dimensional laminar flow first becomes linearly unstable, and what is the nature of this instability? These are the questions we wish to address.

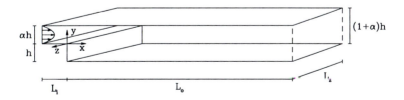

Fig. 6.15. Flow geometry for the backward-facing step. The origin of the coordinate system is at the step edge. We take the ratio of inlet height to step height as $\alpha = 1$, so that the expansion ratio is $1 + \alpha = 2$.

Figure 6.15 illustrates the computational domain under consideration and also serves to define the geometric parameters for the problem. We consider a step of height h and take the edge of the step as the origin of our coordinate system. Fluid arrives from an inlet channel of height αh and flows downstream into an outlet channel of height $(1+\alpha)h$. Here we fix $\alpha = 1$, giving an expansion ratio (outlet to inlet) of $1 + \alpha = 2$. The inflow and outflow lengths L_i and L_o should be large enough that the results are independent of these parameters. At the inlet, $L_i = h$ is sufficient for the range of Reynolds numbers we consider [46, 84]. The required outflow length L_o varies with Reynolds number and must be determined from a proper convergence study. Acceptable values for the range of Re considered here are $15h \leq L_o \leq 55h$ [8]. Finally we take the system to be infinitely large and homogeneous in the spanwise direction, i.e. $L_z = \infty$.

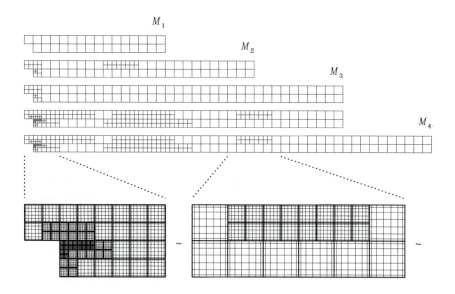

Fig. 6.16. Computational domains for simulating flow over a backward-facing step. Two subsections of mesh M_4 are expanded to show the internal distribution of quadrature points for polynomial order $p = 7$. To simulate a three-dimensional flow the solution is decomposed into M Fourier modes in the periodic spanwise direction, each computed on the same two-dimensional grid [8].

Figure 6.16 shows a collection of nonconforming spectral element grids for simulating this flow [8]. Each grid uses local refinement to isolate the singularity induced by the sharp corner, and to resolve the important recirculation zones in the wake of the step and along the upper wall. The use of local mesh refinement allows high-resolution of the critical regions in this flow

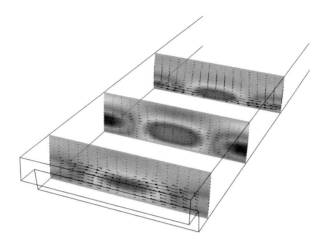

Fig. 6.17. Three-dimensional structure of the leading eigenmode. Contours indicate the strength of the downstream component of the perturbation and vectors indicate the spanwise flow pattern at each downstream plane [8].

along with a large computational domain that pushes the outflow boundary far downstream.

Stability calculations for this flow consist of two parts. First, the steady state solution for a given Re is computed using either time-integration or Newton methods [81]. Second, the relevant bifurcation points along the steady branch of solutions are computed using two- and three-dimensional linear stability analysis based on the iterative methods outlined in Sect. 6.1. The additional parameter for three-dimensional stability calculations is the spanwise wavenumber β of the perturbation. We define $\beta = 2\pi/\lambda$ where λ is the corresponding wavelength.

First consider the three-dimensional stability of the flow. Figure 6.18 shows the neutral stability curve up to $Re = 1000$. Everywhere to the right of the curve the flow has a positive eigenvalue and is linearly unstable. The points were obtained by accurately finding zero crossings of eigenvalue branches (as a function of β) for several Reynolds numbers between 750 and 1000. From Fig. 6.18 it can be seen that the primary linear instability for the backward-facing step occurs very near $Re = 750$. The instability is three dimensional with a streamwise wavenumber $\beta \approx 0.9$.

The three-dimensional structure of the leading eigenmode is shown in Fig. 6.17. The flow visualization is constructed by forming the linear superposition $U + \epsilon u$ of the steady base flow and the computed perturbation field. The structure of the 3D instability represents streamwise vortices that originate in the recirculation zone just downstream of the step. In principle, the flow shown in Fig. 6.17 could be integrated forward in time using the full Navier–

Stokes equations to determine the nonlinear stability of this flow. This is complicated by the presence of a strong convective instability [47] and has not been computed satisfactorily to date.

It is also interesting to look at the two-dimensional stability of this flow. Note that in the limit $\beta \to 0$ the eigenmodes fall into one of two categories, either

$$\hat{\boldsymbol{u}}(x,y) = (\hat{u}(x,y), \hat{v}(x,y), 0), \qquad (6.26)$$

or

$$\hat{\boldsymbol{u}}(x,y) = (0, 0, \hat{w}(x,y)). \qquad (6.27)$$

We shall refer to these as type-I and type-II modes respectively.

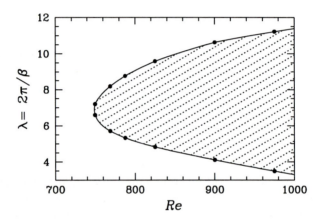

Fig. 6.18. Neutral stability curve for backward-facing step flow. In the shaded region the flow is linearly unstable at the corresponding Reynolds numbers and spanwise wavelengths [8].

Figure 6.19 shows the first three eigenvalues corresponding to two-dimensional modes. At low Reynolds number these modes are, in order of decreasing real part, type-I, type-II, and then again type-I. At $Re \approx 1000$ the second type-I eigenvalue crosses the type-II eigenvalue, but they do not merge because the eigenmodes are of different type. A semi-log plot of the data indicates that the eigenvalues depend exponentially on Re over this range. The corresponding exponential fits are shown in Fig. 6.19. Extrapolation of these fits indicates that the two real eigenvalues would cross at $Re \approx 1350$. It is thus likely that the two eigenvalues join in a complex pair near $Re = 1350$. This is consistent with two-dimensional simulations at $Re = 1350$ which show oscillatory decay to the two-dimensional steady state. Because the exponential fits in Fig. 6.19

will not be valid as the eigenvalues approach one another, it is impossible to estimate what happens at higher Reynolds numbers based on the current data. One possibility is that the two-dimensional linear instability for this flow is a Hopf bifurcation arising from the joining of the two eigenvalue branches [8].

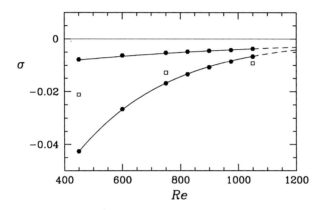

Fig. 6.19. Two-dimensional stability results for the backward-facing step. Solid points and hollow squares denote eigenvalues corresponding to type-I modes and type-II modes, respectively [8].

Cylinder wake For our final example we return the problem of flow past a circular cylinder. The range of Re from about 10 to 1000 shown in Figs. 5.12 and 5.13 represents the entire sequence of states from steady laminar flow to complex turbulent flow for this system. What we wish to understand are the secondary instabilities corresponding to Re_2 and Re_2' and how these instabilities drive the transition to turbulence. Roshko first identified the *transition range* for flow past a circular cylinder as the range of Re where velocity fluctuations become irregular [67]; this is generally quoted at $Re = 150$ to 300. Early flow visualization studies revealed some three-dimensionality in this regime [32, 37], but it was really Williamson who captured the intricate structure of the 3D flow and demonstrated the clear presence of a finite-wavenumber secondary instability [86]. The basic flow patterns consist of two types of 3D vortex shedding now referred to as mode A and mode B. For reasons discussed below these structures are fleeting and can only be captured in pure form on the computer. From this point we proceed in stages, first looking at the linear and nonlinear instabilities that produce these modes, then mechanisms by which they interact to cause transition, and finally some properties of the 'turbulent' flow at higher Re.

Linear stability theory is the natural context for examining the origin of three-dimensionality in the wake [9, 62]. The linear stability problem determines the structure and spatiotemporal symmetry of the global modes and the critical parameter values (Re_2 and Re'_2) where they first become unstable. Once perturbed these modes are self-excited and cause transition to a three-dimensional state. The symmetry of the wake after transition is determined by the spatiotemporal symmetry of the destabilizing global mode.

Computational domains appropriate for simulating the flow past a cylinder were shown in Fig. 5.14. Like the previous example, stability calculations for this flow consist of two parts. First we compute the 2D base flow corresponding to the Karman vortex street by integration the fluid equations until they converge to a *time-periodic* state. Second, we compute the relevant bifurcation points along the 2D time-periodic branch of solutions using three-dimensional Floquet stability analysis.

Figure 6.20 shows the neutral-stability curves for the wake and the two regions of instability that produce modes A and B. These calculations are performed using the stability methods outlined in Sect. 6.1; a detailed explanation is given in [9]. The critical values are $Re_2 \simeq 190$ and $\lambda_2 \simeq 3.96d$ for mode A, $Re'_2 \simeq 260$ and $\lambda'_2 \simeq 0.822d$ for mode B. Note that mode A has a relatively long wavelength that scales on the primary instability wavelength, i.e. the Karman vortex spacing of $\lambda \approx 5d$, while mode B has a relatively short wavelength that presumably scales on the thickness of the separating shear layer. Experimental measurements show exceptional agreement with the predicted maximum growth rate curve for mode A [88]. Measurements for mode B also cluster nicely into the predicted range of unstable wavelengths. Referring back to Figs. 5.12 and 5.13 shows that the critical points for the linear instabilities coincide with the observed transition points in the response of St and C_D. Given the complexity of the system this is outstanding agreement for a non-trivial set of quantities. It is also a triumph for linear stability calculations that reduce the complexity of the full three-dimensional stability problem to a level that can be run on a workstation.

Next we apply the methods for nonlinear stability analysis described in Sect. 6.2 to determine the nonlinear stability of mode A and mode B. As stated, the Landau coefficient in (6.25) can be evaluated from a single time series computed from a full nonlinear calculation. A convenient measure of the amplitude A is the magnitude of the Fourier component corresponding to the 3D perturbation [42]. This analysis indicates that $\alpha = -0.116$ for mode A (subcritical) and $\alpha = 3.92$ for mode B (supercritical). Once these coefficients are known the steady-state amplitudes $|A|$ and $|B|$ can be computed explicitly.[4] Figure 6.21 shows this in the form of a bifurcation diagram for the two instabilities. This figure also includes additional DNS results that

---

[4] Because mode A is subcritical, the coefficient of the next-order term A^5 is necessary to determine saturation. This can be estimated using the same technique applied to determine the Landau constant [42].

verify the validity of the amplitude model near the critical points [41, 42]. Although these results have not yet been confirmed directly by experiment they are consistent with experimental observations. Referring back to Fig. 5.12 we see there is good agreement in the range of hysteresis and the computed frequency drop. The discontinuous drop in shedding frequency is a natural result of the subcritical bifurcation to mode A.

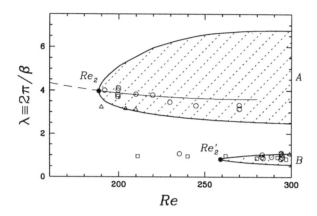

Fig. 6.20. Regions of linear instability for the cylinder wake. neutral curves and critical points (•) are from computations [9]. Open symbols indicate wavelength measurements from various experimental studies [58, 88, 90].

Figure 6.22 shows a visualization of the full nonlinear form modes A and B exhibit at saturation in terms of their streamwise and spanwise components of vorticity. This figure also reveals their distinct space-time symmetries. These symmetries are manifest in the form of a staggered array of streamwise vortices for mode A and an inline array of streamwise vortices for mode B [9, 18, 89]. Several simulations of the three-dimensional flow (all using spectral element methods) have reproduced the essential features observed in experiment and there is now little doubt regarding the qualitative structure of modes A and B [41, 79].

Unfortunately these states are not observed in pure form in the laboratory. In the range $Re \approx 200$ to 260 the natural flow structure may be more appropriately characterized as a mixed A-B state like the one shown in Fig. 6.22c. The relevant facts are the following. Velocity fluctuations exhibit broad-band frequency spectra just beyond the onset of mode A, and mode A is in fact only observed as a transient in the approximate range $Re \approx 180$ to 200. At long times the flow is highly irregular. In contrast to this, mode B is observed with good regularity from $Re \approx 200$ on, and as $Re \to Re_2'$ there is a reasonably well-defined wavelength in the near wake and a sharp peak in the frequency spectrum. However, this peak is superimposed over a broad band of

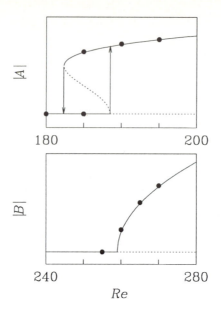

Fig. 6.21. Bifurcation diagrams for (upper) mode A and (lower) mode B. Points (•) indicate results from three-dimensional simulations [41, 42].

frequencies in the background indicative of 'turbulence' farther downstream. From these observations we see that the flow undergoes a fast transition to a state that may be characterized as spatiotemporal (ST) chaos at the *onset* of mode A rather than through a sequence of further bifurcations.

What are the properties of the system that would lead one to expect chaotic behavior? We shall argue this in terms of the spanwise energy spectrum shown in Fig. 6.23, spanwise dimension L, excitation scale l_E, and dissipation scale l_D. ST chaos is a common feature of systems where excitation occurs at a length scale much smaller than the system size but larger than the dissipation scale ($L \gg l_E > l_D$). The excitation scale $l_E \approx \lambda_2$ is fixed by the finite-wavenumber instability of mode A. The subcritical nature of the bifurcation indicates that $l_E > l_D$ at onset. Simulations indicate that the dynamics are time-periodic or quasi-periodic when $L \approx \lambda_2$ so that only one or two mode A instabilities can be excited. When $L \gg \lambda_2$ many A-modes are excited and the simulated flow exhibits ST chaos that is in qualitative agreement with experimental observations. The dynamics in this case are driven by the nonlinear competition between multiple mode A instabilities. This scenario is exactly the Ruelle–Takens–Newhouse (RTN) route to turbulence, a universal route to turbulence in dissipative systems that develop three or more incommensurate modes of oscillation [61, 68]. Finally we close this example with some observations of the 'turbulent' flow that develops beyond the transition regime. If one accepts the definition of a turbulent flow

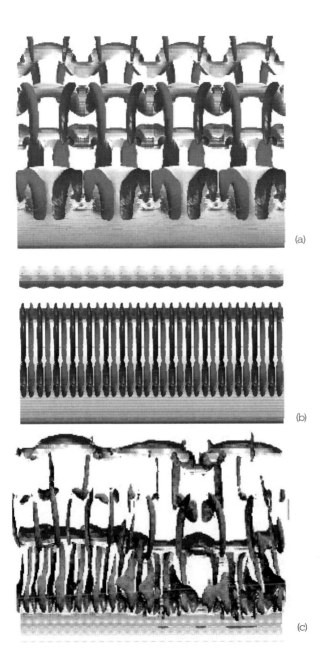

Fig. 6.22. Flow visualization of the three-dimensional vorticity field due to secondary instability in the wake of a circular cylinder: (a) mode A at $Re = 195$, (b) mode B at $Re = 265$, and (c) mixed A-B state at $Re = 265$ [41].

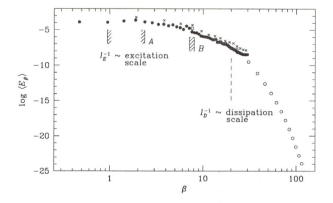

Fig. 6.23. Computed spanwise energy spectrum of the cylinder wake at $Re = 265$, indicating the excitation scale due to mode A and the dissipation scale due to viscosity [41].

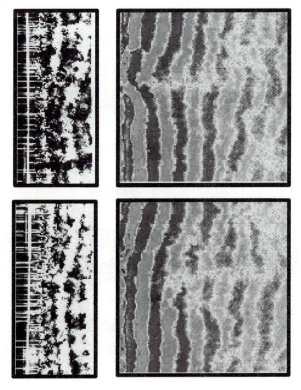

Fig. 6.24. Formation of dislocations in the turbulent cylinder wake: (left) experimental smoke-wire visualization at $Re = 5500$ (Norberg 1992); (right) DNS results at $Re = 1000$ (Henderson 1997).

as being characterized by continuous spatial and temporal spectra, then the cylinder wake is fully turbulent at $Re = 300$. In the classical view further increasing Re pushes the system into the regime of 'featureless' turbulence.

There is at least one additional interesting phenomenon that occurs beyond the transition regime that can be identified as a unique feature of the flow. Figure 6.24 shows a spanwise view of the wake that reveals a set of dislocations in the pattern of vortex shedding. This figure compares both experimental flow visualization and computer simulations with a large spanwise dimension of $L \simeq 25.13d$ [41, 63]. Other experiments of turbulent flow past a cylinder also show evidence of dislocations at Re as high as 10^5 [15]. At high Re these structures develop spontaneously as long as the aspect ratio is sufficiently large.

Dislocation events have a distinct effect on the fluctuation lift and drag. Figure 6.25 shows computed values of C_D and C_L as a function of the spanwise dimension L at $Re = 1000$. In small systems the formation of dislocations is suppressed and the unsteady forces are roughly periodic. In large systems C_L in particular appears in 'bursts.' Minimum values of C_L occur during the formation of a dislocation due to phase differences along the span of the cylinder. This 'bursting' phenomenon is a generic feature of high-Re flow past bluff bodies and is also reported in experimental studies of flow past cylinders and bluff plates [56, 78].

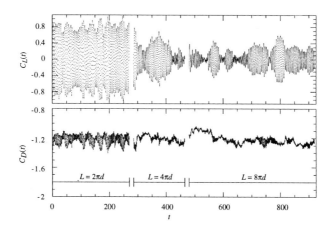

Fig. 6.25. Unsteady lift and drag coefficients for the 'turbulent' flow past a cylinder at $Re = 1000$, illustrating the effect of increasing domain size.

A natural extension of the computational results reported here is to pursue large-eddy simulation (LES) of the turbulent flow at higher Re. A better understanding of the role that large-scale structures play on the overall mixing and dynamics of the flow is certainly necessary for this to succeed. Further experience on the application of high-order methods for LES is also needed. In particular, there are challenges related to issues like proper filtering and formal correctness on locally refined grids.

Acknowledgements

Many people paved the way for the techniques described in these notes by developing the essential framework, theory, and early implementation of spectral element methods. In particular: A. T. Patera, Y. Maday, C. Bernardi, and C. Mavriplis. The author gratefully acknowledges S. J. Sherwin and G. E. Karniadakis for contributing material on triangular and tetrahedral spectral elements, and looks forward to their book on that subject (in preparation). Financial support for this work has been provided primarily by the Office of Naval Research under Grant No. N000-94-1-0793 and by the National Science Foundation under Grant No. CDA-9318145. Most of the 3D computations reported here were performed using machines maintained by the Center for Advanced Computing Research and the JPL High Performance Computing and Communications program at the California Institute of Technology. Most of the 2D computations were performed on the author's Linux PC.

References

1. M. Abramowitz and I. A. Stegun. *Handbook of Mathematical Functions.* Dover, 1970.
2. G. Anagnostou. *Nonconforming Sliding Spectral Element Methods for the Unsteady Incompressible Navier–Stokes Equations.* PhD thesis, M.I.T., February 1991.
3. B. F. Armaly, F. Durst, J. C. F. Pereira, and B. Schönung. Experimental and theoretical investigation of backward-facing step flow. *J. Fluid Mech.*, 127:473–496, February 1983.
4. W. E. Arnoldi. The principle of minimized iterations in the solution of the matrix eigenvalue problem. *Q. Appl. Math.*, 9:17–29, 1951.
5. R. K. Avva. *Computation of the turbulent flow over a backward-facing step using the zonal modeling approach.* PhD thesis, Stanford University, 1988.
6. I. Babuska and M. Suri. The p and h-p versions of the finite element method: Basic principles and properties. Technical report, Institute for Physical Science and Technology, University of Maryland, 1992.
7. I. Babuska and M. Suri. The p and h–p versions of the finite element method, basic principles and properties. *SIAM Review*, 36(4):578–632, 1994.
8. D. Barkley, G. Gomes, and R. D. Henderson. Three-dimensional instability in flow over a backward-facing step. *J. Fluid Mech.*, 1998. *Submitted for publication.*
9. D. Barkley and R. D. Henderson. Three-dimensional Floquet stability analysis of the wake of a circular cylinder. *J. Fluid Mech.*, 319, 1996.
10. D. Barkley and L. S. Tuckerman. Stability analysis of perturbed plane Couette flow. *Phys. Fluids*, 1998. *Submitted for publication.*

11. R. Barrett et al. *Templates* for the solution of linear systems: Building blocks for iterative methods. Text available at http://www.netlib.org, 1994.

12. C. Basdevant, M. Deville, P. Haldenwang, J. M. Lacroix, J. Quazzi, R. Peyret, P. Orlandi, and A. T. Patera. Spectral and finite difference solutions of the Burgers equation. *Computers and Fluids*, 14(1):23–41, 1986.

13. E. R. Benton and G. N. Platzmann. A table of solutions of one-dimensional Burgers equation. *Quart. Appl. Math.*, 29:195–212, 1972.

14. C. Bernardi, Y. Maday, and A. T. Patera. A new nonconforming approach to domain decomposition: the mortar element method. In H. Brezis and J. L. Lions, editors, *Nonlinear Partial Differential Equations and Their Applications*. Pitman and Wiley, 1992.

15. H. M. Blackburn and W. H. Melbourne. The effect of free-stream turbulence on sectional lift forces on a circular cylinder. *J. Fluid Mech.*, 306:267–292, 1996.

16. O. Botella and R. Peyret. Benchmark spectral results on the lid-driven cavity flow. *Computers & Fluids*, 27(4):421–433, 1998.

17. J. P. Boyd. *Chebyshev and Fourier Spectral Methods*. Springer–Verlag, 1989.

18. M. Brede, H. Eckelmann, and D. Rockwell. On secondary vortices in the cylinder wake. *Phys. Fluids*, 8(8):2117–2124, August 1996.

19. J. M. Burgers. A mathematical model illustrating the theory of turbulence. *Adv. Appl. Mech.*, 1:171–199, 1948.

20. C. Canuto, M. Y. Hussaini, A. Quarteroni, and T. A. Zang. *Spectral Methods in Fluid Dynamics*. Springer–Verlag, 2nd edition, 1988.

21. T. F. Chan and T. P. Mathew. Domain decomposition algorithms. *Acta Numerica*, pages 61–143, 1994.

22. D. Chu, R. D. Henderson, and G. E. Karniadakis. Parallel spectral element–Fourier simulation of turbulent flow over riblet-mounted surfaces. *Theoret. Comput. Fluid Dyn.*, 3:219–229, 1992.

23. D. C. Chu and G. E. Karniadakis. A direct numerical simulation of laminar and turbulent flow over streamwise aligned riblets. *J. Fluid Mech.*, 250:1–42, 1993.

24. W. Couzy and M. O. Deville. A fast Schur complement method for the spectral element discretization of the incompressible Navier–Stokes equations. *J. Comput. Phys.*, 116:135, 1995.

25. P. J. Davis and P. Rabinowitz. *Methods of Numerical Integration*. Academic Press, Inc., 1984.

26. M. O. Deville and E. H. Mund. Chebyshev pseudospectral solution of second order elliptic equations with finite element preconditioning. *J. Comput. Phys.*, 60:517–533, 1985.

27. M. Dubiner. Spectral methods on triangles and other domains. *J. Sci. Comp.*, 6:345, 1991.

28. J. P. Johnston E. W, Adams. Effects of the separating shear-layer on the reattachment flow structure part 2: Reattachment length and wall shear-stress. *Exp. Fluids.*, 6(7):493–499, 1988.

29. B. A. Finlayson. *The Method of Weighted Residuals and Variational Principles*. Academic Press, New York, 1972.

30. A. Fortin, M. Jardak, J. J. Gervais, and R. Pierre. Localization of Hopf bifurcations in fluid flow problems. *Int. J. Numer. Methods Fluids*, 24(11):1185–1210, 1997.

31. D. K. Gartling. A test problem for outflow boundary-conditions - flow over a backward-facing step. *Int. J. Numer. Methods Fluids*, 11(7):953–967, 1990.

32. J. H. Gerrard. The wakes of cylindrical bluff bodies at low Reynolds number. *Phil. Trans. R. Soc. Lond.*, 288:351–382, 1978.

33. G. G. Gornowicz. Continuous-field image-correlation velocimetry and its application to unsteady flow over an airfoil. Aeronautical Engineer's Thesis, Caltech, 1997.

34. D. Gottlieb and S. A. Orszag. *Numerical Analysis of Spectral Methods: Theory and Applications.* SIAM, Philadelphia, 1977.

35. L. Greengard and J. Y. Lee. A direct adaptive Poisson solver of arbitrary order accuracy. *J. Comput. Phys.*, 125:415–424, 1996.

36. P. M. Gresho, D. K. Gartling, J. R. Torczynski, K. A. Cliffe, K. H. Winters, T. J. Garratt, A. Spence, and J. W. Goodrich. Is the steady viscous incompressible 2-dimensional flow over a backward-facing step at Re=800 stable? *Int. J. Numer. Methods Fluids*, 17(6):501–541, 1993.

37. F. R. Hama. Three-dimensional vortex pattern behind a circular cylinder. *J. Aeronaut. Sci.*, 24:156, 1957.

38. M. Hammanche and M. Gharib. An experimental study of the parallel and oblique shedding in the wake from circular cylinders. *J. Fluid Mech.*, 232:567, 1991.

39. R. D. Henderson. *Unstructured Spectral Element Methods: Parallel Algorithms and Simulations.* PhD thesis, Princeton University, June 1994.

40. R. D. Henderson. Details of the drag curve near the onset of vortex shedding. *Phys. Fluids*, 7(9), September 1995.

41. R. D. Henderson. Nonlinear dynamics and pattern formation in turbulent wake transition. *J. Fluid Mech.*, 352:65–112, 1997.

42. R. D. Henderson and D. Barkley. Secondary instability in the wake of a circular cylinder. *Phys. Fluids*, 8(6):1683–1685, June 1996.

43. R. D. Henderson and G. Karniadakis. Unstructured spectral element methods for simulation of turbulent flows. *J. Comput. Phys.*, 122(2):191–217, December 1995.

44. T. J. R. Hughes. *The Finite Element Method.* Prentice-Hall, Inc., Englewood Cliffs, New Jersey, 1987.

45. L. Hung. *Direct numerical simulation of turbulent flow over a backward-facing step.* PhD thesis, Stanford University, 1995.

46. L. Kaiktsis, G. E. Karniadakis, and S. A. Orszag. Onset of 3-dimensionality, equilibria, and early transition in flow over a backward-facing step. *J. Fluid Mech.*, 231:501–528, October 1991.

47. L. Kaiktsis, G. E. Karniadakis, and S. A. Orszag. Unsteadiness and convective instabilities in 2-dimensional flow over a backward-facing step. *J. Fluid Mech.*, 321:157–187, 1996.

48. G. E. Karniadakis, M. Israeli, and S. A. Orszag. High-order splitting methods for the incompressible Navier–Stokes equations. *J. Comput. Phys.*, 97(2):414, 1991.

49. G. E. Karniadakis and G. S. Triantafyllou. Three-dimensional dynamics and transition to turbulence in the wake of bluff objects. *J. Fluid Mech.*, 238:1, 1992.

50. A. Knut. *Large eddy simulation of turbulent confined coannular jets and turbulent flow over a backward-facing step (coaxial jet combustor).* PhD thesis, Stanford University, 1995.

51. D. A. Kopriva. Compressible Navier–Stokes computations on unstructured quadrilateral grids by a staggered-grid Chebyshev method. AIAA 98-0133. In *36th Annual Aerospace Sciences Meeting and Exhibit*, Reno, NV, January 1998.

52. D. A. Kopriva. Euler computations on unstructured quarilateral grids by a staggered-grid Chebyshev method. AIAA 98-0132. In *36th Annual Aerospace Sciences Meeting and Exhibit*, Reno, NV, January 1998. AIAA.

53. D. Kosloff and H. Tal-Ezer. Modified Chebyshev pseudospectral methods with $O(N^{-1})$ time step restriction. Technical Report Report 89-71, ICASE, 1989.

54. L. I. G. Kovasznay. Laminar flow behind a two dimensional grid. In *Proc. Cambridge Phil. Society*, 1948.

55. H. O. Kreiss. Numerical methods for solving time-dependent problems for partial differential equations. Technical report, University of Uppsala, Sweden, 1978.

56. D. Lisoski. *Nominally 2-Dimensional Flow About a Normal Flat Plate*. PhD thesis, California Institute of Technology, August 1993.

57. Y. Maday and A. T. Patera. Spectral element methods for the Navier–Stokes equations. *ASME, State of the art surveys in Computational Mechanics*, 1987.

58. H. Mansy, P.-M. Yang, and D. R. Williams. Quantitative measurements of three-dimensional structures in the wake of a circular cylinder. *J. Fluid Mech.*, 270:277–296, 1994.

59. C. Mavriplis. *Nonconforming Discretizations and a Posteriori Error Estimates for Adaptive Spectral Element Techniques*. PhD thesis, M.I.T., February 1989.

60. C. Mavriplis. Adaptive mesh strategies for the spectral element method. *Comput. Methods Appl. Mech. Engrg.*, 116:77–86, 1994.

61. S. Newhouse, D. Ruelle, and F. Takens. Occurence of strange axiom A attractors near quasi periodic flows in T^m, $m \geq 3$. *Commun. Math. Phys.*, 64:35–40, 1978.

62. B. R. Noack and H. Eckelmann. A global stability analysis of the steady and periodic cylinder wake. *J. Fluid Mech.*, 270:297–330, 1994.

63. C. Norberg. Pressure forces on a circular cylinder in cross flow. In *Proc. IUTAM Symp. on Bluff-Body Wakes*, Göttingen, Germany, 1992. Additional photo provided by personal communication (1997).

64. J. T. Oden. Optimal hp-finite element methods. Technical Report TICOM Report 92-09, University of Texas at Austin, 1992.

65. S. A. Orszag and L. C. Kells. Transition to turbulence in plane Poiseuille flow and plane Couette flow. *J. Fluid Mech.*, 96:159, 1980.

66. M. C. Rivara. Selective refinement/derefinement algorithms for sequences of nested triangulations. *Int. J. Numer. Methods Eng.*, 28:2889–2906, 1989.

67. A. Roshko. On the development of turbulent wakes from vortex strects. Technical Report 1191, NACA, 1954.

68. D. Ruelle and F. Takens. On the nature of turbulence. *Commun. Math. Phys.*, 20:167–192, 1971.

69. Y. Saad. Variations on Arnoldi's method for computing eigenelements of large unsymmetric matrices. *Linear Alg. Appl.*, 34:269–295, 1980.

70. J. K. Salmon, M. S. Warren, and G. S. Winckelmans. Fast parallel tree codes for gravitational and fluid dynamical N-body problems. *Int. J. Super. Appl.*, 8(2), 1994.

71. M. F. Schatz, D. Barkley, and H. L. Swinney. Instability in a spatially periodic open flow. *Phys. Fluids*, 7(2):344–358, 1995.

72. K. R. Shariff and R. D. Moser. Two-dimensional mesh embedding for B-spline methods. *J. Comput. Phys.*, 145, 1998.

73. S. J. Sherwin. Hierarchical *hp* finite elements in hybrid domains. *Finite Elements in Analysis and Design*, 27:109–119, 1997.

74. S. J. Sherwin and G. E. Karniadakis. A new triangular and tetrahedral basis for high-order (*hp*) finite-element methods. *Int. J. Numer. Methods*, 38(22):3775–3802, November 1995.

75. S. J. Sherwin and G. E. Karniadakis. A triangular spectral element method – applications to the incompressible Navier–Stokes equations. *Comput. Methods Appl. Mech. Engrg.*, 123(1–4):189–229, June 1995.

76. G. Strang and G. Fix. *An Analysis of the Finite Element Method*. Prentice-Hall, Inc., Englewood Cliffs, New Jersey, 1973.

77. B. Szabo and I. Babuska. *Finite Element Analysis*. John Wiley and Sons, 1991.

78. S. Szepessy and P. W. Bearman. Aspect ratio and end plate effects on vortex shedding from a circular cylinder. *J. Fluid Mech.*, 234:191–217, 1992.

79. M. Thompson, K. Hourigan, and J. Sheridan. Three-dimensional instabilities in the wake of a circular cylinder. *Exp. Therm. Fluid Sci.*, 12:190–196, 1996.

80. A. Tomboulides, S. A. Orszag, and G. E. Karniadakis. Direct and large eddy simulations of axisymmetric wakes. AIAA 93-0546. In *31st Aerospace Sciences Meeting & Exhibit*, Reno, NV, January 1993.

81. L. S. Tuckerman and D. Barkley. Bifurcation analysis for timesteppers. IMA Preprint, April 1998.

82. G. H. Wannier. A contribution to the hydrodynamics of lubrication. *Quart. Appl. Math.*, 8(1), 1950.

83. C. Wieselsberger. Neuere Feststellungen über die Gesetze des Flüssigkeits- und Luftwiderstands. *Phys. Z.*, 22:321–238, 1921.

84. P. T. Williams and A. J. Baker. Numerical simulations of laminar-flow over a 3D backward-facing step. *Int. J. Numer. Methods Fluids*, 24(11):1159–1183, 1997.

85. R. D. Williams. Voxel databases: a paradigm for parallelism with spatial structure. *Concurrency*, 4(8):619–636, 1992.

86. C. H. K. Williamson. The existence of two stages in the transition to three dimensionality of a cylinder wake. *Phys. Fluids*, 31(11):3165–3168, 1988.

87. C. H. K. Williamson. Oblique and parallel modes of vortex shedding in the wake of a circular cylinder at low Reynolds numbers. *J. Fluid Mech.*, 206:579–627, 1989.

88. C. H. K. Williamson. Mode A secondary instability in wake transition. *Phys. Fluids*, 8(6):1680–1682, June 1996.

89. C. H. K. Williamson. Three-dimensional wake transition. *J. Fluid Mech.*, 328:345–407, 1996.

90. J. Wu, J. Sheridan, M. C. Welsh, and K. Hourigan. Three-dimensional vortex structures in a cylinder wake. *J. Fluid Mech.*, 312:201–222, 1996.

91. T. A. Zang. On the rotational and skew-symmetric forms for incompressible flow simulations. *App. Num. Math.*, 7:27–40, 1991.

hp-FEM for Fluid Flow Simulation

Christoph Schwab

Seminar für Angewandte Mathematik, ETH Zürich, Rämistrasse 101,
CH-8092 Zürich, Switzerland

Abstract. We present some mathematical foundations of *hp*-FEM for fluid flow simulation. Particular attention is paid to the mesh-design for viscous, incompressible flow where the regularity of the solution mandates resolution of corner singularities and boundary layers. Stabilized and discontinuous *hp*-FEM for advection dominated and nearly incompressible flows are derived. A new *hp*-adaptive time stepping strategy for spectral accuracy in transient problems is presented.

Table of Contents

1   Introduction

1.1   General Remarks

These lecture notes are intended as an introduction to the subject of hp-Finite Element Methods with particular attention to computational fluid dynamics (CFD) problems. We assume that the reader is familiar with the governing equations of viscous flow, both compressible and incompressible, as well as with the basic facts on hyperbolic systems of conservation laws. Good references on the analysis of the incompressible Navier-Stokes equations are e.g. [71], and for hyperbolic conservation laws with particular attention to numerical methods we mention [32].

What are hp-FEM? There are at present two dominant methodologies in CFD algorithm design, spectral and Finite Difference (FD)/ Finite Volume (FV) methods. Spectral discretizations in fluid dynamics have a long history, see e.g. [17] and the references there. Spectral methods are typically based on subdivisions of the domain in few, rather large elements with high order polynomial discretizations of the field variables. In most cases, the partial differential equations are discretized using collocation in special sets of nodes, mostly the Chebyšev or the Lobatto nodes. In spectral methods, convergence is achieved by raising the *order k of the approximation* rather than by reducing the *meshwidth h*, as is done in finite difference or classical finite element methods. In FD/FV methods, on the contrary, convergence is achieved by refining the mesh, possibly adaptively, and by (adaptively) reducing the order of the scheme near discontinuities using limiters. The resulting numerical schemes are nonlinear, even when applied to linear problems. Unlike FD/FV methods, the convergence order of spectral methods is limited only by the regularity of the solution (loosely speaking by the growth of high order derivatives of the solution provided they exist). There are, however, instances (and we will discuss them) when high derivatives of the solution fail to exist at least in subdomains and in these cases nothing is to be gained by using very high order approximations everywhere [1]

hp-FEM can be viewed as a unification of both ideas – in a sense, they allow the combination of (necessarily anisotropic) local mesh refinement in areas where the exact solution lacks regularity with large, spectral type elements in areas where the solution is smooth.

When to use *hp*-FEM? *hp*-FEM have been successful in applications to structural mechanics, in particular in applications where high accuracy is required and where the solutions lack regularity locally due to corner singularities and/or the presence of small parameters (singular perturbation problems), see e.g. [46], [59]. As we shall see, also in computational fluid dynamics the judicious application of properly designed *hp*-FEM can in many practical situations deliver high resolution and *exponential convergence rates* where either FDM/FVM or spectral methods would only yield algebraic rates.

Let us briefly outline common features and differences between *hp*-FEM and spectral and FD/FV methods. *hp*-FEM share with spectral element and FV methods that arbitrary geometries can be discretized via parametric element maps. Unlike spectral methods, *hp*-FEM allow also for nonuniform distribution of the polynomial degree resp. order of accuracy – for example, not only can the mesh be locally refined near shocks but the order of the method may also be reduced to first order there. This order reduction corresponds to the use of limiters. However, the resulting algorithm is linear for linear problems. This reduction to first order in *hp*-FEM does not entail a loss of overall exponential accuracy if the elements where first order is used are exponentially small. This is typically the case provided we employ *geometric meshes* with a number of refinement levels coupled to the spectral order of the elements. In the small elements supporting the first order discretization,

---

[1] In these cases, the mathematical theory of n-widths indicates that uniform mesh refinement with a low order method will give optimal convergence rates that can at best be matched but not surpassed by spectral methods

all techniques from FDM/FVM for dealing with discontinuous solutions can be brought to bear.

As a rule, *hp*-FEM are based on certain *variational formulations* of the problem under consideration. Discretization is performed by restricting in these formulations the unknown physical fields to finite dimensional subspaces. The design of these subspaces shall be discussed in detail below. In the derivation of *hp*-methods, one assumes first that integrals in the variational formulation are evaluated exactly. This is rarely possible in practice, since for example for curved elements and nonlinearities some form of *numerical quadrature* is an integral part of *hp*-FE algorithms. The resulting, fully discrete methods are in essence *hp-spectral element methods* sharing features of *hp*-FEM (variational formulation, variable polynomial degree/spectral order) and of the traditional spectral methods (collocation of nonlinearities).

The pillars of any convergent numerical algorithm are *stability* and *consistency*. Exponential convergence rates with *hp*-FEM require, as a rule, the proper design of the *hp*-subspaces, i.e. proper choice of the mesh and the degree distribution. Many choices are usually possible. They can be based either upon the dominant solution phenomena or on adaptive strategies. As a rule, *hp*-FEM are most efficient when *highly anisotropic elements* are admitted, e.g. in boundary layers or viscous shock profiles; since *anisotropic adaptive refinements* are to date still not as well developed as isotropic ones, some *a-priori mesh-design* with anisotropic elements should be performed in the appropriate flow regions whenever possible. The use of body-fitted, structured meshes is well-established in CFD and should be kept with *hp*-FEM whenever possible. Note, however, that *hp*-meshes may differ considerably from the ones used with low order methods.

Unlike in solid mechanics, variational formulations of fluid flow problems are usually neither symmetric nor coercive due to dominant transport effects. Therefore *stabilized variational formulations* have to be used to achieve stability of FEM in the presence of advection. We will discuss in detail the most frequently employed formulations such as *Galerkin Least Squares (GLS)* and the *streamline diffusion FEM (SDFEM)* as well as certain *Discontinuous-Galerkin (DG)* methods. Such formulations are well established in CFD, but have to be adapted to accommodate *hp*-FEM.

These lectures aim at the description of *hp*-FEM with particular attention to the formulation of *hp*-schemes for flow problems and their error analysis. Methodologically, we start by describing *hp*-FE discretizations of simple linear diffusion and transport processes, followed by Galerkin schemes for inviscid conservation laws and finally the full, compressible NSE. In each case, we explain carefully the design of meshes and order distributions which are most efficient for the resolution of specific flow phenomena, such as singularities, boundary layers and viscous shock profiles in the context of judiciously chosen model problems. Likewise, the GLS, SDFEM and DG stabilization techniques for convection dominated problems will also be discussed first for such model problems.

These notes are not intended as a mathematical treatise on *hp*-FE theory, they rather try to give a concise overview over variational formulations for *hp*-FEM and theoretical convergence results that are essential for efficient fluid flow simulation. The material presented is biased towards the recent work of the author. Nevertheless, we have tried to give up to date references to related, and particularly computational work. These references, as well as the other articles in the present volume should be consulted for different viewpoints of high order methods.

1.2   Notation

We list some notation which will be used throughout the text. We will denote the physical domain in which the computations will be performed by $\Omega \subset \mathbb{R}^d$ where the dimension $d = 1, 2, 3$ ($d = 1$ will rarely be considered). Partial derivatives with respect to the spatial variables x_i will be denoted by ∂_i and will be understood in the distributional sense, unless stated otherwise. The usual differential operators $\nabla, \Delta, \text{div}$ etc. shall be used and summation over repeated indices is employed. By $L^2(\Omega)$ we denote the usual space of square integrable functions in Ω. By $H^k(\Omega)$, $k \geq 0$, we denote the Sobolev space of functions with kth square integrable derivative in Ω. Evidently, $H^0 = L^2$. By $(\cdot, \cdot)_\Omega$ we denote the L^2 innerproduct over the set Ω, i.e. $(u, v)_\Omega = \int_\Omega uv dx$. $L^2(\Omega)$ is a Hilbert space with inner product $(u, v)_\Omega$ and norm $\|u\| := ((u, u)_\Omega)^{1/2}$. Analogously, $H^k(\Omega)$ is equipped with the innerproduct

$$(u, v)_{k,\Omega} = \sum_{|\alpha| \leq k} (D^\alpha u, D^\alpha v)_\Omega$$

where $\alpha \in \mathbb{N}_0^d$ is a multiindex and D^α is the derivative of order α. The norm $\|u\|_{k,\Omega}$ is defined analogously as in the L^2 case:

$$\|u\|_{k,\Omega} = \left((u, u)_{k,\Omega}\right)^{1/2}$$

Similarly, we define Sobolev spaces and norms on lower dimensional sets, such as e.g. the boundary $\Gamma = \partial\Omega$. The $L^2(\Gamma)$ inner product is just the (Lebesgue) surface integral taken with respect to the surface measure on Γ and we write $(u, v)_\Gamma = \int_\Gamma uv ds$. Spaces of vector valued functions will be denoted by a superscript after the space, i.e.

$$H^k(\Omega)^m = \left[H^k(\Omega)\right]^m$$

denotes the m-fold tensor product of the space $H^k(\Omega)$. Typically, m will denote the number of state variables in the system under consideration.

Throughout, the spectral order of the elements will be denoted by the letter k, elements by K and partitions of Ω into d-dimensional elements by $\mathcal{T}$. The letter $\mathcal{E}$ denotes the set of $d-1$ dimensional, intersections of elements $K, K' \in \mathcal{T}$.

1.3 Governing Equations

Continuum mechanics of a compressible fluid in a domain $\Omega \subset \mathbb{R}^d$ is described by the mass density $\rho : \Omega \to \mathbb{R}$, the velocity field $\mathbf{u} : \Omega \to \mathbb{R}^d$ and the energy $e : \Omega \to \mathbb{R}$. These fields are governed by the (compressible) **Navier-Stokes equations** (NSE) which read (in Eulerian form)

Conservation of Mass

$$\frac{\partial \rho}{\partial t} + \sum_{j=1}^{d} \frac{\partial}{\partial x_j}(\rho u_j) = 0 \,. \tag{1.1}$$

Conservation of Momentum

$$\frac{\partial}{\partial t}(\rho u_i) + \sum_{j=1}^{d} \frac{\partial}{\partial x_j}(\rho u_i u_j + p\delta_{ij}) = \sum_{j=1}^{d} \frac{\partial}{\partial x_j}\tau_{ij} + S_i \tag{1.2}$$

$$\text{for } i = 1, \ldots, d \,.$$

Conservation of Energy

$$\frac{\partial}{\partial t}(\rho e) + \sum_{j=1}^{d} \frac{\partial}{\partial x_j}((\rho e + p)u_j) = \sum_{j=1}^{d} \frac{\partial}{\partial x_j}\left(k\frac{\partial T}{\partial x_j} + \sum_{l=1}^{d}\tau_{jl}u_l\right), \tag{1.3}$$

where $k > 0$ denotes thermal diffusivity, τ is the stress tensor describing the elastic effects in the fluid, e is the internal energy and $\mathbf{S} \in L^2(\Omega)^d$ are given sources.

Part I

Fundamentals of *hp*-FEM

2 Model Problems

We present several scalar model problems modeling diffusive transport of a scalar quantity u which share many features which we will encounter later on also in the context of the Navier-Stokes equations.

Consider linear, diffusive transport of a scalar field $u(\mathbf{x}, t)$ in $\Omega \times (0, T)$ where $\Omega \subset \mathbb{R}^d$ is a bounded domain with piecewise smooth boundary $\Gamma = \partial\Omega$; it is governed by the equation

$$\frac{\partial u}{\partial t} + \operatorname{div} \mathbf{f}(u) + \sigma u = \operatorname{div} \mathbf{q}(\nabla u) + S \quad \text{in } \Omega \times (0, T)\,. \tag{2.1}$$

Here **f** and **q** are the convective resp. diffusive fluxes, and $\sigma \geq 0$ the reaction constraint. We consider here the linear fluxes

$$\mathbf{f}(u) = \boldsymbol{\beta} u \,, \tag{2.2}$$

where $\boldsymbol{\beta} \in L^\infty(\Omega)^d$ is the flux vector and

$$\mathbf{q}(\nabla u) = \mathbf{A}\,\nabla u \tag{2.3}$$

where $\mathbf{A} \in L^\infty(\Omega)^{d\times s}_{\mathrm{sym}}$ is a positive, possibly anisotropic diffusivity matrix satisfying

$$\varepsilon\,\xi^\top\xi \leq \xi^\top\,\mathbf{A}(\mathbf{x})\xi \leq c\varepsilon\,\xi^\top\xi \,\forall \xi \in \mathbb{R}^d, \quad \text{a.e. } \mathbf{x} \in \Omega \tag{2.4}$$

for some $\varepsilon > 0$.

Of particular interest is the case $\mathbf{q} = \varepsilon\mathbf{1}$, whence $\mathbf{q}(\nabla u) = \varepsilon\,\nabla u$ and (2.1) becomes with (2.2), (2.3)

$$\frac{\partial u}{\partial t} + \operatorname{div}(\boldsymbol{\beta} u) + \sigma u = \varepsilon\Delta u + S \quad \text{in } \Omega \times (0,T)\,. \tag{2.5}$$

Here $S \in L^2(\Omega)$ is a source term which we assume time-independent unless stated otherwise.

(2.5) is completed by initial- and boundary conditions. To this end, partition Γ into 2 disjoint parts,

$$\Gamma = \overline{\Gamma_D} \cup \overline{\Gamma_N},\ \Gamma_D \cap \Gamma_N = \emptyset\,.$$

Then we impose initial and boundary conditions

$$u = f \quad \text{on } \Gamma_D\,,$$
$$\mathbf{q}(\nabla u) \cdot \mathbf{n} = g \quad \text{on } \Gamma_N\,, \tag{2.6}$$
$$u(\cdot,0) = u_0 \quad \text{at } t = 0\,,$$

Here **n** is the exterior unit normal vector to Γ. In the following, we will discuss the *hp*-FE discretization of various special cases of (2.5). Since many schemes are based on separate treatment of space and time variables, it is useful to consider first semidiscretization of (2.5) in space. These spatial *hp*-discretizations can be introduced for the steady state case, i.e. for $\frac{\partial u}{\partial t} = 0$ and this is what we will do in the sequel. We consider special cases of (2.5), in particular the reaction-diffusion and the pure advection problem.

While doing so, we will pay particular attention to the singular perturbation character and the variational formulation of the problem – we review classical, mixed, stabilized and the discontinuous Galerkin (DG) formulations. Especially the latter ones are being used with increasing frequency in FE flow simulations (see, e.g., [12,13,18–20,23,26,36,44,49,64–67] but must be complemented by a suitable time stepping scheme. This will be topic of the second part of these notes, however.

The preferable type of discretization depends strongly on the dominant terms in (2.1). We will address several particular cases:

2.1 Reaction-Diffusion

Here $\mathbf{A} = \mathbf{1}$, $\boldsymbol{\beta} = \mathbf{0}$ and $\sigma = 1$ so that (2.5), (2.6) become

$$-\varepsilon \Delta u + u = S \quad \text{in } \Omega, \tag{2.7}$$

$$u = f \text{ on } \Gamma_D, \quad \varepsilon \frac{\partial u}{\partial n} = g \text{ on } \Gamma_N. \tag{2.8}$$

2.2 Convection

We assume that in (2.5) $\boldsymbol{\beta} \in C^1(\overline{\Omega})^d$ and that $\varepsilon = 0$. Then (2.5) becomes, in the steady state case,

$$\boldsymbol{\beta} \cdot \nabla u + (\sigma + \operatorname{div} \boldsymbol{\beta}) u = S \quad \text{in } \Omega. \tag{2.9}$$

This equation is now first order hyperbolic in space and the boundary conditions (2.6) cannot be imposed anymore. It is a model for the continuity equation (1.1). The correct boundary conditions, for which the problem (2.9) is well-posed, are as follows: Define in- and outflow boundaries

$$\Gamma_- = \{ x \in \Gamma : \boldsymbol{\beta}(x) \cdot \mathbf{n}(x) < 0 \}, \quad \Gamma_+ = \{ x \in \Gamma : \boldsymbol{\beta}(x) \cdot \mathbf{n}(x) > 0 \}.$$

and assume that $\Gamma = \bar{\Gamma}_- \cup \bar{\Gamma}_+$. The "inflow" boundary condition for (2.9) is

$$u = f \text{ on } \Gamma_-. \tag{2.10}$$

No boundary conditions can be prescribed on the outflow boundary Γ_+.

2.3 Convection-Diffusion

We observe in (2.9), (2.10) that the vanishing viscosity $\varepsilon \to 0$ in (2.5) has caused a reduction of the order of the equation and the loss of a boundary condition. This is a (very simple) model of the transition from (incompressible) Navier-Stokes to (incompressible) Euler (which is, however, not very well understood at present). The steady state equation (2.5) with $\varepsilon > 0$ and $\boldsymbol{\beta} \neq \mathbf{0}$ is the *convection-diffusion equation*

$$\boldsymbol{\beta} \cdot \nabla u + (\sigma + \operatorname{div} \boldsymbol{\beta}) u = \varepsilon \Delta u + S \quad \text{in } \Omega, \tag{2.11}$$

together with the boundary conditions (2.6).

3   Solution properties

Any stable numerical scheme for the numerical solution of (2.1) - (2.11) will generate solutions u_N which approximate u - for the design of efficient schemes it is therefore necessary to know certain qualitative features of the solutions to be approximated. *hp*-FEM allow for simultaneous mesh-refinement and variation of the polynomial degree and constitute a generalization of both, the standard low order finite-volume / finite-element methods as well of the so-called spectral methods. The large flexibility in *hp*-FEM is most easily used with unstructured, triangular resp. tetrahedral meshes, and high polynomial degree which is best suited for irregular flows with moving features as e.g. the vortex shedding in incompressible flow in the wake of a cylinder. Nevertheless, substantial improvements in accuracy vs. degrees of freedom (and, in particular, exponential convergence) can be realized by using structured meshes in certain subregions of the flow.

In the following, some typical solution features are presented.

3.1   Corner Singularities

Corner singularities are present in 2-dimensional domains whenever

a) the governing equations contain viscosity (i.e. diffusion or elasticity), [48], and the boundary of the domain is not smooth at a point $O \in \partial\Omega$ (even changes in curvature which may not be apparent at first sight excite corner singularities), or

b) when inside a smooth boundary segment the boundary conditions change abruptly, (e.g. P_7 in Figure 3.1). In three dimensional domains, for example in polyhedra, corner singularities arise at *vertices* - in addition, at *edges* so-called *edge-singularities* appear which we discuss below.

Corner singularities are solution components with low regularity which are poorly approximated by low order methods on uniform meshes. In the context of convection-dominated problems, the resulting large approximation error at the corner is transported downstream and maybe responsible for spurious solution features.

We discuss corner singularities in 2-dimensions. Let $\Omega \subset \mathbb{R}^2$ be a polygon with M possibly curved sides Γ_j, cf. Figure 3.1, and vertices P_j, $j = 1, \ldots, M$.

Consider the reaction-diffusion Problem (2.7) with $\varepsilon = 1$ in Ω for smooth source terms S, f and g. We assume that $\overline{\Gamma}_D \cap \overline{D}_N$ coincide with vertices P_j, i.e. each Γ_j is contained in either Γ_D or in Γ_N.

If the source terms are smooth, the solution u of (2.7), (2.8) is also smooth *inside* Ω, but *not at the vertices* P_j. More precisely, for any $s > 0$ the solution u can be decomposed into a smooth part $u_{\mathrm{reg}} \in H^{s+2}(\Omega)$ and *singular functions* $S(r_j, \varphi_j)$:

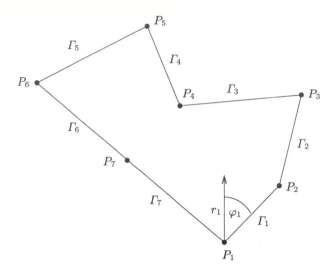

Fig. 3.1. Polygon Ω with vertices P_j

$$u = u_{\mathrm{reg}} + \sum_{j=1}^{M} \chi(r_j) \sum_{k=1}^{K(s)} \sum_{\ell}^{L} a_{k\ell}\, S_{jk\ell}(r_j, \varphi_j) \qquad (3.1)$$

where

$u_{\mathrm{reg}} \in H^{s+2}(\Omega)$,

$\chi(r) > 0$ is a smooth cut-off function,

$\chi \equiv 1$ near zero,

$$S_{jk\ell}(r_j, \varphi_j) = r_j^{\lambda_{jk}} (\log r_j)^\ell\, \Phi_{jk\ell}(\varphi_j) \qquad (3.2)$$

(r_j, φ_j) Polar coordinates at P_j,

$\lambda_{j\ell} > 0$ the singularity exponent,

$\Phi_{j\ell}(\varphi)$ a smooth function of φ.

Notice the dependence of K in (3.1) on s - the smoother u_{reg} is supposed to be, the larger is s and the more terms have to be included into the decomposition. Decomposition (3.1) is by now classical in the theory of elliptic equations - we mention here only [42] and the references there. It is important to note that the $S_{jk\ell}$ and the λ_j do not depend on S, f and on g in (2.7). They only depend on the interior angle of Ω at the vertex P_j, the boundary conditions and on the diffusion operator. Analogous results hold for solutions of (2.11) with $\varepsilon = 1$, since there once again the diffusion part of the operator is equal to $-\Delta u$. The same result holds also for systems, such as for the Stokes-system or the system of linearized elasticity arising in viscous, compressible flow.

3.2   Boundary layers

Other interesting phenomena happen when $\varepsilon \to 0$ in (2.7), (2.11). We see
that formally, at $\varepsilon = 0$, the order of the equation changes: (2.11) becomes the
first order hyperbolic problem (2.9) and (2.7) the "zeroth" order problem

$$u = S \text{ in } \Omega . \tag{3.3}$$

Evidently, for general source terms S, this u will not satisfy the boundary
conditions (2.8) anymore - this (whole or partial) loss of boundary conditions
is typical when the viscosity ε in the system vanishes. As $\varepsilon \to 0$, the solution
u^ε of (2.7) forms steep gradients near $\partial\Omega$. The simplest one-dimensional
problem exhibiting these effects is

$$-\varepsilon u'' + u = 1 \text{ in } (-1,1), \quad u(\pm 1) = 0 . \tag{3.4}$$

We have the exact solution

$$u^\varepsilon(x) = 1 - \frac{\exp(-(1+x)/\sqrt{\varepsilon})}{\exp(1/\sqrt{\varepsilon}) + \exp(-1/\sqrt{\varepsilon})} - \frac{\exp(-(1-x)/\sqrt{\varepsilon})}{\exp(1/\sqrt{\varepsilon}) + \exp(-1/\sqrt{\varepsilon})} \tag{3.5}$$

which is equal to a regular part, u_{reg}, i.e. $S \equiv 1$, up to two terms that
are exponentially decaying off $\partial\Omega$, the so-called (viscous) boundary layers:
i.e. the decomposition $u^\varepsilon = u_{\text{reg}} + u_{b\ell}$. For linear problems with constant
coefficients, viscous boundary layers are always exponential. If the coefficients
are nonconstant or the problem is nonlinear, generally no explicit form of the
layers is known. For nonconstant, analytic coefficients one can show, however,
that *boundary layers with length scale d* satisfy for every n the estimates (see,
e.g., [45] for a proof in the linear, variable coefficient case)

$$|D^n u^\varepsilon_{b\ell}(x)| \leq C K^n \max\{n, 1/\varepsilon\}^n \exp(-b\rho(x)/\varepsilon) \tag{3.6}$$

where $\rho(x) = \text{dist}(x, \partial\Omega)$ is the distance to the boundary and the positive
constants b, C, K are independent of n and d. Evidently, the solution (3.5)
satisfies (3.6).

In two dimensions, *if $\partial\Omega$ is smooth*, an analogous result holds: the solution
u^ε can be decomposed into a *regular part* $u_{\text{reg}}(x)$ (whose derivatives remain
bounded as $\varepsilon \to 0$) and *boundary layers* $u_{b\ell}$ (whose derivatives behave like

$$|D^\alpha u_{b\ell}| \sim O(\varepsilon^{-|\alpha|})$$

as $\varepsilon \to 0$).

Generally, boundary layers of (2.7), (2.11) are special solutions of (2.7)
resp. (2.11) with $S \equiv 0$, but with nonzero data f, g of the forms

$$u_{b\ell} = U(\rho/d(\varepsilon)) \, \Phi(\theta) \tag{3.7}$$

where (ρ, θ) are boundary fitted coordinates near $\partial\Omega$ (see Figure 3.2).

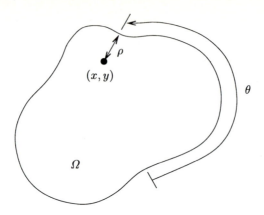

Fig. 3.2. Boundary fitted coordinates (ρ, θ) in Ω

In two dimensions, $0 < s < L$ is the arclength of $\partial\Omega$ and s is the normal distance of a point $P = (x, y)$ to $\partial\Omega$.

The function $U(\cdot)$ in (3.7) is independent of ε and decaying for positive arguments - it is the so-called *boundary layer profile*. In all linear problems, in particular in (2.7), (2.11), the boundary layer profile is exponential, i.e.

$$U(\zeta) = \exp(-\zeta), \quad \zeta \geq 0 . \tag{3.8}$$

The function $\Phi(s)$ in (3.7) is smooth independent of ε and $d(\varepsilon)$, the so-called *length-scale of the layer*, is usually some simple power of ε - in (2.7), it is $d(\varepsilon) = \sqrt{\varepsilon}$, whereas in (2.11) $d(\varepsilon) = \varepsilon$ or $d(\varepsilon) = \sqrt{\varepsilon}$, depending on whether the boundary is characteristic or not. In nonlinear problems, not much is known about decompositions

$$u^\varepsilon = u_{\text{reg}} + u_{b\ell} .$$

For the incompressible Navier-Stokes equations, the profile $U(\zeta)$ in (3.8) is the similarity solution of a nonlinear ODE which again is decaying as ζ tends towards infinity and the length scale is $d(\varepsilon) = Re^{-1/2}$.

3.3   Viscous Shock Profiles

One dimensional case. Consider the scalar conservation law with viscous perturbation in one dimension

$$u_t + f(u)_x = \varepsilon u_{xx} \quad (x, t) \in \mathbb{R} \times \mathbb{R}_+ \tag{3.9}$$

with initial condition

$$u(x, 0) = u_0(x) . \tag{3.10}$$

For $\varepsilon = 0$, $u(x,t)$ in general develops discontinuities in finite time. For $\varepsilon > 0$, these shocks are smeared out - we have a **viscous shock profile**.

Assuming that (3.9) admits a steady asymptotic solution as $t \to \infty$, this solution must satisfy

$$f(\widetilde{u})_x = \varepsilon \widetilde{u}_{xx} \quad \text{in } x \in \mathbb{R} \tag{3.11}$$

$$\lim_{x \to \pm\infty} \widetilde{u}(x) = u^{\pm} \tag{3.12}$$

where $u^{\pm}$ are the left/right states of the shock. We see from (3.11) that $\widetilde{u}(x)$, if it exists, must have the form

$$\widetilde{u}(x) = U((x - x_s)/\varepsilon) \tag{3.13}$$

where x_s is the **shock-location** and $U(\cdot)$ is the **viscous shock-profile**. $U(\xi)$ satisfies the ordinary differential equation

$$U_{\xi\xi} = f(U)_\xi \qquad \xi \in (-\infty, \infty) \tag{3.14}$$

with the boundary conditions

$$\lim_{\xi \to \pm\infty} U(\xi) = u^{\pm} . \tag{3.15}$$

Assuming a solution U of (3.14) exists, this solution will be locally analytic if the flux $f(\cdot)$ is analytic. Moreover, in many cases we have exponential decay of $U(\xi)$ to $u^{\pm}$:

$$|U(\pm\xi) - u^{\pm}| \le C \exp(-b\xi), \; \xi \to \infty . \tag{3.16}$$

See [68], Chapter 24, for more on this.

Consider the viscous Burgers' equation (3.9) where $f(u) = u^2/2$. Here the viscous shock profile developing for initial data

$$u_0(x) = \left\{ \begin{array}{l} \alpha \; x < 0 \\ -\alpha \; x > 0 \end{array} \right\}$$

with $\alpha > 0$ has the form (3.13) with $x_s = 0$ and

$$U(\xi) = -\alpha \tanh(\alpha\xi) \tag{3.17}$$

for some $\alpha > 0$ independent of ε. (3.17) evidently satisfies (3.16) with $u^{\pm} = \mp\alpha$. **We stipulate therefore that viscous shock profiles are internal layers originating in the shock-location x_s.** The viscous shock profiles can be seen as boundary layers at the (generally unknown) free boundary x_s. The viscous shock profile $\widetilde{u}(x)$ in (3.13) is assumed to satisfy an estimate of the form (3.6), i.e. there are $b, C, K > 0$ such that

$$|D^n \widetilde{u}(x)| \le CK^n (\max(n, 1/\varepsilon))^n \exp(-b\rho(|x - x_s|)/\varepsilon) \tag{3.18}$$

for $n = 1, 2, \ldots$ and $x \ne x_s$. The solution $\widetilde{u}(x)$ with U as in the example (3.17) is seen to satisfy condition (3.18).

Higher dimensional case. In dimension $d > 1$, shocks are discontinuities in solutions across possibly curved discontinuity surfaces Σ, which arise in nonlinear, hyperbolic equations. If viscosity is present, the discontinuities will be replaced once more by a viscous shock profile which we assume to have the following generic form: denoting by (s, ρ) coordinates fitted to Σ (see Figure 3.3) the viscous shock profile is of the form

$$u_{\mathrm{sh}}(s, \rho) = C(s)\, u_{b\ell}(|\rho|) \tag{3.19}$$

where $u_{b\ell}(\rho)$ is a boundary layer function satisfying the estimate (3.6) with length scale equal to the viscosity parameter ε and $C(s)$ is smooth (analytic) independent of ε, i.e. $\|D_s^\ell C\|_{L^\infty} \le c\, K^\ell\, \ell!$ for all ℓ, where c, K are independent of ε.

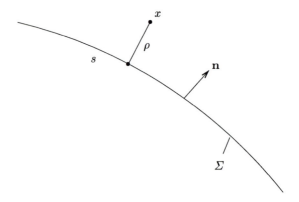

Fig. 3.3. Discontinuity surface Σ and fitted coordinates (s, ρ)

We emphasize that the behavior (3.19) for viscous shock profiles is extrapolated from 1-d, there are, to date, no rigorous regularity results in the nonlinear setting for the solution $u_{\mathrm{sh}}(s, \rho)$ - in particular, the regularity at shock - boundary and shock - shock interaction points in the presence of viscosity is open. We take here the point of view that viscous shock profiles and boundary layers are closely related and that, likewise, the corner singularities and the shock-boundary interaction are of similar nature in that one has low regularity in an $O(\varepsilon)$ neighborhood of the interaction point of a globally relatively smooth (piecewise analytic) solution u.

4   Basic hp FEM

We describe the main components of hp-FEM, beginning with the admissible meshes followed by the function spaces on these meshes. We make provisions for unstructured as well as for patchwise structured meshes since these are very advantageous for the resolution of specific flow phenomena, if they are combined with proper distribution of the polynomial degrees. Many of the meshes used in present day CFD exhibit some structure, such as refinement towards the surfaces and uniform refinement at the trailing edge and corners. We will explain how proper design of meshes and polynomial degree distribution in the *hp*-FEM gives *exponential convergence* for the solution features of the previous section.

4.1   *hp*-FE Spaces on patchwise structured grid

Meshes. Let $\mathcal{P}$ denote a partition of Ω into open patches P which are images of a reference patch $\widehat{P}$ under smooth, bijective maps F_P:

$$\forall P \in \mathcal{P}: \quad P = F_P(\widehat{P}).$$

We assume that $\widehat{P}$ is either the unit cube

$$\widehat{P} = \widehat{Q} := (-1, 1)^d$$

or the unit simplex

$$\widehat{P} = \widehat{S} := \left\{ \hat{x} \in \mathbb{R}^d : \hat{x}_i > 0, \sum_{i=1}^{d} \hat{x}_i < 1 \right\}.$$

The meshes $\mathcal{T}$ are unions of patch meshes $\mathcal{T}_P$ which are constructed in the reference patch $\widehat{P}$ and transported to $P \in \mathcal{P}$ via the patch map F_P. For each P, a patch mesh $\mathcal{T}_P$ is obtained by first subdividing $\widehat{P}$ into triangles resp. quadrilaterals $\widehat{K}$ which are affine equivalent to either $\widehat{Q}$ or $\widehat{S}$; we call this mesh $\widehat{\mathcal{T}}_P$. A mesh $\mathcal{T}_P$ in $P \in \mathcal{P}$ is then obtained by simply mapping $\widehat{\mathcal{T}}_P$ to P using the *patch map F_P*

$$\forall P \in \mathcal{P}: \quad \mathcal{T}_P := \{K \,|\, K = F_P(\widehat{K}), \ \widehat{K} \in \widehat{\mathcal{T}}_P\}. \tag{4.1}$$

The mesh $\mathcal{T}$ in Ω is the collection of all patch meshes, i.e.

$$\mathcal{T} = \bigcup_{P \in \mathcal{P}} \mathcal{T}_P.$$

Note that each element $K \in \mathcal{T}$ is an image of the reference domain $\widehat{P}$ via the *element map F_K*: if $K \in P$ for some $P \in \mathcal{P}$,

$$K = F_K(\widehat{P}), \quad F_K := F_P \circ A_{\widehat{K}} \tag{4.2}$$

where $A_{\widehat{K}} : \widehat{P} \to \widehat{K} \in P$ is affine.

We emphasize that we could choose $A_{\widehat{K}} = id$ and $\mathcal{T}_P = \{P\}$, thereby obtaining the usual parametric elements and arbitrary, unstructured meshes. However, it is advantageous in *hp*-FEM to use structured patch meshes $\mathcal{T}_P$

as e.g. geometric corner refinement, anisotropic boundary layer and edge refinement etc. In what follows, the partition $\mathcal{P}$ and the patch maps $\mathbf{F}_P = \{F_P : P \in \mathcal{P}\}$ shall be fixed, i.e. mesh refinement is performed in $\widehat{P}$.

We call the mesh $\mathcal{T}$ **regular**, if for any two $K, K' \in \mathcal{T}$ the intersection $\overline{K} \cap \overline{K}'$ is either empty or an entire side (more precisely, an entire boundary segment of dimension $0 \le d' < d$ as e.g. a vertex ($d' = 0$), an entire edge ($d' = 1$), an entire side ($d' = 2$) etc.). In order for the mesh $\mathcal{T}$ to be regular, the maps F_P must be **compatible between patches** in the sense that

$$\text{if } \overline{P} \cap \overline{P'} \neq \emptyset : F_P \circ (F_{P'})^{-1}|_{\overline{P} \cap \overline{P'}} = id \text{ on } \overline{P} \cap \overline{P'} . \tag{4.3}$$

The $\mathcal{T}_P$ are **1-irregular**, if they consist of quadratics resp. hexagonal elements with at most one irregular ("hanging") node per side. $\mathcal{T}$ is 1-irregular, if the $\mathcal{T}_P \subset \mathcal{T}$ are either regular or 1-irregular and compatible between patches.

Polynomial subspaces. On the reference element $\widehat{P}$ we define spaces of polynomials of degree $p \ge 0$ as follows:

$$\mathcal{Q}_k = \text{span}\{\hat{x}^\alpha : 0 \le \alpha_i \le k, \ 1 \le i \le d\}$$

$$\mathcal{P}_k = \text{span}\{\hat{x}^\alpha : 0 \le \alpha_i, \ 0 \le \sum_{i=1}^{d} \alpha_i \le k . \tag{4.4}$$

Polynomial subspaces on $\widehat{\mathcal{T}}_P$. Let $\widehat{\mathcal{T}}_P$ be any mesh consisting of patch meshes $\mathcal{T}_P$ and let

$$\mathbf{k} = \{k_K : K \in \mathcal{T}\}$$

be a **polynomial degree vector** on $\mathcal{T}$. The definition of a **discontinuous hp-FE space** is now straightforward: if $\mathbf{F}_\mathcal{P} = \{F_P : P \in \mathcal{P})$ denotes the patch-map vector, we set

$$S^{\mathbf{k},0}(\Omega, \mathcal{T}, \mathbf{F}_\mathcal{P}) := \{u \in L^2(\Omega) : \ u|_K \circ F_K \in \mathcal{Q}_{k_K} \text{ if } K \in \mathcal{T}$$

$$\text{is quadrilateral resp. } u|_K \circ F_K \in \mathcal{P}_{k_K} \text{ if } K \text{ is triangular}\} . \tag{4.5}$$

No interelement continuity is imposed here. If the polynomial degree is uniform, $k_K = k$ for all $K \in \mathcal{T}$, we write $S^{k,0}(\Omega, \mathcal{T}, \mathbf{F}_\mathcal{P})$. If $\mathcal{T}$ and $\mathbf{F}_\mathcal{P}$ are clear from the context, we omit them and write $S^{k,0}(\Omega)$.

Let us now turn to **continuous hp-FE spaces.** Here we assume $\mathcal{T}$ to be either regular or 1-irregular. If the polynomial degrees k_K are uniform, $k_K = k$ for all K, we define for $k \ge 1$

$$S^{k,1}(\Omega, \mathcal{T}, \mathbf{F}_\mathcal{P}) = S^{k,0}(\Omega, \mathcal{T}, \mathbf{F}_\mathcal{P}) \cap H^1(\Omega), \tag{4.6}$$

i.e. interelement continuity is now enforced and the compatibility (4.3) between patches is required. If the polynomial degrees are nonuniform, there are several ways to enforce interelement continuity - assume that $K, K' \in \mathcal{T}$

share a $d - 1$ dimensional set, and that $p_K < p_{K'}$. One can now either enrich the polynomials on K or constrain the polynomials on K'. We adopt with (4.6) the latter approach.

Note that one could even allow anisotropic/nonuniform polynomial degrees **within** an element $K \in \mathcal{T}$ - this becomes important when adaptivity is considered (see [21] and the references there). Definition (4.6) implies that DOFs from K' that are unmatched by those from K are constrained to zero on interfaces $\overline{K} \cap \overline{K'}$.

Basic *hp*-FE Spaces. We introduce the *hp*-FE subspaces $S^{\mathbf{k},\ell}(\Omega, \mathcal{T}, \mathbf{F}_{\mathcal{P}})$, $\ell = 0, 1$, which are basic to the *hp*-FEM; $\ell = 0$ will denote discontinuous functions whereas $\ell = 1$ implies $H^1(\Omega)$ conformity, i.e. full continuity. These are the basic and most frequently used *hp*-spaces.

4.2   Choice of Patch Meshes $\widehat{\mathcal{T}}_P$ in 2-d

Preliminaries. A *mesh* $\mathcal{T}$ on a bounded polygonal patch $P \subseteq \mathbb{R}^2$ is a partition of Ω into disjoint and open quadrilateral and/or triangular elements $\{K\}$ such that $\overline{P} = \cup_{K \in \mathcal{T}} \overline{K}$. The mesh $\mathcal{T}$ is called *regular* if for any two elements $K, K' \in \mathcal{T}$ the intersection $\overline{K} \cap \overline{K'}$ is either empty, a single vertex or an entire side. Otherwise, the mesh $\mathcal{T}$ is called *irregular*. We denote by h_K the diameter of the element K and by ρ_K the diameter of the largest circle inscribed into K. The *meshwidth* h of $\mathcal{T}$ is given by $h = \max_{K \in \mathcal{T}} h_K$. The fraction $\sigma_K := \frac{h_K}{\rho_K}$ is the *aspect ratio* of the cell K. A (regular or irregular) mesh $\mathcal{T}$ is called κ-*shape regular* if there exists $\kappa > 0$ such that

$$\max_{K \in \mathcal{T}} \sigma_K \leq \kappa < \infty. \tag{4.7}$$

$\mathcal{T}$ is called *affine* if each $K \in \mathcal{T}$ is affine equivalent to a reference element $\widehat{P}$ which is either the square $\widehat{Q} = (0, 1)^2$ or the triangle $\widehat{T} = \{(x, y) \colon 0 < x < 1, 0 < y < x\}$, i.e.

$$K = A_K(\widehat{K}), \qquad A_K(\cdot) \text{ affine.}$$

Reference meshes. We introduce now some meshes on the reference elements.

Definition 4.1. Let $n \in \mathbb{N}_0$ and $\sigma \in (0, 1)$. On $\widehat{Q}$, the *(irregular) geometric mesh* $\Delta_{n,\sigma}$ with $n + 1$ *layers* and *grading factor* σ is created recursively as follows: If $n = 0$, $\Delta_\sigma^0 = \{\widehat{Q}\}$. Given $\Delta_{n,\sigma}$ for $n \geq 0$, $\Delta_{n+1,\sigma}$ is generated by subdividing that square $K \in \Delta_{n,\sigma}$ with $0 \in \overline{K}$ into four smaller rectangles by dividing the sides of K in a $\sigma : (1 - \sigma)$ ratio. The *(regular) geometric mesh* $\widetilde{\Delta}_{n,\sigma}$ is obtained from $\Delta_{n,\sigma}$ by removing the hanging nodes as indicated in Figure 4.1.

In Figure 4.1 the geometric meshes are shown for $n = 3$ and $\sigma = 0.5$. Clearly, $\Delta_{n,\sigma}$ is an irregular affine mesh, it has so-called *hanging nodes* while $\widetilde{\Delta}_{n,\sigma}$ is regular. The elements of the geometric mesh $\Delta_{n,\sigma}$ are numbered as in Figure 4.1, i.e.

$$\Delta_{n,\sigma} = \{\Omega_{11}\} \cup \{\Omega_{ij} : 1 \le i \le 3, 2 \le j \le n+1\}. \tag{4.8}$$

The elements Ω_{1j}, Ω_{2j} and Ω_{3j} constitute the *layer j*.

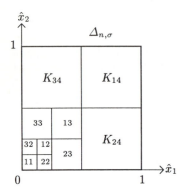 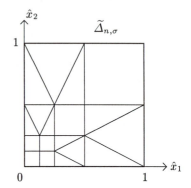

Fig. 4.1. The geometric meshes $\Delta_{n,\sigma}$ and $\widetilde{\Delta}_{n,\sigma}$ with $n = 3$ and $\sigma = 0.5$.

Remark 4.2. On the reference triangle $\widehat{T}$, $\Delta_{n,\sigma}$ and $\widetilde{\Delta}_{n,\sigma}$ can be defined in a similar way. $\Delta_{n,\sigma}$ is depicted in Figure 4.3.

Definition 4.3. Let $\mathcal{T}_x$ be an arbitrary mesh on $I = (0,1)$, given by a partition of I into subintervals $\{K_x\}$. On $\widehat{Q}$, the *boundary layer mesh* $\Delta_{\mathcal{T}_x}$ is the product mesh

$$\Delta_{\mathcal{T}_x} = \{K : K = K_x \times I, K_x \in \mathcal{T}_x\}.$$

Figure 4.2 shows a typical boundary layer mesh. We emphasize that any $\mathcal{T}_x$ is allowed, in particular, rectangles of arbitrary high aspect ratio can be used such that boundary layer meshes are not κ-uniform.

Definition 4.4. Let $n \in \mathbb{N}_0$ and $\sigma \in (0,1)$. Let $\mathcal{T}_{n,\sigma}$ be the one dimensional geometric mesh refined towards 0 given by a partition of $I = (0,1)$ into subintervals $\{I_j\}_{j=1}^{n+1}$ where

$$I_j = (x_{j-1}, x_j) \text{ with } x_0 = 0 \text{ and}$$
$$x_j = \sigma^{n+1-j}, \, j = 1, \dots, n+1.$$

On $\widehat{Q}$, the *geometric tensor product mesh* $\Delta_{n,\sigma}^2$ is then given by $\mathcal{T}_{n,\sigma} \otimes \mathcal{T}_{n,\sigma}$, i.e.

$$\Delta_{n,\sigma}^2 = \{I_j \times I_k : I_j \in \mathcal{T}_{n,\sigma}, I_k \in \mathcal{T}_{n,\sigma}\}.$$

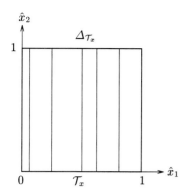 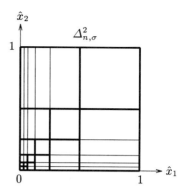

Fig. 4.2. Boundary layer mesh and geometric tensor product mesh on $\widehat{Q}$.

The tensor product mesh $\Delta^2_{n,\sigma}$ contains anisotropic rectangles with arbitrary large aspect ratio. For the proof of the inf-sup conditions ahead, it is important that $\Delta^2_{n,\sigma}$ can be understood as the geometric mesh $\Delta_{n,\sigma}$ into which appropriately scaled versions of boundary layer meshes $\Delta_{\mathcal{T}_x}$ are inserted to remove the hanging nodes. A geometric tensor product mesh is shown in Figure 4.2 with $n = 5$ and $\sigma = 0.5$. The underlying geometric mesh $\Delta_{n,\sigma}$ is indicated by bold lines.

Remark 4.5. As before, $\Delta^2_{n,\sigma}$ can also be defined on the reference triangle $\widehat{T}$. This is shown in Figure 4.3. On the reference square $\widehat{Q}$ we can even admit mixtures of geometric tensor product meshes and geometric meshes $\Delta_{n,\sigma}$ or $\widetilde{\Delta}_{n,\sigma}$ as illustrated in Figure 4.4. They are denoted by $\Delta^m_{n,\sigma}$ and $\widetilde{\Delta}^m_{n,\sigma}$. Of course, other mixtures are imaginable.

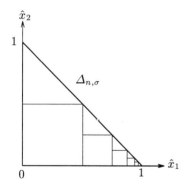 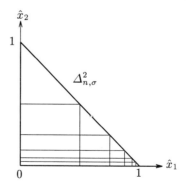

Fig. 4.3. The meshes $\Delta_{n,\sigma}$ and $\Delta^2_{n,\sigma}$ on the reference triangle $\widehat{T}$.

Admissible patch meshes $\mathcal{T}$.

Definition 4.6. An affine mesh $\widehat{\mathcal{T}}$ on $\widehat{P}$ is called a $(\mathcal{F}, \mathcal{T}_m, \kappa)$-mesh if

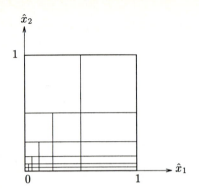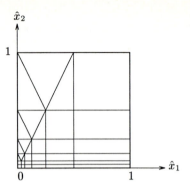

Fig. 4.4. The meshes $\Delta_{n,\sigma}^m$ and $\tilde{\Delta}_{n,\sigma}^m$ with $n = 5$ and $\sigma = 0.5$.

1. $\mathcal{T}_m$ is an affine mesh which is coarser than $\mathcal{T}$ and κ-uniform for some $\kappa > 0$. The elements of $\mathcal{T}_m$ are called *macro-elements* and $\mathcal{T}_m$ is the *macro-element mesh* of $\mathcal{T}$.

2. $\mathcal{F}$ is a nonempty family of affine reference meshes on the reference square $\widehat{Q}$ or the reference triangle $\widehat{T}$.

3. The restriction $\mathcal{T}_K := \mathcal{T}|_K$ of $\mathcal{T}$ to any macro-element $K \in \mathcal{T}_m$ is given by $\mathcal{T}_K = F_K(\widehat{\mathcal{T}})$ for some $\widehat{\mathcal{T}}$ in $\mathcal{F}$ where F_K is the affine mapping between $\widehat{K}$ and K.

A $(\mathcal{F}, \mathcal{T}_m, \kappa)$-mesh is thus obtained from the κ-uniform mesh $\mathcal{T}_m$ by refining some or all elements with the strategies given by the family $\mathcal{F}$. In the simple case where

$$\mathcal{F} = \left\{ \{\widehat{Q}\}, \{\widehat{T}\} \right\}$$

the notion "$(\mathcal{F}, \mathcal{T}_m, \kappa)$-mesh" reduces to the already introduced notion of κ-uniform affine meshes consisting of quadrilaterals and/or triangles and the notion of "macro-elements" becomes unnecessary. We are mainly interested in the family

$$\mathcal{F}^\sigma = \left\{ \Delta_{n,\sigma}, \tilde{\Delta}_{n,\sigma}, \Delta_{T_x}, \Delta_{n,\sigma}^2, \Delta_{n,\sigma}^m, \tilde{\Delta}_{n,\sigma}^m, \right.$$
$$\left. \{\widehat{Q}\}, \{\widehat{T}\} : n \in \mathbb{N}_0, \, \mathcal{T}_x \text{ arbitrary} \right\} \tag{4.9}$$

for $\sigma \in (0, 1)$ fixed. Here, $\Delta_{n,\sigma}$ and $\Delta_{n,\sigma}^2$ is understood as a mesh on $\widehat{Q}$ or $\widehat{T}$. Alternatively, one could consider $\tilde{\Delta}_{n,\sigma}$ as a part of the macro-element mesh $\mathcal{T}_m$ and put only the irregular patches into the family $\mathcal{F}^\sigma$. If $\mathcal{T}$ contains no triangles, $\mathcal{F}^\sigma$ can be reduced to

$$\mathcal{F}^\sigma = \left\{ \Delta_{n,\sigma}, \Delta_{n,\sigma}^2, \Delta_{T_x}, \Delta_{n,\sigma}^m, \{\widehat{Q}\} : n \in \mathbb{N}_0, \, \mathcal{T}_x \text{ arbitrary} \right\} \tag{4.10}$$

where $\Delta_{n,\sigma}$ and $\Delta_{n,\sigma}^2$ have now to be meshes on $\widehat{Q}$. We call a $(\mathcal{F}^\sigma, \mathcal{T}_m, \kappa)$-mesh shortly (κ, σ)-*mesh* where we choose the reduced family $\mathcal{F}^\sigma$ if $\mathcal{T}$ contains no triangles.

(κ, σ)-meshes are a quite general class of possibly highly irregular meshes. They are well suited for the effective resolution of boundary layer and corner singularity phenomena. Typically, mesh-patches from $\mathcal{T}_m$ near the boundary of the domain are partitioned anisotropically using $\Delta_{\mathcal{T}_x}$-meshes to approximate boundary layers. Patches near corners are geometrically refined towards the corner with the meshes $\Delta_{n,\sigma}$ or $\Delta_{n,\sigma}^2$. This takes into account boundary layers as well as the singular behaviour of the solution near a corner. In the interior of the domain a simple κ-uniform mesh can be used. Some examples of (κ, σ)-meshes are shown in Figure 4.5 and 4.6.

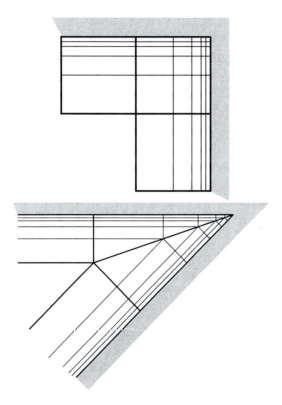

Fig. 4.5. Geometric (κ, σ)-boundary layer meshes near convex corners.

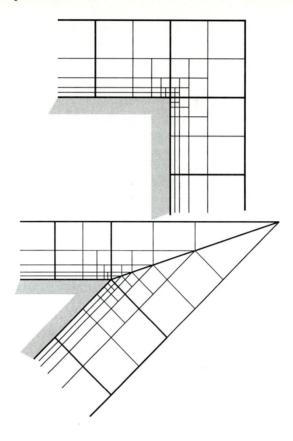

Fig. 4.6. Geometric (κ, σ)-boundary layer meshes near reentrant corners.

4.3 hp-spaces on $\widehat{\mathcal{T}}$

We introduce the hp-FE spaces investigated later on. Therefore, let $\widehat{\mathcal{T}}$ be an affine mesh on $\widehat{P}$. With each element $K \in \mathcal{T}$ we associate a polynomial degree k_K. All degrees are stored in a degree vector

$$\mathbf{k} = \{k_K : K \in \mathcal{T}\}. \tag{4.11}$$

We define spaces of continuous and discontinuous piecewise polynomial functions, respectively, by

$$S^{\mathbf{k},1}(\widehat{P}, \widehat{\mathcal{T}}) := \left\{ \begin{array}{l} u \in H^1(\widehat{P}) : u|_K \in \\ \left\{ \begin{array}{l} \mathcal{Q}_{k_K}(\widehat{Q}) \text{ if } K \text{ is a} \\ \qquad \text{quadrilateral} \\ \mathcal{P}_{k_K}(\widehat{T}) \text{ if } K \text{ is a triangle} \end{array} \right\} \forall K \in \widehat{\mathcal{T}} \end{array} \right. \tag{4.12}$$

and

$$S^{\mathbf{k},0}(\widehat{P},\widehat{\mathcal{T}}) := \begin{cases} p \in L^2(\widehat{P}) : p|_K \in \\ \qquad \begin{cases} \mathcal{Q}_{k_K}(\widehat{Q}) \text{ if } K \text{ is a} \\ \qquad\qquad\qquad \text{quadrilateral} \\ \mathcal{P}_{k_K}(\widehat{T}) \text{ if } K \text{ is a triangle} \end{cases} \end{cases} \forall K \in \widehat{\mathcal{T}} . \qquad (4.13)$$

We set further

$$S_0^{\mathbf{k},1}(\widehat{P},\widehat{\mathcal{T}}) = S^{\mathbf{k},1}(\widehat{P},\widehat{\mathcal{T}}) \cap H_0^1(\widehat{P}) ,$$
$$S_0^{\mathbf{k},0}(\widehat{P},\widehat{\mathcal{T}}) = S^{\mathbf{k},0}(\widehat{P},\widehat{\mathcal{T}}) \cap L_0^2(\widehat{P}) .$$

If the polynomial degree is constant throughout the mesh $\widehat{\mathcal{T}}$ (i.e. $k_K = k \,\forall \widehat{K} \in \widehat{\mathcal{T}}$), we use the shorthand notations $S^{k,1}(\widehat{P},\widehat{\mathcal{T}})$ and $S^{k,0}(\widehat{P},\widehat{\mathcal{T}})$.

5 *hp*-Error Estimates

We present *hp*-error estimates with particular attention to the approximation of boundary layers, corner singularities and viscous shock profiles as discussed above. It is well-known that *hp* and spectral methods achieve exponential convergence rates for smooth (analytic) solutions ([17,58]). Exponential convergence of *hp*-FEM for boundary layers, corner singularities and viscous shock profiles, however, requires the combination of structured patch meshes $\mathcal{T}_P$ with the proper polynomial degree distribution **k**. *hp*-FEM are robust in the sense that exponential convergence holds even under certain changes in the mesh and the degree distribution which we will indicate in each case. This makes the results presented here relevant in practice.

5.1 Basic error estimates

One dimensional *hp*-approximation. We cite some approximation results from [58]. To this end, we set $\widehat{I} = (-1,1)$ and denote by $\|u\|_{k,\widehat{I}}$ resp. $|u|_{k,\widehat{I}}$ the $H^k(\widehat{I})$ norm resp. seminorm on $\widehat{I}$. Denote further $S^p(\widehat{I})$ the polynomials of degree p on $\widehat{I}$. Then we have

Theorem 5.1. *Let $u_0 \in H^{k+1}(\widehat{I})$ for some $k \geq 0$. Then, for every $p \geq 1$, there exists $s_0 = \pi_p u_0 \in S^p(\widehat{I})$ such that*

$$\|u_0' - s_0'\|_{0,\widehat{I}}^2 \leq \frac{(p-s)!}{(p+s)!} |u_0|_{s+1,\widehat{I}}^2 \qquad (5.1)$$

for any $0 \leq s \leq \min(p,k)^2$ and such that

$$\|u_0 - s_0\|_{0,\widehat{I}}^2 \leq \frac{1}{p(p+1)} \frac{(p-t)!}{(p+t)!} |u_0|_{t+1,\widehat{I}}^2 \qquad (5.2)$$

---

[2] Interpreting the factorials in terms of Gamma functions and the norms as interpolation norms for fractional indices

for any $0 \leq t \leq \min(p,k)$. Moreover, we have

$$s_0(\pm 1) = u_0(\pm 1) . \tag{5.3}$$

For the proof, we refer e.g. to [58]. We emphasize that in (5.1), (5.2) the dependence of the error on the polynomial degree p as well as on the regularities s, t of u is completely explicit. Such results cannot be obtained by Taylor's theorem and its generalizations which are common in the analysis of low order FEM.

Corollary 5.2. *The projector π_p in Theorem 5.1 is bounded as follows:*

$$\|(\pi_p u)'\|_{0,\widehat{I}} \leq 2\|u'\|_{0,\widehat{I}} \tag{5.4}$$

$$\|\pi_p u\|_{0,\widehat{I}} \leq \|u\|_{0,\widehat{I}} + \frac{1}{\sqrt{p(p+1)}} \|u'\|_{0,\widehat{I}} \tag{5.5}$$

for all $p \geq 1$ and every $u \in H^1(\widehat{I})$ where $C > 0$ is independent of p.

Proof. (5.1) with $s = 0$ implies (5.4) since

$$\|s_0'\|_{0,\widehat{I}} \leq \|s_0' - u_0'\|_{0,\widehat{I}} + \|u_0'\|_{0,\widehat{I}} \leq 2\|u_0'\|_{0,\widehat{I}} .$$

(5.2) with $t = 0$ implies (5.5) since

$$\|s_0\|_{0,\widehat{I}} \leq \|s_0 - u_0\|_{0,\widehat{I}} + \|u_0\|_{0,\widehat{I}}$$
$$\leq \|u_0\|_{0,\widehat{I}} + \frac{1}{\sqrt{p(p+1)}} \|u_0'\|_{0,\widehat{I}} .$$
$\square$

Approximation on quadrilaterals. Higher dimensional approximation results can be obtained from Theorem 5.1 by tensor product construction. We denote by $\pi_p^i u_0$ the one-dimensional projector in Theorem 5.1 applied to u_0 as function of the i th coordinate alone and perform the error analysis for $d = 2$.

Let $\widehat{Q} = (-1,1)^2$ and denote by $\hat{\gamma}_i, i = 1,2,3,4$ the sides of $\widehat{Q}$ as shown in Figure 5.1.

Theorem 5.3. *(Reference Element Approximation) Let $\widehat{Q} = (-1,1)^2$ as in Figure 5.1 and $u_0 \in H^{k+1} = (\widehat{Q})$ for some $k \geq 1$. Let $\Pi_p = \pi_p^1 \pi_p^2$ denote the tensor product projector. Then there holds:*

$$\pi_p u_0 = u_0 \quad \text{at the vertices of } \widehat{Q} , \tag{5.6}$$

$$\pi_p u_0|_{\hat{\gamma}_i} = \begin{cases} \pi_p^1(u_0|_{\hat{\gamma}_i}) & \text{if } i \text{ is odd}, \\ \pi_p^2(u_0|_{\hat{\gamma}_i}) & \text{if } i \text{ is even}. \end{cases} \tag{5.7}$$

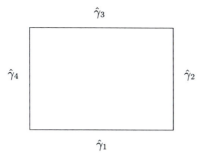

Fig. 5.1. $\widehat{Q}$ and the notation for the sides

There hold the **error estimates**

$$\|\nabla(u_0 - \pi_p u_0)\|_{0,\widehat{Q}}^2 \le 2 \frac{(p-s)!}{(p+s)!} \{\|\partial_1^{s+1} u_0\|_{0,\widehat{Q}}^2 + \|\partial_2^{s+1} u_0\|_{0,\widehat{Q}}^2\} +$$

$$\frac{2}{p(p+1)} \frac{(p-s+1)!}{(p+s-1)!} \{\|\partial_1^s \partial_2 u_0\|_{0,\widehat{Q}}^2 + \|\partial_1 \partial_2^s u_0\|_{0,\widehat{Q}}^2\} \qquad (5.8)$$

$$\|(u_0 - \Pi_p u_0)\|_{0,\widehat{Q}}^2 \le$$

$$\frac{2}{p(p+1)} \frac{(p-s)!}{(p+s)!} \left(\|\partial_1^{s+1} u_0\|_{0,\widehat{Q}}^2 + 2\|\partial_2^{s+1} u_0\|_{0,\widehat{Q}}^2\right) +$$

$$\frac{4}{p^2(p+1)^2} \frac{(p-s+1)!}{(p+s-1)!} \|\partial_1 \partial_2^s u\|_{0,\widehat{Q}}^2 \qquad (5.9)$$

for any $0 \le s \le \min(p, k)$.

Proof. We prove (5.9). It holds

$$\|u - \Pi_p u\|_{0,\widehat{Q}}^2 \le 2\|u - \pi_p^1 u\|_{0,\widehat{Q}}^2 + 2\|\pi_p^1(u - \pi_p^2 u)\|_{0,\widehat{Q}}^2 .$$

For the first term we use the bound (5.2), resulting in

$$\|u - \pi_p^1 u\|_{0,\widehat{Q}}^2 \le \frac{1}{p(p+1)} \frac{(p-s)!}{(p+s)!} \|\partial_1^{s+1} u\|_{0,\widehat{Q}}^2 .$$

For the second term, (5.5) and (5.1), (5.2) give

$$\|\pi_p^1(u - \pi_p^2 u)\|_{0,\widehat{Q}}^2 \le 2\|u - \pi_p^2 u\|_{0,\widehat{Q}}^2 + \frac{2}{p(p+1)} \|\partial_1(u - \pi_p^2 u)\|_{0,\widehat{Q}}^2$$

$$\le \frac{2}{p(p+1)} \frac{(p-t)!}{(p+t)!} \|\partial_2^{t+1} u\|_{0,\widehat{Q}}^2 +$$

$$\frac{2}{p^2(p+1)^2} \frac{(p-r)!}{(p+r)!} \|\partial_1 \partial_2^{r+1} u\|_{0,\widehat{Q}}^2 .$$

Selecting $t = s$ and $r = s - 1$ gives (5.9). The proof of (5.8) is analogous.   □

Approximation on quadrilateral meshes with hanging meshes. Consider now a patch $P \in \mathcal{P}$ with mesh $\mathcal{T}_P$ and corresponding reference mesh $\widehat{\mathcal{T}}_P$ in $\widehat{P}$. We assume that all $K \in \mathcal{T}_P$ are quadrilateral, possibly with hanging nodes.

Theorem 5.4. *(Discontinuous Approximation) Let $P \in \mathcal{P}$ with quadrilateral, possibly 1-irregular mesh $\mathcal{T}_P$ of shape-regular elements and polynomial degree distribution $\mathbf{p}$. For all $K \in \mathcal{T}_P$ let $u|_K \in H^{k_K+1}(K)$ for some $k_K \geq 1$ and define $\Pi u \in S^{\mathbf{p},0}(P, \mathcal{T}_P)$ elementwise by*

$$(\Pi u)|_K \circ F_P := \Pi_{p_K}(u|_K \circ F_P) \quad \forall K \in \mathcal{T}_P$$

with Π_p as in Theorem 5.3.

Then there holds the estimate for $0 \leq s_K \leq \min(p_K, k_K)$

$$\|u - \Pi u\|_{0,P}^2 \leq$$

$$C \sum_{K \in \mathcal{T}_P} \left(\frac{h_K}{2}\right)^{2s_K+2} \frac{1}{p_K(p_K+1)} \Phi(p_K, s_K)|\hat{u}|_{s_K+1,\widehat{K}}^2 \tag{5.10}$$

where $\hat{u} = u \circ F_P$, $K = F_P(\widehat{K})$ and where

$$\Phi(p,s) := \frac{(p-s)!}{(p+s)!} + \frac{1}{p(p+1)} \frac{(p-s+1)!}{(p+s-1)!}, \quad 0 \leq s \leq p.$$

Further, there holds

$$\|\nabla(u - \Pi u)\|_{0,P}^2 \leq C \sum_{K \in \mathcal{T}_P} \left(\frac{h_K}{2}\right)^{2s_K} \Phi(p_K, s_K)|\hat{u}|_{s_K+1,\widehat{K}}^2. \tag{5.11}$$

The constant $C > 0$ depends only on F_P, but is independent of h_K, p_K and s_K.

Proof. The L^2-estimate (5.10) follows immediately by a change of variables and a scaling argument from Theorem 5.3.

For the gradient estimate, we observe that

$$\|\nabla(u - \Pi_u)\|_{0,P}^2 \leq C(F_P)(\|(u - \Pi u) \circ F_P\|_{0,\widehat{P}}^2 + \|\widehat{\nabla}((u - \Pi u) \circ F_P)\|_{0,\widehat{P}}^2).$$

For the first term we use (5.10), for the second one we use (5.8), after scaling to the reference element:

$$\|\nabla((u - \Pi u) \circ F_P)\|_{0,\widehat{P}}^2$$

$$= \sum_{\widehat{K} \in \widehat{\mathcal{T}}_P} \|\partial_1((u - \Pi u) \circ F_P)\|_{0,\widehat{K}}^2 + \|\partial_2((u - \Pi u) \circ F_P)\|_{0,\widehat{K}}^2$$

$$= \sum_{\substack{\widehat{K} \in \widehat{\mathcal{T}}_P \\ i=1,2}} h_{1,K}\, h_{2,K} \|\partial_i(I - \Pi_{p_K})u \circ F_P \circ A_K\|_{0,\widehat{P}}^2$$

$$\overset{(5.8)}{\leq} 2(h_K)^2 \sum_{\widehat{K} \in \widehat{\mathcal{T}}_P} \left\{ \frac{(p_K - s_K)!}{(p_K + s_K)!} \left(\|\hat{\partial}_1^{s_K+1} u_{0,K}\|_{0,\widehat{P}}^2 + \|\hat{\partial}_2^{s_K+1} u_{0,K}\|_{0,\widehat{P}}^2 \right) \right.$$

$$\left. + \frac{1}{p_K(p_K+1)} \frac{(p_K - s_K + 1)!}{(p_K + s_K - 1)!} \left(\|\hat{\partial}_1^{s_K} \hat{\partial}_2\, u_{0,K}\|_{0,\widehat{P}}^2 + \|\hat{\partial}_1 \hat{\partial}_2^{s_K}\, u_{0,K}\|_{0,\widehat{P}}^2 \right) \right\}$$

where

$$u_{0,K} := u \circ F_P \circ A_K = \hat{u} \circ A_K, \quad K \in \mathcal{T}_P .$$

Affine scaling from $\widehat{P}$ to $\widehat{K} \in \widehat{\mathcal{T}}_P$ gives the assertion.   □

The error bounds in Theorem 5.4 simplify for uniform p.

Corollary 5.5. *(Uniform order estimate)*
Assume that $\hat{u} := u \circ F_P \in H^{k+1}(\widehat{P})$ and that for all $K \in \mathcal{T}_P$

$$p_K = p, \quad s_K = s, \quad 0 \leq s \leq \min(p, k) .$$

Then there holds for $\Pi u \in S^{p,0}(P, \mathcal{T}_P)$ and $\hat{u} := u \circ F_P$

$$\|u - \Pi u\|_{0,P}^2 \leq C \frac{1}{p(p+1)}\, \Phi(p, s) \sum_{K \in \mathcal{T}_P} \left(\frac{h_K}{2} \right)^{2s+2} |\hat{u}|_{s+1,\widehat{K}}^2 \tag{5.12}$$

and

$$\|\nabla(u - \Pi u)\|_{0,P}^2 \leq C \Phi(p, s) \sum_{K \in \mathcal{T}_P} \left(\frac{h_K}{2} \right)^{2s} |\hat{u}|_{s+1,K}^2 . \tag{5.13}$$

Here C depends only on the patch mapping F_P but not on s, p, h_K.

Remark 5.6. (Anisotropic error estimates)

We note in passing that the above error estimate assumed the shape regularity of the $\widehat{K}$ merely for convenience - in fact the explicit error bounds in Theorem 5.3 and 5.4 above could be easily generalized to anisotropic element shapes (with edge-lengths h_{1K} and h_{2K}) and even to anisotropic polynomial degrees p_{1K}, p_{2K}, say. Error bounds explicit in these parameters can be deduced by inspecting the proofs of the above theorems.

Theorem 5.4 addressed only discontinuous approximations; it turns out, however, that also continuous, piecewise polynomial approximations can be obtained.

Theorem 5.7. *(Continuous approximations)*

Let $\Omega \subset \mathbb{R}^2$ and let $P \in \mathcal{P}$ with a 1-irregular mesh consisting of shape regular quadrilaterals K of diameter h_K. Let the polynomial degree be uniform, $p_K = p$. Let $u|_K \in H^{k_K+1}(K)$ for some $k_K \geq 1$ and let $u \in H^2(P)$.

Then there exists a projector $\widetilde{\Pi}u \in S^{p,1}(P, \mathcal{T}_P)$ such that the error bounds (5.12), (5.13) hold, with a possibly different value of C.

Proof. If $\mathcal{T}_P$ does not contain hanging nodes, $\mathcal{T}_P$ is regular and we take $\widetilde{\Pi} = \Pi$ in Theorem 5.4. Since Π was constructed elementwise, the properties (5.10), (5.11) together with the assumption that $u \in H^2(P)$ give the continuity of Πu in $\overline{P}$.

Consider now that $\mathcal{T}_P$ contains hanging nodes. A typical situation in the reference mesh $\widehat{\mathcal{T}}_P$ is shown in Figure 5.2 where the elements have been scaled to unit size for convenience.

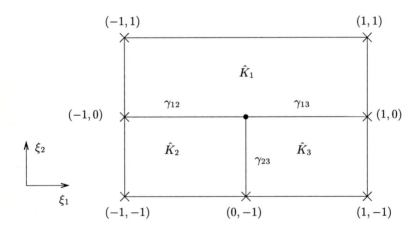

Fig. 5.2. Hanging node ● and adjacent elements

Since $u \in H^2(P)$, $u \in C^0(\overline{P})$. By (5.6), $u - \Pi u$ vanishes at the points × in Figure 5.2. Denote by $[u - \Pi u]_{ij}$ the jump of $u - \Pi u$ across γ_{ij}. By (5.9), the jump of $\varPi u$ across γ_{23} is zero. Since $u \in C^0(\overline{P})$, $[u - \varPi u]_{ij} = -[\varPi u]_{ij}$. Further, $[\varPi u]_{ij} \in P_p(\gamma_{ij})$.

We now construct a trace-lifting of $[\varPi u]$ across $\gamma_{12} \cup \gamma_{13}$ as follows: We set

$$V(\xi) = -(\xi_2 + 1) \begin{cases} [\varPi u]_{12}(\xi_1) & \text{on } \widehat{K}_2 , \\ [\varPi u]_{13}(\xi_1) & \text{on } \widehat{K}_3 . \end{cases}$$

Since $[\varPi u]_{23} = 0$, V is continuous on $\overline{\widehat{K}_2 \cup \widehat{K}_3}$ and

$$\|\nabla V\|_{L^2(\widehat{K}_2 \cup \widehat{K}_3)} \leq C \, \|\, [\varPi u]\,\|_{H^{\frac{1}{2}}(\gamma_{12} \cup \gamma_{13})} \tag{5.14}$$

where C is independent of p. By the trace theorem and since $u \in C^0(\bigcup_1^3 \overline{\widehat{K}_i})$, we have

$$\begin{aligned}
\|\, [\varPi u]\,\|_{H^{\frac{1}{2}}(\gamma_{12} \cup \gamma_{13})} &= \|\, [u - \varPi u]\,\|_{H^{\frac{1}{2}}(\gamma_{12} \cup \gamma_{13})} \\
&\leq \|\, (u - \varPi u)_+\|_{H^{\frac{1}{2}}(\gamma_{12} \cup \gamma_{13})} \\
&\quad + \|\, (u - \varPi u)_-\|_{H^{\frac{1}{2}}(\gamma_{12} \cup \gamma_{13})} \\
&\leq C \sum_{i=1}^{3} \|u - \varPi u\|_{H^1(\widehat{K}_i)} .
\end{aligned} \tag{5.15}$$

where $()_\pm$ denote traces from $\xi_2 > 0$ and $\xi_2 < 0$, respectively. By construction $V + \varPi u$ is continuous on and across γ_{12} and γ_{13}.

We define

$$\widetilde{\varPi} u := \begin{cases} \varPi u & \text{on } \widehat{K}_1 \\ V_1 + \varPi u & \text{on } \widehat{K}_2 \cup \widehat{K}_3 . \end{cases}$$

Then, on $\widehat{K} := \overline{\widehat{K}_1 \cup \widehat{K}_2 \cup \widehat{K}_3}$,

$$\|\nabla(u - \widetilde{\varPi} u)\|_{0,\widehat{K}} \leq \|\nabla V\|_{0,\widehat{K}_2 \cup \widehat{K}_3} + \sum_{i=1}^{3} \|\nabla(u - \varPi u)\|_{0,\widehat{K}_i} .$$

Using (5.10), (5.11) we get

$$\|(\nabla u - \widetilde{\varPi} u)\|_{0,\widehat{K}}^2 \leq C \sum_{i=1}^{3} \|(\nabla u - \varPi u)\|_{0,\widehat{K}_i}^2 \tag{5.16}$$

where $C > 0$ is independent of p.

If the $\widehat{K}_i$ are not of unit size, we may scale the estimate (5.13) without incurring h-powers. Since

$$\tilde{\Pi}u|_{\partial\widehat{K}} = \Pi u_{\partial\widehat{K}} \,,$$

further liftings in the presence of additional hanging nodes on $\partial\widehat{K}$ can be performed in the adjacent element patches, resulting in the error bounds (5.12), (5.13) with a larger C.   □

5.2   Corner singularities

Corner singularities are present in polygons and polyhedra whenever the governing equations contain second order, viscous terms, but also appear in certain inviscid problems (see, eg. Figure 12 in the article [19]). A recent reference is [33], [42] where further references can be found. We address the hp-FE approximation of corner singularities – although these singularities have very low regularity at the corner, exponential convergence results are nevertheless possible. To present ideas in the simplemost setting, we start in dimension one (where corner singularities do not arise in practice), continue in the 2-d case and comment finally on the 3-d case, where 2 types of singularities, *edge and vertex singularities*, must be distinguished.

One dimensional case. In $I = (0, 1)$ a typical corner singularity function is given by

$$s(x) = g(x)r^\lambda \tag{5.17}$$

where $r(x) = |x|$ and $g(x)$ is analytic in $[0, 1]$. The *singularity exponent* λ is not an integer and it must hold that

$$\lambda > 1/2$$

to ensure that $s(x)$ has finite energy, i.e. that $\|s\|_{1,I} < \infty$. Typically, λ is small. For example, for $\lambda < 3/2$ the singular function $s(x) \notin H^2(I)$ and finite difference/ finite volume methods on uniform meshes can not even achieve first order convergence in $H^1(I)$. Likewise, spectral methods which approximate $s(x)$ on I by increasing the polynomial degree k will produce low algebraic convergence rates such as (see, e.g. [58])

$$\|s - s_k\|_{\ell,I} \le Ck^{-(2(\lambda-\ell)+1)}, \quad k = 1, 2, ..., \ \ell = 0, 1 \,. \tag{5.18}$$

Nevertheless, $s(x)$ is analytic on the set $(0, 1]$, so the low rate (5.18) is caused solely by the point singularity at $x = 0$. hp-FEM exploit this piecewise analyticity as follows.

Consider the sequence of *geometric meshes* $\mathcal{T}^{n,\sigma}$ with n layers and geometric grading factor $\sigma, 0 < \sigma < 1$, and polynomial degrees k_K shown in Figure 8. Notice that here the grading factor $\sigma = 0.5$ and that the number n of refinements is proportional to the maximal polynomial degree $k_{\max} = \max\{k_K : K \in \mathcal{T}^{n,\sigma}\}$ in the mesh: As n increases, mesh and polynomial orders change simultaneously. We have the following hp-approximation result:

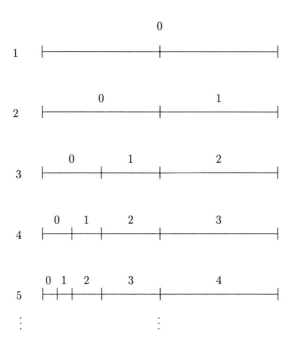

Fig. 5.3. Geometric mesh and polynomial degree distribution **k** for root singularity at $x = 0$ (discontinuous polynomials)

Theorem 5.8. *Consider the root singularity $s(x)$ in (5.17) defined in $I = (0, 1)$. Let $\mathcal{T}^{n,\sigma} = \{K_j^n : j = 1, \ldots, n\}$, $K_1^n = (0, \sigma^{n-1})$, $K_j^n = (\sigma^{n-j+1}, \sigma^{n-j})$, $j = 2, \ldots, n$ be the geometric mesh with n layers and grading factor $0 < \sigma < 1$ as in Figure 4.1. Let the degree distribution $\mathbf{k}\,(n) = \{k_j^n : j = 1, \ldots, n\}$ satisfy $k_j^n \geq \mu(j-1)$ for some $\mu = \mu(\sigma) > 0$ sufficiently large. Then for every n there exists a (possibly discontinuous) polynomial*

$$s_n(x) \in S_n := S^{\mathbf{k}(n),0}(I, \mathcal{T}^{n,\sigma})$$

satisfying the error bound

$$\|s - s_n\|_{L^2(I)} \leq Ce^{-bn} \leq Ce^{-b\sqrt{N}} \tag{5.19}$$

where $N = \dim(S_n) = O(n^2)$ and C, b are independent of n (but depend on σ and α).

If $k_j^n \geq \mu j$, (5.19) still holds with an $s_n(x) \in S^{\mathbf{k}(n),1}(I, \mathcal{T}^{n,\sigma})$.

If the polynomial degree is uniform, i.e. $k_j^n = k = n$ for all n, then (5.19) still holds, with possibly different constants b and C.

For a proof, we refer for example to [58].

Theorem 5.8 shows that by judicious combination of mesh $\mathcal{T}$ and degree vector $\mathbf{k}$, exponential convergence can be achieved. Mesh refinement or order increase alone yield only algebraic convergence rates. Similar results hold also when the pointwise error is of interest.

Two dimensional case. Consider a polygon $\Omega \subset \mathbb{R}^2$ as shown in Figure 3.1. A corner singular function $S(r_j, \varphi_j)$ at vertex P_j is as in (3.2). To simplify the notation, we may assume that $P_j = O$ and that $r(x) = r_j(x) = |x|$. Then there holds again an exponential convergence result.

Theorem 5.9. *Let Ω be a polygonal domain containing the origin O as a vertex and let $u_{\text{sing}} = S(r, \varphi)$ be a singular function as in (3.2). Let $0 < \sigma < 1$ and $\{\mathcal{T}^{n,\sigma}\}_n$ be a sequence of geometric meshes refined towards O with n layers (see, e.g. Figure 5.3) and grading factor σ, $0 < \sigma < 1$. Let the polynomial degree k be uniform and proportional to the number of layers, i.e. $k \sim n$. Then, for every n exists a continuous, piecewise polynomial function $u_n(x) \in S^{k,1}(\Omega, \mathcal{T}^{n,\sigma})$ such that*

$$\|S(r, \varphi) - u_n\|_{H^1(\Omega)} \leq Ce^{-bn} = Ce^{-bN^{1/3}} \qquad (5.20)$$

where $b, C > 0$ are independent of $N = \dim(S^{k,1}(\Omega, \mathcal{T}^{n,\sigma}))$, the number of degrees of freedom of $S^{k,1}(\Omega, \mathcal{T}^{n,\sigma})$.

For a proof, we refer for example to [34], [58].

Remark 5.10. We emphasize that uniform polynomial degree k is not necessary - it suffices in fact to allow $k = 2$ in the element abutting at O, and to let k_K increase linearly with the number of elements $K' \in \mathcal{T}^{n,\sigma}$ between K and O.

Remark 5.11. Theorems 5.8 and 5.9 give exponential convergence for *any* $0 < \sigma < 1$. There arises the question for the optimal σ. In one dimension, one can show that $\sigma_{\text{opt}} = (\sqrt{2} - 1)^2 = 0.17\ldots$ is optimal *regardless of the strength of the singularity*. In two dimensions, no analytical result is known, but also here geometric meshes with grading $\sigma \approx \sigma_{\text{opt}}$ outperform meshes with other values of σ, see also Figure 10.11 below.

If ω contains more than one vertex as e.g. in Figure 3.1, (5.20) still holds if at each vertex (reentrant or not) a geometric mesh patch $\widehat{\mathcal{T}}^{n,\sigma}$ is used.

Remark 5.12. For (5.20) to hold, it is not necessary that the domain Ω is a straight sided polygon. The same result holds also for curved domains, see [2].

Remark 5.13. In three dimensions, at vertices the construction is analogous, whereas at edges of polyhedra $\Omega \subset \mathbb{R}^3$ the geometric mesh refinement is anisotropic towards the edge. The resulting geometric meshes contain in the vicinity of edges the so-called "needle elements" of aspect ratio $1 : \sigma^p$ - this is necessary to achieve exponential convergence in three dimensional polyhedra. Geometric refinement towards the edge with κ-uniform meshes will **not** give exponential convergence rates (see [3] and Remark 5.18 for more).

5.3 *hp*-Boundary layer resolution

Analogous to corner singularities, *hp*-FEM can deal very effectively with boundary layers and viscous shock profiles as introduced in Sections 3.2 and 3.3. Here, we collect the main mesh design principles and convergence results for the *hp*-FEM for these problems (see also the references [59], [61], [45] for proofs and further details).

Boundary layers are, like corner singularities, essentially one-dimensional phenomena; therefore, we first address the *hp*-FEM for boundary layers in one dimension.

One dimensional results. On the interval $I = (0,1)$, consider a *boundary layer function with length-scale $d > 0$* satisfying the estimates (3.6). A typical example is the (ubiquitous) exponential boundary layer $u_{b\ell}^d(x) = \exp(-x/d)$. For the *hp*-FEM, we have the following result [59].

Theorem 5.14. *In $I = (0,1)$, consider the exponential boundary layer function*

$$u_{b\ell}^d(x) = \exp(-x/d) \, .$$

For $0 < \kappa < 4/e$, $0 < d \leq 1$, and $k = 1, 2, \ldots$ let $\mathcal{T}_k$ be a sequence of meshes defined by

$$
\begin{aligned}
\mathcal{T}_{b\ell}^k &= \{(0, \kappa k d), \ (\kappa k d, 1)\} \ \text{if} \ \kappa k d < 1 \, , \\
\mathcal{T}_d^k &= \{(0, 1)\} \qquad\qquad\quad \text{if} \ \kappa k d \geq 1 \, .
\end{aligned}
\tag{5.21}
$$

Let the polynomial degree be uniform and equal to k; then for every $k \in \mathbb{N}$ exists $u_k^d \in S^{k,1}(I, \mathcal{T}_{b\ell}^k)$ such that the following error estimates hold:

$$\|u_{b\ell}^d - u_k^d\|_{L^2(I)} \leq Cd^{1/2} \exp(-bk) \, ,$$

$$\|u_{b\ell}^d - u_k^d\|_{H^1(I)} \leq Cd^{-1/2} \exp(-bk) \, ,$$

$$\|u_{b\ell}^d - u_k^d\|_{L^\infty(I)} \leq C \ \exp(-bk) \, .$$

Here $b, C > 0$ are constants which are independent of d, k, but depend on κ.

We see that in the presence of boundary layers, *2 elements are sufficient for robust exponential resolution of boundary layers* in the context of the *hp*-FEM. Note, however, that the size of the smaller element is crucial - it must be proportional to kd; the precise value need not be achieved and the constant C does not depend sensitively on κ, as the results in Figure 5.4 show. In figures 5.5 - 5.7, we see the comparison of various finite element methods in terms of the error vs. the number of degrees of freedom. Low order methods with uniform meshes as well as spectral methods on a fixed mesh are clearly inferior to low order methods on judiciously refined meshes which in turn are inferior to the *hp*-FEM, especially at very small values of d.

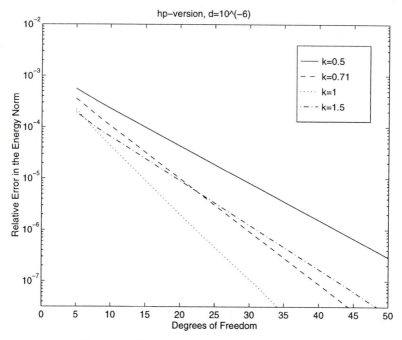

Fig. 5.4. The dependence on the parameter κ.

We see here the comparison in approximability in the "Energy" norm

$$\|u\|_E := d|u|_{H^1(I)} + \|u\|_{L^2(I)}$$

for various methods - here $I = (-1, 1)$ and u^d was as in (3.6). We clearly see the superiority of the *hp*-FEM over all other approaches. In particular, for small values of d the only way to get high accuracy in the layer at a reasonable number of DOF is the *hp*-FEM.

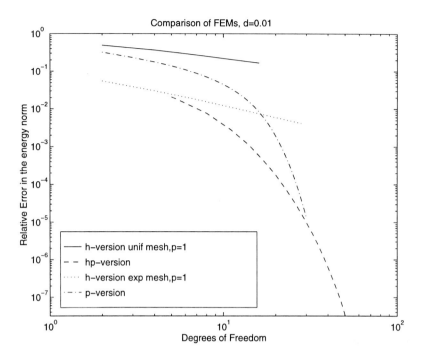

Fig. 5.5. Comparison of various methods, $d = 10^{-2}$.

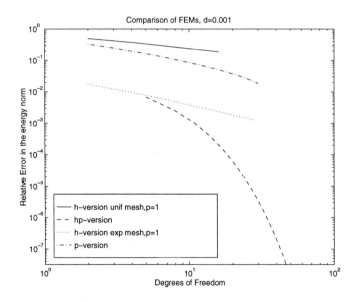

Fig. 5.6. Comparison of various methods, $d = 10^{-3}$.

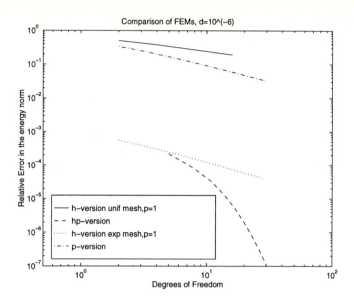

Fig. 5.7. Comparison of various methods, $d = 10^{-6}$.

Let us still comment further on theorem 5.14. Two items seem to limit the generality of the result: the explicit form of $u_{b\ell}^d$ and the specific knowledge of the parameter d for the mesh design. In fact, both prerequisites can be relaxed. We have [45,59].

Theorem 5.15. *Let the boundary layer function $u_{b\ell}^d$ on $I = (0,1)$ be as in (3.6). Define for $k = 1, 2, 3, \dots$ the mesh $\mathcal{T}_k$ as in (5.21) above. Then there exists $u_k^d \in S^{k,1}(I, \mathcal{T}_k)$ on I such that, for $0 < \kappa \leq \kappa_0$,*

$$\|u_{b\ell}^d - u_k^d\|_{L^\infty(I)} + \kappa p d \|(u_{b\ell}^d - u_k^d)'\|_{L^\infty(I)} \leq C e^{-b\kappa k}$$

where $b, C > 0$ are independent of d, κ and k.
Moreover,

$$u_k^d(0) = u_{b\ell}^d(0), \quad u_k^d(1) = u_{b\ell}^d(1) .$$

If the length scale ε of the boundary layer is not known explicitly, or if several length scales $d_1, d_2, \dots, d_\ell$ are present, these scales must be known explicitly in order to construct the hp-boundary layer meshes. Moreover, the FE-subspaces are not *hierarchic*, since at every k-increase the meshes are changed. This is overcome once more by means of geometric meshes.

Theorem 5.16. *On $I = (0,1)$ consider a boundary layer $u_{b\ell}^d$ of length scale d as in (3.6). Let $n \in \mathbb{N}$ be fixed and consider in I the geometric mesh $\mathcal{T}^{n,\sigma}$ with n layers and grading factor σ. Assume **scale resolution**, i.e. that*

$$\sigma^L \leq cd \tag{5.22}$$

for some $c > 0$. Then there are $C, \tau > 0$ such that for every $k \in \mathbb{N}$ there is $u_k^d \in S^{k,1}(I, \mathcal{T}^{n,\sigma})$ with

$$\|u_{b\ell}^d - u_k^d\|_{L^\infty(I)} + d\,\|(u_{b\ell}^d - u_k^d)'\|_{L^\infty(I)} \le C e^{-\tau k} . \tag{5.23}$$

For a proof, see [45].

As compared to Theorem 5.15, we have an additional condition (5.22); in the context of *hp*-FEM, scale resolution is not a very severe condition, since geometric mesh refinement allows to resolve extremely small scales with few layers. For example, let $d = 10^{-10}$ and $\sigma = 0.1$. Then $L = 10$ layers will suffice in (5.22). More generally, we get scale resolution provided that $L = O(\log_\sigma(d))$, a weak requirement if compared to uniform mesh refinement necessary for low order elements; even adaptive low order elements will require considerably more DOF (in terms of small d) to resolve the small scales.

Two dimensional results (smooth domain). The previous results on one-dimensional *hp*-boundary layer resolution apply immediately to boundary layers of the form (5.5). The main idea is now to use a *tensor product mesh* with *anisotropic element* that are aligned with the layer. The following figure shows in detail this construction.

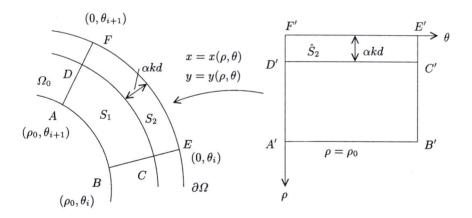

Fig. 5.8. Boundary-fitted elements in Ω_0.

If we now look at the components of (3.7), we see that the boundary layer effect is still only a one-dimensional one, in the direction of ρ (the functions $\tau_i(\theta)$ being smooth). Hence we may define boundary-fitted elements (as shown in Figure 5.8) on Ω_0. We do this by dividing $\partial\Omega$ into subintervals (θ_i, θ_{i+1}), $1 \le i \le m-1$, $\theta \in \partial\Omega$ and drawing the inward normal at θ_i, $1 \le i \le m$, of length ρ_0. Then the points (ρ_0, θ_i) are connected by the curve $\rho = \rho_0$.

Each curvilinear quadrilateral $S = ABEF$ is then further subdivided into two elements S_1 and S_2 by the curve $\rho = \kappa kd$, according to the prescription in the previous section. Looking at $ABEF$ in the (ρ, θ) coordinates then gives two rectangular elements $\widehat{S}_1 = A'B'C'D'$ and $\widehat{S}_2 = D'C'E'F'$ as shown in Figure 5.8. The local polynomial space on S_i, $i = 1, 2$ is then defined (using the notation $v(x, y) = \widehat{v}(\rho, \theta)$ for $(x, y) = (x(\rho, \theta), y(\rho, \theta))$) by

$$Q_k(S_i) = \{v(x, y) : \widehat{v}(\rho, \theta) \in Q_k(\widehat{S}_i)\}.$$

Note that the basis functions we use are polynomials in (ρ, θ) instead of in (x, y).

Consider the local approximation of (3.7) over the space

$$V_k(S) = \left\{v \in C^0(S) : v|_{S_i} \in Q_k(S_i)\right\}.$$

The function τ_i being smooth, is approximated exponentially by a piecewise polynomial $\tau_i^k(\theta)$ of degree k. The boundary layer function $\exp(-\alpha\rho/d)$ is approximated at an exponential rate by a piecewise polynomial $v(\rho)$, of degree $k - q$, as in Theorem 5.7. Then, for q fixed, k large enough, we obtain by a simple tensor product argument

$$\left\| \xi(\rho, \theta) - \sum_{i=0}^{q} \tau_i^k(\theta) \rho^i v(\rho) \right\|_{E, d, S} \leq Cd^{1/2} \exp(-bk) \tag{5.24}$$

so that the local approximation in the energy norm is the same as that in the one-dimensional case.

Remark 5.17. So far, we considered only boundary-fitted meshes. Analogous results are also valid for more general, properly refined triangulations at the boundary [46].

Similar arguments apply also for the other results, Theorems ((5.15)) and 5.16, if they are combined with high order polynomial approximation on large elements *along* the layer/front.

Remark 5.18. (on anisotropic refinement) We emphasize here that anisotropic mesh refinement is a conditio-sine-qua-non for the robust exponential convergence of hp-FEM in the presence of boundary layers and edge singularities in polyhedra; isotropic refinement will not suffice, since e.g. in shape regular geometric meshes the number of elements (and therefore the number of degrees of freedom) will increase exponentially (see Figure 5.9).

Boundary layer-corner singularity interaction. The above remarks apply only for a smooth boundary resp. near a smooth boundary segment. For flow problems with small viscosity or the reaction-diffusion equation (2.7) with small diffusion constant ε in a polygon. Here boundary layers appear

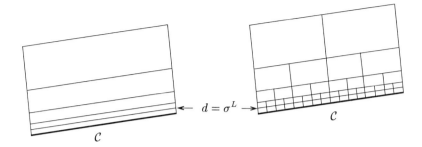

Fig. 5.9. Geometric isotropic and anisotropic refinement towards curve $\mathcal{C}$

near smooth boundary segments, corner singularities near vertices and corner-layers in the transition region between corner and boundary segment. All these effects are resolved by hp-FEM based on the (k, σ)-meshes in Section 4 (see in particular Figures 8-9 in Chapter 4). We conjecture that the solution of (2.7) in polygonal domains for $0 < \varepsilon \leq 1$ can be approximated robustly at an exponential rate. This is corroborated also by numerical experiments.

In the convection-diffusion problem (2.11) the additional difficulty arises that the dominant transport terms propagate the effect of corner singularities into the domain along characteristics. For positive $\varepsilon > 0$ at vertices, the typical corner singularities arise which generate so-called characteristic boundary layers along characteristics. For piecewise analytic data, the singular support of the solution contains characteristic lines which changes the length-scale of the layers associated with these lines. Schematically, this is shown in Figure 5.10. At the outflow boundary

$$\Gamma_+ = \{x \in \Gamma : \boldsymbol{\beta} \cdot \mathbf{n}(x) > 0\}$$

we have an **outflow boundary layer** of width $O(\varepsilon)$, whereas along the characteristic sets

$$\mathcal{C} := \{x \in \Omega : \dot{x}(s) = \boldsymbol{\beta}(x(s)), \ x(0) = P_i, \ i = 1, \ldots M\} \,,$$

i.e. the union of integral curves (contained in Ω) of the advection field $\boldsymbol{\beta}(x)$ through the vertices $O(\sqrt{\varepsilon})$, so-called **parabolic layers** arise. Notice that the corner singularities at inflow vertices $P_i \in \Gamma_- := \Gamma \backslash \Gamma_+$ (we assume here that Γ does not contain characteristic segments) influence these layers; their precise regularity is, even in the linear, $2 - d$ case, still under investigation.

In Figure 5.10, the lines in $\mathcal{C}$ are straight since the field β is constant. In general, these lines are curved for variable $\boldsymbol{\beta} = \boldsymbol{\beta}(x)$ and the hp-mesh design *must* be anisotropic and geometric towards $\mathcal{C}$ in order to achieve exponential convergence (see Remark 5.18).

Similar remarks apply also to the viscous shock profiles introduced in Section 3.3.

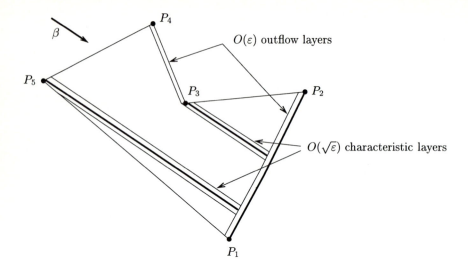

Fig. 5.10. Convection-Diffusion problem (2.11) in a polygon Ω-length scales of the layers

Part II

hp discretization techniques

The convergence of any numerical method is based upon *consistency of the approximation* and upon *stability of the discretization*. We have seen in the first part that *hp*-FEM can achieve *exponential approximation rates* for typical flow viscous features; this requires the combination and simultaneous variation of polynomial order and strong, possibly anisotropic, mesh-refinement. There arises the question on how to stably discretize CFD problems with *hp*-approximations. This part of the notes deals with the most important discretization techniques for such problems. All discretizations are based on some form of Galerkin projection upon the *hp*-subspaces. This methodology is well-established in solid mechanics where stable variational principles for most problems are readily available. In CFD we have to deal in particular with strongly advection dominated problems for which the usual Galerkin type discretizations do not exhibit good stability properties. To ensure robustness, we must therefore resort to non-standard – from the point of view of symmetry – discretization techniques for the viscous terms such as finite-volume, discontinuous Galerkin methods or in particular the stabilized Galerkin schemes, i.e. streamline-diffusion FEM (SDFEM) and the Galerkin-Least Squares (GLS) techniques. The presentation of these techniques is the purpose of the second part of the notes.

6   Reaction-Diffusion

We consider the discretization of the problem (2.7), (2.8). Several FE discretizations are presented, each based on specific *variational formulation* of (2.7), (2.8). Of particular interest are discontinuous approximations which can be used with the discontinuous Galerkin technique for first order problems, see Sections 7 and 12 ahead.

6.1   Standard (continuous) Discretization

We assume first that the Dirichlet data f in (2.8) is zero and introduce the space

$$H_D^1(\Omega) := \{u \in H^1(\Omega) : u = 0 \text{ on } \Gamma_D\} . \tag{6.1}$$

The variational function of (2.7), (2.8) with general $\mathbf{A}(x)$ is

$$u \in H_D^1(\Omega) : \ a(u,v) = \ell(v) \quad \forall v \in H_D^1(\Omega) \tag{6.2}$$

where the forms are defined by

$$a(u,v) := \varepsilon(\nabla v, \mathbf{A}\nabla u)_\Omega + (v,u)_\Omega ,$$
$$\ell(v) \quad := (S,v)_\Omega + (g,v)_{\Gamma_N} .$$

The form $a(\cdot,\cdot)$ is *symmetric* and *coercive*, i.e.

$$a(u,u) \geq \|u\|_\varepsilon^2 := \varepsilon \|\nabla u\|_{0,\Omega}^2 + \|u\|_{0,\Omega}^2 > 0 \tag{6.3}$$

if $u \neq 0$, due to (2.4).

The discretization of (6.2) is obtained by restricting u and v to FE subspaces of $H_D^1(\Omega)$: in order to achieve it, subspaces of continuous, piecewise polynomial functions must be chosen. We have

$$u_{\text{FE}} \in S_D^{\mathbf{k},1}(\Omega, \mathcal{T}, \mathbf{F}_\mathcal{P}) :$$
$$a(u_{\text{FE}}, v) = \ell(v) \quad \forall v \in S^{\mathbf{k},1}(\Omega, \mathcal{T}), \mathbf{F}_\mathcal{P}) . \tag{6.4}$$

Here

$$S_D^{\mathbf{k},1} := S^{\mathbf{k},1} \cap H_D^1 .$$

Let $N = \dim\{S_D^{\mathbf{k},1}\}$ and $\{\varphi_i : i = 1, \dots, N\}$ be a basis for $S_D^{\mathbf{k},1}$. Then (6.4) is equivalent to the linear system

$$\mathbf{\Lambda x} = \boldsymbol{\ell}$$

where the entries of the diffusion-stiffness matrix are given by

$$A_{ij} = a(\varphi_i, \varphi_j) = a(\varphi_j, \varphi_i), \ 1 \leq i,j \leq N$$

and the entries of the load vector ℓ are $\ell_j = \ell(\varphi_j)$.

The diffusion matrix is symmetric and positive definite and must be evaluated by numerical quadrature of sufficiently high order, in particular if the elements are curved, see [41], [65], [66], [67] and [46] for quadrature techniques and error estimates.

6.2   Mixed discretization

The continuity of the FE solution u_{FE} in (6.4) is restrictive - in connection with finite volume methods for convection dominated problems or discontinuous Galerkin methods it is desirable to admit *discontinuous approximations* for u_{FE}. To this end, the variational formulation (6.2) must be changed. We write

$$-\operatorname{div}\mathbf{q}(\nabla u) + \sigma u = S \text{ in } \Omega, \tag{6.5}$$

where the flux $\mathbf{q}$ is given by

$$\mathbf{q}(\nabla u) = \mathbf{A}\nabla u \text{ in } \Omega. \tag{6.6}$$

We get the (dual) **mixed variational formulation**:
find $u \in L^2(\Omega)$ and $\mathbf{q} \in H(\operatorname{div}, \Omega)$ such that $\mathbf{n} \cdot \mathbf{q} = g$ on Γ_N and

$$\begin{aligned}
(v, \sigma u)_\Omega - (v, \operatorname{div}\mathbf{q})_\Omega &= (S, v) && \forall v \in L^2(\Omega), \\
(u, \nabla \cdot \mathbf{p})_\Omega + (\mathbf{p}, \mathbf{q})_\Omega &= (f, \mathbf{n} \cdot \mathbf{p})_{\Gamma_D} && \forall \mathbf{p} \in H(\operatorname{div}, \Omega).
\end{aligned} \tag{6.7}$$

Here $H(\operatorname{div}, \Omega)$ is defined as follows:

$$H(\operatorname{div}, \Omega) := \{\mathbf{q} \in L^2(\Omega)^d : \operatorname{div}\mathbf{q} \in L^2(\Omega)\}, \tag{6.8}$$

where the divergence is understood in the weak sense.

The mixed FE discretization of (6.7) is based on subspaces $S^{\mathbf{k},0}(\Omega, \mathcal{T}, \mathbf{F}_{\mathcal{P}}) \subset L^2(\Omega)$ and $S^{\mathbf{k}}_{\operatorname{div}}(\Omega, \mathcal{T}, \mathbf{F}_{\mathcal{P}}) \subset H(\operatorname{div}, \Omega)$; now u_{FE} can be discontinuous, but d components of the flux $\mathbf{q}$ must be discretized; the finite element fluxes $\mathbf{q}_{\mathrm{FE}}$ must have a continuous normal component across element interfaces, but their tangential component(s) may be discontinuous.

The linear system corresponding to (6.7) has the form (for constant σ)

$$\begin{pmatrix} \sigma\mathbf{M} & -\mathbf{B} \\ \mathbf{B}^\top & \mathbf{C} \end{pmatrix} \begin{pmatrix} \mathbf{u} \\ \mathbf{q} \end{pmatrix} = \begin{pmatrix} \mathbf{S} \\ \mathbf{f} \end{pmatrix} \tag{6.9}$$

where $\mathbf{M}$ is the L^2-mass matrix of u, $\mathbf{C}$ is the mass-matrix of $\mathbf{q}$ and $\mathbf{B}$, $\mathbf{B}^\top$ correspond to the nonsymmetric forms of (6.7). In the conjunction with an explicit time-stepping strategy, the spectrum of the matrix in (6.9) is of interest. We have

$$\begin{aligned}
\begin{pmatrix} \mathbf{u} \\ \mathbf{q} \end{pmatrix}^\top \begin{pmatrix} \sigma\mathbf{M} & -\mathbf{B} \\ \mathbf{B}^\top & \mathbf{C} \end{pmatrix} &= \begin{pmatrix} \mathbf{u} \\ \mathbf{q} \end{pmatrix} \\
&= \sigma\mathbf{u}^\top \mathbf{M}u - \mathbf{u}^\top \mathbf{B}\mathbf{q} + \mathbf{q}^\top \mathbf{B}^\top \mathbf{u} + \mathbf{q}^\top \mathbf{C}\mathbf{q} \\
&= \sigma\mathbf{u}^\top \mathbf{M}u + \mathbf{q}^\top \mathbf{C}\mathbf{q} > 0
\end{aligned} \tag{6.10}$$

if $\sigma > 0$, $\mathbf{u} \neq 0$, $\mathbf{q} \neq 0$, i.e. the matrix has eigenvalues with positive real part, if $\sigma > 0$, so that this discretization is dissipative.

If $\sigma = 0$, stability of (6.9) is not guaranteed in general. In this case, we must require a compatibility condition of the spaces $S^{\mathbf{k},0}$ and $S^{\mathbf{k}}_{\mathrm{div}}$, the so-called discrete *inf-sup condition*:

$$\forall 0 \neq u \in S^{\mathbf{k},0} : \quad \sup_{0 \neq \mathbf{q} \in S^{\mathbf{k}}_{\mathrm{div}}} \frac{(u, \mathrm{div}\, \mathbf{q})}{\|\mathrm{div}\, \mathbf{q}\|_{0,\Omega}} \geq \gamma \|u\|_{0,\Omega} \qquad (6.11)$$

for some $\gamma > 0$.

An example of an element family satisfying (6.11) is the so-called discrete **Raviart-Thomas family** (see [16], Chapter III for more).

6.3   Mortar-Discretization

The mixed discretization has the disadvantage that for each component u of the flow field d additional fluxes must be discretized leading to a large number of unknowns. Another approach is to use discontinuous u and to penalize the interelement jumps by Lagrange-Multipliers on the element interfaces, leading to the so-called **Mortar Element Method (MEM)**. Some relevant references are [11], [8] and, for the *hp*-MEM, [63].

We describe the MEM for the model problem

$$-\mathrm{div}\, \mathbf{q}(\nabla u) + \sigma u = S \quad \text{in} \ \Omega, \qquad (6.12)$$

$$\begin{aligned} u = 0 \quad &\text{on} \ \Gamma_D, \\ \mathbf{n} \cdot \mathbf{q}(\nabla u) = g \quad &\text{on} \ \Gamma_N. \end{aligned} \qquad (6.13)$$

Here $\sigma \geq 0$ and the flux $\mathbf{q}(\nabla u)$ is as in (6.6). Let $\mathcal{T}$ be a mesh in Ω built out of regular patch meshes $\mathcal{T}_P$, which are possibly irregular across patch interfaces for $K, K' \in \mathcal{T}$ with intersection $\Gamma_{KK'}$ of positive $d-1$ dimensional measure. In the MEM, we use the standard variational formulation (6.2) with discontinuous $u, v \in S^{\mathbf{k},0}_D(\Omega, \mathcal{T}, \mathbf{F}_P)$.

The bilinear form $a(\cdot, \cdot)$ must be reinterpreted then, since the $H^1(\Omega)$ norm is not defined for $u, v \in S^{\mathbf{k},0}_D$.

Broken Bilinear Forms and Spaces. We reformulate therefore (6.2) for piecewise H^1-functions on the partition $\mathcal{T}$ and set

$$H^1(\Omega, \mathcal{T}) := \{u \in L^2(\Omega) : u|_K \in H^1(K) \ \ \forall K \in \mathcal{T}\}, \qquad (6.14)$$

equipped with the broken seminorm and norm

$$|u|^2_{1,\Omega,\mathcal{T}} := \sum_{K \in \mathcal{T}} \|\nabla u\|^2_{0,K}, \quad \|u\|^2_{1,\Omega,\mathcal{T}} = \sum_{K \in \mathcal{T}} \|u\|^2_{1,K}. \qquad (6.15)$$

To generalize (6.2) to $u, v \in S_D^{\mathbf{k},0}$, we also introduce the *broken bilinear form*: for $u, v \in H^1(\Omega, \mathcal{T})$, set

$$a_{\mathcal{T}}(u, v) := \sum_{K \in \mathcal{T}} \varepsilon (\nabla v, \mathbf{A} \nabla u)_{0,K} + (v, u)_{0,K} = \sum_{K \in \mathcal{T}} a_K(u, v) . \tag{6.16}$$

Variational formulation in broken spaces. We derive an analog to (6.2) in broken spaces. Let $K \in \mathcal{T}$ be any element. Multiplying the equation

$$-\mathrm{div} \; \mathbf{q}(\nabla u) + \sigma u = S \; \text{ in } \; K$$

by $v \in H^1(K)$ and integrating by parts on K gives

$$\varepsilon (\nabla v, \mathbf{A} \nabla u)_{0,K} + \sigma(u, v)_{0,K} = (S, v)_{0,K} + (v, \mathbf{n} \cdot \mathbf{q}_K)_{0,\partial K} . \tag{6.17}$$

To get a variational formulation of (6.2), we sum (6.17) over all $K \in \mathcal{T}$, giving

$$a_{\mathcal{T}}(u, v) = (S, v)_{0,\Omega} + \sum_{K \in \mathcal{T}} (v, \mathbf{n}_K \cdot \mathbf{q}_K)_{0,\partial K} \tag{6.18}$$

where $\mathbf{n}_K$ is the exterior unit normal to $K \in \mathcal{T}$ and $\mathbf{q}_K = \mathbf{q}|_K$. Denote by the *Skeleton* S_{int} the union of all element intersections of positive $d - 1$ dimensional measure.

$$S_{\mathrm{int}} = \left\{ e = \overline{K} \cap \overline{K'} \in \mathcal{E} : K, K' \in \mathcal{T}, \int_e ds > 0 \right\} \tag{6.19}$$

and set

$$S_D := \{ e \in \mathcal{E} : e \subset \Gamma_D \} \tag{6.20}$$

where $\mathcal{E}$ is the set of all $d - 1$ dimensional element boundary segments.

For the exact solution u of (6.2), the fluxes $\mathbf{n}_K \cdot \mathbf{q}(\nabla u)$ are continuous across edges $e \in S_{\mathrm{int}}$. The MEM for (6.12), (6.13) consists in enforcing the vanishing of the jumps of u across $e \in S_{\mathrm{int}}$ as follows: find $u \in H_D^1(\Omega, \mathcal{T})$, $\mu \in M$ such that

$$\begin{aligned} a_{\mathcal{T}}(u, v) + b_{\mathcal{T}}(v, \mu) &= (S, v)_{0,\Omega} + (g, v)_{0,\Gamma_N} \quad &&\forall v \in H^1(\Omega, \mathcal{T}), \\ b_{\mathcal{T}}(u, \lambda) \quad\quad &= 0 \quad &&\forall \lambda \in M . \end{aligned} \tag{6.21}$$

Here

$$b_{\mathcal{T}}(u, \lambda) := \sum_{e \in S_{\mathrm{int}}} ([u], \lambda)_{0,e}$$

and $[u]$ denotes the jump of $u \in H^1(\Omega, \mathcal{T})$ across $e \in S_{\mathrm{int}}$. The Mortar space M is a multiplier space contained in $\prod_{e \in S_{\mathrm{int}}} H^{-1/2}(e)$. Notice that by (6.18), if u is smooth, the mortar μ in (6.21) will give the canonical flux $\mathbf{n}_K \cdot \mathbf{q}(\nabla u)$ on $e \in S_{\mathrm{int}}$. Note also that (6.21) has saddle point form, similarly to the mixed formulation (6.7).

It is crucial for the stability of (6.21) that $b_\mathcal{T}(\cdot, \cdot)$ satisfies a suitable inf-sup condition; this is indeed the case, see e.g. [8].

The *finite element discretization of* (6.21) is as usual:
find $u_{\text{FE}} \in S_D^{\mathbf{k},0}(\Omega, \mathcal{T}, \mathbf{F}_\mathcal{P})$, $\mu_{\text{FE}} \in M^{\mathbf{k},0}(\Omega, \mathcal{T})$:

$$
\begin{aligned}
a_\mathcal{T}(u_{\text{FE}}, v) + b_\mathcal{T}(v, \mu_{\text{FE}}) &= (S, v)_{0,\Omega} + (g, v)_{0,\Gamma_N} \quad \forall v \in S_D^{\mathbf{k},0} \\
b_\mathcal{T}(u_{\text{FE}}, \lambda) &= 0 \quad\quad\quad\quad\quad\quad \forall \lambda \in M_\mathcal{T}^{\mathbf{k}} .
\end{aligned}
\tag{6.22}
$$

Here the additional mortar space $M_\mathcal{T}^{\mathbf{k}}$ enters, similarly to the flux-subspaces $S_{\text{div}}^{\mathbf{k}} \subset H(\text{div}, \Omega)$ in (6.7). Several choices for $M_\mathcal{T}^{\mathbf{k}}$ are possible. However, care must be taken that the forms $b_\mathcal{T}(\cdot, \cdot)$ satisfy a discrete inf-sup condition

$$
\inf_{\lambda \in M_\mathcal{T}^{\mathbf{k}}} \sup_{v \in S^{\mathbf{k},0}} \frac{b_\mathcal{T}(v, \lambda)}{\|v\|_{1,\Omega,\mathcal{T}} \|\lambda\|_M} \geq \gamma(\mathcal{T}, \mathbf{k}) > 0
\tag{6.23}
$$

holds. For uniform degree k, the mortar space

$$
M_\mathcal{T}^k = \{\lambda \in L^2(\mathcal{S}_{\text{int}}) : \lambda|_e \in \mathcal{P}_{k-1}(e)\}
\tag{6.24}
$$

has been shown (for a fixed patch mesh $\widehat{\mathcal{T}}_\mathcal{P}$ allowing in particular also geometric meshes) in [63] to have an inf sup constant $\gamma(\mathcal{T}, k) \geq C(\sigma) k^{-3/4}$ in two dimensions.

The usual theory of mixed methods (see e.g. [16]) implies then quasi optimal error bounds for u as well as for the fluxes μ_{FE}.

Remark 6.1. Notice that the degree K of the mortar space $M_\mathcal{T}^k$ in (6.24) is one less than k in the domain - the lowest degree admissible is hence $k = 1$; no variant of the MEM is known which admits $k = 0$ in the elements. In comparison with the mixed formulation (6.7), the mortar method involves less additional degrees of freedom - only fluxes on interfaces must be discretized, rather than fluxes in the elements. Nevertheless, the mortar approach still involves more DOF than the conforming method (6.4).

Implementation without fluxes. It is possible to eliminate the mortar μ_{FE}, λ from (6.22), thereby reducing the number of unknowns. The idea is to restrict u_{FE} and $v \in S_D^{\mathbf{k},0}$. If, for example, $[u_{\text{FE}}] = 0$ and $[v] = 0$ on $\mathcal{S}_{\text{int}}$, so that u_{FE} and v are continuous, $b_\mathcal{T}$ in (6.22) vanishes and we get again the symmetric formulation (6.4) (since then $u_{\text{FE}}, v \in S_D^{\mathbf{k},1}$) i.e. nothing new.

A second possibility not enforcing interelement continuity is use (6.24) and to restrict u_{FE}, v to

$$
S_D^{k,\delta} = \Big\{ u \in S_D^{k,0} : \forall e_{K,K'} \in \mathcal{S}_{\text{int}} \; \forall \varphi \in \mathcal{P}_{k-1}(e_{KK'}) :
$$
$$
\int_{e_{KK'}} (u|_K - u|_{K'}) \varphi \, ds = 0 \Big\}
\tag{6.25}
$$

resp. more generally

$$S_D^{k,\delta} = \left\{ u \in S_D^{k,0} : \forall e_{KK'} \in \mathcal{S}_{\text{int}} \;\; \forall \varphi \in M_{\mathcal{T}}^k : \right.$$
$$\left. \int_{e_{KK'}} (u|_K - u|_{K'}) \varphi|_{e_{KK'}} \, ds = 0 \right\} \tag{6.26}$$

where $e_{KK'} = \overline{K} \cap \overline{K'}$ for $K, K' \in \mathcal{T}$.

We observe that on any $e_{K,K'}$ the jump $[u]$ belongs to $\mathcal{P}_k(e_{K,K'})$. The orthogonality

$$\int_{e_{K,K'}} [u]\varphi \, ds \quad \forall \varphi \in \mathcal{P}_{k-1}(e_{K,K'}) \tag{6.27}$$

consists in $k = \dim \mathcal{P}_{k-1}(e_{K,K'})$ constraints which are linear combinations of the side degrees of freedom of $u|_K$ and $u|_{K'}$. The condensed stiffness matrix $\mathbf{A}$ can be written in the form

$$\widehat{\mathbf{A}} = \mathbf{Q}^{\top} \operatorname{diag}\{\mathbf{A}_K : K \in \mathcal{T}\} \mathbf{Q} \tag{6.28}$$

where A_K are elemental stiffness matrices corresponding to $a_K(u, v)$ in (6.16) and the matrices $\mathbf{Q}$ contain the coefficients of the constraints (6.16). In iterative solvers for $\widehat{\mathbf{A}}\mathbf{x} = \mathbf{b}$, (6.28) is never formed explicitly and, in particular, the element stiffness matrices $\mathbf{A}_K$ could reside on different processors during the iterations.

It can be proved that the bilinear form $a_{\mathcal{T}}(u, v)$ is coercive on $S_D^{k,\delta}$ and hence the matrix $\widehat{\mathbf{A}}$ is positive-definite [11], i.e. the mortar discretization (6.22) preserves dissipativity. Note however, that the coercivity constant resulting from the proof in [11] depends on the triangulation in an unspecified way.

Remark 6.2. We emphasize that the MEM presented here differs from the one considered in [8], [9], [11], in that we allow here discontinuities on each edge whereas the cited works treated the MEM as a variant of the domain decomposition method where the number of subdomains coupled by the mortar is fixed and mesh refinement with conforming elements takes place within the subdomains. Clearly, the formulation presented here is more general and closely related to FEM with Lagrangean Multipliers resp. to the global element method.

Remark 6.3. Finally, we remark that the MEM with eliminated fluxes coincides at least in one case with a known method: consider on $\mathcal{T}$ consisting of triangles the space $S^{1,0}(\Omega, \mathcal{T})$ of piecewise linear, discontinuous functions. Choosing the mortar space $M_{\mathcal{T}}^0$ of piecewise constants on the edges, (6.28) implies that the averages of the jumps of $u \in S^{1,\delta}$ over each edge must vanish – this element is just the Crouzeix-Raviart element. Here the matrix $\widehat{\mathbf{A}}$ is coercive independent of the meshwidth.

6.4   Discontinuous Galerkin Method for second order problems

The DGFEM allows to discretize diffusion problems with discontinuous shape functions without extra unknowns due to fluxes or multipliers. The stiffness matrix is nonsymmetric but positive semidefinite which is desirable for explicit time stepping schemes.

Derivation from the Mortar Method. Closely related to the MEM is the Discontinuous Galerkin (DG) method for the problem (2.7), (2.8). It can be derived as follows: consider (6.21). Adding the equations, we get: find $(u, \mu) \in H_D^1(\Omega, \mathcal{T}) \times M$ such that

$$B_{\mathcal{T}}(u, \mu; v, \lambda) = \ell(v, \lambda) \quad \forall (v, \lambda) \in H_D^1(\Omega, \mathcal{T}) \times M \tag{6.29}$$

where we set

$$B_{\mathcal{T}}(u, \mu; v, \lambda) := a_{\mathcal{T}}(u, v) + b_{\mathcal{T}}(v, \mu) + b_{\mathcal{T}}(u, \lambda),$$
$$\ell(v, \lambda) \qquad := (S, v)_{0,\Omega} + (g, v)_{0, \Gamma_N}$$

and where M is a suitable mortar space.

(6.29) is equivalent to (6.21) and its discretization:
find $(u_{\mathrm{FE}}, \mu_{\mathrm{FE}}) \in V_{\mathcal{T}}^{\mathbf{k}}$ such that

$$B_{\mathcal{T}}(u_{\mathrm{FE}}, \mu_{\mathrm{FE}}) = \ell(v, \lambda) \quad \forall (v, \lambda) \in V_{\mathcal{T}}^{\mathbf{k}} \tag{6.30}$$

where

$$V_{\mathcal{T}}^{\mathbf{k}} := S_D^{\mathbf{k},0}(\Omega, \mathcal{T}) \times M^{\mathbf{k},0}(\Omega, \mathcal{T}) .$$

The discontinuous Galerkin FEM consists in eliminating a-priori the multipliers μ_{FE} and λ in (6.30) by the flux-averages: on $e_{K,K'} \subset \mathcal{S}_{\mathrm{int}}$, set

$$\mu_{\mathrm{FE}} = \frac{1}{2} \left(\mathbf{q}_K \cdot n_K + \mathbf{q}_{K'} \cdot n_K \right) =: \langle \mathbf{q}(\nabla u) \cdot n_e \rangle \tag{6.31}$$

where n_e is, for example, the exterior unit normal n_K to the element K with higher index (any other, fixed, choice of n_K would do). Analogously, we select

$$\lambda = -\langle \mathbf{q}(\nabla v) \cdot n_e \rangle \quad \text{on } e \in \mathcal{S}_{\mathrm{int}} \tag{6.32}$$

and get the DGFEM: find $u_{\mathrm{DG}} \in S_D^{\mathbf{k},0}$ such that

$$\begin{aligned}
B_{\mathrm{DG}}(u_{\mathrm{DG}}, v) = &\sum_K a_K(u_{\mathrm{DG}}, v) \\
&+ \sum_{e \in \mathcal{S}_{\mathrm{int}}} \int_e ([u_{\mathrm{DG}}]\langle \mathbf{q}(\nabla v) \cdot n_e \rangle - [v]\langle \mathbf{q}(\nabla u_{\mathrm{DG}}) \cdot n_e \rangle) \, ds
\end{aligned} \tag{6.33}$$

for all $v \in S_D^{\mathbf{k},0}$. Here $\mathbf{q}(\nabla u) = \varepsilon \mathbf{A} \nabla u$, cf. (6.2).

The minus sign in (6.32) is crucial - taking there plus gives a symmetric bilinear form which is, however, *indefinite* - this property is very bad for explicit time stepping schemes.

In contrast, the form $B_{DG}(\cdot,\cdot)$ in (6.33) is nonsymmetric, but positive semidefinite, i.e.

$$\forall u \in H^1(\Omega,\mathcal{T}) \quad B_{DG}(u,u) = \sum_K a_K(u,u) \geq 0 , \qquad (6.34)$$

i.e. (the real parts of) the eigenvalues of the corresponding stiffness matrix A_{DG} are nonnegative and the DG discretization (6.33) of the diffusion operator will be dissipative in an explicit time-stepping scheme, an observation due to Oden and Baumann [49].

Stability of the DG-method. The stability of (6.33) is, to some extent, an open problem. We prove here

Proposition 6.4. *Assume that $\mathcal{T}$ is a quasi-uniform, shape-regular mesh on Ω of meshwidth h and that there exists $c > 0$ such that*

$$\forall K \in \mathcal{T} \quad \forall u \in H^1(K) : \ a_K(u,u) \geq c \, \|u\|_{1,K}^2 \qquad (6.35)$$

and

$$\forall K \in \mathcal{T} \quad \forall u,v \in H^1(K) : \ a_K(u,v) \leq c^{-1} \, \|u\|_{1,K} \|v\|_{1,K} . \qquad (6.36)$$

Define further on $H^1(\Omega,\mathcal{T})$ the broken norm

$$\|u\|_{\mathcal{T}} := \left(\sum_K \|u\|_{1,K}^2 \right)^{1/2} . \qquad (6.37)$$

Then there holds

$$\inf_{0\neq u\in S^{k,0}} \ \sup_{0\neq v\in S^{k,0}} \ \frac{B_{DG}(u,v)}{\|u\|_{\mathcal{T}} \, \|v\|_{\mathcal{T}}} \geq \gamma > 0 \qquad (6.38)$$

and

$$|B_{DG}(u,v)| \leq C(k+1)h^{-1} \ \|u\|_{\mathcal{T}} \, \|v\|_{\mathcal{T}} . \qquad (6.39)$$

where $C,\gamma > 0$ are independent of h and of k; they depend only on the shape-regularity of the elements.

Proof.

1) Given $u \in S^{k,0} \subset H^1(\Omega,\mathcal{T})$, select $v_u = u$. Then $\|v_u\|_{\mathcal{T}} = \|u\|_{\mathcal{T}}$ and

$$
\begin{aligned}
B_{DG}(u,v_u) &= B_{DG}(u,u) = \sum_K a_K(u,u) \\
&\overset{(6.35)}{\geq} c \sum_K \|u\|_{H^1(K)}^2 = c \, \|u\|_{\mathcal{T}}^2 .
\end{aligned}
$$

2) Let $u, v \in S^{k,0}$. Then from (6.33) with $\varepsilon = 1$ we get

$$
\begin{aligned}
|B_{\mathrm{DG}}(u,v)| \ &\leq \ \sum_K |a_K(u,v)| + \sum_e \||[u]|\|_{0,e} \|\langle n_e \cdot \mathbf{A}\nabla v\rangle\|_{0,e} \\
&\quad + \sum_e \||[v]|\|_{0,e} \|\langle n_e \cdot \mathbf{A}\nabla u\rangle\|_{0,e} \\
&\stackrel{(6.35)}{\leq} \ c^{-1} \sum_K \|u\|_{1,K}\|v\|_{1,K} \\
&\quad + C \sum_e h^{-1/2}\|u\|_{1,K \cup K'} (k+1)h^{-1/2}\|v\|_{1,K \cup K'} \\
&\leq \ c^{-1}\|u\|_{\mathcal{T}}\|v\|_{\mathcal{T}} + C(k+1)h^{-1}\|u\|_{\mathcal{T}}\|v\|_{\mathcal{T}}
\end{aligned}
$$

where we used the trace inequality

$$
\|u\|_{0,e}^2 \leq C(\|\nabla u\|_{0,K}\|u\|_{0,K} + h^{-1}\|u\|_{0,K}^2)
$$

and, by (6.36),

$$
\|n_e \cdot \mathbf{A}\nabla v\|_{0,e} \leq \|\mathbf{A}\nabla v\|_{0,e} \leq C(k+1)h^{-1/2}\|\nabla v\|_{0,K} .
$$

$\square$

Remark 6.5. Note that (6.35) rules out the case when we have pure diffusion, i.e. the Laplacean. Then $a_K(u,u) = \int_K |\nabla u|^2 \, dx$ and (6.35) is violated. Moreover, in this case

$$
B_{\mathrm{DG}}(u,u) = \sum_K \int_K |\nabla u|^2 \, dx = 0 \iff u \in S^{0,0}(\Omega, \mathcal{T}) , \tag{6.40}
$$

i.e. the bilinear form B_{DG} has a large kernel. Note also that for diffusion problems resulting from implicit time discretization, (6.35) is usually satisfied, see Section 11 below for more.

Remark 6.6. In one dimension an inf-sup condition (6.38) and continuity (6.32) with constants independent of h and k holds [49] even in the absence of an absolute term. In our case, the *hp*-error estimates of Section 5 apply with a loss of $(k+1)h^{-1}$.

Remark 6.7. The form a_K in assumption (6.35) is as in (6.16), with $\varepsilon = 1$. Nevertheless, the argument in the proof goes through also for $\varepsilon < 1$ if in the definition (6.37) of the norm $\|\circ\|_{H^1(K)}$ is replaced by $\varepsilon |\circ|_{1,K} + \|\circ\|_{0,K}$.

Remark 6.8. In terms of computational efficiency, the DGFEM (6.33) has numerous advantages over, e.g. the schemes in 6.2 and 6.3. For example, since continuity is only weakly enforced, there is no need to code interelement constraints any more. This, in turn, allows to modify the definition (4.5) of the FE space in that the FE space on element $K \in \mathcal{T}$ need not be defined in terms of parametric element mappings F_K. Rather, we can in the DGFEM, adopt the definition

$$S^{k,0}(\Omega, \mathcal{T})\{u \in L^2(\Omega) : \ u|_K \in \mathcal{P}_{k_K} \text{ for } K \in \mathcal{T}\},$$

i.e. the FE-spaces may be defined in local carthesian coordinates. Moreover, even if K is a quadrilateral element, the local approximation space may be $\mathcal{P}_k$ rather than $\mathcal{Q}_k$.

Remark 6.9. We have seen in Proposition 6.4 that in the hp-DGFEM we must generally expect a loss of optimal convergence. It can be shown, however, that this loss of convergence orders can be overcome by a stabilization via penalization of the interelement jumps - a device going back to J. Nitsche in 1971. There, the bilinear form $B_{\mathrm{DG}}(u, v)$ in (6.33) is modified by an additional term to

$$B_{\mathrm{DG},\gamma}(u, v) := B_{\mathrm{DG}}(u, v) + \sum_{e \in \mathcal{S}_{\mathrm{int}}} \gamma_e \int_e [u][v] \, ds$$

where $\gamma_e > 0$ is a stabilization parameter to be selected. The resulting method has the advantage to be defined also for $k = 0$, i.e., for piecewise constants. Judicious choice of γ_e allows to recover optimal convergence rates in the diffusive case ([37], Section 4). The price to be paid by the penalization of the interelement jumps is a) increased stiffness and condition number of the discrete problem and b) loss of elementwise conservation property.

7   Convection

Contrary to the reaction-diffusion case, the convection problem (2.9) and the continuity equation (1.1) are first order, hyperbolic equations. Consequently, the variational formulation underlying the hp-FEM will not be symmetric any more and a standard Galerkin approach as in the reaction-diffusion case is well known to have poor stability properties. This parallels the classical instability of central differencing for the linear advection equation. To obtain stable discretizations, some sort of *stabilization* must be introduced into the variational formulation. We will discuss the following devices: a) streamline diffusion techniques and b) discontinuous Galerkin approximations.

The streamline diffusion method was introduced by Hughes and Johnson and their coworkers in the early 80ies in order to combat instabilities of C^0-FEM for advection-dominated flows [36], [38], [39]. It consists in replacing the test function v in the Galerkin scheme by $v + \delta_K \mathcal{L}v$ where $\mathcal{L}$ is the advection operator. The parameter δ must be chosen in terms of the discretization parameters, i.e. the meshwidth and, in hp-FEM, also in terms of the elemental polynomial degree k_K. This so-called *stabilization parameter* is at the disposal of the analyst and can be adjusted in specific computations, but for each element $K \in \mathcal{T}$ there is a coupling to h_K and k_K which ensures the optimal convergence rates for first order problems, both in h and k.

7.1   Model convection problem

Let Ω be a bounded curved polyhedral domain in $\mathbb{R}^d$, $d \geq 2$. Given that $\mathbf{a} = (a_1, \ldots, a_d)$ is a d-component vector function defined on $\overline{\Omega}$ with $a_i \in C^1(\overline{\Omega})$, $i = 1, \ldots, d$, we define the following subsets of $\Gamma = \partial\Omega$

$$\Gamma_- = \{x \in \Gamma : \mathbf{a}(x) \cdot \mathbf{n}(x) < 0\}, \quad \Gamma_+ = \{x \in \Gamma : \mathbf{a}(x) \cdot \mathbf{n}(x) > 0\},$$

where $\mathbf{n}(x)$ denotes the unit outward normal vector to Γ at $x \in \Gamma$. It is assumed here implicitly that in these definitions x ranges only through those points of Γ at which $\mathbf{n}(x)$ is defined; consequently, Γ_- and Γ_+ are not necessarily connected subsets of Γ.

For the sake of simplicity, we shall suppose that Γ is non-characteristic in the sense that $\overline{\Gamma}_- \cup \overline{\Gamma}_+ = \Gamma$.

The convection problem (2.9) takes the form

$$\begin{cases} \mathcal{L}u \equiv \mathbf{a} \cdot \nabla u + bu = S \text{ in } \Omega, \\ \qquad\qquad\qquad u = f \text{ on } \Gamma_-. \end{cases} \tag{7.1}$$

for some $b \in C(\overline{\Omega})$, $S \in L^2(\Omega)$, $f \in L^2(\Gamma_-)$.

This problem has a unique weak solution $u \in L^2(\Omega)$ with $\mathbf{a} \cdot \nabla u \in L^2(\Omega)$ and the boundary condition satisfied as an equality in $[H_{00}^{1/2}(\Gamma_-)]'$.

In the next two subsections we shall formulate the *hp*-streamline diffusion and *hp*-discontinuous Galerkin finite element approximation of (7.1).

7.2   The *hp*-SDFEM

The *hp*-SDFEM approximation of (7.1) is defined as follows: find $u_{\mathrm{SD}} \in S^{\mathbf{k},1}$ such that

$$\begin{aligned} B_{\mathrm{SD}}(u_{\mathrm{SD}}, v) &:= (\mathcal{L}u_{\mathrm{SD}}, v + \delta\mathcal{L}v) + (u_{\mathrm{FE}}, v)_{\Gamma_-} \\ &= F_{\mathrm{SD}}(v) = (S, v + \delta\mathcal{L}v) + (f, v)_{\Gamma_-} \quad \forall v \in S^{\mathbf{k},1} \end{aligned} \tag{7.2}$$

where δ is a positive piecewise constant function defined on the mesh triangulation $\mathcal{T}$.

In (7.2), $(\cdot, \cdot)$ denotes the inner product of $L_2(\Omega)$, and

$$(w, v)_{\Gamma_-} = \int_{\Gamma_-} |\mathbf{a} \cdot \mathbf{n}| \, wv \, ds,$$

with analogous definition of $(\cdot, \cdot)_{\Gamma_+}$ and associated norms $\|\cdot\|_{\Gamma_-}$ and $\|\cdot\|_{\Gamma_+}$.

The stability of the *hp*-SDFEM is expressed in the next lemma.

Lemma 7.1. *Suppose that there exists a positive constant c_0 such that*

$$b(x) - \frac{1}{2} \nabla \cdot \mathbf{a}(x) \geq c_0, \quad x \in \overline{\Omega}. \tag{7.3}$$

Then u_{SD} obeys for $\delta \geq 0$ the bound

$$\|\sqrt{\delta} \mathcal{L} u_{SD}\|^2 + c_0 \|u_{SD}\|^2 + \|u_{SD}\|_{\Gamma_+}^2 + \frac{1}{2} \|u_{SD}\|_{\Gamma_-}^2$$
$$\leq \|\sqrt{\delta} S\|^2 + \frac{1}{c_0} \|S\|^2 + 2\|f\|_{\Gamma_-}^2.$$

Proof. Select $v = u_{SD}$ in (7.2) and note that

$$(\mathcal{L} u_{SD}, u_{SD}) + (u_{SD}, u_{SD})_{\Gamma_-}$$
$$= \left(\left(b - \frac{1}{2} \nabla \cdot a \right) u_{SD}, u_{SD} \right) + \frac{1}{2} \|u_{SD}\|_{\Gamma_+}^2 + \frac{1}{2} \|u_{SD}\|_{\Gamma_-}^2. \tag{7.4}$$

Applying (7.3) here and using the Cauchy Schwarz inequality on the right-hand side in (7.2) with $v = u_{SD}$, the result follows. □

We observe that the bound in Lemma 7.1 controls the L^2-norm of the discrete solution as well as some derivatives of it in the advection direction, *provided $\delta > 0$. We see that $\delta = 0$ gives only L^2-stability.*

Now we turn to the error analysis of (7.2). We begin by decomposing

$$u - u_{SD} = (u - \Pi u) + (\Pi u - u_{SD})$$
$$\equiv \eta + \xi, \tag{7.5}$$

where Πu is a suitable projection of u into $S^{k,1}$; the choice of the projector Π will be deferred until later. The key is a bound on ξ in terms of η; the final error bound on $u - u_{SD}$ will then follow from bounds on the projection error η in Section 7.4 below.

Lemma 7.2. *Assuming that (7.3) holds, and that $u \in H^1(\Omega)$, we have*

$$\|\sqrt{\delta} \mathcal{L} \xi\|^2 + \|c\xi\|^2 + \frac{1}{2} \|\xi\|_{\Gamma_+}^2 + \|\xi\|_{\Gamma_-}^2$$
$$\leq \|\sqrt{\delta} \mathcal{L}\eta - \frac{1}{\sqrt{\delta}} \eta\|^2 + 4\|c\eta\|^2 + 2\|\eta\|_{\Gamma_+}^2, \tag{7.6}$$

where $c \in C(\overline{\Omega})$ is defined by

$$c^2(x) = b(x) - \frac{1}{2} \nabla \cdot \mathbf{a}(x), \quad x \in \overline{\Omega}. \tag{7.7}$$

For the proof, we refer to [36].

7.3   Discontinuous-Galerkin *hp*-FEM

Given that K is an element in the partition $\mathcal{T}$, we denote by ∂K the union of open faces of K. This is non-standard notation in that ∂K is a subset of the boundary of K. Let $x \in \partial K$ and suppose that $\mathbf{n}(x)$ denotes the unit outward normal vector to ∂K at x. With these conventions, we define the inflow and outflow parts of ∂K, respectively by

$$\partial_- K = \{x \in \partial K \,:\, \mathbf{a}(x) \cdot \mathbf{n}(x) < 0\}, \ \ \partial_+ K = \{x \in \partial K \,:\, \mathbf{a}(x) \cdot \mathbf{n}(x) \geq 0\} \,.$$

For each $K \in \mathcal{T}$ and any $v \in H^1(K)$ we denote by v_+ the interior trace of v on ∂K (taken from within K). Now consider an element K such that the set $\partial_- K \backslash \Gamma_-$ is nonempty; then for each $x \in \partial_- K \backslash \Gamma_-$ (with the exception of a set of $(d-1)$ dimensional measure zero) there exists a unique element K', depending on the choice of x, such that $x \in \partial_+ K'$. This is illustrated in Figure 7.1.

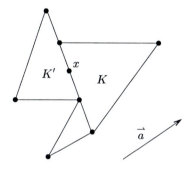

Fig. 7.1. A point x such that $x \in \partial_- K$ and $x \in \partial_+ K'$

Now suppose that $v \in H^1(K)$ for each $K \in \mathcal{T}$. If $\partial_- K \cap \Gamma_-$ is nonempty for some $K \in \mathcal{T}$, then we can also define the outer trace v_- of v on $\partial_- K \backslash \Gamma_-$ relative to K as the inner trace v_+ relative to those elements K' for which $\partial_+ K'$ has intersection with $\partial_- K \backslash \Gamma_-$ of positive $(d-1)$-dimensional measure. We also introduce the jump of v across $\partial_- K \backslash \Gamma_-$:

$$[\,v\,] = v_+ - v_- \,.$$

Let $\delta \in H^1(K)$ for each $K \in \mathcal{T}$, and suppose that δ is positive on each $K \in \mathcal{T}$. Typically, δ is chosen to be constant on each $K \in \mathcal{T}$, although we shall require this for now.

Suppose that $v, w \in H^1(K)$ for each $K \in \mathcal{T}$. We define

$$
\begin{aligned}
B_{\mathrm{DG}}(w, v) = \sum_K \int_K \mathcal{L}w \cdot (v + \delta\mathcal{L}v)\, dx \\
- \sum_K \int_{\partial_- K \backslash \Gamma_-} (\mathbf{a} \cdot \mathbf{n})[w]\, v_+ ds - \sum_K \int_{\partial_- K \cap \Gamma_-} (\mathbf{a} \cdot \mathbf{n}) u_+ v_+\, ds
\end{aligned}
\tag{7.8}
$$

and put

$$
\ell_{\mathrm{DG}}(v) = \sum_K \int_K f \cdot (v + \delta\mathcal{L}v)dx - \sum_K \int_{\partial_- K \cap \Gamma_-} (\mathbf{a} \cdot \mathbf{n})\, g v_+\, ds .
\tag{7.9}
$$

Note that the term $\delta\mathcal{L}v$ in (7.8), (7.9) is a stabilization parameter, for $\delta = 0$, we get the usual discontinuous Galerkin method. The hp-DGFEM approximation of (7.1) is defined as follows: find $u_{\mathrm{DG}} \in S^{k,0}$ such that

$$
B_{\mathrm{DG}}(u_{\mathrm{DG}}, v) = \ell_{\mathrm{DG}}(v) \quad \forall v \in S^{k,0} .
\tag{7.10}
$$

Next we study the stability of the discrete problem (7.10).

Lemma 7.3. *Suppose that there exists a positive constant c_0 such that (7.3) holds. Then u_{DG} obeys the bound*

$$
\begin{aligned}
\sum_K \|\sqrt{\delta}\mathcal{L}u_{\mathrm{DG}}\|_K^2 + c_0\|u_{\mathrm{DG}}\|_K^2 + \sum_K \|u_{\mathrm{DG}}^+ - u_{\mathrm{DG}}^-\|_{\partial_- K \backslash \Gamma_-}^2 \\
+ \sum_K \|u_{\mathrm{DG}}^+\|_{\partial_+ K \cap \Gamma_+}^2 + \frac{1}{2} \sum_K \|u_{\mathrm{DG}}\|_{\partial_- K \cap \Gamma_-}^2 \\
\le \sum_K \|\sqrt{\delta}f\|_K^2 + \frac{1}{c_0} \sum_K \|f\|_K^2 + 2 \sum_K \|g\|_{\partial_- K \cap \Gamma_-}^2 .
\end{aligned}
\tag{7.11}
$$

The proof can be found in [36].

Remark 7.4. This bound is analogous to the estimate (7.3) for the hp-SDFEM.

We now discuss the error analysis of hp-DGFEM. We write

$$
\begin{aligned}
u - u_{\mathrm{DG}} = (u - \Pi u) + (\Pi u - u_{\mathrm{DG}}) \\
\equiv \eta + \xi
\end{aligned}
\tag{7.12}
$$

where Πu is a suitable projection of u into $S^{k,0}$, to be chosen below. There is an analog of Lemma 7.2:

Lemma 7.5. *Assuming that (7.3) holds and $u \in H^1(K)$ for each $K \in \mathcal{T}$.*
We have that

$$\sum_K \|\sqrt{\delta}\mathcal{L}\xi\|_K^2 + \sum_K \|c\,\xi\|_K^2 + \sum_K \|\xi^+\|_{\partial_- K \cap \Gamma_-}^2$$

$$+ \frac{1}{2}\sum_K \|\xi^+\|_{\partial_+ K \cap \Gamma_+}^2 + \frac{1}{2}\sum_K \|\xi^+ - \xi^-\|_{\partial_- K \setminus \Gamma_-}^2$$

$$\leq \sum_K \|\sqrt{\delta}\mathcal{L}\eta - \frac{1}{\sqrt{\delta}}\,\eta\|_K^2 + 4\sum_K \|c\,\eta\|_K^2 \tag{7.13}$$

$$+ 2\sum_K \|\eta^+\|_{\partial_+ K \cap \Gamma_+}^2 + \sum_K \|\eta^-\|_{\partial_- K \setminus \Gamma_-}^2 \;.$$

The proof follows by elementary manipulation and we refer to [36] for details.

7.4 *hp*-Error Analysis of the DG- and the SDFEM

In this section, we shall construct the *hp*-approximation projector Π in the error estimates (7.5) and derive *hp*-error bounds for the *hp*-SDFEM as well as for the *hp*-DGFEM introduced in the previous section. The bounds are explicit in h and p and in the regularities of the solution and allow to deduce in particular exponential convergence estimates for piecewise analytic solutions.

We are now in position to present error estimates for both, the SD- and the DGFEM. We shall use the following mesh dependent norm defined by

$$\||u\||_{\mathrm{DG}}^2 := \sum_{K \in \mathcal{T}} \Big\{ \|\sqrt{\delta}\mathcal{L}u\|_K^2 + \|cu\|_K^2 + \|u^+\|_{\partial_- K \cap \Gamma_-}^2$$

$$+ \frac{1}{2}\|u^+\|_{\partial_+ K \cap \Gamma_+}^2 + \frac{1}{2}\|u^+ - u^-\|_{\partial_- K \setminus \Gamma_-}^2 \Big\} . \tag{7.14}$$

Notice that for the SDFEM, the last term vanishes. Here is our main error estimate for the *hp*-DGFEM.

Theorem 7.6. *(Convergence rate of the hp-DGFEM)*
Let $\Omega \subset \mathbb{R}^2$ and $\mathcal{T}, \mathcal{P}$ be as in Section 2 with (possibly irregular) patch meshes $\mathcal{T}_P$, $P \in \mathcal{P}$, consisting of shape-regular quadrilateral elements of degree $p_K \geq 1$. Select the stabilization parameters δ_K according to

$$\delta_K = h_K/k_K \quad \text{for all } K \in \mathcal{T} . \tag{7.15}$$

Then

$$\||u - u_{\mathrm{DG}}\||_{\mathrm{DG}}^2 \leq C \sum_K \left(\frac{h_K}{2}\right)^{2s_K+1} \frac{\Phi(k_K, s_K)}{k_K} \,|\hat{u}|_{s_K+1,\widehat{K}}^2 \tag{7.16}$$

where $C > 0$ depends only on elemental shape regularity, and on the coefficients a, b, but is independent of k_K, s_K, h_K and where $\Phi(p, s)$ is as in (5.10).

Proof. (7.12) and Lemma 7.5 imply

$$
\begin{aligned}
\|\|u - u_{\mathrm{DG}}\|\|_{\mathrm{DG}} \ &\leq\ \|\|\eta\|\|_{\mathrm{DG}} + \|\|\xi\|\|_{\mathrm{DG}} \\
&\overset{(7.13)}{\leq}\ \|\|\eta\|\|_{\mathrm{DG}} + \Big(\sum_K \|\delta^{\frac{1}{2}} \mathcal{L}\eta - \delta^{-\frac{1}{2}}\eta\|_K^2 \Big)^{\frac{1}{2}} + 2 \Big(\sum_K \|c\eta\|_K^2 \Big)^{\frac{1}{2}} \\
&\quad + \sqrt{2} \Big(\sum_K \|\eta_+\|_{\partial_+ K \cap \Gamma_+}^2 \Big)^{\frac{1}{2}} + \Big(\sum_K \|\eta_-\|_{\partial_- K \backslash \Gamma_-}^2 \Big)^{\frac{1}{2}} \\
&\leq\ \Big(\sum_K \|\delta^{\frac{1}{2}} \mathcal{L}\eta\|_K^2 \Big)^{\frac{1}{2}} + \Big(\sum_K \|c\eta\|_K^2 \Big)^{\frac{1}{2}} \\
&\quad + \Big(\sum_K \|\eta_+\|_{\partial_- K \cap \Gamma_-}^2 \Big)^{\frac{1}{2}} + \frac{1}{\sqrt{2}} \Big(\sum_K \|\eta^+\|_{\partial_+ K \cap \Gamma_+}^2 \Big)^{\frac{1}{2}} \\
&\quad + \frac{1}{\sqrt{2}} \Big(\sum_K \|\eta^+ - \eta^-\|_{\partial_- K \backslash \Gamma_-}^2 \Big)^{\frac{1}{2}} \\
&\quad + \Big(\sum_K \|\delta^{\frac{1}{2}} \mathcal{L}\eta\|_K + \|\delta^{-\frac{1}{2}}\eta\|_K \Big)^{\frac{1}{2}} + 2 \Big(\sum_K \|c\eta\|_K^2 \Big)^{\frac{1}{2}} \\
&\quad + \sqrt{2} \Big(\sum_K \|\eta_+\|_{\partial_+ K \cap \Gamma_+}^2 \Big)^{\frac{1}{2}} + \Big(\sum_K \|\eta_-\|_{\partial_- K \backslash \Gamma_-}^2 \Big)^{\frac{1}{2}} \\
&\leq\ C \Big\{ \sum_K \delta_K^{\frac{1}{2}} \|\nabla\eta\|_K^2 + \delta_K^{\frac{1}{2}} \|\eta\|_K^2 + \|\eta\|_K^2 + \delta_K^{-\frac{1}{2}} \|\eta\|_K^2 \\
&\quad + \sum_K \|\eta^+\|_{\partial_+ K \cap \Gamma_+}^2 + \|\eta^+\|_{\partial_- K \cap \Gamma_-}^2 \\
&\quad + \|\eta^-\|_{\partial_- K \backslash \Gamma_-}^2 + \|\eta^+\|_{\partial_- K \backslash \Gamma_-}^2 \Big\}^{\frac{1}{2}} \\
&=\ C(A + B)^{\frac{1}{2}} .
\end{aligned}
$$

where C depends on (a, b).

We select $\eta = u - \Pi u$ with Π as in Theorem 5.4. This gives the bound

$$
A \leq \sum_K \Big(\frac{h_K}{2} \Big)^{2s_K} \Phi(k_K, s_K)(\delta_K + \delta_K^{-1} h_K^2 k_K^{-2}) |\hat{u}|_{s_K+1, \widehat{K}}^2 .
$$

To bound B, we must estimate $\|\eta\|_{0, \partial K}^2$. We use the inequality

$$
\|\eta\|_{L^2(\partial K)}^2 \leq C \left(\|\nabla\eta\|_K \|\eta\|_K + h_K^{-1} \|\eta\|_K^2 \right) \quad \forall K \in \mathcal{T}
$$

and obtain the bound

$$
\begin{aligned}
B &\leq C \sum_K \left(\frac{h_K}{2}\right)^{s_K} \Phi(p_K, s_K)^{\frac{1}{2}} \cdot \left(\frac{h_K}{2}\right)^{s_K+1} \Phi(p_K, s_K)^{\frac{1}{2}} \, p_K^{-1} |\hat{u}|^2_{s_K+1, \widehat{K}} \\
&\quad + h_K^{-1} \left(\frac{h_K}{2}\right)^{2s_K+2} \Phi(p_K, s_K) \, p_K^{-2} |\hat{u}|^2_{s_K+1, \widehat{K}} \\
&= C \sum_K \left(\frac{h_K}{2}\right)^{2s_K+1} p_K^{-1} \cdot \Phi(p_K, s_K)(1 + p_K^{-1}) |\hat{u}|^2_{s_K+1, \widehat{K}} \, .
\end{aligned}
$$

Selecting δ_K as in (7.15) concludes the proof. $\square$

An analogous error estimate holds true for the *hp*-SDFEM.

Theorem 7.7. *(Convergence rate of the hp-SDFEM)*
Let $\Omega \subset \mathbb{R}^2$ and $\mathcal{T}, \mathcal{P}$ be as in Section 2 with a 1-irregular mesh consisting of shape-regular quadrilateral elements of degree $p_K \geq 1$. Select the stabilization parameter δ_K as in (7.15).
Then there holds the error estimate

$$
\||u - u_{\mathrm{SD}}\||^2_{\mathrm{SD}} \leq C \sum_K \left(\frac{h_K}{2}\right)^{2s_K+1} \frac{\Phi(k_K, s_K)}{k_K} |\hat{u}|^2_{s_K+1, K} \tag{7.17}
$$

where

$$
\||u\||^2_{\mathrm{SD}} := \|\sqrt{\delta}\mathcal{L}u\|^2_0 + \|cu\|^2_0 + \frac{1}{2} \|u\|^2_{0,\Gamma_+} + \|u\|^2_{\Gamma_-}
$$

and

$$
0 \leq s_K \leq k_K \quad \forall K \in \mathcal{T}, \ \hat{u} = u \circ F_P \ \text{if} \ K \in \mathcal{T}_P
$$

and $\Phi(k, s)$ *is as in (5.10).*

Let us discuss special cases of the above, general error bounds.

Remark 7.8.

1) If $k_K = k$ is fixed, and $h_K = h \to 0$, (7.16) is optimal in h.
2) As s is fixed, $k_K = k \to \infty$, Stirling's formula implies

$$
\Phi(k, s) \leq C(s) \, k^{-2s}
$$

and

$$
\||u - u_{\mathrm{DG}}\||^2_{\mathrm{DG}} \leq C \sum_K \left(\frac{h_K}{k_K}\right)^{2s_K+1} \|\hat{u}\|_{s_K+1, \widehat{K}} \, .
$$

The bound (7.16) is optimal also in k, improving upon [12], [13].
3) If u is patchwise analytic, we have the bounds

$$
\forall K \in \mathcal{T} \ \exists d_K > 1, \ C > 0 \quad \forall s > 0: \ |\hat{u}|_{s, \widehat{K}} \leq C(d_K)^s s! \tag{7.18}
$$

In this case, we get the exponential convergence estimate [36]

$$|||u - u_{\mathrm{DG}}|||_{\mathrm{DG}}^2 \leq C \sum_K \left(\frac{h_K}{2}\right)^{2s_K+1} (k_K)^2 \exp(-2b_K k_K) \,.$$

By Theorem 7.7 an analogous bound holds also for the hp-SDFEM on quadrilateral, possibly 1-irregular meshes.

8   Convection-Diffusion

Based on the discretizations of the diffusion operator in Section 6 and of the advection problem in Section 7, it is now easy to derive discretizations of the convection-diffusion problem

$$\mathcal{L}_\varepsilon u = -\varepsilon \Delta u + \mathbf{a}(x) \cdot \nabla u + b(x)u = S \text{ in } \Omega \,, \tag{8.1}$$

$$u = 0 \text{ on } \partial\Omega \,. \tag{8.2}$$

Here the viscosity $\varepsilon \in (0,1]$, $f \in L^2(\Omega)$ and $\mathbf{a}(x)$, $b(x)$ are assumed to belong to $C^1(\overline{\Omega})$ and to satisfy (7.3).

8.1   Standard Galerkin discretization

The standard Galerkin discretization of (8.1) reads:
find $u \in H_0^1(\Omega)$ such that

$$\begin{aligned} B_\varepsilon(u,v) &:= \varepsilon(\nabla u, \nabla v)_\Omega \\ &+(\mathbf{a}\cdot\nabla u + bu, v)_\Omega = (S,v)_\Omega \quad \forall v \in H_0^1(\Omega) \,. \end{aligned} \tag{8.3}$$

The Galerkin finite element discretization of (8.3) reads:
find $u_{\mathrm{FE}} \in S_0^{\mathbf{k},1}(\Omega, \mathcal{T})$ such that

$$B_\varepsilon(u_{\mathrm{FE}}, v) = (S, v)_\Omega \quad \forall v \in S_0^{\mathbf{k},1}(\Omega, \mathcal{T}) \,. \tag{8.4}$$

Condition (7.3) guarantees the solvability of (8.3), (8.4), since, for $u \in S_0^{\mathbf{k},1}(\Omega, \mathcal{T})$ it holds

$$\begin{aligned} B_\varepsilon(u,u) &= \varepsilon \|\nabla u\|^2 + \int_\Omega (\mathbf{a}\cdot\nabla u + bu)\, u \, dx \\ &= \varepsilon \|\nabla u\|^2 + \int_\Omega \left(b - \frac{1}{2}\nabla\cdot\mathbf{a}\right) |u|^2 \, dx \\ &\geq \varepsilon \|\nabla u\|^2 + c_0 \|u\|^2 \\ &\geq \min(1, c_0) \|u\|_\varepsilon^2 \end{aligned} \tag{8.5}$$

where we used the formula

$$\int_\Omega u\mathbf{a}\cdot\nabla u\,dx = -\int_\Omega u^2\nabla\cdot\mathbf{a} - \int_\Omega u\mathbf{a}\cdot\nabla u + \int_{\partial\Omega} u^2\mathbf{a}\cdot\mathbf{n}\,ds\ .$$

We see that (8.4) is stable in the $\|\circ\|_\varepsilon$-norm, whence it follows that

$$\|u - u_{\mathrm{FE}}\|_\varepsilon \le C\ \|u - v\|_\varepsilon \quad \forall v \in S_0^{\mathbf{k},1}(\Omega,\mathcal{T})\ . \tag{8.6}$$

The Galerkin FEM (8.4) without stabilization converges therefore optimally, *provided the FE-space $S_0^{\mathbf{k},1}(\Omega,\mathcal{T})$ resolves the fine scales of the solution* (such as boundary layers, eddies, fronts etc.). If this is not the case, the Galerkin FEM (8.4) is prone to *pollution*, i.e. a local underresolution of fine solution scales triggers oscillations which spread throughout the domain Ω.

To prevent this, *stabilized schemes* must be used (in fact, the main impetus for the development of stabilized methods has come from the inability of the FEM to resolve all small scales of the flow). We present here two stabilization techniques, the *hp*-SDFEM and the *hp*-DGFEM.

8.2   Streamline-Diffusion FEM

Formulation and main properties. The *hp*-SDFEM discretization of (8.3) reads: find $u_{\mathrm{FE}} \in S_0^{\mathbf{k},1}(\Omega,\mathcal{T})$ such that

$$B_{\mathrm{SD}}(u_{\mathrm{FE}},v) = F_{\mathrm{SD}}(v) \quad \forall v \in S_0^{\mathbf{k},1}(\Omega,\mathcal{T})\ . \tag{8.7}$$

Here the bilinear form and the right hand side include the so-called stabilization terms: for $u \in S_0^{\mathbf{k},1}(\Omega,\mathcal{T})$, we have with $\mathcal{L}_\varepsilon$ as in (8.1)

$$B_{\mathrm{SD}}(u,v) := B_\varepsilon(u,v) + \sum_{K\in\mathcal{T}}\delta_K\int_K (\mathcal{L}_\varepsilon u)(\mathcal{L}_0 v)\,dx\ , \tag{8.8}$$

$$F_{\mathrm{SD}}(v) := (S,v)_\Omega + \sum_{K\in\mathcal{T}}\delta_K\int_K S(\mathcal{L}_0 v)\,dx\ . \tag{8.9}$$

Remark 8.1. At first sight, it would appear that the stabilization terms in (8.8), (8.9) require, for positive ε, $H^2(K)$-regularity of u. This is not so - all that is required for B_{SD}, F_{SD} to make sense is that $\mathcal{L}_\varepsilon u \in L^2(K)$, and this is satisfied for the exact solution if S in (8.1) belongs to $L^2(K)$ for all $K \in \mathcal{T}$.

Remark 8.2. The SDFEM formulation is fully consistent, i.e. for any value of ε and δ, the exact solution of (8.1) satisfies (8.7). Adding the stabilization terms on the right hand sides of (8.8), (8.9) therefore does not alter the problem to be discretized.

Remark 8.3. As in the pure convection case, the SDFEM contains free parameters δ_K at our disposal; for $\delta_K = 0$, (8.7) reduces to (8.4), $\delta_K > 0$ will imply stabilizations. δ_K needs to be selected in dependence on k_K as well as on the element shape - this will be explained below. Proper choice of δ_K is crucial for good performance.

Remark 8.4. We see that for $\varepsilon = 0$ the SDFEM (8.7) becomes (7.2). All properties are shown below for the SDFEM.

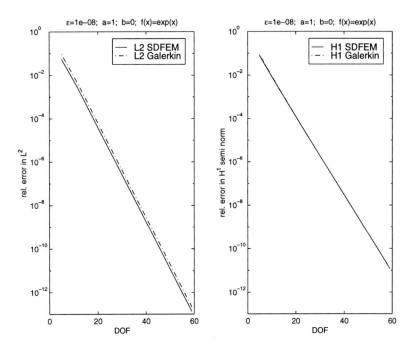

Fig. 8.1. L^2 and energy performance of "two-element mesh" for Galerkin FEM and SDFEM, $\varepsilon = 10^{-8}$;

Stability. As we pointed out in Section 5.3, the solution of (8.1) exhibits for small $\varepsilon > 0$, boundary layers and hp-FEM will not give exponential convergence uniform in ε if unstructured, shape regular meshes are used (no layers are present for $\varepsilon = 0$, i.e. for the pure convection problem). We therefore address now the choice of the parameters δ_K in (8.8), (8.9) and the stability of the method. We assume that the mesh $\mathcal{T}$ in (8.7) is given in terms of patches $P \in \mathcal{P}$, regular patch maps $F_P : \widehat{K} \to P$ and allow patch meshes $\mathcal{T}_P$ with anisotropic quadrilateral elements, of the type introduced in Section 5.3. Then there holds:

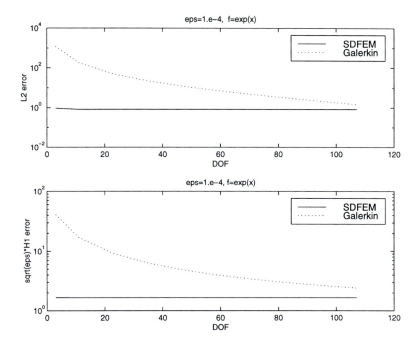

Fig. 8.2. L^2 and energy performance of p version on uniform mesh with $h = 0.5$ for Galerkin FEM and SDFEM, $\varepsilon = 10^{-4}$

Theorem 8.5. *Let the mesh $\mathcal{T}$ consist of shape regular triangles of diameter h_K or of possibility anisotropic quadrilaterals with sidelengths $h_{K,\max}$ and $h_{K,\min}$, respectively (no bound on the aspect ratio $h_{K,\max}/h_{K,\min}$ is assumed). Then there exists $\delta_0 > 0$ independent of h_K, k and of the aspect ratio, such that for all $0 < \delta < \delta_0$ the choice*

$$\delta_K = \frac{\delta}{k_K^2} \cdot \frac{h_{K,\max}\, h_{K,\min}}{\sqrt{h_{K,\max}^2 + h_{K,\min}^2}} \tag{8.10}$$

will render the hp-SDFEM (8.7) stable independent of the aspect ratio, i.e. it holds

$$\frac{1}{2}\,\|u\|_{\mathrm{SD}}^2 \leq B_{\mathrm{SD}}(u, u) \quad \forall u \in S_0^{\mathbf{k},1}(\Omega, \mathcal{T}) \tag{8.11}$$

where the norm $\|\circ\|_{\mathrm{SD}}$ is defined by

$$\|u\|_{SD}^2 := \varepsilon\,\|\nabla u\|_{0,\Omega}^2 + \|u\|_{0,\Omega}^2 + \sum_{K \in \mathcal{T}} \delta_K \|\mathcal{L}_0 u\|_{0,K}^2\;.$$

For the proof, we refer to [28].

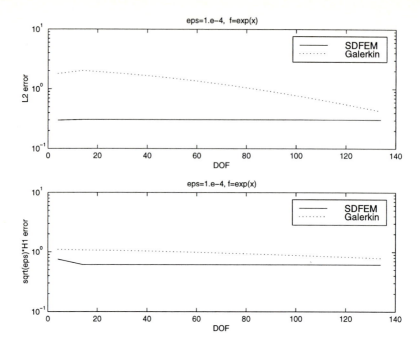

Fig. 8.3. L^2 and energy performance of p version on uniform mesh with $h = 0.5 +$ small element size ε, $\varepsilon = 10^{-4}$

Note that for shape-regular elements $h_{K,\mathrm{max}} = h_{K,\mathrm{min}} = h_K$ and (8.10) becomes simply

$$\delta_K = \delta\, h_K / k_K^2 , \qquad (8.12)$$

which should be compared with the choice (7.15): we see that the appearance of the viscous terms changes the weight δ_K from h_K / k_K to h_K / k_K^2, at least as far as the stability analysis is concerned.

Remark 8.6. The previous Theorem applies in particular also to the (κ, σ)-geometric boundary layer meshes shown in Figures 8 and 9.

We shall see in Section 10 below how stabilized formulations like (8.8), (8.9) can also be used in the computation of incompressible fluid flow.

Computational Experiments. In this section, we illustrate the performance of the hp-SDFEM (8.7) with numerical examples for 1-d convection-dominated problems. All findings which we report below are mathematically explained in detail in [47]. Our aims in these numerical experiments are

1. to illustrate the theoretical results obtained above, in particular the ability of the hp-FEM to resolve very narrow fronts and layers, leading to the asymptotic exponential convergence with few degrees of freedom;

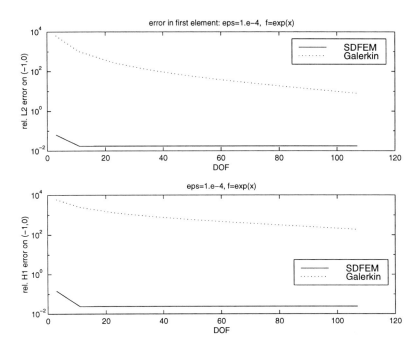

Fig. 8.4. L^2 and H^1 performance on the first element of p version on uniform mesh with $h = 0.5$, $\varepsilon = 10^{-4}$

2. to compare *hp*-SDFEM and *hp*-Galerkin FEM in the preasymptotic phase, i.e., if the small scales of the solution are not resoled. In particular, we will see that the appropriate choice of mesh sequences lead to robust exponential convergence on compact subsets for the *hp*-SDFEM.

We consider two types of problems, a standard advection-dominated problem and a turning point problem which satisfies the crucial assumption (7.3).

The boundary layer case. Let us first consider the problem

$$-\varepsilon u'' + au' = e^{\omega x}, \qquad u(\pm 1) = 0, \qquad \omega = 1, \qquad a = 1 \qquad (8.13)$$

The exact solution has a boundary layer at the outflow boundary $x = 1$ and is given by

$$u_\varepsilon = \frac{e^{\omega x}}{\omega(a - \omega\varepsilon)} + \frac{c}{a} + \tilde{c}e^{-a(1-x)/\varepsilon} \qquad (8.14)$$

$$c = \frac{ae^{-\omega}}{\omega(\omega\varepsilon - a)} \frac{1 - e^{2\omega}e^{-2a/\varepsilon}}{1 - e^{-2a/\varepsilon}} = O(1),$$

$$\tilde{c} = \frac{e^{-\omega}(e^{2\omega} - 1)}{\omega(\omega\varepsilon - a)(1 - e^{-2a/\varepsilon})} = O(1). \qquad (8.15)$$

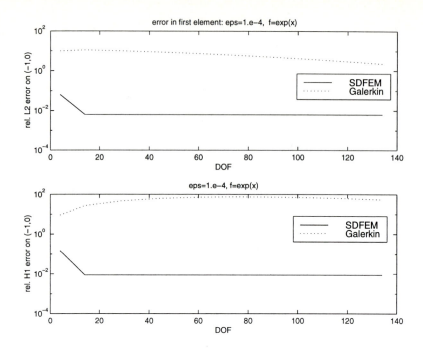

Fig. 8.5. L^2 and H^1 performance on the first element of p version on uniform mesh with $h = 0.5 +$ small element size ε, $\varepsilon = 10^{-4}$

Note that both $\|u_\varepsilon\|_{L^2(\Omega)}$ and $\||u_\varepsilon\||$ are $O(1)$ independently of ε.

Global SDFEM performance. We present numerical results for the SD-FEM. In order to illustrate the robustness of the SDFEM with respect to the weights $(\delta_i)_{i=1}^N$ noted in Section 8.2 we choose the weights $(\delta_i)_{i=1}^N$ of the SDFEM as

$$\delta_i = \begin{cases} \frac{1}{2} h_i & \text{if } \varepsilon k^2/h_i \leq \frac{1}{4} \\ 0 & \text{otherwise .} \end{cases}$$

We point out that numerical results are practically identical if the choice is made.

In our first series of numerical experiments, we resolve the boundary layer with the two-element mesh of (5.21) with $\kappa = 1$. Fig. 8.1 compares the behavior of the Galerkin and the SDFEM in the L^2 norm and the energy norm $\||\cdot\||$ (which is $\sqrt{\varepsilon} \cdot |\cdot|_{H^1(\Omega)}$) for $\varepsilon = 10^{-8}$ where the order k ranges from 1 to 27. The theory of [47] yields robust exponential convergence in the energy norm for the SDFEM as well as the Galerkin FEM on this two element mesh. This exponential convergence is visible in the bottom figure of Fig. 8.1.

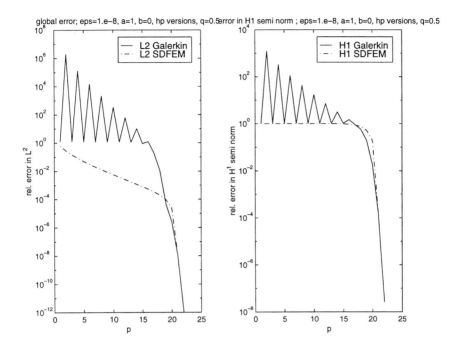

Fig. 8.6. L^2 and energy performance for "hp"-mesh; SDFEM; $\sigma = 0.5$, $\varepsilon = 10^{-8}$

Furthermore, for the SDFEM, we have robust exponential convergence in L^∞ and thus in L^2 (cf. the top figure of Fig. 8.1); we also observe robust exponential convergence in L^2 for the standard Galerkin FEM, Fig. 8.1. We note that the qualitative behavior of the schemes is comparable although the error of the *hp*-SDFEM is slightly smaller than that of the Galerkin FEM for this problem.

We conclude that the two-element mesh scheme is able to resolve the boundary layer at the outflow boundary and that no stabilization is required in this case.

Our next experiment is geared towards getting insight in the behavior of the Galerkin method and the SDFEM if the boundary layer has not been resolved. To that end, we consider the performance of the *p* version on a *uniform* mesh with $h = 0.5$ (i.e., 4 elements). Here, the order k ranges from 1 to 27 and $\varepsilon = 10^{-4}$. Fig. 8.2 shows the behavior in the L^2 and the energy norm $\|\|\cdot\|\|$. The error in the *hp*-SDFEM is considerably smaller than that of the Galerkin method, but the rate of convergence SDFEM is very poor also—in the energy norm, no convergence can be observed!

Finally, Fig. 8.3 shows the performance of a uniform mesh ($h = 0.5$) augmented by one small element of size ε in the outflow boundary layer (i.e., the mesh given by the nodes $\{-1, -0.5, 0, 0.5, 1-\varepsilon, 1\}$). As to be expected, insert-

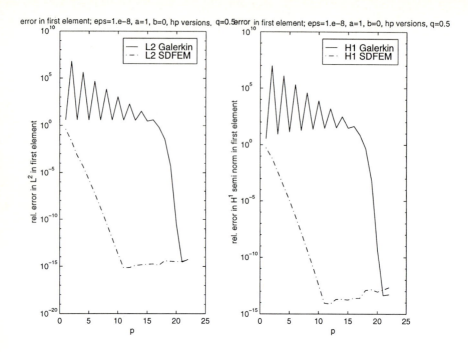

Fig. 8.7. L^2 and H^1 performance on first element $(-1,0)$ for "hp"-mesh; SDFEM; $\sigma = 0.5$, $\varepsilon = 10^{-8}$

ing one small element of size ε greatly alleviates the problems of the standard Galerkin method (cf. [47] for a detailed analysis). Comparing Fig. 8.2 with Fig. 8.3, the error of the Galerkin FEM is reduced by two orders of magnitude. Nevertheless, both the Galerkin method and the SDFEM yield poor rates of convergence as the p version on a mesh with one small element of size ε near the boundary cannot resolve the boundary layer properly. Hence, comparing the results with those in Fig. 8.1, we see that the proper element length εk at the boundary is essential for the boundary layer (compare Theorem 5.15) resolutions as well as for robust exponential convergence.

Local p-SDFEM performance — pollution. We have just seen that the pure p version Galerkin FEM and SDFEM have poor convergence properties if the error is measured in a global norm such as the L^2 or the $|||\cdot|||$ norm. The performance was not substantially improved by inserting one small element of size ε in the layer. The local behavior of the pure p-version SDFEM is investigated in Figs. 8.4, 8.5 by plotting the relative L^2 and H^1 errors in the first element $I_1 = (-1,0)$ for a uniform mesh with $h = 0.5$ and a uniform mesh with $h = 0.5$ that is augmented by one small element of size ε in the layer. Although the SDFEM, which suppresses spurious oscillations, is much

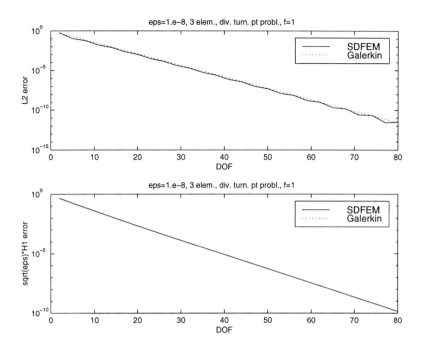

Fig. 8.8. L^2 and energy error for turning pt. problem, $a = 1$, $\varepsilon = 10^{-8}$, 3 elem.

more accurate (1% in both L^2 and H^1 on $(-1, -0.5)$) than the Galerkin FEM, we see that increasing the order k does not reduce the error. We conclude that the pure *p*-version of both the Galerkin FEM and the SDFEM are prone to *pollution*, i.e., the error introduced by not resolving the boundary layer affects strongly the accuracy achievable in the whole computational domain.

Local SDFEM performance on special mesh sequences. Our next numerical example shows that the *hp*-SDFEM leads to robust exponential convergence on compact subsets not containing the layers if an increase of the polynomial degree is combined with a mesh refinement towards the layer. We therefore consider the following scheme: For a grading factor $\sigma \in (0, 1)$ let

$$k_0 \in \mathbb{N} \quad \text{be the smallest integer s.t.} \quad \sigma^{k_0} < k_0 \varepsilon$$

and let for each polynomial degree k a geometrically refined mesh with p layers be given by the points

$$\{-1, 1, 1 - \sigma^i \,|\, i = 0, \dots, \min(k, k_0)\}. \tag{8.16}$$

On such meshes, we will consider as trial spaces the space $S_0^{k,1}(\mathcal{T})$ (cf. Fig. 8.10). We note that such mesh sequences would typically be generated

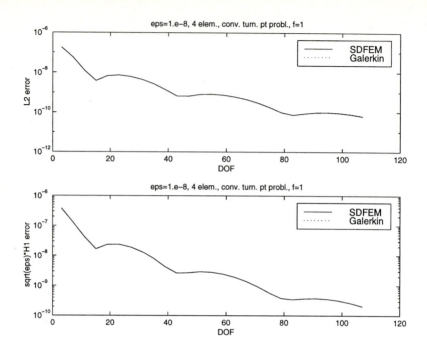

Fig. 8.9. L^2 and energy error for turning pt. problem, $a = -1$, $\varepsilon = 10^{-8}$, 4 elem.

by adaptive schemes that locate and try to resolve the layers. It can be shown using ideas of [40,75] (cf. [47] for the details) that the hp-SDFEM converges robustly and exponentially on compact subsets of Ω for such mesh sequences:

Theorem 8.7. Let $a = 1$, $b = 0$, $\sigma \in (0,1)$, $\xi \in (-1,1)$ be fixed. For $k \in \mathbb{N}$ consider the meshes $\mathcal{T}$ defined by the nodes (8.16). Assume that the weights $(\delta_i)_{i=1}^{N}$ are of the form (8.12). Then there are constants C, $b > 0$ independent of ε, k such that

$$\|u_\varepsilon - u_{SD}\|_{H^1(-1,\xi)} \le Ce^{-bk}, \quad k = 1, 2, \ldots$$

The refinement factor σ is chosen in the following experiments as $\sigma = 1/2$ and the weights $(\delta_i)_{i=1}^{N}$ are given in both cases by

$$\delta_i = \begin{cases} \dfrac{1}{4}\dfrac{h_i}{k} & \text{if } \varepsilon < \dfrac{1}{4}\dfrac{h_i}{k} \\ 0 & \text{otherwise .} \end{cases} \tag{8.17}$$

Again, we point out that choosing the weights $(\delta)_{i=1}^{N}$ as in (8.10), (8.12)) leads to similar numerical results. For $\varepsilon = 10^{-8}$ and k going from 1 to 22. Figs. 8.6- 8.7 show the performance of the SDFEM in comparison with the Galerkin FEM. Fig. 8.6 depicts their behavior in global norms (L^2 and energy norm) whereas Fig. 8.7 shows the relative error (measured in the L^2 and H^1 norm) in the first element $I_1 = (-1, 0)$. Fig. 8.6 illustrates once more that both Galerkin FEM and *hp*-SDFEM do not lead to convergence in the energy norm until the layer is resolved, that is, $\sigma^k \approx \varepsilon k$ (for $\sigma = 0.5$ and $\varepsilon = 10^{-8}$ this happens for $k \approx 22$). The behavior of the Galerkin FEM is, however, completely different from that of the *hp*-SDFEM if the error on the first element $I_1 = (-1, 0)$ is of interest (cf. Fig. 8.7). The Galerkin FEM is highly prone to *pollution*: The local error in I_1 cannot be controlled until k is so large that the smallest element in the layer has width $\sigma^k \approx k\varepsilon$. In contrast to this, the SDFEM is *pollution-free* as robust exponential convergence on the compact subset $(-1, 0)$ can be achieved according to Theorem 8.7 and in fact is visible in Fig. 8.7.

Turning point problems. Let us now consider a problem with a turning point at $x = 0$. We consider

$$-\varepsilon u_\varepsilon'' + ax u_\varepsilon' + u_\varepsilon = 1, \text{ on } (-1, 1), \; a = \pm 1 \tag{8.18}$$

$$u_\varepsilon(\pm 1) = 0 \tag{8.19}$$

In the case $a = 1$, the exact solution has boundary layers at both endpoints ± 1; for $a = -1$, the exact solution exhibits an internal layer at the turning points $x = 0$. The exact solutions are given by

$$u_\varepsilon(x) = 1 - \exp\left\{(x^2 - 1)/(2\varepsilon)\right\} \text{ for } a = 1 \tag{8.20}$$

$$u_\varepsilon(x) = 1 - cx \operatorname{erf}\left(x/\sqrt{2\varepsilon}\right) - \sqrt{2/\pi} \, c\sqrt{\varepsilon} \exp\left\{-x^2/(2\varepsilon)\right\}$$
$$\text{for } a = -1$$

$$c := \left(\operatorname{erf}\left(1/\sqrt{2\varepsilon}\right) + \sqrt{2\varepsilon/\pi} \exp\left(-1/(2\varepsilon)\right)\right)^{-1} \approx 1 \tag{8.21}$$
$$\text{for small } \varepsilon$$

$$\operatorname{erf}(x) := \frac{2}{\sqrt{\pi}} \int_0^x \exp\left(-t^2\right) dt, \; \operatorname{erf}(x) \to 1 \text{ for } x \to \infty$$

Equation (8.18) satisfies the crucial assumption (7.3) and the fact that the coefficient a is a polynomial allows us to modify the arguments as to accommodate the case of (8.18) as well. For the SDFEM we use the weights (8.10), i.e.

$$\delta_i = \frac{1}{4} \frac{h_i}{k^2}.$$

The solution given by (8.20) (i.e., the case $a = 1$) has two boundary layers at both endpoints with length scale $O(\varepsilon)$. The structure of the boundary layers is essentially of the form analyzed in Section 3.2 so that the approximation results with the "two-element" meshes introduced apply. In fact, a "three-element" mesh consisting of two small elements of size $k\varepsilon$ at the boundary points and one large element in the middle (that is, the mesh is given by the points $\{-1, -1 + k\varepsilon, 1 - k\varepsilon, 1\}$) is well-suited to resolve the layers in both the Galerkin as well as the SDFEM (cf. Figs. 8.8 where $\varepsilon = 10^{-8}$).

In the case $a = -1$, the solution is given by (8.21) and has an internal layer of width $O(\sqrt{\varepsilon})$. Again, the "two-element" mesh in Theorem 5.14 can be applied successfully for the approximation of the internal layer if at least one element of size $O(k\sqrt{\varepsilon})$ is introduced at the turning point $x = 0$. Figs. 8.9 show the performance of the Galerkin FEM and the SDFEM for a "four-element" mesh based on the points $\{-1, -k\sqrt{\varepsilon}, 0, k\sqrt{\varepsilon}, 1\}$ and $\varepsilon = 10^{-8}$. Although the error graphs do not behave monotonically, the overall convergence of the "four-element" hp-SDFEM shows exponential convergence rates.

Conclusions on hp-SDFEM for convection-diffusion. From our numerical experiments we conclude that some mesh refinement in the layer is indispensable for proper performance (in a global norm) of both, hp Galerkin FEM and hp SDFEM; in this case, both methods perform comparably well. If, however, the length scales of the solution are not completely resolved, the hp-SDFEM is considerably more robust than the Galerkin FEM in the sense that it effectively suppresses spurious oscillations in the pre-asymptotic range of convergence, and that its asymptotic convergence rate is very close to that of the best approximation.

A successful strategy in more complicated settings will therefore combine mesh adaptation at low p with SDFEM stabilization in the preasymptotic range in order to locate the layers/fronts. Once the layers/fronts are located, our mesh design principles based on the "two-element" mesh can be successfully applied to resolve the layers.

In this pre-asymptotic range, when the layers/fronts are still to be located by some adaptive scheme the hp-SDFEM leads already to robust exponential convergence on compact subsets "upstream". The pure Galerkin FEM on the other hand does not produce reliable results anywhere in the computational domain until the layer is resolved.

We emphasize that we investigated here only one-dimensional linear problems where very precise regularity properties of the solution are available. The stability analysis of the hp-SDFEM, however, did not exploit these properties so that similar findings will likely hold also in two- and three-dimensional situations. The main conclusion which can be drawn from the numerical experiments is that localized small scale phenomena of viscous flow can be resolved with moderate computational by hp-FE discretizations and that the hp-SDFEM can perform very satisfactorily in an adaptive environment.

Proof. If $\int_{\Gamma_D} ds > 0$, there exists $C(\Omega, \Gamma_D) > 0$ such that *Korn's inequality* holds

$$\forall \mathbf{u} \in H^1(\Omega, \Gamma_D)^d : \ \|\mathbf{D}(\mathbf{u})\|_{L^2(\Omega)} \geq C \, \|\mathbf{u}\|_{H^1(\Omega)} . \qquad (9.10)$$

Writing $\alpha = \frac{1}{d}$ trace $(\mathbf{D}(\mathbf{u}))$ and defining the deviatoric part $\mathbf{D}_0(\mathbf{u}) := \mathbf{D}(u) - \alpha \, \mathbf{1}$, we get

$$\begin{aligned}
E(\mathbf{u}, \mathbf{u}) &= \gamma \, \|\text{div } \mathbf{u}\|_\Omega^2 + 2\mu(\mathbf{D}(\mathbf{u}), \, \mathbf{D}(\mathbf{u}))_\Omega \\
&= 2\mu \|D_0(\mathbf{u})\|_\Omega^2 + (2\mu + \gamma d) \, d\|\alpha\|_\Omega^2
\end{aligned}$$

where we used that

$$\|\mathbf{D}(\mathbf{u})\|_{L^2(\Omega)}^2 = (\mathbf{D}(\mathbf{u}), \, \mathbf{D}(\mathbf{u}))_\Omega = \|\mathbf{D}_0(\mathbf{u})\|_\Omega^2 + d \, \|\alpha\|_\Omega^2 .$$

Hence we get

$$E(\mathbf{u}, \mathbf{u}) \geq \min(2\mu, 2\mu + \gamma d)\|\mathbf{D}(\mathbf{u})\|_\Omega^2 .$$

Korn's inequality and (9.5) imply the assertion. □

Remark 9.2. In the case of a monatomic gas, $\mu > 0$ and

$$\gamma + 2\mu/d = 0$$

(Stokes' relation). The dissipativity of div $\boldsymbol{\tau}(\mathbf{u})$ in (1.2) is then not clear.

The FE discretizations of (9.9) are analogous to those of Section 6 and we present them here.

9.3 Standard continuous discretization

It reads: find $\mathbf{u}_{\text{FE}} \in S_D^{\mathbf{k},1}(\Omega, \mathcal{T}, \mathbf{F}_\mathcal{P})^d$ such that

$$E(\mathbf{u}_{\text{FE}}, \mathbf{v}) = F(\mathbf{v}) \quad \forall v \in S_D^{\mathbf{k},1}(\Omega, \mathcal{T}, \mathbf{F}_\mathcal{P})^d . \qquad (9.11)$$

9.4 Dual mixed formulation

Again, to accommodate discontinuous velocities $\mathbf{u}_{\text{FE}}$, a mixed formulation is useful. Now the flux is simply the stress tensor $\boldsymbol{\tau}(\mathbf{u})$. The mixed form of (9.1), (9.7), (9.8) reads

$$\begin{aligned}
-\text{div } \boldsymbol{\tau} &= \mathbf{S} && \text{in } \Omega , \\
\boldsymbol{\tau} &= \mathbf{C}\mathbf{D}(\mathbf{u}) && \text{in } \Omega , \\
\mathbf{u} &= \mathbf{0} && \text{on } \Gamma_D , \\
\boldsymbol{\tau}(\mathbf{u})\mathbf{n} &= \mathbf{g} && \text{in } \Gamma_N ,
\end{aligned} \qquad (9.12)$$

and in weak form: find $\boldsymbol{\tau} \in \mathbf{H}(\mathrm{div}, \varOmega)$, $\mathbf{u} \in L^2(\varOmega)^d$ such that, assuming (9.5) holds,

$$(\mathbf{C}^{-1}\boldsymbol{\tau}, \boldsymbol{\sigma})_\varOmega + (\mathbf{u}, \mathrm{div}\,\boldsymbol{\tau})_\varOmega = 0 \quad \forall \boldsymbol{\sigma} \in \mathbf{H}(\mathrm{div}, \varOmega),$$
$$(\mathbf{v}, \mathrm{div}\,\boldsymbol{\sigma})_\varOmega = (\mathbf{S}, \mathbf{v})_\varOmega + (\mathbf{g}, \mathbf{v})_{\varGamma_N} \quad \forall \mathbf{v} \in L^2(\varOmega)^d .$$

(9.13)

Here $\mathbf{H}(\mathrm{div}, \varOmega) := \{\boldsymbol{\tau} \in L^2(\varOmega)_{\mathrm{sym}}^{d\times d} : \mathrm{div}\,\boldsymbol{\tau} \in L^2(\varOmega)^d\}$ and $\mathbf{C}^{-1}$ is the inverse of the elasticity compliance tensor.

The construction of finite elements which are $\mathbf{H}(\mathrm{div}, \varOmega)$ conforming and stable for (9.13) is delicate. Some $2-d$ examples can be found in [16], Chapter VII.2. Note that in $\mathbb{R}^d$ in the mixed formulation (9.13) $d(d+1)/2$ additional fields have to be discretized; for elasticity problems with discontinuous $\mathbf{u}_{\mathrm{FE}}$, the incentive to consider mortar resp. DG-FEM is therefore even higher than for scalar advection-diffusion problems. Moreover, the discretization techniques for the scalar case carry over to large extent. We therefore do not recommend discrete versions of (9.13) in fluid flow simulations, and turn to the mortar and DG methods.

9.5   Mortar Discretization

Basic Discretization. We use the notation of Section 6.3 and proceeding analogously we arrive at: find $\mathbf{u} \in H_D^1(\varOmega, \mathcal{T})^d$, $\boldsymbol{\mu} \in M^d$ such that $\forall \mathbf{v} \in H^1(\varOmega, \mathcal{T})^d$, $\forall \boldsymbol{\lambda} \in M^d$

$$E_\mathcal{T}(\mathbf{u}, \mathbf{v}) + \varPhi_\mathcal{T}(\mathbf{v}, \boldsymbol{\mu}) = (\mathbf{S}, \mathbf{v})_\varOmega + (\mathbf{g}, \mathbf{v})_{\varGamma_N}$$
$$\varPhi_\mathcal{T}(\mathbf{u}, \boldsymbol{\lambda}) = 0 .$$

(9.14)

Here the broken bilinear forms are given by

$$E_\mathcal{T}(\mathbf{u}, \mathbf{v}) := \sum_K E_K(\mathbf{u}, \mathbf{v}) = \sum_K (\mathbf{C}\mathbf{D}(\mathbf{u}), \mathbf{D}(\mathbf{v}))_K ,$$

$$\varPhi_\mathcal{T}(\mathbf{u}, \boldsymbol{\lambda}) := \sum_{e \in \mathcal{S}_{\mathrm{int}}} ([\mathbf{u}], \boldsymbol{\lambda})_e .$$

The discretization of (9.14) is analogous to (6.22): find $\mathbf{u}_{\mathrm{FE}} \in S_D^{k,0}(\varOmega, \mathcal{T})^d$ and $\boldsymbol{\mu}_{\mathrm{FE}} \in M^{k,0}(\varOmega, \mathcal{T})^d$ such that $\forall \mathbf{v} \in S_D^{k,0}(\varOmega, \mathcal{T})^d$, $\forall \boldsymbol{\lambda} \in M^{k,0}(\varOmega, \mathcal{T})^d$

$$E_\mathcal{T}(\mathbf{u}_{\mathrm{FE}}, \mathbf{v}) + \varPhi_\mathcal{T}(\mathbf{v}, \boldsymbol{\mu}) = (\mathbf{S}, \mathbf{v})_\varOmega + (\mathbf{g}, \mathbf{v})_{\varGamma_N}$$
$$\varPhi_\mathcal{T}(\mathbf{u}_{\mathrm{FE}}, \boldsymbol{\lambda}) = 0 .$$

(9.15)

The structure of the linear system corresponding to (9.15) is analogous (6.9).

Implementation without fluxes. Assume that the polyhedron $\Omega \subset \mathbb{R}^d$ is partitioned into a regular triangulation $\mathcal{T}$ of simplicial elements K. We consider (9.1), (9.7) and (9.8) and assume the polynomial degrees $\mathbf{k}$ are uniform and equal $k \geq 2$. Let $\mathcal{S}_{\text{int}}$ denote the set of all $d - 1$ dimensional simplices e_{KK}, which are interfaces of $K, K' \in \mathcal{T}$. Then we have, analogous to (6.25),

$$(\mathcal{S}_D^{k,\delta})^d = \Big\{ \mathbf{u} \in \mathcal{S}_D^{k,0}(\Omega, \mathcal{T})^d :$$

$$\int_{e_{KK'}} [\mathbf{u}] \cdot \boldsymbol{\varphi} \, ds = 0 \quad \forall \boldsymbol{\varphi} \in \mathcal{P}^{k-1}(e_{KK'})^d \Big\} . \tag{9.16}$$

The matrix $\widehat{\mathbf{A}}$ is the stiffness matrix of the form $E_\mathcal{T}(\mathbf{u}, \mathbf{v})$ on $(\mathcal{S}_D^{k,\delta})^d \times (\mathcal{S}_D^{k,\delta})^d$. And it holds that $\widehat{\mathbf{A}}$ is symmetric, positive definite, if $k \geq 2$ and if (9.5) holds.

To see it, assume that $0 = E_\mathcal{T}(\mathbf{u}, \mathbf{u})$ for some $\mathbf{0} \neq \mathbf{u} \in (\mathcal{S}_D^{k,\delta})^d$. Hence we get

$$0 = E_\mathcal{T}(\mathbf{u}, \mathbf{u}) \geq \min\{2\mu, 2\mu + \gamma d\} \sum_{K \in \mathcal{T}} \|D(\mathbf{u})\|_K^2$$

which implies that $u|_K$ is a rigid body motion, i.e.

$$\forall K \in \mathcal{T} \;\; \exists \mathbf{A}_K = -\mathbf{A}_K^\top, \; \mathbf{b}_K : \mathbf{u}|_K = \mathbf{A}_K \mathbf{x} + \mathbf{b}_K .$$

If $\overline{K} \cap \Gamma_D \neq \emptyset$, then $\mathbf{u}|_K = \mathbf{0}$ in these K.

Further, $\mathbf{u} = \mathbf{0}$ in the remaining elements $K \in \mathcal{T}$, since

$$\forall e_{KK'} \in \mathcal{S}_{\text{int}} : \int_{e_{KK'}} [\mathbf{u}] \cdot \boldsymbol{\varphi} \, ds = 0$$

for every $\boldsymbol{\varphi} \in P_1(e_{KK'})^d$, due to $k \geq 2$, and since $[\mathbf{u}]$ is linear on $e_{KK'}$. Therefore $E_\mathcal{T}(\mathbf{u}, \mathbf{u}) = 0 \implies \mathbf{u} = \mathbf{0}$ and $\widehat{\mathbf{A}}$ is positive definite, hence this discretization of viscous stresses is dissipative. Notice that in the scalar case in Section 6.3 the above argument works even for $k \geq 1$, since the "rigid body motions" are piecewise constant then.

As in Section 6.3, the *hp*-MEM can be implemented without the fluxes. To this end, we evaluate the broken bilinear form $E_\mathcal{T}(\mathbf{u}, \mathbf{v})$ on the constrained space $(\mathcal{S}_D^{\mathbf{k},\delta})^d$, resulting in a symmetric, positive definite matrix $\widehat{\mathbf{A}}$, i.e. giving rise to a dissipative term, provided (9.5) holds.

9.6 Discontinuous Galerkin discretization

From (9.15) it is now straightforward to derive the DG-discretization of (9.1); as for the diffusion problems in Section 6.3, we replace in $\Phi_\mathcal{T}(\mathbf{v}, \boldsymbol{\mu})$ on each edge $e \in \mathcal{S}_{\text{int}}$ (9.14) the multiplier $\boldsymbol{\mu}$ in the saddle point form by the flux average, i.e.

$$\boldsymbol{\mu}|_e = \langle \boldsymbol{\tau}(\mathbf{u}) \, \mathbf{n} \rangle, \quad \boldsymbol{\lambda}|_e = -\langle \boldsymbol{\tau}(\mathbf{v}) \, \mathbf{n} \rangle \tag{9.17}$$

where $\mathbf{n}$ denotes the unit normal vector perpendicular to the interface e, resulting in: find $\mathbf{u}_{FE} \in S_D^{k,0}(\Omega, \mathcal{T}, \mathbf{F}_{\mathcal{P}})^d$ such that

$$E_{DG}(\mathbf{u}_{FE}, \mathbf{v}) = (\mathbf{S}, \mathbf{v})_\Omega + (\mathbf{g}, \mathbf{v})_{\Gamma_N} \qquad \forall \mathbf{v} \in S^{k,0}(\Omega, \mathcal{T}, \mathbf{F}_{\mathcal{P}})^d, \qquad (9.18)$$

where we defined

$$
\begin{aligned}
E_{DG}(\mathbf{u}, \mathbf{v}) = & \sum_{K \in \mathcal{T}} E_K(\mathbf{u}, \mathbf{v}) \\
& + \sum_{e \in \mathcal{S}_{int}} \int_e \left\{ \langle \tau(\mathbf{v})\mathbf{n} \rangle [\mathbf{u}] - [\mathbf{v}] \langle \tau(\mathbf{u})\mathbf{n} \rangle \right\} ds .
\end{aligned}
\qquad (9.19)
$$

And we have once more the positive semidefiniteness

$$E_{DG}(\mathbf{u}, \mathbf{u}) \geq 0$$

and $E_{DG}(\mathbf{u}, \mathbf{u}) = 0$ if and only if $\mathbf{u}|_K$ is a rigid body motion. It is at present open if (9.19) satisfies a discrete inf-sup stability condition.

10   Incompressibility

10.1   Basic Equations

For an incompressible medium, $\gamma \to \infty$ in (9.3) thereby imposing in (1.2) the *incompressibility constraint*

$$\operatorname{div} \mathbf{u} = 0 \text{ in } \Omega. \qquad (10.1)$$

This constraint changes the momentum equation (1.2) to

$$\frac{\partial}{\partial t}(\rho u_i) + \sum_{j=1}^d \frac{\partial}{\partial x_j}(\rho u_i u_j) + \frac{\partial p}{\partial x_i} = \mu \Delta u_i + S_i, \ i = 1, \dots, d \qquad (10.2)$$

if $\mu > 0$ is constant. The system (1.1), (10.1), (10.2) constitutes the inhomogeneous, incompressible NSE.

If in addition $\rho = \rho_0 = \text{const in } \Omega$, it follows that

$$\rho_0 \frac{\partial u_i}{\partial t} + \rho_0 \nabla \cdot (\mathbf{u}\, u_i) + \nabla p = \mu \Delta u_i + S_i, \ i = 1, \dots, d$$

or, upon the rescaling

$$\nu \leftarrow \mu/\rho_0, \quad p \leftarrow p/\rho_0, \quad \mathbf{S} \leftarrow \frac{1}{\rho_0} \mathbf{S}, \qquad (10.3)$$

$$\frac{\partial u_i}{\partial t} + \nabla \cdot (\mathbf{u} u_i) + \nabla p = \nu \Delta u_i + S_i, \ i = 1, \dots, d. \qquad (10.4)$$

We remark that the energy equation is now absent and that the function p is a Lagrange multiplier for the constraint (10.1). We shall not dwell upon the derivation of (10.2). We note, however, that (10.1) generally causes difficulties for a FE discretization which will also appear at large, but finite values of γ in (9.3). Stable FE discretizations for (10.1), (10.2) promise also robust performance for (1.2), and (9.3) as $\gamma \to \infty$.

Once again, we focus on the space discretization of (10.1) - (10.4). To this end, we consider the steady case ($\partial/\partial t = 0$). Linearizing around $\mathbf{u} = \mathbf{w}$ with div $\mathbf{w} = 0$ yields in (10.4) the *Oseen-equations*

$$-\nu \Delta \mathbf{u} + \mathbf{w} \cdot \nabla \mathbf{u} + \nabla p = \mathbf{S} \text{ in } \Omega,$$
$$\nabla \cdot \mathbf{u} = 0 \text{ in } \Omega. \tag{10.5}$$

If, in addition, $\mathbf{w} = \mathbf{0}$, we get the *Stokes-equations*

$$-\nu \Delta \mathbf{u} + \nabla p = \mathbf{S} \text{ in } \Omega,$$
$$\nabla \cdot \mathbf{u} = 0 \text{ in } \Omega. \tag{10.6}$$

Both, (10.5) and (10.6), are completed by *no-slip boundary conditions*

$$\mathbf{u} = \mathbf{0} \text{ on } \partial\Omega. \tag{10.7}$$

10.2   Variational formulation of the Stokes problem

Consider first (10.6) and assume $\mathbf{S} \in L^2(\Omega)^d$. The discretization of the incompressibility constraint (10.1) can be done in 2 ways:

a) incorporation into the space, i.e. we look for $\mathbf{u} \in J_0 := \{\mathbf{u} \in H_0^1(\Omega)^d : \nabla \cdot \mathbf{u} = 0 \text{ in } L^2(\Omega)\}$. It is generally difficult to construct FE subspaces of J_0,

b) enforcement of (10.1) via Lagrange multiplier p:
find $\mathbf{u} \in H_0^1(\Omega)^d$, $p \in L_0^2(\Omega)$ such that

$$\nu(\nabla \mathbf{u}, \nabla \mathbf{v})_\Omega - (p, \nabla \cdot \mathbf{v})_\Omega = (\mathbf{S}, \mathbf{v})_\Omega \quad \forall \mathbf{v} \in H_0^1(\Omega)^d,$$
$$(\nabla \cdot \mathbf{u}, q)_\Omega = 0 \quad\quad \forall q \in L_0^2(\Omega). \tag{10.8}$$

Here $L_0^2(\Omega) = \{q \in L^2(\Omega) : (q, 1)_\Omega = 0\}$.

This is now a mixed problem and the FE discretization of (10.7) can be based on the standard spaces $S^{k,\ell}$ of Section 4.

10.3   FE-discretization of the Stokes problem

Stability. Let $\mathbf{V}_N \subset H_0^1(\Omega)^d$, $M_N \subset L_0^2(\Omega)$ be any pair of finite-dimensional spaces. The Galerkin discretization of (10.6), (10.7) reads:

find $\mathbf{u}_N \in \mathbf{V}_N$, $p_N \in M_N$ such that

$$\nu(\nabla\mathbf{u}_N, \nabla\mathbf{v})_\Omega - (p_N, \nabla \cdot \mathbf{v})_\Omega = (\mathbf{S}, \mathbf{v})_\Omega \quad \forall \mathbf{v} \in \mathbf{V}_N,$$
$$(\nabla \cdot \mathbf{u}_N, q)_\Omega \qquad\qquad\qquad = 0 \qquad \forall q \in M_N. \tag{10.9}$$

In principle, we may choose for $\mathbf{V}_N$, M_N the hp-FE subspaces of Section 4. However, the pair $\mathbf{V}_N$, M_N must satisfy the *discrete inf-sup stability condition*

$$\inf_{0\neq q \in M_N} \sup_{0\neq u \in \mathbf{V}_N} \frac{(q, \nabla \cdot \mathbf{u})_\Omega}{\|q\|_0 \, \|\nabla\mathbf{u}\|_0} \geq \gamma_N > 0 \tag{10.10}$$

where γ_N is the inf-sup (or stability) constant. (10.10) ensures stability of the approximation and precludes in particular spurious pressure modes. Natural choices for $\mathbf{V}_N$, M_N, such as

$$\mathbf{V}_N = S_0^{k,1}(\Omega, \mathcal{T})^d, \quad M_N = S_0^{k-1,0}(\Omega, \mathcal{T})$$

$k \geq 1$, generally fail (10.10): One must choose $(\mathbf{V}_M, M_N)$ carefully.

Divergence stable elements on shape regular meshes. Let us present various choices of stable hp-spaces. To this end assume that all element mappings F_K are *affine* and that $\mathcal{T}$ is *shape-regular*. Then the spaces $S^{k,\ell}(\Omega, \mathcal{T})$ are determined by the polynomial spaces $\widehat{\mathbf{V}}_K$, $\widehat{M}_K$ on the reference element $\widehat{K}$

$$S_0^{k,0}(\Omega, \mathcal{T}) = \{q \in L_0^2(\Omega) : q \circ F_K \in \widehat{M}_K\}, \tag{10.11}$$

$$S_0^{k,1}(\Omega, \mathcal{T})^d = \{\mathbf{u} \in H_0^1(\Omega)^d : \mathbf{u} \circ F_K \in \widehat{\mathbf{V}}_K\}. \tag{10.12}$$

In the following table we list some pairs $\widehat{\mathbf{V}}_K$, $\widehat{M}_K$ and the mathematically established bounds on the inf-sup constant γ_N in (10.10). We assume shape regular, possibly non-quasiuniform meshes

$\widehat{\mathbf{V}}_K$	$\widehat{M}_K$	γ_N	$\widehat{K}$	
$\mathcal{Q}_k$	$\mathcal{Q}_{k-2}$	$\mathcal{O}(k^{-d/2})$	$\widehat{Q}$	(10.13a)
$\mathcal{Q}_k$	$\mathcal{P}_{k-1}$	$\mathcal{O}(1)$	$\widehat{Q}$	(10.13b)
$\mathcal{P}_k$	$\mathcal{P}_{k-2}$	$\mathcal{O}(k^{-3})$	$\widehat{T}$	(10.13c)

(10.13a) and (10.13b) are sharp and hold in two and three dimensions. We remark that the bound (10.13c), proved in [58], [60] is suboptimal and only valid in two dimensions. If used on a shape regular, possibly geometric mesh, the velocity-pressure combinations (10.13) give in (10.10) inf-sup constants γ_N which are independent of the element sizes h_K, $K \in \mathcal{T}$.

Stable elements on (κ, σ)-boundary layer meshes. The situation is different on geometric, affine (κ, σ)-boundary layer meshes (cf. Section 4.2 and Figures 8 and 9) containing long rectangles. Here the combination $\mathcal{Q}_k \times \mathcal{Q}_{k-2}$ is stable independent of the aspect ratio [62], [55].

Theorem 10.1. *Let $\Omega \subset \mathbb{R}^2$ be a polygon and $\mathcal{T}$ be an affine (κ, σ) geometric boundary layer mesh. Let in (10.10) for $k = 2, 3, \ldots$*

$$\mathbf{V}_N = S_0^{k,1}(\Omega, \mathcal{T})^d, \quad M_N = S_0^{k,0}(\Omega, \mathcal{T})$$

with element spaces (10.13). Then (10.10) holds with $\gamma_N \geq C k_{\max}^{-3}$ if $\mathcal{T}$ contains triangles and $\gamma_N \geq C k_{\max}^{-1/2}$ otherwise. Here $C > 0$ is independent of k and of the aspect ratio of the rectangles (it depends only on κ and σ).

No divergence stable, high order and high aspect ratio triangular element family is known to date.

10.4   GLS stabilized *hp*-FEM for the Stokes problem

The divergence stability (10.10) imposed the use of different polynomial orders for velocity pressure approximations. Equal order spaces are not divergence stable. There is, however, a GLS (Galerkin Least Squares) approach due to Hughes and Franca which allows a) to circumvent (10.10) and b) to use equal order approximations for $\mathbf{V}_N$ and M_N. We show here an *hp*-extension of this approach. Select

$$\mathbf{V}_N = S_0^{k,1}(\Omega, \mathcal{T})^d, \quad M_N = S_0^{k,1}(\Omega, \mathcal{T}) \tag{10.14}$$

with equal elemental polynomial degrees. Then the *hp*-GLS FEM for (10.7) reads: find $(\mathbf{u}_{\mathrm{GLS}}, p_{\mathrm{GLS}}) \in \mathbf{V}_N \times M_N$ such that

$$B_\alpha(\mathbf{u}_{\mathrm{GLS}}, p_{\mathrm{GLS}}; \mathbf{v}, q) = F_\alpha(\mathbf{v}, q) \quad \forall (\mathbf{v}, q) \in \mathbf{V}_N \times M_N, \tag{10.15}$$

where $\alpha > 0$ is a parameter independent of $\mathbf{k}$ and h_K, and

$$B_\alpha(\mathbf{u}, p; \mathbf{v}, q) := \nu(\nabla \mathbf{u}, \nabla \mathbf{v}) - (p, \nabla \cdot \mathbf{v}) - (\nabla \cdot \mathbf{u}, q)$$
$$- \alpha \sum_{K \in \mathcal{T}} \frac{h_K^2}{k_K^4} (-\nu \Delta \mathbf{u} + \nabla p, -\nu \Delta \mathbf{v} + \nabla q)_K$$
$$F_\alpha(\mathbf{v}, q) := (\mathbf{S}, \mathbf{v}) - \alpha \sum_{K \in \mathcal{T}} \frac{h_K^2}{k_K^4} (\mathbf{S}, \nu \Delta \mathbf{v} + \nabla q)_K.$$

Notice that $\alpha = 0$ gives the (unstable) Galerkin-formulation (10.9) (just add the equations there). Note also that the GLS formulation (10.15) is fully consistent - inserting the exact solution $(\mathbf{u}, p)$, we see that the GLS terms disappear, for any value of α.

We have

Theorem 10.2. *[54], Let $\Omega \subset \mathbb{R}^2$ be a polygon and $\mathcal{T}$ be a shape-regular mesh. Then there exists $C > 0$ independent of α, $\mathbf{k}$ and h_K such that B_α in (10.15) is stable, more precisely that*

$$\sup_{\substack{0 \neq v \in \mathbf{V}_N \\ 0 \neq q \in M_N}} \frac{B_\alpha(\mathbf{u}, p; \mathbf{v}, q)}{(\|\mathbf{u}\|_{1,\Omega}^2 + \|p\|_{0,\Omega}^2)^{1/2}(\|\mathbf{v}\|_{1,\Omega}^2 + \|q\|_{0,\Omega}^2)^{1/2}} \geq C \frac{\alpha}{k_{\max}^4}$$

for all $0 \neq \mathbf{u} \in \mathbf{V}_N$, $0 \neq p \in M_N$.

Remark 10.3. The above result does not allow for anisotropic rectangles - it does allow, however, curved elements, i.e., nonaffine patch maps F_P. GLS stabilization on *curved, anisotropic* meshes is open at present. For more information on GLS methods, we refer to [38] and the references there.

Remark 10.4. The original GLS methods were developed for $k = 1$, so that the domain integrals in B_α would simplify. The evaluation of second order derivatives of high order polynomials in stabilization terms in B_α is costly in the element stiffness matrix evaluation of (10.15).

10.5 Numerical experiments

Implementational details. Here we present some numerical results, taken from [29], to show: a) that hp-FEM give exponential convergence even if the solution has singularities and b) to compare the pure Galerkin approach with divergence stable elements with the GLS approach and equal order hp-interpolation. Accordingly, we compare

The Galerkin formulation (GFEM): Let

$$\mathbf{V}_N = S_0^{k,1}(\mathcal{T})^2, \qquad M_N = S_0^{k-2,0}(\mathcal{T}).$$

The GFEM is to find $(\mathbf{u}_N, p_N) \in \mathbf{V}_{N,0} \times M_{N,0}$ such that

$$B_0(\mathbf{u}_N, p_N; \mathbf{v}, q) = F_0(\mathbf{v}, q) \text{ for all } (\mathbf{v}, q) \in \mathbf{V}_N \times M_N.$$

The Galerkin Least Squares formulation (GLSFEM):
Let $\alpha > 0$ and
$$\mathbf{V}_N = S_0^{k,1}(\mathcal{T})^2, \qquad M_{N,0} = S^{k,1}(\mathcal{T}).$$

The GLSFEM is to find $(\mathbf{u}_N, p_N) \in \mathbf{V}_N \times M_N$ such that

$$B_\alpha(\mathbf{u}_N, p_N; \mathbf{v}, q) = F_\alpha(\mathbf{v}, q) \text{ for all } (\mathbf{v}, q) \in \mathbf{V}_N \times M_N.$$

Note that we consider a continuous pressure approximation in the GLS-FEM while the pressure is discontinuously interpolated in the GFEM. This choice has been made since it points out the principal advantages of implementing GLSFEM: In the GLSFEM velocity and pressure degrees of freedom

are treated in exactly the same way. For the GFEM implementational difficulties arise if one enforces different polynomial degrees for the velocity and
the pressure and different interelement continuity requirements for $\mathbf{u}_N$ and
p_N.

Our *hp*-FE implementation for the Stokes problem is based on HP90, a
flexible FE code for general elliptic problems in Fortran 90 [21]. HP90 allows
for isotropic and anisotropic mesh refinements, both *h*- and *p*-refinements. In
particular, *h*-refinements can lead to irregular meshes with hanging nodes.
HP90 is designed to handle such meshes and enforces the appropriate continuity requirements by constraining these irregular nodes. We refer to [21],
[50] for a detailed description of the constraining procedure.

In our numerical examples we use quadrilateral finite elements to discretize the domain Ω. Implementationally, the elemental polynomial degrees
k_K are further split into edge and internal degrees that can vary within the
element, i.e. k_K is to be understood as the vector $k_K = \{k_K^1, k_K^2, k_K^3, k_K^4, k_K^5\}$.
Here k_K^i, $i = 1, \ldots, 4$, is the polynomial degree on the *i*-th edge, and k_K^5 the
polynomial degree in the interior of the element. The nodes $\hat{a}_K^1, \ldots, \hat{a}_K^9$ correspond to k_K, where $\hat{a}_K^1, \ldots, \hat{a}_K^4$ denote the vertex nodes, $\hat{a}_K^5, \ldots \hat{a}_K^8$ the
mid-side nodes and $\hat{a}_K^9$ is the middle node. This is shown schematically in
Figure 10.1 for the reference square $\widehat{Q} = (0,1)^2$. The shape functions that are
associated with the nodes $\hat{a}_K$ of the reference element are the nodal based
Lagrange shape functions but other shape functions can be used as well (cf.
[21]).

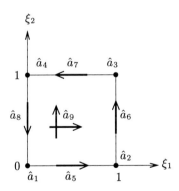

Fig. 10.1. Quadrilateral reference element $\widehat{Q}$ with nodes $(\hat{a}^1, \ldots, \hat{a}^9)$.

In the case of the GLS method we need to interpret the reference element
$\widehat{Q}$ as a vector valued reference element, i.e. we use the shape functions and
degrees of freedom (dof) that correspond to $\widehat{Q}$ to approximate each velocity
component and the pressure. The least-squares stabilization term in B_α in
(10.15) involves second derivatives and therefore we also need the second

derivatives of the reference element shape functions $\varphi(\xi_1, \xi_2)$ with respect to the physical coordinates $(x_1, x_2) = F_K(\xi_1, \xi_2)$. In the case of an affine element mapping F_K, the chain rule gives

$$\frac{\partial^2 \varphi}{\partial x_i^2} = \frac{\partial^2 \varphi}{\partial \xi_1^2} \left(\frac{\partial \xi_1}{\partial x_i} \right)^2 + \frac{\partial^2 \varphi}{\partial \xi_2^2} \left(\frac{\partial \xi_2}{\partial x_i} \right)^2. \tag{10.16}$$

But the terms $\partial \xi_j / \partial x_i$ are not constant in the case of a general (e.g. bilinear) element mapping which leads to

$$\frac{\partial^2 \varphi}{\partial x_i^2} = \frac{\partial^2 \varphi}{\partial \xi_1^2} \frac{\partial \xi_1}{\partial x_i} + \frac{\partial \varphi}{\partial \xi_1} \frac{\partial^2 \xi_1}{\partial x_i^2} + \frac{\partial^2 \varphi}{\partial \xi_2^2} \frac{\partial \xi_2}{\partial x_i} + \frac{\partial \varphi}{\partial \xi_2} \frac{\partial^2 \xi_2}{\partial x_i^2}. \tag{10.17}$$

The terms $\partial \xi_j / \partial x_i$ and $\partial^2 \xi_j / \partial x_i^2$ are rational functions and can thus not be integrated exactly, but the use of a higher order integration rule reduces the error in the element computations. Nevertheless, the element computations are completely standard and for an element K the local element stiffness matrix $E_{K,\alpha}$ and load vector $F_{K,\alpha}$ result in an element system of equations that is of the well known form

$$E_{K,\alpha} \begin{bmatrix} \mathbf{u} \\ p \end{bmatrix} = \begin{bmatrix} A_\alpha & 0 & B_{1,\alpha}^\top \\ 0 & A_\alpha & B_{2,\alpha}^\top \\ B_{1,\alpha} & B_{1,\alpha} & \alpha M \end{bmatrix} \begin{bmatrix} u_1 \\ u_2 \\ p \end{bmatrix} = \begin{bmatrix} F_{1,\alpha} \\ F_{2,\alpha} \\ 0 \end{bmatrix} \tag{10.18}$$

where $\mathbf{u} = (u_1, u_2)$, A_α, $B_{1,\alpha}$, $B_{2,\alpha}$ as well as M correspond to the usual velocity and pressure combinations in (10.13) and $X^\top$ it the transpose of X.

In the context of geometric refinements with irregular nodes we have to modify $E_{K,\alpha}$ in order to account for these irregular nodes. HP90 is designed to enforce the appropriate constraints automatically on the local element stiffness matrix and load vector. This procedure [21] results in a modified local stiffness matrix $\tilde{E}_{K,\alpha}$ that corresponds to the actual globally existing dof. This modified matrix $\tilde{E}_{K,\alpha}$ can then be assembled to obtain the global stiffness matrix.

In the case of the G method the situation is somewhat more complicated due to the different approximation orders for the velocities and the pressure and additionally the pressure being discontinuous. Here we use the shape functions of order k on $\hat{Q}$ to approximate the velocity components and the shape functions of order $k - 2$ to approximate the pressure. The dof for the velocity components are interpreted in the standard way but the pressure dof are now all interpreted as dof that belong to bubble shape functions, although the shape functions of order $k - 2$ contain vertex and side shape functions. It is obvious that the number of bubble shape functions of order k on $\hat{Q}$ is exactly the same as the total number of shape functions of order $k - 2$. This motivates to interprete $\hat{Q}$ as a vector valued reference element with two components for the vertex and side dof and three components for the bubble dof and the shape functions being chosen as described above. The element computations for the G method are then again standard but we have to consider the unusual element definition. The local element stiffness matrix E_K is of a form similar to (10.18) with $\alpha = 0$. We further emphasize that we do not need the second derivatives of the shape functions to compute E_K. For an irregular mesh, we again have to modify E_K to a local matrix $\widetilde{E}_K$ that corresponds to global dof. But now we apply the constraints only to the velocity components because the pressure may be discontinuous across element boundaries. The element matrices $\widetilde{E}_K$ are then assembled in principle in the usual way but the non standard element definition requires a generalization of the assembling procedure to account for the presence of continuous and discontinuous field variables.

In both the G & GLS method we have to enforce Dirichlet boundary conditions that correspond to the boundary values of the exact solutions. The standard procedure very often used in practice is to interpolate the boundary data at equidistant points, but this procedure is known to be numerically instable for higher approximation orders. In connection with higher order methods interpolation at the Gauss Lobatto points is better suited (cf. [21]). We enforce the Dirichlet data for the G & GLS method in exactly the same way at the element level.

Although we apply Dirichlet boundary conditions to the velocity components, the global stiffness matrix is not invertible in both formulations, because the constant pressure mode is still not eliminated. To obtain invertibility of the global system we fix the pressure at one dof. Then the global system can be solved and we only have to postprocess the pressure so that the mean value is zero, i.e. so that the pressure is an element of $L_0^2(\Omega)$.

Numerical results for G & GLS *hp* FEM. In the following we first describe the two model problems that we use. Both model problems have exact solutions and therefore allow for a numerical convergence study. These two exact solutions have significantly different characteristics, i.e. one solution is smooth and the other one has a corner singularity at the reentrant corner. These two model problems are well suited for a comparison of the G- and GLS- *hp*-FEM.

In our numerical results we present always the relative errors that we obtained with our *hp*-FE implementation. We show only the errors for the first velocity component (the results for the second one being completely similar) and the pressure. The velocity error is computed in the H^1-norm and the pressure error in the L^2-norm. In order to be consistent with the pressure being in L_0^2, we subtract the mean value from the exact pressure p and the numerical pressure p_N , i.e. we subtract terms of the form

$$\bar{p} = \frac{1}{|\Omega|} \int_\Omega p \, dx, \qquad \bar{p}_N = \frac{1}{|\Omega|} \int_\Omega p_N \, dx , \qquad (10.19)$$

and the relative error in the pressure is computed as

$$\frac{\|(p-\bar{p})-(p_N-\bar{p}_N)\|_{L^2(\Omega)}}{\|p-\bar{p}\|_{L^2(\Omega)}}. \tag{10.20}$$

The relative H^1-error in the velocity components is computed in the standard way. We remark finally that the Gauss integration rule that we use to compute the errors is of significantly higher order than the integration rule in the element computations.

Model problems. In our model problems we consider the Stokes equation (10.6), (10.7) with viscosity $\nu = 1$ in the L-shaped domain Ω shown in Figure 10.2. Such domains appear also in the backward facing step flow problem or in the so-called 4:1 contraction problem. On Ω we use geometric meshes $\mathcal{T}^{n+1,\sigma}$ with $n+1$ layers. Such a mesh (with irregular nodes) is shown for a grading factor $\sigma = 0.5$ in Figure 10.2.

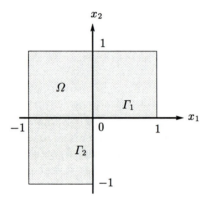

Fig. 10.2. L-shaped domain Ω and a geometric mesh on Ω.

We use two exact solutions $(\mathbf{u}_1, p_1)$ and $(\mathbf{u}_2, p_2)$, the first one exhibiting corner singularity phenomena at the reentrant corner 0, the second one being analytical in $\overline{\Omega}$ (including the corners). In polar coordinates (r, φ) at the origin the first exact solution is given by

$$\mathbf{u}_1(r,\varphi) = r^\lambda \begin{pmatrix} (1+\lambda)\sin(\varphi)\Psi(\varphi) + \cos(\varphi)\Psi'(\varphi) \\ \sin(\varphi)\Psi'(\varphi) - (1+\lambda)\cos(\varphi)\Psi(\varphi) \end{pmatrix}, \tag{10.21}$$

$$p_1 = -r^{\lambda-1}[(1+\lambda)^2\Psi'(\varphi) + \Psi'''(\varphi)]/(1-\lambda) \tag{10.22}$$

with

$$\Psi(\varphi) = \sin((1+\lambda)\varphi)\cos(\lambda\omega)/(1+\lambda) - \cos((1+\lambda)\varphi) - \\ \sin((1-\lambda)\varphi)\cos(\lambda\omega)/(1-\lambda) + \cos((1-\lambda)\varphi),$$

$$\omega = \frac{3\pi}{2}.$$

The exponent λ is the smallest positive solution of

$$\sin(2\lambda\omega) + \lambda\sin(2\omega) = 0, \tag{10.23}$$

which is $\lambda \approx 0.5444838205973307$. This solution satisfies the homogeneous Stokes equation, i.e. $-\Delta\mathbf{u}_1 + \nabla p_1 = 0$ in Ω, and we have $\mathbf{u}_1 = 0$ on the segments Γ_1, Γ_2 shown in Figure 10.2. We emphasize that $(\mathbf{u}_1, p_1)$ is analytical in $\overline{\Omega} \backslash \{0\}$, but $\nabla\mathbf{u}_1$ and p_1 are singular at the origin. Especially, $\mathbf{u}_1 \notin H^2(\Omega)^2$ and $p_1 \notin H^1(\Omega)$. This first solution reflects perfectly the typical (singular) behavior of solutions of the Stokes equations near reentrant corners and is generic (compare with (3.1)).

The second exact solution we use is somehow artificial, since it is analytic in $\overline{\Omega}$ (including the corners). In practice, one can not expect solutions to behave so nicely at reentrant corners. Nevertheless, smooth solutions arise for example in smooth domains and it is hence reasonable to validate the numerical performance for such exact solutions too. We take

$$\mathbf{u}_2(x,y) = \begin{pmatrix} -\exp(x)[y\cos(y) + \sin(y)] \\ \exp(x)y\sin(y) \end{pmatrix}, \tag{10.24}$$

$$p_2 = 2\exp(x)\sin(y). \tag{10.25}$$

As above, $-\Delta\mathbf{u}_2 + \nabla p_2 = 0$.

Choice of stabilization parameter α. Theorem 10.2 guarantees stability of the GLSFEM as long as the parameter α remains in a range $0 < \alpha < \alpha_{max}$. α_{max} is independent of the element sizes h_K and the approximations orders k_K and is essentially determined by the best constant C for which the inverse inequality

$$\|\nabla\varphi\|_{L^2(\widehat{K})} \le Ck^2\|\varphi\|_{L^2(\widehat{K})} \tag{10.26}$$

holds on the reference element $\widehat{K}$ for all polynomials $\varphi \in S^k(K)$ and all $k \in \mathbb{N}$ (cf. [54]). In one dimension the best constant C in (10.26) is explicitly known and equal to $3\sqrt{2}$ (if $\widehat{K} = (-1,1)$). In two space dimensions this best constant seems not to be available, but we expect it to be of about the same order. In addition, one may ask whether this upper bound α_{max} is just an artefact of the stability proof or whether it can really be observed in practice. On the other hand, we expect the GLSFEM to become instable as α approaches 0. In fact, for $\alpha = 0$ the G- and GLS-discretization coincide and it is well known that the Galerkin method is instable for velocity and pressure spaces of the same polynomial order.

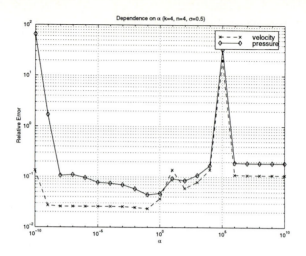

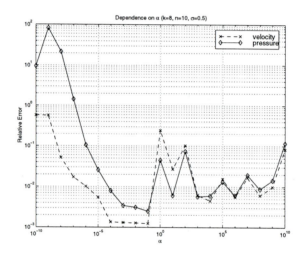

Fig. 10.3. Dependence of the relative error on stabilization parameter α.

We addressed these questions numerically by varying α in a large range. We considered two configurations for the model problem (10.21)-(10.22) in the L-shaped domain, the first one being $k = 4$, $n = 4$ and $\sigma = 0.5$, the second one $k = 8$, $n = 10$ and $\sigma = 0.5$, where k is the polynomial degree and n, σ determine the geometric mesh $\mathcal{T}^{n,\sigma}$ with $n + 1$ layers and grading factor σ. In Figure 10.3 the relative errors of the first velocity component and the pressure are plotted for these two configurations against α ranging from 10^{-10} to 10^{10}. The error curves become oscillatory for increasing α. The "existence" of an upper bound α_{max} can not be answered affirmatively with absolute certainty. Anyway, the performance of the GLSFEM is rather poor in the range $\alpha \geq 10^0$. But the deterioration of the GLS-scheme as α approaches zero can indeed be observed: The errors begin to grow and finally explode for $\alpha \leq 10^{-5}$. In this range the velocities are still more or less accurate but the obtained pressures become strongly oscillatory. This phenomena (already mentioned in [38]) is to be expected since the pressure terms are in fact the terms that are stabilized. We see, however, that good results are obtained for $\alpha \in (10^{-5}, 10^0)$ which depend weakly on the particular value of α in this range (see Figure 10.4). We conclude that *in practice the precise value of* α *is not critical to the accuracy, as long as the dependence on* h_K *and* k_K *are accounted for properly.*

In all our numerical results that follow we use $\alpha = 0.1$.

Numerical experiments for the smooth solution. In Figure 10.5 we present convergence rates for the $h-$ and $p-$version G & GLSFEM that we obtained by approximating the smooth solution (10.24)-(10.25) to the Stokes problem. In the $h-$version we use uniform meshes and expect algebraic convergence rates.

The approximation order for the velocity is choosen to be cubic and this implies a linear approximation of the pressure in the G method. We start with 3 elements in the L-shaped domain and uniformly h-refine the mesh. Note that the meshwidth h is given by $CN^{\frac{1}{2}}$, where N is the number of dof. It is evident from Figure 10.5 that the $h-$version yields algebraic convergence of order 2 for the G method. For the GLS method the $h-$version convergence rate is 3, which is optimal.

Since the exact solution (10.24)-(10.25) is analytic in $\overline{\Omega}$, we expect exponential convergence of the $p-$version. We start again with a 3 element mesh and increase the polynomial approximation order k from 3 to 8 for the velocity. Here, we have $p \approx N^{\frac{1}{2}}$. The convergence rates displayed in Figure 10.5 indicate the exponential convergence of the G & GLS FEM for this smooth solution.

Numerical experiments for the singular solution. In this section we present numerical results for the first solution (10.21)-(10.22). We recall that the solution has a singularity at the reentrant corner. Therefore, it is necessary to perform mesh refinements towards the singularity in order to capture its singular behavior. In Figures 10.6 to 10.9 we present convergence rates that correspond to meshes of affine elements that have been refined geometrically towards the reentrant corner with grading factor $\sigma = 0.5$. An example of such a mesh is displayed in Figure 10.2. This mesh contains $l = 8$ layers of elements, which have been generated by successively refining 3 initial elements. The irregular nodes in this mesh are constrained automatically by HP90.

In Figures 10.6 and 10.7 we show the performance of the $p-$version FEM (resp. the spectral method) by fixing a grid with l layers, and increasing the polynomial approximation order k from 3 to 8. As to be expected, the graphs indicate algebraic rates of convergence which in fact are very close to the a-priori bound of $k^{0.5-2\lambda} \approx N^{0.25-\lambda}$, where λ is the constant in (10.23). This a-priori bound is optimal in view of [4] and the fact that the inf-sup constant γ_N in (10.9) is $Ck^{-0.5}$ in the G method for the elements chosen here [69].

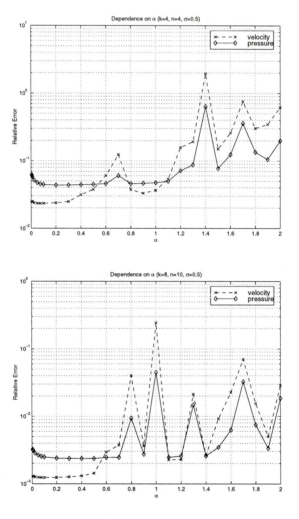

Fig. 10.4. Dependence of the relative error on α .

In Figure 10.7 the same plot is depicted for the GLSFEM and shows a convergence similar to the GFEM. This indicates that the dependence of the inf-sup constant on the approximation order in Theorem 10.2 is probably suboptimal.

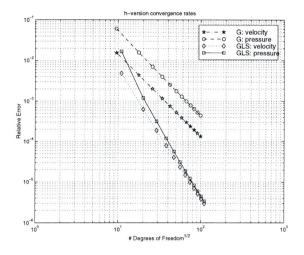

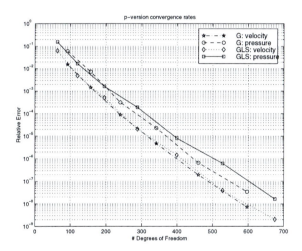

Fig. 10.5. *h*− and *p*−version G & GLS FEM convergence rates.

For the *hp*−version we show the convergence rates in Figures 10.8 and 10.9. Here we do not only vary the polynomial degree but also the grid, i.e. the number of layers in the mesh. We do this with respect to the parameter μ where

$$l = \lfloor \mu \cdot k \rfloor. \tag{10.27}$$

Again, here l is the number of layers and k the polynomial degree. The *hp* convergence rates for various parameters μ indicate the exponential convergence of the *hp*−version, as expected and predicted by Theorems 5.8 and 5.9.

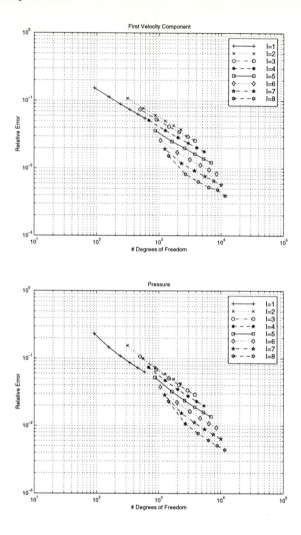

Fig. 10.6. $p-$version GFEM convergence rates for geometric meshes with hanging nodes.

The affine geometric meshes with hanging nodes are obtained by bisecting elements in the middle, which results in a mesh grading factor of $\sigma = 0.5$. With this σ we obtain reliable results, but recall from Remark 5.11 that $\sigma = 0.5$ is not optimal. To study the dependence on σ, we use geometric meshes with variable order elements that have bilinear element mappings. An example mesh with geometric refinement toward the reentrant corner with $\sigma = 0.3$ and 8 layers of elements is shown in Figure 10.10 (the elements in the layers at the reentrant corner are so small that they are not visible in Figure 10.10).

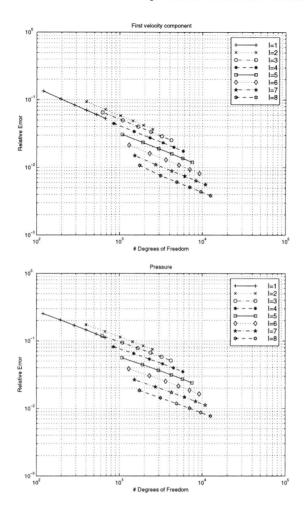

Fig. 10.7. *p*−version GLSFEM conv. rates for geometric meshes with hanging nodes.

We demonstrate the dependence of the GFEM performance on the geometric mesh grading in Figure 10.11. The *hp*−version GFEM is converging exponentially for all values of σ on these geometric meshes in accordance with Theorem 5.9. Further, the performance is best for $\sigma = 0.15$ and $\sigma = 0.2$, which are very close to the optimal σ in one dimension (see Remark 5.11). In particular, for $\sigma = 0.5$ the error is about one order of magnitude larger than for the optimal grading factor 0.15. The best result with $\sigma = 0.5$ is obtained with $N \approx 5000$ while for $\sigma = 0.15$ the same accuracy is already obtained with 1500 dof. This underlines the importance of refining towards the singularity with the grading factor $\sigma = 0.15$.

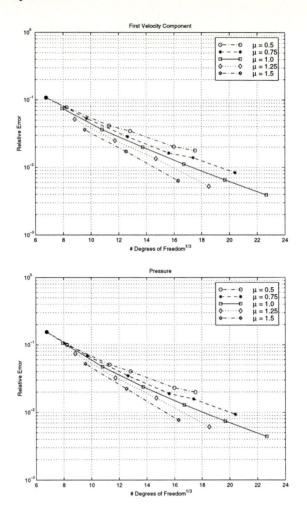

Fig. 10.8. hp−version GFEM conv. rates for geometric meshes with hanging nodes.

10.6   Almost incompressibility

We return to the elasticity problem (9.1) and assume that

$$\overline{\mu} \geq \mu \geq \underline{\mu} > 0, \quad \gamma \gg 1, \tag{10.28}$$

i.e. the medium is almost incompressible. Both, mixed and stabilized methods, are able to handle the limiting, incompressible case $\gamma \to \infty$ and are, in fact, *robust with respect to* γ.

In what follows, we assume that

$$\varepsilon := 2\mu/\gamma \in [0,1]. \tag{10.29}$$

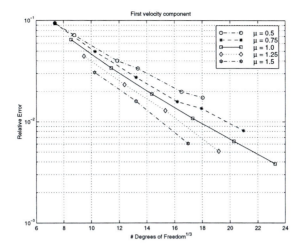

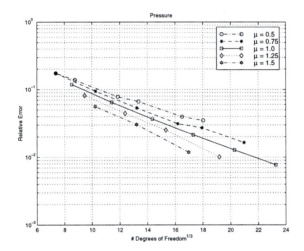

Fig. 10.9. *hp*−version GLSFEM conv. rates for geometric meshes with hanging nodes.

For $\varepsilon > 0$, introduce in (9.1), (9.3) a new variable p by

$$\varepsilon p := -\nabla \cdot \mathbf{u} \qquad (10.30)$$

and obtain the *saddle-point form* of (9.1), (9.3)

$$-2\operatorname{div}(\mu \mathbf{D}(\mathbf{u})) + 2\mu\nabla p = \mathbf{S} \text{ in } \Omega \,,$$
$$\nabla \cdot \mathbf{u} + \varepsilon p = 0 \text{ in } \Omega \,. \qquad (10.31)$$

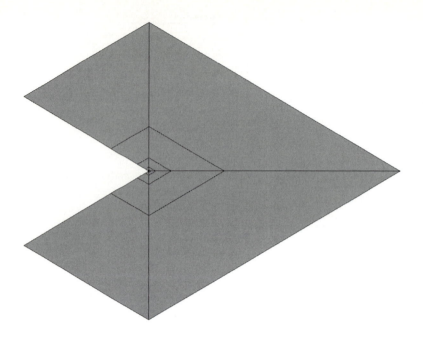

Fig. 10.10. Geometric mesh with 8 layers of elements.

Mixed hp-FEM. The boundary conditions are

$$\mathbf{u} = \mathbf{0} \text{ on } \Gamma_D ,$$
$$2\mu(\mathbf{D}(\mathbf{u}) - p\,\mathbf{1})\mathbf{n} = \mathbf{g} \text{ on } \Gamma_N . \tag{10.32}$$

The variational formulation of (10.31), (10.32) reads:
find $(\mathbf{u}, p) \in H_D^1(\Omega)^d \times L^2(\Omega)$ such that $\forall \mathbf{v} \in H_D^1(\Omega)^d$

$$2(\mu\mathbf{D}(\mathbf{u}), \mathbf{D}(\mathbf{v}))_\Omega - 2(\mu p, \nabla \cdot \mathbf{v})_\Omega = (\mathbf{S}, \mathbf{v})_\Omega + (\mathbf{g}, \mathbf{v})_{\Gamma_N} , \tag{10.33}$$

$$(\nabla \cdot \mathbf{u}, q)_\Omega + (\varepsilon p, q)_\Omega = 0 \quad \forall q \in L^2(\Omega) . \tag{10.34}$$

The discretization proceeds by choosing subspaces $\mathbf{V}_N \subset H_D^1(\Omega)^d$, $M_N \subset L^2(\Omega)$: find $\mathbf{u}_N, p_N \in \mathbf{V}_N \times M_N$ such that (10.33), (10.34) hold for all $\mathbf{v}, q \in \mathbf{V}_N \times M_N$.

Theorem 10.5. *Let $(\mathbf{V}_N, M_N)$ satisfy the discrete inf-sup condition (10.10), with constant $\gamma_N > 0$. Then the bilinear form*

$$B(\mathbf{u}, p; \mathbf{v}, q) := \{2(\mu\mathbf{D}(\mathbf{u}), \mathbf{D}(\mathbf{v}))_\Omega$$
$$- (p, \nabla \cdot \mathbf{v})_\Omega + (\nabla \cdot \mathbf{u}, q)_\Omega + (\varepsilon p, q)_\Omega\} \tag{10.35}$$

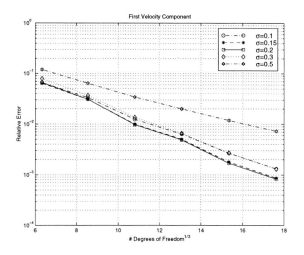

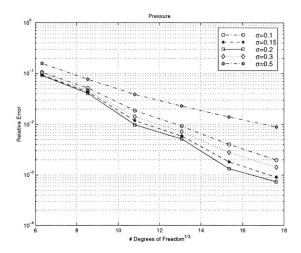

Fig. 10.11. $hp-$version GFEM convergence rates for geometric meshes with $\mu = 1$ and varying σ.

satisfies, for any $\varepsilon \geq 0$,

$$\inf_{\substack{0 \neq \mathbf{u} \in \mathbf{V}_N \\ 0 \neq p \in M_N}} \sup_{\substack{0 \neq \mathbf{v} \in \mathbf{V}_N \\ 0 \neq q \in M_N}} \frac{B(\mathbf{u}, p; \mathbf{v}, q)}{|||(\mathbf{u}, p)||| \; |||(v, q)|||} \geq C \mu \, \gamma_N^2 \tag{10.36}$$

where $C > 0$ is independent of ε, N, μ and

$$|||(u, p)|||^2 := \|\nabla u\|^2 + \|p\|^2 \, .$$

Proof. Let $0 \neq \mathbf{u} \in \mathbf{V}_N$, $0 \neq p \in M_N$. W.l.o.g. assume $\mu = 1/2$. Then

$$B(\mathbf{u}, p; \mathbf{u}, p) = \|\mathbf{D}(\mathbf{u})\|^2 + \|\sqrt{\varepsilon}\, p\|^2 \geq C_K \|\nabla u\|^2$$

by Korn's inequality (9.10), for any $\varepsilon \geq 0$.

By (10.10), for any $0 \neq p \in M_N$ there exists $\mathbf{w}_p \in \mathbf{V}_N$ such that:

$$\|\nabla \mathbf{w}_p\| = \|p\|, \quad -(p, \nabla \cdot \mathbf{w}_p, p) \geq \gamma_N \|p\|^2 .$$

Hence, for any $\theta > 0$ we get

$$\begin{aligned}
B(\mathbf{u}, p; \mathbf{w}_p, 0) &\geq \gamma_N \|p\|^2 - \|\mathbf{D}(\mathbf{u})\|\, \|\mathbf{D}(\mathbf{w}_p)\| \\
&\geq \gamma_N \|p\|^2 - C \|\nabla \mathbf{u}\|\, \|\nabla \mathbf{w}_p\| \\
&= \gamma_N \|p\|^2 - C \|\nabla \mathbf{u}\|\, \|p\| \\
&\geq \gamma_N \|p\|^2 - \frac{C}{2\theta} \|\nabla u\|^2 - \frac{C\theta}{2} \|p\|^2 \\
&= (\gamma_N - C\theta/2) \|p\|^2 - \frac{C}{2\theta} \|\nabla u\|^2 .
\end{aligned}$$

Let $\delta > 0$ and put $\mathbf{v} := \mathbf{u} + \delta \mathbf{w}_p$, $q = p$. Then

$$\begin{aligned}
B(\mathbf{u}, p; \mathbf{v}, q) &= B(\mathbf{u}, p; \mathbf{u}, p) + \delta B(\mathbf{u}, p; \mathbf{w}_p, 0) \\
&\geq (C_K - \delta C/2\theta) \|\nabla u\|^2 + \delta(\gamma_N - C\theta/2) \|p\|^2 .
\end{aligned}$$

Pick $\theta = \gamma_N/C$ and $\delta = C_K \gamma_N/C^2$ to get

$$\begin{aligned}
B(\mathbf{u}, p; \mathbf{v}, q) &\geq \frac{C_K}{2} \|\nabla u\|^2 + \frac{C_K \gamma_N}{C^2} \cdot \frac{\gamma_N}{2} \|p\|^2 \\
&\geq \frac{C_K}{2} \min(1, \gamma_N^2/C^2) \, \|\|(u, p)\|\|^2 .
\end{aligned}$$

Since $0 < \gamma_N \leq \bar{\gamma}$, there is a constant $\widehat{C} > 0$ independent of ε, N

$$\begin{aligned}
\|\|(\mathbf{v}, q)\|\| &\leq \|\|(\mathbf{u}, p)\|\| + \delta \|\nabla \mathbf{w}_p\| = \|\|(\mathbf{u}, p)\|\| + C_K \gamma_N C^{-2} \|p\| \\
&\leq \widehat{C} \|\|(\mathbf{u}, p)\|\| .
\end{aligned}$$

$\square$

Choosing $\mathbf{V}_N$ *and* M_N *as in the Stokes problem, for example, for* $k_K \geq 2$, $K \in \mathcal{T}$,

$$\mathbf{V}_N = S_D^{\mathbf{k},1}(\Omega, \mathcal{T})^d, \quad M_N = S^{\mathbf{k}-2,0}(\Omega, \mathcal{T}) \tag{10.37}$$

with $\mathcal{T}$ *denoting a geometric boundary layer mesh, gives discretizations of* (10.33), (10.34) *which are uniformly stable as* $\varepsilon \to 0$ *(i.e. as* $\gamma \to \infty$*). In* particular, the conditioning of the stiffness matrix corresponding to (10.33), (10.34), which, for constant γ and μ reads,

$$\begin{pmatrix} 2\mu\,\mathbf{A} & -2\mu\,\mathbf{B}^\top \\ 2\mu\,\mathbf{B} & \varepsilon\,\mathbf{M} \end{pmatrix} \tag{10.38}$$

is independent of ε and there holds the stability inequality

$$2\underline{\mu}(\|\mathbf{u}_N\|_{1,\Omega} + \|p_N\|_{0,\Omega}) \leq C\,(\|\mathbf{S}\|_{0,\Omega} + \|\mathbf{g}\|_{0,\Gamma_N}) \tag{10.39}$$

where $C > 0$ is independent of $\varepsilon \geq 0$ (it depends only on the inf-sup constant γ_N in (10.10)); for a proof we refer for example to [16], Chapter II. Note that for $\varepsilon > 0$ the variable p in (10.33), (10.34) is *not* related to the hydrostatic pressure in (1.2).

GLS-stabilized *hp*-FEM. The *hp*-FEM for (10.33), (10.34) requires again elements of different order for $\mathbf{V}_N$ and M_N if robustness w.r. to $\varepsilon = 2\mu/\gamma$ is to hold. Equal order elements cannot achieve robustness. The remedy is again a GLS stabilization: find $(\mathbf{u}_N, p_N) \in \mathbf{V}_N \times M_N$ such that

$$B_\alpha(\mathbf{u}_N, p_N; \mathbf{v}, q) = F_\alpha(\mathbf{v}, q) \quad \forall (\mathbf{v}, q) \in \mathbf{V}_N \times M_N \tag{10.40}$$

where $0 < \alpha \leq \alpha_0$ is a stabilization parameter and (see (10.15))

$$B_\alpha(\mathbf{u}, p; \mathbf{v}, q) :=$$
$$\{2(\mu\,\mathbf{D}(\mathbf{u}), \mathbf{D}(\mathbf{v}))_\Omega - (p, \nabla \cdot \mathbf{v})_\Omega - (\nabla \cdot \mathbf{u}, q)_\Omega - (\varepsilon p, q)_\Omega\}$$
$$-\alpha \sum_{K \in \mathcal{T}} \frac{h_K^2}{k_K^4} (-2(\nabla \cdot \mu\mathbf{D}(\mathbf{u}) + \nabla p)_K, -2(\nabla \cdot \mu\mathbf{D}(\mathbf{v}) + \nabla q))_K \tag{10.41}$$

and

$$F_\alpha(\mathbf{v}, q) := (\mathbf{S}, \mathbf{v})_\Omega + (\mathbf{g}, \mathbf{v})_{\Gamma_N}$$
$$-\alpha \sum_{K \in \mathcal{T}} \frac{h_K^2}{k_K^4} (\mathbf{S}, -2\mu(\nabla \cdot \mathbf{D}(\mathbf{v}) + \nabla q))_K . \tag{10.42}$$

Remark 10.6. The triangulation $\mathcal{T}$ in (10.40) must be shape regular, no GLS stabilization for anisotropic elements is known.

Remark 10.7. In (10.41), (10.42), we assumed that $\mu|_K = $ const. for all $K \in \mathcal{T}$.

Remark 10.8. Note that (10.41) is fully consistent - inserting the exact solution $(\mathbf{u}, p)$, the least squares terms cancel.

The formulation (10.41) is stable uniformly in ε:

Theorem 10.9. *There is $\alpha_0 > 0$ independent of h_K, k_K such that for $0 \leq \alpha \leq \alpha_0$ and all $0 \leq \varepsilon \leq 1$ it holds for*

$$\mathbf{V}_N = S_D^{\mathbf{k},1}(\Omega, \mathcal{T})^d, \quad M_N = S^{\mathbf{k},1}(\Omega, \mathcal{T})$$

(equal order, continuous elements):

$$\inf_{\substack{0 \neq \mathbf{u} \in \mathbf{V}_N \\ 0 \neq p \in M_N}} \sup_{\substack{0 \neq \mathbf{v} \in \mathbf{V}_N \\ 0 \neq q \in M_N}} \frac{B_\alpha(\mathbf{u}, p; \mathbf{v}, q)}{(\|\mathbf{u}\|_1^2 + (1+\varepsilon)\|p\|_0^2)^{1/2}(\|\mathbf{v}\|_1^2 + (1+\varepsilon)\|q\|_0^2)^{1/2}} \geq C\alpha$$

where $C > 0$ depends only on μ and on the shape-regularity of $\mathcal{T}$.

The proof is a slight modification of [27] and omitted here.

10.7 Advection dominated compressible (elastic) flow

So far, we discretized only the "elliptic" part of (1.2), incompressible or elastic. Now we include advection terms of the left hand side of (1.2) into the problem and consider the *compressible Oseen-equations*:

$$-\operatorname{div} \boldsymbol{\tau}(\mathbf{u}) + \mathbf{w} \cdot \nabla \mathbf{u} = \mathbf{S} \text{ in } \Omega, \tag{10.43}$$

$$\mathbf{u} = \mathbf{0} \text{ on } \partial\Omega. \tag{10.44}$$

where again, with ε as in (10.29) and μ, γ constant,

$$\boldsymbol{\tau}(\mathbf{u}) = 2\mu \left\{ \mathbf{D}(\mathbf{u}) + \frac{1}{\varepsilon} \nabla \cdot \mathbf{u} \right\}. \tag{10.45}$$

For $\varepsilon = 0$ we obtain the incompressible limit, i.e. the Oseen equations (10.5). Mixed boundary conditions like (10.32) can also be posed instead of (10.44); for ease of notation we develop the methods for (10.44).

Advection stabilized mixed hp-FEM. In (10.43) we have 2 effects: a) advection dominance, i.e. $|\mathbf{w}|$ large, and b) near incompressibility, i.e. $\varepsilon \to 0$. We handle the latter by adopting a mixed formulation as in (10.31), and the former by GLS stabilization as in the hp-SDFEM in Section 8. We will see that the resulting method is stable independent of the advection size and the incompressibility constraint on high aspect ratio (κ, σ)-boundary layer meshes.

Using (10.30) (note that $p = -\varepsilon^{-1} \nabla \cdot u$ is not related to the pressure in (1.2)), we get in (10.42) with (10.45) the system

$$-2 \operatorname{div} \mu \, \mathbf{D}(\mathbf{u}) + 2 \nabla \mu \, p + \mathbf{w} \cdot \nabla \mathbf{u} = \mathbf{S} \text{ in } \Omega, \tag{10.46}$$

$$\nabla \cdot \mathbf{u} + \varepsilon p = 0 \text{ on } \Omega. \tag{10.47}$$

The stabilized saddle point formulation reads: find $\mathbf{u} \in \mathbf{V}_N$, $p \in M_N$ such that

$$B_\alpha(\mathbf{u}, p; \mathbf{v}, q) = F_\alpha(\mathbf{v}, q) \quad \forall (\mathbf{v}, q) \in \mathbf{V}_N \times M_N \tag{10.48}$$

where $\mathbf{V}_N$ and M_N is a pair of stable spaces for the Stokes Problem, as e.g.

$$\mathbf{V}_N = S_0^{k,1}(\Omega, \mathcal{T})^d, \quad M_N = S^{k-2,0}(\Omega, \mathcal{T}) \tag{10.49}$$

and we define, for a parameter $\alpha > 0$ and with δ_K as in (8.10), the forms

$$B_\alpha(\mathbf{u}, p; \mathbf{v}, q) := 2(\mu \mathbf{D}(\mathbf{u}), \mathbf{D}(\mathbf{v}))_\Omega - (p, \nabla \cdot \mathbf{v})_\Omega$$
$$+ (\nabla \cdot \mathbf{u}, q)_\Omega + (\varepsilon p, q)_\Omega + \frac{1}{2} \{(\mathbf{w} \cdot \nabla \mathbf{u}, \mathbf{v})_\Omega - (\mathbf{u}, \mathbf{w} \cdot \nabla \mathbf{v})_\Omega\} \tag{10.50}$$
$$+ \alpha \sum_K \delta_K (-2(\operatorname{div} \mu \mathbf{D}(\mathbf{u}) - \mu \nabla p) + \mathbf{w} \cdot \nabla \mathbf{u}, \mathbf{w} \cdot \nabla \mathbf{v})_K$$

$$F_\alpha(\mathbf{v}, q) := (\mathbf{S}, \mathbf{v})_\Omega + \alpha \sum_K \delta_K (\mathbf{S}, \mathbf{w} \cdot \nabla \mathbf{v})_K . \tag{10.51}$$

Remark 10.10. In (10.49), (10.50) the stabilization is fully consistent once more, notice, however, that now only the advection term $\mathbf{w} \cdot \nabla \mathbf{v}$ has been stabilized. Stabilization of the incompressibility condition is not needed if either $\varepsilon = 1$ or if the pair $(\mathbf{V}_N, M_N)$ is stable for the Stokes Problem, as e.g. (10.49). In particular, by the stability of (10.37) in (10.33), (10.34) on geometric boundary layer meshes, and with the choice (8.10) of the δ_K, (10.48) is stable on geometric boundary layer meshes also for advection dominated flow. By Theorem 10.9, (10.48), (10.49) will also work for the Oseen problem (10.5), i.e. for $\varepsilon = 0$.

Remark 10.11. In (10.49), (10.50) we used divergence stable mixed elements and stabilized the method only toward the advection term $\mathbf{w} \cdot \nabla \mathbf{u}$. One can, however, also include additional stabilization to accommodate equal order elements for velocity and pressure, i.e. stabilize also against divergence instability. This is done in [30], [74].

11 *hp*-time-stepping

All discretizations considered so far addressed the spatial parts of (1.1) - (1.3), ignoring the time derivative altogether. Here we address the time-discretization of (1.1) - (1.3). We semidiscretize the system (1.1) - (1.3) in time, thereby reducing it to a sequence of nonlinear, convection dominated elliptic-hyperbolic systems in space which are of the type considered above. Thus, our approach is, in a sense, complementary to the usual method of lines. Many time stepping schemes have been proposed in the literature based on schemes from initial-value ODEs and we do not want to survey them here (see eg. [35]). We merely observe here that all of the schemes in [35] are based on

Taylor expansions in time and yield error estimates of order $O(\Delta t^r)$, $r \geq 1$, for solutions smoothly depending on t, as the time step $\Delta t \to 0$; examples are the classical Runge-Kutta methods or the multistep-methods. In none of these methods, error bounds that are explicit in the order k are usually available and if so, they do not allow to deduce spectral convergence for smooth solutions. The error analysis of low order methods, in particular for viscous, incompressible flow, has reached some maturity by now (see [51] and the references there).

Here, we present new hp-time-stepping approaches based on a hp-DGFEM in time [57], [62]. The methods are single step schemes which allow arbitrary variation in order r as well as in the time step Δt. Conceptually, this is reminiscent of the Runge-Kutta-Fehlberg approach to initial value ODEs. However, there are important differences. The hp-DGFEM converge as the order $r \to \infty$ and the time step $\Delta t > 0$ is fixed. They give spectral accuracy in transient problems with smooth time-dependence and, in conjunction with geometric meshes and variable order in time, give exponential convergence for parabolic evolution problems with piecewise analytic (in time) solutions (which arise, e.g. at $t = 0$ for incompatible initial data or for piecewise analytic forcing terms).

Moreover, they are unconditionally stable for parabolic problems *independent of the spatial discretization*. This is crucial, since hp-FEM in space require highly anisotropic meshes for efficient resolution of layers and fronts which tend to produce very stringent CFL limitations in explicit schemes. The underlying variational structure of hp-DGFEM allows moreover for a-posteriori error estimation and adaptivity.

We proceed as follows: we first elaborate on the hp-DGFEM and the hp-SDFEM for first order hyperbolic equations as e.g. (1.1). It turns out that the analysis in Chapter 7 applies directly here as well. Next, we present the hp-DG time stepping technique from [57,62] for (systems of nonlinear) parabolic initial value problems. Finally, we apply this technique to some parabolic model evolution problems and discuss convergence results as well as implementation issues.

11.1 hp-FEM for first order transient, hyperbolic problems

In a bounded domain $\Omega \subset \mathbb{R}^d$ and for $0 < t < T$, consider the unsteady linear advection problem

$$\frac{\partial u}{\partial t} + \mathbf{a} \cdot \nabla u + bu = S \quad \text{in } (0, T) \times \Omega \tag{11.1}$$

where $\mathbf{a} \in C^1(\overline{\Omega})^d$, $b \in C(\overline{\Omega})$ and $S \in L^2(\Omega)$. This is the transient variant of (7.1) and, in fact, a special case of it: we put

$$\hat{x} := (t, x) \in Q := (0, T) \times \Omega \subset \mathbb{R}^{d+1},$$

$$\hat{\mathcal{L}}u := \hat{\mathbf{a}} \cdot \widehat{\nabla} u + bu$$

where $\hat{\mathbf{a}} := (1, a_1, a_2, \ldots, a_d)^\top$, $\widehat{\nabla} = (\partial_t, \partial_1, \ldots, \partial_d)$. Then (11.1) takes the form (7.1), and the initial condition

$$u(\cdot, 0) = u_0 \quad \text{in} \quad \Omega \tag{11.2}$$

becomes simply an "inflow" boundary condition on $\Omega \times \{t = 0\}$. We may therefore discretize now (11.1), (11.2) in $\mathbb{R}^{d+1}$ as proposed in Section 7 - the resulting method will allow, in fact, arbitrary combinations of space and time meshes and orders, if the *hp*-DGFEM in Section 7.3 is used. For example, Fig. 11.1 shows a possible mesh in $d = 1$: here $Q = (0, T) \times (0, 1)$.

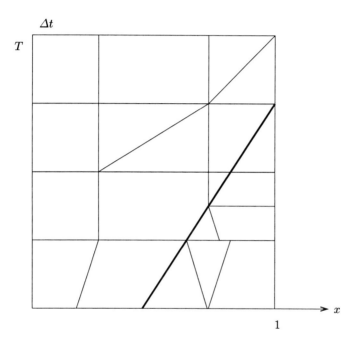

Fig. 11.1. Space time mesh for *hp*-DGFEM

Notice that the element boundaries need not be aligned with the (x, t) axes - this is essential if propagating perturbations arising in hyperbolic equations are to be tracked accurately with large time steps (see the bold line in Fig. 11.1).

Note also that we still kept in Fig. 11.1 time levels - the first order problems can be solved explicitly by propagating information with the flow $\hat{\mathbf{a}}$ through the elements. It should also be clear from Section 7 that space and time orders can be varied independently here. We shall not go into detailed error estimates here.

We next present the DG(r) scheme for nonlinear initial value ODEs. This is of independent interest also for high order MOL discretizations. In the following subsection we address then the combined space-time discretization of parabolic problems.

11.2   The DG(r)-FEM for nonlinear initial value problems

Let $J = [t_0, t_0 + T]$ for $T > 0$. Let $f : J \times \mathbb{R}^M \to \mathbb{R}^M$ be continuous and $\mathbf{u}_0 \in \mathbb{R}^K$ be given. Consider the IVP

$$\mathbf{u}'(t) = \mathbf{f}(t, \mathbf{u}(t)), \quad t \in J, \ \mathbf{u}(t_0) = \mathbf{u}_0 . \tag{11.3}$$

We assume that $\mathbf{f}(t, \mathbf{u})$ is uniformly Lipschitz continuous w.r. to $\mathbf{u}$, i.e. $\mathbf{f}$ satisfies

$$\|\mathbf{f}(t, \mathbf{u}) - \mathbf{f}(t, \mathbf{v})\| \leq L \|\mathbf{u} - \mathbf{v}\| \quad \mathbf{u}, \mathbf{v} \in \mathbb{R}^M, t \in J . \tag{11.4}$$

(11.4) implies that (11.3) admits a unique solution $\mathbf{u}(t) \in C^1(J; \mathbb{R}^<)$.

Let $\mathcal{M}$ denote a partition of J into N timesteps $t_0 < t_1 < \cdots < t_{N-1} < t_N = t_0 + T$ and set $\Delta t_n := t_n - t_{n-1}$, $n = 1, \ldots, N$, $\Delta t = \max\{\Delta t_n : 1 \leq n \leq N\}$. Let $\varphi : J \to \mathbb{R}^M$ be a piecewise continuous function on $\mathcal{M}$. Then we define the one-sided limits

$$\varphi_n^\pm := \lim_{0 < s \to 0} \varphi(t_n \pm s), \quad 0 \leq n \leq N - 1,$$

and the jumps

$$[\varphi]_n = \varphi_n^+ - \varphi_n^- .$$

On the time-mesh $\mathcal{M}$, we introduce

$$C_b^0(\mathcal{M}; \mathbb{R}^M) := \{\varphi : J \to \mathbb{R}^M \ |\varphi|_{I_n} \in C_b^0(I_n; \mathbb{R}^M)\}$$

of $\mathbb{R}^M$-valued, piecewise continuous and bounded functions. The IVP (11.3) admits the following variational formulation: find $\mathbf{u} \in C_b^0(\mathcal{M}; \mathbb{R}^M)$ such that, for all $\varphi \in C_b^0(\mathcal{M}; \mathbb{R}^M)$,

$$\sum_{n=1}^N \int_{I_n} \langle \mathbf{u}'(t) - \mathbf{f}(t, \mathbf{u}(t)), \varphi(t) \rangle \, dt +$$
$$\sum_{n=2}^N \langle [\mathbf{u}]_{n-1}, \varphi_{n-1}^+ \rangle + \langle \mathbf{u}_0^+, \varphi_0^+ \rangle = \langle \mathbf{u}_0, \varphi_0^+ \rangle . \tag{11.5}$$

DG(r)-discretization. We associate with each time interval I_n a polynomial degree $r_n \geq 0$ and combine these degrees in the vector $\mathbf{r} = \{r_n\}_{n=1}^N$. Define the subspace

$$\mathcal{V}^{\mathbf{r}}(\mathcal{M}; \mathbb{R}^M) = \{\varphi : J \to \mathbb{R}^M \ |\varphi|_{I_n} \in \mathcal{P}^{r_n}(I_n; \mathbb{R}^M), 1 \leq n \leq N\} . \tag{11.6}$$

As before, if $r_n = r$ for all n, we write $\mathcal{V}^r(\mathcal{M}; \mathbb{R}^M)$. Then the DG(**r**)-method reads: find $\mathbf{U} \in \mathcal{V}^{\mathbf{r}}(\mathcal{M}; \mathbb{R}^M)$ such that

$$
\begin{aligned}
&\sum_{n=1}^{N} \int_{I_n} \langle \mathbf{U}'(t) - \mathbf{f}(t, \mathbf{U}(t)), \ \boldsymbol{\varphi}(t) \rangle \, dt + \\
&\sum_{n=2}^{N} \langle [\mathbf{U}]_{n-1}, \boldsymbol{\varphi}_{n-1}^+ \rangle + \langle \mathbf{U}_0^+, \boldsymbol{\varphi}_0^+ \rangle = \langle \mathbf{u}_0, \boldsymbol{\varphi}_0^+ \rangle
\end{aligned} \tag{11.7}
$$

for all $\boldsymbol{\varphi} \in \mathcal{V}^{\mathbf{r}}(\mathcal{M}; \mathbb{R}^M)$.

Notice that (11.7) is only apparently global - owing to the discontinuity of the $\boldsymbol{\varphi}$, (11.5) amounts to solving successively on I_n, $n = 1, 2, 3, \ldots$ the problems

$$
\int_{I_n} \langle \mathbf{U}'(t) - \mathbf{f}(t, \mathbf{U}(t)), \ \boldsymbol{\varphi}(t) \rangle \, dt + \langle \mathbf{U}_{n-1}^+, \boldsymbol{\varphi}_{n-1}^+ \rangle = \langle \mathbf{U}_{n-1}^-, \boldsymbol{\varphi}_{n-1}^+ \rangle , \tag{11.8}
$$

for all $\boldsymbol{\varphi} \in \mathcal{P}^{r_n}(I_n; \mathbb{R}^M)$.

In each timestep, this is a system of $M(r_n + 1)$ nonlinear equations for the polynomial coefficients of $\mathbf{U}|_{I_n}$.

To solve it, we propose the *fixed point iteration*:

Let $\widetilde{\mathbf{U}} \in \mathcal{P}^{r_n}(I_n, \mathbb{R}^M)$ be given. Then $\mathbf{U} = T\widetilde{\mathbf{U}}$ is the solution of the nonlinear problem: $\forall \boldsymbol{\varphi} \in \mathcal{P}^{r_n}(I_n : \mathbb{R}^M)$:

$$
\begin{aligned}
&\int_{I_n} \langle \mathbf{U}'(t), \boldsymbol{\varphi}(t) \rangle \, dt + \langle \mathbf{U}_{n-1}^+, \boldsymbol{\varphi}_{n-1}^+ \rangle \\
&= \langle \mathbf{U}_{n-1}^-, \boldsymbol{\varphi}_{n-1}^+ \rangle + \int_{I_n} \langle \mathbf{f}(t, \widetilde{\mathbf{U}}(t)), \boldsymbol{\varphi}(t) \rangle \, dt
\end{aligned} \tag{11.9}
$$

A fixed point $\widetilde{\mathbf{U}} = T\widetilde{\mathbf{U}}$ of (11.9) solves (11.8). We have [57]:

Theorem 11.1. *Let* $r_n \geq 0$ *be arbitrary and assume the CFL-condition*

$$
\Delta l = \max\{\Delta t_n : n = 1, \ldots, N\} < \sqrt{\frac{6}{5}} \, L^{-1}. \tag{11.10}
$$

Then (11.8) *has a unique solution and the fixed point iteration*

$$
\widetilde{\mathbf{U}}^{\ell+1} = T\widetilde{\mathbf{U}}^\ell, \quad \widetilde{\mathbf{U}}^0 = \mathbf{U}_{n-1}^-
$$

converges.

Using a more sophisticated iteration (eg. Newton's Method), larger time steps are allowable. The error $\mathbf{u} - \mathbf{U}$ can be estimated as follows:

Theorem 11.2. *There is* $c > 0$ *independent of* **r** *and* $\mathcal{M}$ *and* K *such that the DG(**r**) solution* $\mathbf{U} \in \mathcal{V}^r(\mathcal{M}; \mathbb{R}^M)$ *of* (11.7) *satisfies.*

$$
\begin{aligned}
\|\mathbf{u} - \mathbf{U}\|_{L^2(I_n, \mathbb{R}^M)} &\leq c(1 + LT \exp(cLT))^{\frac{1}{2}} \\
&\max_{1 \leq i \leq N} \|\mathbf{u} - \mathbf{V}\|_{L^2(I_i, \mathbb{R}^M)} .
\end{aligned} \tag{11.11}
$$

If the time steps Δt_n are monotonically increasing,

$$\|\mathbf{u} - \mathbf{U}\|_{L^2(J,\mathbb{R}^M)} \le c(1 + LT \, \exp(cLT))^{\frac{1}{2}} \|\mathbf{u} - \mathbf{V}\|_{L^2(J;\mathbb{R}^M)} \, . \qquad (11.12)$$

Here $\mathbf{V} \in \mathcal{V}^\mathbf{r}(\mathcal{M}; \mathbb{R}^M)$ is the interpolant on I_n defined by $\mathbf{V}_n^- := \mathbf{u}(t_n)$, and

$$\int_{I_n} \langle \mathbf{V}, \varphi \rangle \, dt = \int_{I_n} \langle \mathbf{u}, \varphi \rangle \, dt \quad \forall \varphi \in \mathcal{P}^{r_n}(I_n; \mathbb{R}^M) \, . \qquad (11.13)$$

Remark 11.3. The results hold also when $\mathbb{R}^M$ is replaced by a Hilbert-space with norm $\| \circ \|$ and inner product $\langle \cdot, \cdot \rangle$.

11.3   DG(r)-FEM for abstract initial boundary value problems

We will generalize the *hp* DG-FEM to abstract parabolic equations, including convection dominated diffusion and viscous, incompressible flows.

Abstract Setting. Let X, H be complex, separable Hilbert spaces, $X \hookrightarrow H$ with dense injection and norms $\| \cdot \|_X$ and $\| \cdot \|_H$, respectively. Denote the scalar product on H by $(\cdot, \cdot)_H$ and identify H and H^*, the antidual of H. We get the Gelfand triple

$$X \hookrightarrow H \cong H^* \hookrightarrow X^* \qquad (11.14)$$

and write $(\cdot, \cdot)_{X^* \times X}$ for the $X^* \times X$ duality pairing and $\| \cdot \|_{X^*}$ for the norm in X^*. Typically, for viscous flow, we have $H = L^2(\Omega)$ and $X = H_0^1(\Omega)$.

Let $J = (a, b)$ be a time interval. Then the weak time derivative of an X^*-valued distribution $u \in \mathcal{D}'(J; X^*)$ is

$$\int_J (\dot{u}, v)_{X^* \times X} \varphi(t) dt = - \int_J (u(t), v)_H \dot{\varphi}(t) dt \qquad (11.15)$$

for all $v \in X, \varphi \in \mathcal{D}(J)$. This time derivative has the following properties:

$$u \in L^2(J; X), \ \dot{u} \in L^2(J; X^*) \Longrightarrow$$
$$\qquad\qquad\qquad\qquad\qquad\qquad\qquad\qquad (11.16\mathrm{a})$$
$$u \in C([a, b]; H) \ u, v \in L^2(J; X), \dot{u}, \dot{v} \in L^2(J; X^*) \Longrightarrow$$

$$(u(t), v(t))_H - (u(s), v(s))_H$$
$$= \int_s^t (\dot{u}, v)_{X^* \times X} d\tau + \int_s^t (u, \dot{v})_{X^* \times X} d\tau \qquad (11.16\mathrm{b})$$

$$\|u(t)\|_H^2 - \|u(s)\|_H^2 = 2\mathrm{Re} \int_s^t (\dot{u}, u)_{X^* \times X} d\tau \quad \forall s, t \in [a, b] \, . \qquad (11.16\mathrm{c})$$

On the Gelfand triple (11.14) we introduce the *spatial sesquilinear form*

$$a : X \times X \to \mathbb{C} .$$

We call the (possibly nonsymmetric) form $a(\cdot, \cdot)$ (α, β)-elliptic, if

$$|a(u,v)| \quad \leq \alpha \|u\|_X \|v\|_X \quad \forall u, v \in X ,$$

$$\operatorname{Re} a(u,u) \geq \beta \|u\|_X^2 \qquad \forall u \in X .$$

The form $a(\cdot, \cdot)$ corresponds to a weak formulation of the differential operator $L : X \to X^*$ which could be any of the operators previously considered.

Let now $0 < T < \infty$ and $J = (0, T)$. Consider the abstract evolution problem:
given $g \in L^2(J; X^*)$, $u_0 \in H$, find

$$\dot{u}(t) + Lu(t) = g(t) \tag{11.17a}$$

$$u(0) = u_0 . \tag{11.17b}$$

The weak formulation of (11.17) is:
find $u \in L^2(J; X) \cap H^1(J; X)$, $u(0) = u_0$, such that

$$-\int_J (u(t), v)_H \, \dot{\varphi}(t) dt + \int_J a(u, v) \varphi(t) dt$$

$$= \int_J (g(t), v)_{X^* \times X} \varphi(t) dt \quad \forall v \in X, \quad \forall \varphi \in \mathcal{D}(J). \tag{11.18}$$

The DG(r) method. We discretize (11.18) in time, reducing it to a sequence of spatial problems involving the form $a(\cdot, \cdot)$ which can be discretized using the *hp*-FEM in Chapters 1-10.

Let $\mathcal{M}$ be a partition of $J = (0, T)$ into N time intervals

$$I_n = [t_{n-1}, t_n], \quad 1 \leq n \leq N, \quad 0 = t_0 < t_1 < \cdots < t_N = T .$$

and set $\Delta t_n := t_n - t_{n-1} = |I_n|$.

Set further $\Delta t = \max_n \Delta t_n$. For $u : J \to X$, define the one-sided limits

$$u_n^\pm := \lim_{0 < s \to 0} u(t_n \pm s), \quad 0 \leq n \leq N - 1$$

and $u_N^- = \lim_{0 < s \to 0} u(T - s)$. For $1 \leq n \leq N - 1$, set $[u]_n = u_n^+ - u_n^-$ and introduce on $\mathcal{M} = \{I_n\}_{n=1}^N$ the space

$$C_b^0(\mathcal{M}; X) = \{u : J \to X \,|\, u \in C_b^0(I_n; X)\} .$$

By integration by parts in time and elementary algebra, we obtain

Proposition 11.4. *A weak solution* $u \in L^2(J; X) \cap H^1(J; X^*)$ *of* (11.17) *satisfies*

$$\sum_{n=1}^{N} \int_{I_n} (\dot{u} + Lu, v)_{X^* \times X} + \sum_{n=2}^{N} ([u]_{n-1}, v_{n-1}^+)_H + (u_0^+, v_0^+)_H =$$

$$(u_0, v_0^+)_H + \sum_{n=1}^{N} \int_{I_n} (g, v)_{X^* \times X} dt \tag{11.19}$$

for all $v \in C_b^0(\mathcal{M}; X)$.

Let now
$$\mathcal{P}^r(I; X) = \{p : I \to X : p(t) = \sum_{j=0}^{r} t^j x_j, x_j \in X\}$$

and
$$\mathcal{V}^r(\mathcal{M}; X) = \{u : J \to X : u|_{I_n} \in \mathcal{P}^{r_n}(I_n; X), 1 \le n \le N\} .$$

Then the DG(r) method reads:
find $U \in \mathcal{V}^r(\mathcal{M}, X)$ such that

$$B(U, V) := \sum_{n=1}^{N} \int_{I_n} (\dot{U}, V)_{X^* \times X} dt + \sum_{n=1}^{N} \int_{I_n} (LU, V)_{X^* \times X} dt +$$

$$\sum_{n=2}^{N} ([U]_{n-1}, V_{n-1}^+)_H + (U_0^+, V_0^+)_H \tag{11.20}$$

$$= (u_0, V_0^+)_H + \sum_{n=1}^{N} (g(t), V)_{X^* \times X} dt$$

for all $V \in \mathcal{V}^r(\mathcal{M}; X)$.

Once again, (11.20) can be solved recursively. On each I_n, we get an elliptic system for $r_n + 1$ unknown fields in X. The structure of this system and the algorithmic complexity of its solution crucially depend on the basis functions chosen in time.

Let I be any time interval and let $\{\hat{\varphi}_i\}_{i=0}^r$ and $\{\hat{\psi}_i\}_{i=0}^r$ be two bases of $\mathcal{P}^r(-1, 1)$, and denote by φ_i, ψ_i their transported variants on I.
Then

$$\frac{d\varphi_i}{dt} = \frac{2}{\Delta t} \frac{d\hat{\varphi}_i}{d\hat{t}}, \quad \frac{d\psi_i}{dt} = \frac{2}{\Delta t} \frac{d\hat{\psi}_i}{d\hat{t}} . \tag{11.21}$$

We introduce the matrices

$$\widehat{A}_{ij} = \widehat{A}_{ij}^1 + \widehat{A}_{ij}^2 = \int_{-1}^{1} \hat{\varphi}_j' \, \hat{\psi}_i \, d\hat{t} + \hat{\varphi}_j(-1)\hat{\varphi}_i(-1) \tag{11.22}$$

$$\widehat{B}_{ij} = \int_{-1}^{1} \hat{\varphi}_j \hat{\varphi}_i d\hat{t} \tag{11.23}$$

Then $U, V \in \mathcal{P}^r(I, X)$ can be written as

$$U = \sum_{j=0}^{r} U_j \varphi_j, \quad V = \sum_{i=0}^{r} V_i \psi_i, \quad U_j, V_i \in X .$$

The problem (11.20) in time interval I_n is equivalent to the elliptic system: find $U_j \in X$ such that

$$\sum_{j=0}^{r_n} \widehat{A}_{ij}(U_j, V_i) + \frac{\Delta t}{2} \widehat{B}_{ij} a(U_j, V_i) = \frac{\Delta t}{2} \hat{f}_i^1 + \hat{f}_i^2, \quad i = 0, ..., r_n. \tag{11.24}$$

Here we defined, for $V \in X$,

$$\hat{f}_i^1(V) := \left(V, \int_I g \psi_i dt \right) \quad \text{and} \quad \hat{f}_i^2 := (U_{n-1}^-, V) \, \hat{\psi}_i(-1) .$$

Remark 11.5. If $r_n = 0$, (11.20) corresponds to backward Euler and if $r_n = 1$, analogs of the Crank-Nicolson scheme are obtained.

Spectral decoupling. The convection-diffusion system (11.24) is very costly to solve, particularly in three space dimensions and for $r_n > 0$. In practice, it is therefore very important that (11.24) can, in fact, be decoupled into r_n independent equations of the same type. This is achieved by a clever selection of the basis functions φ_i, ψ_i in (11.23). Assume that we have, for $r > 0$, an $(r+1) \times (r+1)$ matrix $\mathbf{M}$ such that

$$\mathbf{M}^{-1} \widehat{\mathbf{A}} \, \mathbf{M} = \mathrm{diag}\{\sigma_i\}_{i=0}^r, \quad \mathbf{M}^{-1} \widehat{\mathbf{B}} \mathbf{M} = \mathbf{1} . \tag{11.25}$$

Then, changing from the unknowns U_i to $\widehat{U}_i$ by

$$\widehat{U}_i = \sum_{j=0}^{r} (\mathbf{M}^{-1})_{ij} \, U_j ,$$

(11.24) decouples and becomes:
for $j = 0, \ldots r$, find $\widehat{U}_j \in X$ such that

$$a(\widehat{U}_j, \widehat{V}_j) + \frac{2\sigma_j}{\Delta t} (\widehat{U}_j, V_i) = \tilde{f}_j^1 + \frac{2}{\Delta t} \tilde{f}_j^2 \tag{11.26}$$

for all $\widehat{V}_j \in X$, where the $\tilde{f}_j^k$ are certain linear combinations of the $\hat{f}_i^k$. We observe that (11.26) is an elliptic system with an additional mass matrix added, completely analogous to the system resulting from the backward Euler scheme for $r = 0$.

There remains the question, if $\mathbf{M}$ in (11.25) can be found for $r > 0$. This is theoretically open. In practice, up to $r = 50$, such matrices can be found.

They are, in all cases, complex as are the σ_j, nevertheless, the additional gain by decoupling the system (11.24) is worth the use of complex arithmetic, especially in dimension $d = 3$.

Note also that [62]

$$|\sigma_r| \sim r^2 \quad \text{as} \;\; r \to \infty \,,$$

so that the problems (11.26) are singularly perturbed as $\Delta t \to 0$ or as $r \to \infty$.

hp-error estimates. With the hp-DGFEM, exponential convergence rates for the time stepping scheme (11.20) can be achieved. We present here one result from [57]. The starting point is the following abstract error estimate, valid for any $\mathbf{r}$ and $\mathcal{M}$.

Theorem 11.6. *Let $u \in L^2(J, X) \cap H^1(J; X^*)$ be the solution of (11.18) for an elliptic operator L and let $U \in \mathcal{V}^{\mathbf{r}}(\mathcal{M}, X)$ be the discrete solution of the DGFEM (11.20).*

Let $Iu \in \mathcal{V}^{\mathbf{r}}(\mathcal{M}, X)$ be the interpolant of U defined on each time interval I_n by $1 \leq n \leq N$, by

$$\int_{I_n} (u - Iu, \varphi)_{X^* \times S} dt = 0 \quad \forall \varphi \in \mathcal{P}^{r_n - 1}(J; X), \;\; (u - Iu)_n^- = 0 \;\; \text{in} \;\; X \,.$$

Then there holds the error estimate

$$\|u - U\|_{L^2(J;X)}^2 \leq 2\left(1 + \frac{\alpha}{\beta}\right) \|u - Iu\|_{L^2(J;X)}^2 \,.$$

11.4   An example: Heat-equation

DG - discretization. Here $L = -\Delta$ and $X = H_0^1(\Omega)$, $H = L^2(\Omega)$. On a generic time interval $I = (a, b)$, $\Delta t = b - a$, we have to solve: find $U \in \mathcal{P}^r(I; H_0^1(\Omega))$ such that

$$\int_I \{(\dot{U}, V) + (\nabla U, \nabla V)\} dt + (U(a), V(a)) =$$

$$\int_I (g(t), V) dt + (U_a^-, V(a)) \quad \forall V \in \mathcal{P}^r(I : H_0^1(\Omega)) \,. \tag{11.27}$$

Here $\{\hat{\varphi}_i\}_{i=0}^r$ and $\{\hat{\psi}_i\}_{i=0}^r$ are two bases of $\mathcal{P}^r(-1, 1)$, and we denote by φ_i, ψ_i their transported variants on I. Then

$$\frac{d\varphi_i}{dt} = \frac{2}{\Delta t} \frac{d\hat{\varphi}_i}{d\hat{t}}, \quad \frac{d\psi_i}{dt} = \frac{2}{\Delta t} \frac{d\hat{\psi}_i}{d\hat{t}} \,. \tag{11.28}$$

Again, $U, V \in \mathcal{P}^r(I, H_0^1(\Omega))$ can be written as

$$U = \sum_{j=0}^r U_j \varphi_j, \quad V = \sum_{i=0}^r V_i \psi_i, \quad U_j, V_i \in H_0^1(\Omega) \,.$$

Inserting this into (11.27) yields:
find $\{U_j\}_{j=0}^r \subset H_0^1(\Omega)$ such that

$$\sum_{i,j=0}^r \left\{ \left[\int_I \varphi_j' \psi_i dt + \varphi_j(a) \psi_i(a) \right] (U_j, V_i) + \left[\int_I \varphi_j \psi_i dt \right] (\nabla U_j, \nabla V_i) \right\}$$
$$= \sum_{i=0}^r \left\{ \left(V_i, \int_I g \psi_i dt \right) + (U_a^-, V_i) \psi_i(a) \right\} \tag{11.29}$$

for all $\{V_i\}_{i=0}^r \subset H_0^1(\Omega)$.

We introduce the matrices $\widehat{A}_{ij}$, $\widehat{B}_{ij}$ as in (11.22), (11.23).

Then problem (11.29) is the elliptic system:
find $\{U_j\}_{j=0}^r \subset H_0^1(\Omega)$ such that

$$\sum_{j=0}^r \left\{ \widehat{A}_{ij} U_j - \frac{\Delta t}{2} \widehat{B}_{ij} \Delta U_j \right\} = \frac{\Delta t}{2} \hat{f}_i^1 + \hat{f}_i^2, \quad i = 0, \dots, r \tag{11.30}$$

where, for $V \in H_0^1(\Omega)$,

$$\hat{f}_i^1(V) := \left(V, \int_I g \hat{\psi}_i dt \right), \quad \hat{f}_i^2(V) = (U_a^-, V) \hat{\psi}_i(-1) .$$

Decoupling. The work for the solution of the coupled system (11.30) is substantial, in particular in three space dimensions. Using the simultaneous diagonalization of the matrices A and B, however, there exists M such that

$$\mathbf{M}^{-1} \widehat{\mathbf{A}} \, \mathbf{M} = \operatorname{diag}\{\sigma_i\}_{i=0}^r, \quad \mathbf{M}^{-1} \widehat{\mathbf{B}} \, \mathbf{M} = \mathbf{1}$$

with complex σ_i, however.

Then, changing bases $\widehat{U}_i = M_{ij}^{-1} U_j$, (11.30) decouples into $r + 1$ scalar problems:

$$-\frac{\Delta t}{2} \Delta \widehat{U}_i + \sigma_i \widehat{U}_i = \frac{\Delta t}{2} \tilde{f}_i^1 + \tilde{f}_i^2 \iff$$
$$-\Delta \widehat{U}_i + \frac{2\sigma_i}{\Delta t} \widehat{U}_i = \tilde{f}_i^1 + \frac{2}{\Delta t} \tilde{f}_i^2, \quad i = 0, \dots, r . \tag{11.31}$$

More generally, in the context of (11.20), we get the $r+1$ decoupled problems:
find $\widehat{U}_i \in X$ such that

$$a(\widehat{U}_i, V) + \frac{2\sigma_i}{\Delta t} (\widehat{U}_i, V)_H = \tilde{f}_i^1 + \frac{2}{\Delta t} \tilde{f}_i^2 \quad \forall V \in X . \tag{11.32}$$

We see that we must solve in each timestep I_n altogether $r_n + 1$ independent elliptic systems of reaction-diffusion type discussed in Section 6 with same principal part and different right hand side. This can be done in parallel when each system is assigned to one processor. Notice also that (11.32) is, for small Δt or large r, singularly perturbed, regardless of the presence of small viscosity effects in $a(\cdot, \cdot)$. Problem (11.32) is now in the form considered in Sections 1-10, and any of the techniques there can be used for space discretization.

hp-error estimates. Combining the abstract a-priori error estimate Theorem 11.6 with time regularity of the heat equation, we obtain exponential convergence, even if the initial data u_0 does not satisfy any compatibility condition.

Theorem 11.7. *Consider the heat equation*

$$u_t - \Delta u = g \text{ in } \Omega \times (0, T) \quad u = 0 \text{ on } \partial\Omega \times (0, T)$$

with initial data $u_0 \in H^\theta(\Omega) := (L^2(\Omega), H^1(\Omega))_{\theta, 2}$ for some $0 < \theta < 1/2$, and analytic right hand side g satisfying

$$\|g^{(\ell)}(t)\|_{L^2(\Omega)} \le C\ell! \, d^\ell \quad t \in [0, T], \ \ell \in \mathbb{N}_0 \,.$$

Discretize it in time using the hp-DGFEM on a geometric mesh $\mathcal{M}_{n,\sigma}$ with n layers and grading factor $0 < \sigma < 1$ with degrees r_i satisfying, for some $\mu > 0$,

$$r_1 = 0, \ r_j \ge \lfloor \mu j \rfloor, \ j = 2, \ldots, n \,.$$

Then the semidiscrete solution U obtained from (11.27) satisfies

$$\|u - U\|^2_{L^2(J; H^1_0(\Omega))} \le C \exp(-bn) \le C \exp(-bM^{1/2})$$

where M denotes the number of spatial problems to be solved.

Remark 11.8. We emphasize that no compatibility is required for the initial data for Theorem 11.7 to hold. Analogous results hold also in the abstract setting of Section 11.3. if the operator L is the infinitesimal generator of an analytic semigroup [62].

References

1. I. Babuška and A.K. Aziz: Survey lectures on the mathematical foundations of the finite element method, in: The mathematical foundations of the finite element method, A.K.Aziz and I. Babuška (Eds.), Academic Press, (1972).
2. I. Babuška and B.Q. Guo, The hp-version of the finite element method for domains with curved boundaries, SIAM J. Numer. Anal. **25** (1988), 837-861.
3. I. Babuška and B.Q. Guo, Approximation properties of the hp-version of the finite element method, Comp. Meth. Appl. Mech. Engg. **133** (1996), 319-346.

4. I. Babuška and M. Suri, The *p* and *hp*-FEM - a survey. SIAM Review **36** (1994), 578-632.

5. C.E. Baumann, A new *hp*-adaptive Discontinuous Galerkin Finite Element Method for Computational FLuid Dynamics, Dissertation, Texas Inst. Comp. & Appl. Math., UT Austin, TX 78712 (1997).

6. C.E. Baumann and J.T. Oden, A discontinuous *hp* Finite Element method for the Navier-Stokes equations, 10th Int. Conf. on Finite Elements in Fluid Tucson, Arizona, Jan 5-8, 1998.

7. C.E. Baumann and J.T. Oden, A discontinuous *hp* Finite Element Method for the Euler Equations of gas dynamics, 10th Int. Conf. on Finite Elements in Fluid Tucson, Arizona, Jan 5-8, 1998.

8. F. Ben Belgacem, The Mortar Finite Element Method with Lagrange Multipliers, Preprint, (1993).

9. F. Ben Belgacem and Y. Maday, The mortar finite element method for three dimensional finite elements, RAIRO Anal. Numérique **31** (1997), 289-302.

10. C. Bernardi and Y. Maday, Approximations spectrales de problèmes aux limites elliptiques, Springer-Verlag Paris, (1992).

11. C. Bernardi, Y. Maday and A.T. Patera, Domain decomposition by the mortar element method, in *Asymptotic and Numerical Methods for PDEs with Critical Parameters*, H.G. Kaper and M. Garbey (Eds.), (1993), 269-286.

12. K. Bey and J.T. Oden, *hp*-Version discontinuous Galerkin methods for hyperbolic conservation laws, Comp. Math. Appl. Mech. Engg. **133** (1996), 259-286.

13. K. Bey, J.T. Oden and A. Patra, *hp*-Version discontinuous Galerkin Methods for hyperbolic conservation laws, Comp. Meth. Appl. Mech. Engg. **133** (1996), 259-286.

14. D. Braess, W. Dahmen, Stability estimates of the Mortar Finite Element Method for 3-Dimensional Problems, Report 162, IGPM, RWTH Aachen, (1998).

15. J. Bramble, J. Pasciak, J. Wang and J. Xu, Convergence estimates for multigrid algorithms without regularity assumptions, Math. Comp. **37** (1991), 23-45.

16. F. Brezzi and M. Fortin, Mixed and Hybrid Finite Element Methods, Springer Verlag, (1991).

17. C. Canuto, M.Y. Hussaini, A. Quarteroni and T.A. Zhang, Spectral Methods in fluid dynamics, Springer-Verlag, (1988).

18. M. Cayco, L. Foster and H. Swann, On the convergence rate of the cell-discretization algorithm for solving elliptic problems, Math. Comp. **64** (1995), 1397-1419.

19. B. Cockburn, Discontinuous Galerkin Methods for Convection Dominated Problems, Lecture Notes of the NASA/VKI Summer School on High Order Discretization Methods in Comp. Fluid Dynamics, 14-25 September 1998.

20. B. Cockburn, S. Hou and C. Shu, TVB Runge-Kutta local projection discontinuous Galerkin finite elements for conservation laws IV: The multidimensional case, Math. Comp. **54** (1990), 545 ff.

21. L. Demkowicz, K. Gerdes, C. Schwab, A. Bajer and T. Walsh, *HP*90 - A general and flexible Fortran 90 *hp*-FE code, Report 97-17, Seminar for Applied Mathematics, ETH-Zürich (in press in Computing and Visualization in Science)

22. L. Demkowicz, J.T. Oden, W. Rachowicz and O. Hardy, Toward a universal *hp*-adaptive FE strategy, I, Comp. Meth. Appl. Mech. Engg. **77** (1989), 77-112.

23. L. Demkowicz, J.T. Oden, W. Rachowicz and O. Hardy, An hp-Taylor-Galerkin method for the compressible Euler equations, Comp. Meth. Appl. Mech. Engg. **86** (1991), 363-396.
24. K.D. Devine and J.E. Flaherty, Parallel adaptive hp-refinement techniques for conservation laws, Appl. Num. Math. **20** (1996), 367-386.
25. M. Feistauer and C. Schwab, Coupling of an Interior Navier-Stokes Problem with an Exterior Oseen Problem, Report 98-01, Seminar for Applied Mathematics, ETH Zürich, Switzerland (submitted).
26. J.E. Flaherty, R.M. Roy, M.S. Shepard, B.K. Szymanski, J.D. Teresco and L.H. Ziantz, Adaptive local refinement with octree load balancing for the parallel solution of three-dimensional conservation laws, J. parall. Computing **47** (1997), 139-152.
27. L. Franca and R. Stenberg, Error analysis of some Galerkin least squares methods for the elasticity equations, SIAM J. Numer. Anal. **28** (1991), 1680-1697.
28. K. Gerdes, M. Melenk, D. Schötzau and C. Schwab, Fully discrete hp-spectral FEM (in preparation).
29. K. Gerdes and D. Schötzau, hp-FEM for incompressible fluid flow - stable and stabilized, Report 97-18, Seminar for Applied Mathematics, ETH Zürich, Switzerland (submitted).
30. P. Gervasio and F. Saleri, Stabilized spectral element approximation for the Navier-Stokes equations, Numer. Meth. Part. Diff. Eq. **14** (1998), 115-141.
31. R. Girault and P.A. Raviart, Finite Elements for Navier-Stokes Equations, Springer-Verlag, (1986).
32. H. Godlewski and P.A. Raviart, Numerical Approximation of Conservation laws, Springer Verlag, New York, Heidelberg, (1996).
33. P. Grisvard, Singularities in boundary value problems, Masson Publ. and Springer Verlag, Paris and New York (1992).
34. B.Q. Guo and I. Babuška, The hp-Version of the Finite Element Method, Comp. Mech. **1** (1986), 21 ff.
35. E. Hairer and G. Wanner, *Solving ordinary differential equations II*, Springer Series in Computational Mathematics 14, Springer Verlag, (1991).
36. P. Houston, C. Schwab and E. Süli, Stabilized hp-FEM for first order hyperbolic problems, Report, Oxford University Computing Lab, (1998).
37. P. Houston, C. Schwab and E. Süli, hp-Discontinuous Galerkin FEM for Connection-Diffusion Problems, in preparation, (1999).
38. T.J.R. Hughes, L.P. Franca and M. Balestra. A new finite element formulation for computational fluid dynamics: V. Circumventing the Babuška-Brezzi condition: A stable Petrov-Galerkin formulation of the Stokes problem accommodating equal-order interpolations. *Computer Methods in Applied Mechanics and Engineering* **59** (1986), 85-99.
39. C. Johnson and J. Saranen, Streamline diffusion methods for the incompressible Euler and Navier-Stokes Equations, Math. Comp **47** (1986), 1-18.
40. C. Johnson, A. Schatz and L. Wahlbin, Crosswind Smear and Pointwise Errors in the Streamline Diffusion Finite Element Method, Math. Comp. **49** (1987), 25-38.
41. K. Korzak and T. Patera, An isoparametric spectral element method for solution of the Navier-Stokes equations in complex geometry, J. Comp. Phys. **62** (1985) 361-382.
42. V. Kozlov, V.G. Mazya and J. Rossmann, Elliptic boundary value problems in domains with point singularities. AMS, Providence, Rhode Island, (1997).

43. P.L. Lions, Mathematical topics in fluid mechanics, parts I and II, Lecture Series in Mathematics and its Applications, Oxford University Press (1996) and (1998).

44. I. Lomtev and G.E. Karniadakis, Simulations of Viscous Supersonic flows on unstructured *hp*-meshes AIAA paper 97-0754, presented at 35th Aerospace Sciences Meeting & Exhibit Jan. 1997.

45. J.M. Melenk, On the robust exponential convergence of *hp* finite element methods for problems with boundary layers, IMA J. Numer. Anal. **17** (1997), 577-601.

46. J.M. Melenk and C. Schwab, *hp*-FEM for reaction-diffusion problems, SIAM J. Numer. Anal. **35** (1998), 1520-1557.

47. J. M. Melenk and C. Schwab, The *hp* Streamline Diffusion Finite Element Method for Convection Dominated Problems in one Space Dimension, Report 98-10, Seminar for Applied Mathematics, ETH Zürich, Switzerland, (1998).

48. H.K. Moffatt, Viscous and resistive eddies near a sharp corner, J. Fluid Mech. **18** (1964).

49. J.T. Oden, I. Babuška and C.E. Baumann, A discontinuous *hp* Finite Element method for diffusion problems, Journal of Comp. Physics **146** (1998), 491-519.

50. J.T. Oden, L. Demkowicz, W. Rachowicz and O. Hardy, Toward a Universal $h - p$ Adaptive Finite Element Strategy. Part 1: Constrained Approximation and Data Structure, Computer Methods in Applied Mechanics and Engineering **77** (1989), 79-112.

51. A. Prohl, Projection and quasi-compressibility methods for solving the incompressible Navier-Stokes equations, B.G. Teubner Publ. Stuttgart, (1997).

52. A. Quarteroni and A. Valli, Numerical Approximation of Partial Differential Equations, Springer Series in Comp. Mathematics **23**, Springer Verlag Heidelberg, (1994).

53. W. Rachowicz, An anisotropic *h*-adaptive finite element method for compressible Navier-Stokes equations, Comp. Meth. Appl. Mech. Engg. (1997).

54. D. Schötzau, K. Gerdes and C. Schwab, Galerkin Least-squares *hp*-FEM for the Stokes problem, C.R. Acad. Sci. Paris **326** (1998), 249-254.

55. D. Schötzau and C. Schwab, Mixed *hp*-FEM on anisotropic meshes, Math. Meth. Mod. Appl. Sci. **8**, No. 5 (1998), 787-820.

56. D. Schötzau, C. Schwab and R. Stenberg, Mixed *hp*-FEM on anisotropic meshes II: Hanging nodes and tensor products of boundary layer meshes, Report 97-14, Seminar for Applied Mathematics, ETH Zürich, Switzerland (submitted to Numerische Mathematik).

57. D. Schötzau and C. Schwab, Time discretization of parabolic problems by the *hp*-version of the discontinuous Galerkin FEM, Report 99-04, Seminar for Applied Mathematics, (1999).

58. C. Schwab, *p*- and *hp*-FEM, Oxford University Press, (1998).

59. C. Schwab and M. Suri, The *p* and $h - p$ version of the finite element method for problems with boundary layers" Math. Comp. **65** (1996), 1403-1429.

60. C. Schwab and M. Suri, Mixed *hp*-FEM for Stokes and Non-Newtonian flow, Report 97-19, Seminar for Applied Mathematics, ETH Zürich, Switzerland (submitted).

61. C. Schwab, M. Suri and C.A. Xenophontos, The *hp*-Finite Element Method for problems in mechanics with boundary layers, Report 96-20, Seminar for Applied Mathematics, ETH Zürich, in press in Comp. Meth. Appl. Mech. Engg. **157** (1998), 311-333.

62. D. Schötzau, hp-DGFEM for parabolic evolution problems, Doctoral Dissertation ETHZ (to appear 1999).
63. P. Seshayer and M. Suri, Uniform hp convergence results for the mortar element method, Preprint, Dept. of Math. and Stat., Univ. Maryland Baltimore County, Math. Comp. (to appear), (1999).
64. M. S. Shepard and J.E. Flaherty, A straightforward procedure to construct shape functions for variable p-order meshes, Comp. Meth. Appl. Mech. Engg. **147** (1997), 209-233.
65. S. Sherwin, Hierarchical hp-finite elements in hybrid domains, Finite Elements in Analysis and Design **27** (1997), 109-119.
66. S. Sherwin and G.E. Karniadakis, A new triangular and tetrahedral basis for high-order (hp) Finite Element Methods, Int. J. Num. Meth. Engg. **38** (1995), 3775-3802.
67. S. Sherwin and G.E. Karniadakis, Tetrahedral hp Finite Elements: Algorithms and flow simulations, J. Comp. Phys. **124** (1996), 14-45.
68. J. Smoller, Shock waves and reaction-diffusion equations (2nd Ed.), Springer-Verlag, New-York, (1995).
69. R. Stenberg and M. Suri, Mixed hp Finite Element Methods for problems in elasticity and Stokes flow. *Numer. Math.* **72** (1996), 367-389.
70. B.A. Szabo and I. Babuška, *Finite Element Analysis*, Wiley Publ. New York (1991).
71. R. Temam, Navier-Stokes equations, North-Holland Publ., (1983).
72. V. Thomee, Galerkin Finite Element Methods for parabolic equations, Springer Series in Computational Mathematics, (1997).
73. L. Tobiska and G. Lube, A modified streamline diffusion method for solving the stationary Navier-Stokes equations, Numer. Math. **59** (1991), 13-29.
74. L. Tobiska and R. Verfürth: Analysis of a streamline diffusion finite element method for the Stokes and Navier-Stokes equations, SIAM J. Numer. Anal. **33** (1996). 107-127.
75. G. Zhou and R. Rannacher, Pointwise superconvergence of the SDFEM, Num. Meth. PDEs **12** (1996), 123-145.

High Order ENO and WENO Schemes for Computational Fluid Dynamics

Chi-Wang Shu[*]

Division of Applied Mathematics, Brown University, Providence, Rhode Island 02912, USA. E-mail: shu@cfm.brown.edu

Abstract. In these lectures we present the basic ideas and recent development in the construction, analysis, and implementation of ENO (Essentially Non-Oscillatory) and WENO (Weighted Essentially Non-Oscillatory) schemes and their applications to computational fluid dynamics. ENO and WENO schemes are high order accurate finite difference or finite volume schemes designed for problems with piecewise smooth solutions containing discontinuities. The key idea lies at the approximation level, where a nonlinear adaptive procedure is used to automatically choose the locally smoothest stencil, hence avoiding crossing discontinuities in the interpolation procedure as much as possible. ENO and WENO schemes have been quite successful in computational fluid dynamics and other applications, especially for problems containing both shocks and complicated smooth solution structures, such as compressible turbulence simulations and aeroacoustics.

Table of Contents

[*] Research of the author was partially supported by NSF grants DMS-9500814, DMS-9804985, ECS-9627849 and INT-9601084, ARO grant DAAG55-97-1-0318, NASA Langley grant NAG-1-2070, and AFOSR grant F49620-96-1-0150.

1   Introduction

We are concerned in these lectures about high order finite difference and finite volume schemes and their applications to computational fluid dynamics. These are schemes based on interpolations of discrete data, mostly by using algebraic polynomials. The foundation of such interpolation is in the approximation theory, that a wider interpolation stencil yields a higher order of accuracy, provided the function being interpolated is smooth inside the stencil. Traditional finite difference and finite volume methods are based on fixed stencil interpolations. For example, to obtain an interpolation for cell i to third order accuracy, the information of the three cells $i-1$, i and $i+1$ can be used to build a second order interpolation polynomial. In other words, one always looks one cell to the left, one cell to the right, plus the center cell itself, regardless of where in the domain one is situated. This works well for globally smooth problems. The resulting scheme is linear for linear PDEs, hence stability can be easily analyzed by Fourier transforms (for the uniform grid periodic case). However, fixed stencil interpolation of second or higher order accuracy is necessarily *oscillatory* near a discontinuity, see Fig. 3.1, left, in Sect. 3. Such oscillations, which are called the Gibbs phenomena in spectral methods, do not decay in magnitude when the mesh is refined. It is a nuisance to say the least for practical calculations, and often leads to numerical instabilities in nonlinear problems containing discontinuities.

Earlier attempts to eliminate or reduce such spurious oscillations near discontinuities were mainly based on two approaches: explicit artificial viscosity and limiters. The first approach was to add an artificial viscosity. This could be tuned so that it was large enough near the discontinuity to suppress, or at least reduce the oscillations, but was small elsewhere to maintain high-order accuracy. One disadvantage of this approach is that fine tuning of the parameters controlling the artificial viscosity is problem dependent. The second approach was to apply limiters to eliminate the oscillations. In effect, one reduced the order of accuracy of the interpolation near the discontinuity (e.g. by reducing the slope of a linear interpolant, or by using a linear rather than a quadratic interpolant near the shock). By carefully designing such limiters, the TVD (total variation diminishing) property could be achieved for one dimensional nonlinear scalar problems or linear systems, and maximum norm stability can be achieved for multi dimensional scalar problems. Also, there is usually no free parameters in the limiters to tune. One disadvantage of this approach is that accuracy necessarily degenerates to first order near *smooth* extrema. This could be fixed by using the TVB (total variation bounded) modifications to the limiter in Shu [85] and Cockburn and Shu [18], but such modifications are not self-similar. We will not discuss the method of adding explicit artificial viscosity or the TVD limiters in these lectures. We refer the readers to the books by Sod [96], LeVeque [66] and Godlewski and Raviart [35], and the references listed therein.

ENO (Essentially Non-Oscillatory) schemes were first introduced by Harten, Engquist, Osher and Chakravarthy in 1987 [47]. Their paper now has become a classic and has been quoted numerous times. The Journal of Computational Physics decided to republish it as part of the journal's celebration of its 30th birthday [88].

The ENO idea proposed in [47] seems to be the first successful attempt to obtain a self similar (i.e. no mesh size dependent parameter), uniformly high order accurate, yet essentially non-oscillatory interpolation (i.e. the magnitude of the oscillations decays as $O(\Delta x^k)$ where k is the order of accuracy) for piecewise smooth functions. The generic solution for hyperbolic conservation laws is in the class of piecewise smooth functions. The reconstruction in [47] is a natural extension of an earlier second order version of Harten and Osher [46]. In [47], Harten, Engquist, Osher and Chakravarthy investigated different ways of measuring local smoothness to determine the local stencil, and developed a hierarchy that begins with one or two cells, then adds one cell at a time to the stencil from the two candidates on the left and right, based on the size of the two relevant Newton divided differences. Although there are other reasonable strategies to choose the stencil based on local smoothness, such as comparing the magnitudes of the highest degree divided differences among all candidate stencils and picking the one with the least absolute value, experience seems to show that the hierarchy proposed in [47] is the most robust for a wide range of grid sizes, Δx, both *before* and inside the asymptotic regime.

As one can see from the numerical examples in [47] and in later papers, ENO schemes are indeed uniformly high order accurate and resolve shocks with sharp and monotone (to the eye) transitions. ENO schemes are especially suitable for problems containing both shocks and complicated smooth flow structures, such as those occurring in shock interactions with a turbulent flow and shock interaction with vortices.

Since the publication of the original paper of Harten, Engquist, Osher and Chakravarthy [47], the original authors and many other researchers have followed the pioneer work, improving the methodology and expanding the area of its applications. ENO schemes based on point values and TVD Runge-Kutta time discretizations, which can save computational costs significantly for multi space dimensions, were developed in Shu and Osher [89], [90]. Biasing in the stencil choosing process to enhance stability and accuracy were developed in Fatemi, Jerome and Osher [31] and in Shu [87]. Finite volume ENO schemes based on a staggered grid and Lax-Friedrichs formulation were given in Bianco, Puppo and Russo [9]. Weighted ENO (WENO) schemes were developed, using a convex combination of all candidate stencils instead of just one as in the original ENO, Liu, Osher and Chan [69] for 1D, Jiang and Shu [55] for multi dimensional finite difference formulation with improved accuracy, Friedrich [32] for multi dimensional finite volume formulation, Hu and Shu [49], [50] for multi dimensional finite volume formulation with improved accuracy, and Levy, Puppo and Russo [67] for 1D finite volume based on a staggered grid and Lax-Friedrichs formulation. ENO schemes based on other than polynomial building blocks were constructed in Iske and Soner [52] and in Christofi [17]. Sub-cell resolution and artificial compression to sharpen contact discontinuities were studied in Harten [44], Yang [105], Shu and Osher [90] and in Jiang and Shu [55]. Multidimensional ENO schemes based on general triangulation were developed in Abgrall [1]. ENO and WENO schemes for Hamilton-Jacobi type equations were designed and applied in Osher and Sethian [78], Osher and Shu [79], Lafon and Osher [62] and in Jiang and Peng [57]. ENO schemes using one-sided Jocobians for field by field decomposition, which improves the robustness for calculations of systems, were discussed in Donat and Marquina [28]. Combination of ENO with multiresolution ideas was pursued in Bihari and Harten [10]. Combination of ENO with spectral method using a domain decomposition approach was carried out in Cai and Shu [11]. On the application side, ENO and WENO have been successfully used to simulate shock turbulence interactions, Shu and Osher [90], Shu, Zang, Erlebacher, Whitaker and Osher [91], and Adams and Shariff [2]; to the direct simulation of compressible turbulence, Shu, Zang, Erlebacher, Whitaker and Osher [91], Walsteijn [102], and Ladeinde, O'Brien, Cai and Liu [61]; to relativistic hydrodynamics equations in Dolezal and Wong [27]; to shock vortex interactions and other gas dynamics problems in Casper and Atkins [15], Erlebacher, Hussaini and Shu [30], and in Jiang and Shu [55]; to incompressible flow problems in E and Shu [29] and Harabetian, Osher and Shu [40]; to viscoelasticity equations with fading memory in Shu and Zeng [92]; to semi-conductor device simulation in Fatemi, Jerome and Osher [31] and Jerome and Shu [53], [54]; to image processing in Osher and Sethian [78], Sethian [84], and Siddiqi, Kimia and Shu [93]; etc. This list is definitely incomplete and perhaps biased by the author's own research experience, but one can already see that ENO and WENO have been applied quite extensively in many different fields. Most of the problems solved by ENO and WENO schemes are of the type in which solutions contain both strong shocks and rich smooth region structures. Lower order methods usually have difficulties for such problems and it is thus attractive and efficient to use high order stable methods such as ENO and WENO to handle them.

Today the study and application of ENO and WENO schemes are still very active. We expect the schemes and the basic methodology to be developed further and to become even more successful in the future.

In these lectures we present the basic ideas and recent development in the construction, analysis, and implementation of ENO and WENO schemes and their applications to computational fluid dynamics. For readers interested in coding the methods, sample codes are available from the author.

2   Reconstruction and Approximation in One Dimension

This section gives the necessary background information about polynomial interpolation and approximation in one space dimension.

Given a grid

$$a = x_{\frac{1}{2}} < x_{\frac{3}{2}} < ... < x_{N-\frac{1}{2}} < x_{N+\frac{1}{2}} = b, \tag{2.1}$$

We define cells, cell centers, and cell sizes by

$$I_i \equiv \left[x_{i-\frac{1}{2}}, x_{i+\frac{1}{2}} \right], \qquad x_i \equiv \frac{1}{2} \left(x_{i-\frac{1}{2}} + x_{i+\frac{1}{2}} \right),$$
$$\Delta x_i \equiv x_{i+\frac{1}{2}} - x_{i-\frac{1}{2}}, \qquad i = 1, 2, ..., N. \tag{2.2}$$

We denote the maximum cell size by

$$\Delta x \equiv \max_{1 \leq i \leq N} \Delta x_i. \tag{2.3}$$

2.1   Reconstruction from Cell Averages

The first approximation problem we will face, in solving hyperbolic conservation laws using cell averages (finite volume schemes, see Sect. 4.1), is the following *reconstruction* problem [47].

Problem 2.1. One dimensional reconstruction.

Given the cell averages of a function $v(x)$:

$$\bar{v}_i \equiv \frac{1}{\Delta x_i} \int_{x_{i-\frac{1}{2}}}^{x_{i+\frac{1}{2}}} v(\xi)\, d\xi, \qquad i = 1, 2, ..., N, \qquad (2.4)$$

find a polynomial $p_i(x)$, of degree at most $k - 1$, for each cell I_i, such that it is a k-th order accurate approximation to the function $v(x)$ inside I_i:

$$p_i(x) = v(x) + O(\Delta x^k), \quad x \in I_i, \quad i = 1, ..., N. \qquad (2.5)$$

In particular, this gives approximations to the function $v(x)$ at the cell boundaries

$$v_{i+\frac{1}{2}}^- = p_i(x_{i+\frac{1}{2}}), \quad v_{i-\frac{1}{2}}^+ = p_i(x_{i-\frac{1}{2}}), \qquad i = 1, ..., N \qquad (2.6)$$

which are k-th order accurate:

$$v_{i+\frac{1}{2}}^- = v(x_{i+\frac{1}{2}}) + O(\Delta x^k), \quad v_{i-\frac{1}{2}}^+ = v(x_{i-\frac{1}{2}}) + O(\Delta x^k), \quad i = 1, ..., N. \quad (2.7)$$

$\square$

The polynomial $p_i(x)$ in Problem 2.1 can be replaced by other simple functions, such as trigonometric polynomials. See Sect. 8.3.

We will not discuss boundary conditions in this section. We thus assume that $\bar{v}_i$ is also available for $i \le 0$ and $i > N$ if needed.

In the following we describe a procedure to solve Problem 2.1.

Given the location I_i and the order of accuracy k, we first choose a "stencil", based on r cells to the left, s cells to the right, and I_i itself if $r, s \ge 0$, with $r + s + 1 = k$:

$$S(i) \equiv \{I_{i-r}, ..., I_{i+s}\}. \qquad (2.8)$$

There is a unique polynomial of degree at most $k - 1 = r + s$, denoted by $p(x)$ (we will drop the subscript i when it does not cause confusion), whose cell average in each of the cells in $S(i)$ agrees with that of $v(x)$:

$$\frac{1}{\Delta x_j} \int_{x_{j-\frac{1}{2}}}^{x_{j+\frac{1}{2}}} p(\xi)\, d\xi = \bar{v}_j, \qquad j = i - r, ..., i + s. \qquad (2.9)$$

This polynomial $p(x)$ is the k-th order approximation we are looking for, as it is easy to prove (2.5), see the discussion below, as long as the function $v(x)$ is smooth in the region covered by the stencil $S(i)$.

For solving Problem 2.1, we also need the approximations to the values of $v(x)$ at the cell boundaries, (2.6). Since the mappings from the given cell averages $\overline{v}_j$ in the stencil $S(i)$ to the values $v^-_{i+\frac{1}{2}}$ and $v^+_{i-\frac{1}{2}}$ in (2.6) are linear, there exist constants c_{rj} and $\tilde{c}_{rj}$, which depend on the left shift r of the stencil $S(i)$ in (2.8), on the order of accuracy k, and on the cell sizes Δx_j in the stencil S_i, but *not* on the function v itself, such that

$$v^-_{i+\frac{1}{2}} = \sum_{j=0}^{k-1} c_{rj}\overline{v}_{i-r+j}, \qquad v^+_{i-\frac{1}{2}} = \sum_{j=0}^{k-1} \tilde{c}_{rj}\overline{v}_{i-r+j}. \tag{2.10}$$

We note that the difference between the values with superscripts $\pm$ at the same location $x_{i+\frac{1}{2}}$ is due to the possibility of different stencils for cell I_i and for cell I_{i+1}. If we identify the left shift r not with the cell I_i but with the point of reconstruction $x_{i+\frac{1}{2}}$, i.e. using the stencil (2.8) to approximate $x_{i+\frac{1}{2}}$, then we can drop the superscripts $\pm$ and also eliminate the need to consider $\tilde{c}_{rj}$ in (2.10), as it is clear that

$$\tilde{c}_{rj} = c_{r-1,j}.$$

We summarize this as follows: given the k cell averages

$$\overline{v}_{i-r}, \quad \ldots, \quad \overline{v}_{i-r+k-1},$$

there are constants c_{rj} such that the reconstructed value at the cell boundary $x_{i+\frac{1}{2}}$:

$$v_{i+\frac{1}{2}} = \sum_{j=0}^{k-1} c_{rj}\overline{v}_{i-r+j}, \tag{2.11}$$

is k-th order accurate:

$$v_{i+\frac{1}{2}} = v(x_{i+\frac{1}{2}}) + O(\Delta x^k). \tag{2.12}$$

To understand how the constants $\{c_{rj}\}$ are obtained, as well as how the accuracy property (2.5) is proven, we look at the primitive function of $v(x)$:

$$V(x) \equiv \int_{-\infty}^{x} v(\xi)\, d\xi, \tag{2.13}$$

where the lower limit $-\infty$ is not important and can be replaced by any fixed number. Clearly, $V(x_{i+\frac{1}{2}})$ can be expressed by the cell averages of $v(x)$ using (2.4):

$$V(x_{i+\frac{1}{2}}) = \sum_{j=-\infty}^{i} \int_{x_{j-\frac{1}{2}}}^{x_{j+\frac{1}{2}}} v(\xi)\, d\xi = \sum_{j=-\infty}^{i} \overline{v}_j \Delta x_j, \tag{2.14}$$

thus with the knowledge of the cell averages $\{\bar{v}_j\}$ we also know the primitive function $V(x)$ at the cell boundaries exactly. If we denote the unique polynomial of degree at most k, which interpolates $V(x_{j+\frac{1}{2}})$ at the following $k+1$ points:

$$x_{i-r-\frac{1}{2}}, \quad \cdots, \quad x_{i+s+\frac{1}{2}}, \tag{2.15}$$

by $P(x)$, and denote its derivative by $p(x)$:

$$p(x) \equiv P'(x), \tag{2.16}$$

then it is easy to verify (2.9):

$$
\frac{1}{\Delta x_j} \int_{x_{j-\frac{1}{2}}}^{x_{j+\frac{1}{2}}} p(\xi)\,d\xi = \frac{1}{\Delta x_j} \int_{x_{j-\frac{1}{2}}}^{x_{j+\frac{1}{2}}} P'(\xi)\,d\xi
$$

$$
= \frac{1}{\Delta x_j} \left(P(x_{j+\frac{1}{2}}) - P(x_{j-\frac{1}{2}}) \right)
$$

$$
= \frac{1}{\Delta x_j} \left(V(x_{j+\frac{1}{2}}) - V(x_{j-\frac{1}{2}}) \right)
$$

$$
= \frac{1}{\Delta x_j} \left(\int_{-\infty}^{x_{j+\frac{1}{2}}} v(\xi)\,d\xi - \int_{-\infty}^{x_{j-\frac{1}{2}}} v(\xi)\,d\xi \right)
$$

$$
= \frac{1}{\Delta x_j} \int_{x_{j-\frac{1}{2}}}^{x_{j+\frac{1}{2}}} v(\xi)\,d\xi
$$

$$
= \bar{v}_j, \qquad j = i - r, ..., i + s,
$$

where the third equality holds because $P(x)$ interpolates $V(x)$ at the points $x_{j-\frac{1}{2}}$ and $x_{j+\frac{1}{2}}$ whenever $j = i - r, ..., i + s$. This implies that $p(x)$ is the polynomial we are looking for. Standard approximation theory (see any elementary numerical analysis book) tells us that

$$P'(x) = V'(x) + O(\Delta x^k), \qquad x \in I_i.$$

This is the accuracy requirement (2.5).

Now let us look at the practical issue of how to obtain the constants $\{c_{rj}\}$ in (2.11). For this we could use the Lagrange form of the interpolation polynomial:

$$P(x) = \sum_{m=0}^{k} V(x_{i-r+m-\frac{1}{2}}) \cdot \prod_{\substack{l=0 \\ l \neq m}}^{k} \frac{x - x_{i-r+l-\frac{1}{2}}}{x_{i-r+m-\frac{1}{2}} - x_{i-r+l-\frac{1}{2}}}. \tag{2.17}$$

For easier manipulation we subtract a constant $V(x_{i-r-\frac{1}{2}})$ from (2.17), and use the fact that

$$\sum_{m=0}^{k} \prod_{\substack{l=0 \\ l \neq m}}^{k} \frac{x - x_{i-r+l-\frac{1}{2}}}{x_{i-r+m-\frac{1}{2}} - x_{i-r+l-\frac{1}{2}}} = 1,$$

to obtain:

$$P(x) - V(x_{i-r-\frac{1}{2}}) =$$

$$\sum_{m=0}^{k} \left(V(x_{i-r+m-\frac{1}{2}}) - V(x_{i-r-\frac{1}{2}}) \right) \prod_{\substack{l=0 \\ l \neq m}}^{k} \frac{x - x_{i-r+l-\frac{1}{2}}}{x_{i-r+m-\frac{1}{2}} - x_{i-r+l-\frac{1}{2}}} \quad (2.18)$$

Taking derivative on both sides of (2.18), and noticing that

$$V(x_{i-r+m-\frac{1}{2}}) - V(x_{i-r-\frac{1}{2}}) = \sum_{j=0}^{m-1} \bar{v}_{i-r+j} \Delta x_{i-r+j}$$

because of (2.14), we obtain

$$p(x) = \sum_{m=0}^{k} \sum_{j=0}^{m-1} \bar{v}_{i-r+j} \Delta x_{i-r+j} \left(\frac{\sum_{\substack{l=0 \\ l \neq m}}^{k} \prod_{\substack{q=0 \\ q \neq m,l}}^{k} \left(x - x_{i-r+q-\frac{1}{2}} \right)}{\prod_{\substack{l=0 \\ l \neq m}}^{k} \left(x_{i-r+m-\frac{1}{2}} - x_{i-r+l-\frac{1}{2}} \right)} \right).$$

$$(2.19)$$

Evaluating the expression (2.19) at $x = x_{i+\frac{1}{2}}$, we finally obtain

$$v_{i+\frac{1}{2}} = p(x_{i+\frac{1}{2}})$$

$$= \sum_{j=0}^{k-1} \Delta x_{i-r+j} \, \bar{v}_{i-r+j} \sum_{m=j+1}^{k} \frac{\sum_{\substack{l=0 \\ l \neq m}}^{k} \prod_{\substack{q=0 \\ q \neq m,l}}^{k} \left(x_{i+\frac{1}{2}} - x_{i-r+q-\frac{1}{2}} \right)}{\prod_{\substack{l=0 \\ l \neq m}}^{k} \left(x_{i-r+m-\frac{1}{2}} - x_{i-r+l-\frac{1}{2}} \right)},$$

i.e. the constants c_{rj} in (2.11) are given by

$$c_{rj} = \Delta x_{i-r+j} \sum_{m=j+1}^{k} \frac{\sum_{\substack{l=0 \\ l \neq m}}^{k} \prod_{\substack{q=0 \\ q \neq m,l}}^{k} \left(x_{i+\frac{1}{2}} - x_{i-r+q-\frac{1}{2}} \right)}{\prod_{\substack{l=0 \\ l \neq m}}^{k} \left(x_{i-r+m-\frac{1}{2}} - x_{i-r+l-\frac{1}{2}} \right)}. \quad (2.20)$$

Although there are many zero terms in the inner sum of (2.20) when $x_{i+\frac{1}{2}}$ is a node in the interpolation, we will keep this general form so that it applies also to the case where $x_{i+\frac{1}{2}}$ is not an interpolation point.

For a nonuniform grid, one would want to pre-compute the constants $\{c_{rj}\}$ as in (2.20), for $0 \leq i \leq N$, $-1 \leq r \leq k-1$, and $0 \leq j \leq k-1$, and store them before solving the PDE.

For a uniform grid, $\Delta x_i = \Delta x$, the expression for c_{rj} does not depend on i or Δx any more:

$$c_{rj} = \sum_{m=j+1}^{k} \frac{\sum_{\substack{l=0 \\ l \neq m}}^{k} \prod_{\substack{q=0 \\ q \neq m,l}}^{k} (r-q+1)}{\prod_{\substack{l=0 \\ l \neq m}}^{k} (m-l)}. \qquad (2.21)$$

We list in Table 2.1 the constants c_{rj} in this uniform grid case (2.21), for order of accuracy between $k = 1$ and $k = 6$.

Table 2.1. The constants c_{rj} in (2.21).

k	r	j=0	j=1	j=2	j=3	j=4	j=5
1	-1	1					
	0	1					
2	-1	3/2	-1/2				
	0	1/2	1/2				
	1	-1/2	3/2				
3	-1	11/6	-7/6	1/3			
	0	1/3	5/6	-1/6			
	1	-1/6	5/6	1/3			
	2	1/3	-7/6	11/6			
4	-1	25/12	-23/12	13/12	-1/4		
	0	1/4	13/12	-5/12	1/12		
	1	-1/12	7/12	7/12	-1/12		
	2	1/12	-5/12	13/12	1/4		
	3	-1/4	13/12	-23/12	25/12		
5	-1	137/60	-163/60	137/60	-21/20	1/5	
	0	1/5	77/60	-43/60	17/60	-1/20	
	1	-1/20	9/20	47/60	-13/60	1/30	
	2	1/30	-13/60	47/60	9/20	-1/20	
	3	-1/20	17/60	-43/60	77/60	1/5	
	4	1/5	-21/20	137/60	-163/60	137/60	
6	-1	49/20	-71/20	79/20	-163/60	31/30	-1/6
	0	1/6	29/20	-21/20	37/60	-13/60	1/30
	1	-1/30	11/30	19/20	-23/60	7/60	-1/60
	2	1/60	-2/15	37/60	37/60	-2/15	1/60
	3	-1/60	7/60	-23/60	19/20	11/30	-1/30
	4	1/30	-13/60	37/60	-21/20	29/20	1/6
	5	-1/6	31/30	-163/60	79/20	-71/20	49/20

From Table 2.1, we would know, for example, that

$$v_{i+\frac{1}{2}} = -\frac{1}{6}\overline{v}_{i-1} + \frac{5}{6}\overline{v}_i + \frac{1}{3}\overline{v}_{i+1} + O(\Delta x^3).$$

2.2 Conservative Approximation to the Derivative from Point Values

The second approximation problem we will face, in solving hyperbolic conservation laws using point values (finite difference schemes, see Sect. 4.2), is the following problem in obtaining high order *conservative* approximation to the derivative from point values [89,90].

Problem 2.2. One dimensional conservative approximation.

Given the point values of a function $v(x)$:

$$v_i \equiv v(x_i), \qquad i = 1, 2, ..., N, \tag{2.22}$$

find a numerical flux function

$$\hat{v}_{i+\frac{1}{2}} \equiv \hat{v}(v_{i-r}, ..., v_{i+s}), \qquad i = 0, 1, ..., N, \tag{2.23}$$

such that the flux difference approximates the derivative $v'(x)$ to k-th order accuracy:

$$\frac{1}{\Delta x_i}\left(\hat{v}_{i+\frac{1}{2}} - \hat{v}_{i-\frac{1}{2}}\right) = v'(x_i) + O(\Delta x^k), \quad i = 0, 1, ..., N. \tag{2.24}$$

□

We again ignore the boundary conditions here and assume that v_i is available for $i \leq 0$ and $i > N$ if needed.

The solution of this problem is essential for the high order conservative schemes based on point values (finite difference) rather than on cell averages (finite volume).

This problem looks quite different from Problem 2.1. However, we will see that there is a close relationship between these two. *We assume that the grid is uniform*, $\Delta x_i = \Delta x$. This assumption is, unfortunately, essential in the following development.

If we can find a function $h(x)$, which may depend on the grid size Δx, such that

$$v(x) = \frac{1}{\Delta x}\int_{x-\frac{\Delta x}{2}}^{x+\frac{\Delta x}{2}} h(\xi)d\xi, \tag{2.25}$$

then clearly

$$v'(x) = \frac{1}{\Delta x}\left[h\left(x + \frac{\Delta x}{2}\right) - h\left(x - \frac{\Delta x}{2}\right)\right],$$

hence all we need to do is to use

$$\hat{v}_{i+\frac{1}{2}} = h(x_{i+\frac{1}{2}}) + O(\Delta x^k) \tag{2.26}$$

to achieve (2.24). We note here that it would look like an $O(\Delta x^{k+1})$ term in (2.26) is needed in order to get (2.24), due to the Δx term in the denominator. However, in practice, the $O(\Delta x^k)$ term in (2.26) is usually smooth, hence the difference in (2.24) would give an extra $O(\Delta x)$, just to cancel the one in the denominator.

It is not easy to approximate $h(x)$ via (2.25), as it is only implicitly defined there. However, we notice that the known function $v(x)$ is the cell average of the unknown function $h(x)$, so to find $h(x)$ we just need to use the *reconstruction* procedure described in Sect. 2.1. If we take the primitive of $h(x)$:

$$H(x) = \int_{-\infty}^{x} h(\xi)d\xi, \tag{2.27}$$

then (2.25) clearly implies

$$H(x_{i+\frac{1}{2}}) = \sum_{j=-\infty}^{i} \int_{x_{j-\frac{1}{2}}}^{x_{j+\frac{1}{2}}} h(\xi)d\xi = \Delta x \sum_{j=-\infty}^{i} v_j. \tag{2.28}$$

Thus, given the point values $\{v_j\}$, we "identify" them as cell averages of another function $h(x)$ in (2.25), then the primitive function $H(x)$ is exactly known at the cell interfaces $x = x_{i+\frac{1}{2}}$. We thus use the same reconstruction procedure described in Sect. 2.1, to get a k-th order approximation to $h(x_{i+\frac{1}{2}})$, which is then taken as the numerical flux $\hat{v}_{i+\frac{1}{2}}$ in (2.23).

In other words, if the "stencil" for the flux $\hat{v}_{i+\frac{1}{2}}$ in (2.23) is the following k points:

$$x_{i-r}, \ ..., \ x_{i+s}, \tag{2.29}$$

where $r + s = k - 1$, then the flux $\hat{v}_{i+\frac{1}{2}}$ is expressed as

$$\hat{v}_{i+\frac{1}{2}} = \sum_{j=0}^{k-1} c_{rj} v_{i-r+j}, \tag{2.30}$$

where the constants $\{c_{rj}\}$ are given by (2.21) and Table 2.1.

From Table 2.1 we would know, for example, that if

$$\hat{v}_{i+\frac{1}{2}} = -\frac{1}{6}v_{i-1} + \frac{5}{6}v_i + \frac{1}{3}v_{i+1},$$

then

$$\frac{1}{\Delta x}\left(\hat{v}_{i+\frac{1}{2}} - \hat{v}_{i-\frac{1}{2}}\right) = v'(x_i) + O(\Delta x^3).$$

We emphasize again that, unlike in the reconstruction procedure in Sect. 2.1, here the grid *must* be uniform: $\Delta x_j = \Delta x$. Otherwise, it can be proven that no choice of constants c_{rj} in (2.30) (which may depend on the local grid sizes but not on the function $v(x)$) could make the conservative approximation to the derivative (2.24) higher than second order accurate ($k > 2$). The proof is a simple exercise of Taylor expansions. Thus, the high order finite difference (third order and higher) discussed in these lecture notes can apply only to uniform or smoothly varying grids.

Because of this equivalence of obtaining a conservative approximation to the derivative (2.23)-(2.24) and the reconstruction problem discussed in Sect. 2.1, we will only need to consider the reconstruction problem in the following sections.

2.3   Fixed Stencil Approximation

By fixed stencil, we mean that the left shift r in (2.8) or (2.29) is *the same for all locations* i. Usually, for a globally smooth function $v(x)$, the best approximation is obtained either by a central approximation $r = s - 1$ for even k (here central is relative to the location $x_{i+\frac{1}{2}}$), or by a one point upwind biased approximation $r = s$ or $r = s - 2$ for odd k. For example, if the grid is uniform $\Delta x_i = \Delta x$, then a central 4th order reconstruction for $v_{i+\frac{1}{2}}$, in (2.11), is given by

$$v_{i+\frac{1}{2}} = -\frac{1}{12}\bar{v}_{i-1} + \frac{7}{12}\bar{v}_i + \frac{7}{12}\bar{v}_{i+1} - \frac{1}{12}\bar{v}_{i+2} + O(\Delta x^4),$$

and the two one point upwind biased 3rd order reconstructions for $v_{i+\frac{1}{2}}$ in (2.11), are given by

$$v_{i+\frac{1}{2}} = -\frac{1}{6}\bar{v}_{i-1} + \frac{5}{6}\bar{v}_i + \frac{1}{3}\bar{v}_{i+1} + O(\Delta x^3)$$

$$\text{or} \quad v_{i+\frac{1}{2}} = \frac{1}{3}\bar{v}_i + \frac{5}{6}\bar{v}_{i+1} - \frac{1}{6}\bar{v}_{i+2} + O(\Delta x^3).$$

Similarly, a central 4th order flux (2.30) is

$$\hat{v}_{i+\frac{1}{2}} = -\frac{1}{12}v_{i-1} + \frac{7}{12}v_i + \frac{7}{12}v_{i+1} - \frac{1}{12}v_{i+2},$$

which gives

$$\frac{1}{\Delta x}\left(\hat{v}_{i+\frac{1}{2}} - \hat{v}_{i-\frac{1}{2}}\right) = v'(x_i) + O(\Delta x^4),$$

and the two one point upwind biased 3rd order fluxes (2.30) are given by

$$\hat{v}_{i+\frac{1}{2}} = -\frac{1}{6}v_{i-1} + \frac{5}{6}v_i + \frac{1}{3}v_{i+1}$$

$$\text{or} \quad \hat{v}_{i+\frac{1}{2}} = \frac{1}{3}v_i + \frac{5}{6}v_{i+1} - \frac{1}{6}v_{i+2},$$

which gives

$$\frac{1}{\Delta x}\left(\hat{v}_{i+\frac{1}{2}} - \hat{v}_{i-\frac{1}{2}}\right) = v'(x_i) + O(\Delta x^3).$$

Traditional central and upwind schemes, either finite volume or finite difference, can be derived by these fixed stencil reconstructions or flux differenced approximations to the derivatives.

3   ENO and WENO Reconstruction and Approximation in One Dimension

In the previous section we are mainly concerned with the approximation result when the stencil is chosen and fixed. In this section we will mainly discuss the issue of how to choose the stencils.

For solving hyperbolic conservation laws, we are interested in the class of piecewise smooth functions. These are functions which have as many derivatives as the scheme calls for, everywhere except for at finitely many isolated points. At these finitely many discontinuity points, the function $v(x)$ and its derivatives are assumed to have finite left and right limits. Such functions are "generic" for solutions to hyperbolic conservation laws, in the sense that in applications we mostly encounter such functions.

For such piecewise smooth functions, the order of accuracy we refer to in these lecture notes are *formal*, that is, it is defined as whatever accuracy determined by the local truncation error in the *smooth regions* of the function. This is the tradition taken in the literature when discussing about discontinuous solutions.

If the function $v(x)$ is only piecewise smooth, a fixed stencil approximation described in Sect. 2.3 may not be adequate near discontinuities. Fig. 3.1 (left) gives the 4-th order (piecewise cubic) interpolation with a central stencil for the step function, i.e. the polynomial approximation inside the interval $[x_{i-\frac{1}{2}}, x_{i+\frac{1}{2}}]$ interpolates the step function at the four points $x_{i-\frac{3}{2}}, x_{i-\frac{1}{2}}, x_{i+\frac{1}{2}}, x_{i+\frac{3}{2}}$. Notice the obvious over/undershoots for the cells near the discontinuity.

These oscillations (termed *the Gibbs Phenomena* in spectral methods) happen because the stencils, as defined by (2.15), actually contain the discontinuous cell for x_i close enough to the discontinuity. As a result, the approximation property (2.5) is no longer valid in such stencils.

3.1   ENO Approximation

A closer look at Fig. 3.1 (left) motivates the idea of "adaptive stencil", namely, the left shift r changes with the location x_i. The basic idea is to avoid including the discontinuous cell in the stencil, if possible.

To achieve this effect, we need to look at the Newton formulation of the interpolation polynomial.

We first review the definition of the Newton divided differences. The 0-th degree divided differences of the function $V(x)$ in (2.13)-(2.14) are defined by:

$$V[x_{i-\frac{1}{2}}] \equiv V(x_{i-\frac{1}{2}}); \tag{3.1}$$

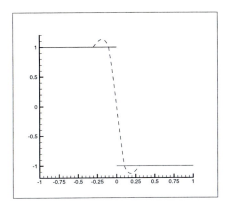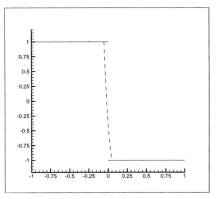

Fig. 3.1. Fixed central stencil cubic interpolation (left) and ENO cubic interpolation (right) for the step function. Solid: exact function; Dashed: interpolant piecewise cubic polynomials.

and in general the j-th degree divided differences, for $j \geq 1$, are defined inductively by

$$V[x_{i-\frac{1}{2}}, ..., x_{i+j-\frac{1}{2}}] \equiv \frac{V[x_{i+\frac{1}{2}}, ..., x_{i+j-\frac{1}{2}}] - V[x_{i-\frac{1}{2}}, ..., x_{i+j-\frac{3}{2}}]}{x_{i+j-\frac{1}{2}} - x_{i-\frac{1}{2}}}. \qquad (3.2)$$

Similarly, the divided differences of the cell averages $\overline{v}$ in (2.4) are defined by

$$\overline{v}[x_i] \equiv \overline{v}_i; \qquad (3.3)$$

and in general

$$\overline{v}[x_i, ..., x_{i+j}] \equiv \frac{\overline{v}[x_{i+1}, ..., x_{i+j}] - \overline{v}[x_i, ..., x_{i+j-1}]}{x_{i+j} - x_i}. \qquad (3.4)$$

We note that, by (2.14),

$$V[x_{i-\frac{1}{2}}, x_{i+\frac{1}{2}}] = \frac{V(x_{i+\frac{1}{2}}) - V(x_{i-\frac{1}{2}})}{x_{i+\frac{1}{2}} - x_{i-\frac{1}{2}}} = \overline{v}_i, \qquad (3.5)$$

i.e. the 0-th degree divided differences of $\overline{v}$ are the first degree divided differences of $V(x)$. We can then write the divided differences of $V(x)$ of first degree and higher in terms of $\overline{v}$, using (3.5) and (3.2), thus completely avoid the computation of V.

The Newton form of the k-th degree interpolation polynomial $P(x)$, which interpolates $V(x)$ at the $k+1$ points (2.15), can be expressed using the divided differences (3.1)-(3.2) by

$$P(x) = \sum_{j=0}^{k} V[x_{i-r-\frac{1}{2}}, ..., x_{i-r+j-\frac{1}{2}}] \prod_{m=0}^{j-1} \left(x - x_{i-r+m-\frac{1}{2}} \right). \qquad (3.6)$$

We can take the derivative of (3.6) to get $p(x)$ in (2.16):

$$p(x) = \sum_{j=1}^{k} V[x_{i-r-\frac{1}{2}}, ..., x_{i-r+j-\frac{1}{2}}] \sum_{m=0}^{j-1} \prod_{\substack{l=0 \\ l \neq m}}^{j-1} \left(x - x_{i-r+l-\frac{1}{2}} \right). \qquad (3.7)$$

Notice that only first and higher degree divided differences of $V(x)$ appear in (3.7). Hence by (3.5), we can express $p(x)$ completely by the divided differences of $\bar{v}$, without any need to reference $V(x)$.

Let us now recall an important property of divided differences:

$$V[x_{i-\frac{1}{2}}, ..., x_{i+j-\frac{1}{2}}] = \frac{V^{(j)}(\xi)}{j!}, \qquad (3.8)$$

for some ξ inside the stencil: $x_{i-\frac{1}{2}} < \xi < x_{i+j-\frac{1}{2}}$, *as long as the function* $V(x)$ *is smooth in this stencil.* If $V(x)$ is discontinuous at some point inside the stencil, then it is easy to verify that

$$V[x_{i-\frac{1}{2}}, ..., x_{i+j-\frac{1}{2}}] = O\left(\frac{1}{\Delta x^j}\right). \qquad (3.9)$$

Thus the divided difference is a measurement of the smoothness of the function inside the stencil.

We now describe the ENO idea by using (3.6). Suppose our job is to find a stencil of $k+1$ consecutive points, which must include $x_{i-\frac{1}{2}}$ and $x_{i+\frac{1}{2}}$, such that $V(x)$ is "the smoothest" in this stencil comparing with other possible stencils. We perform this job by breaking it into steps, in each step we only add one point to the stencil. We thus start with the two point stencil

$$\tilde{S}_2(i) = \{x_{i-\frac{1}{2}}, x_{i+\frac{1}{2}}\}, \qquad (3.10)$$

where we have used $\tilde{S}$ to denote a stencil for the primitive function V. Notice that the stencil $\tilde{S}$ for V has a corresponding stencil S for $\bar{v}$ through (3.5), for example (3.10) corresponds to a single cell stencil

$$S(i) = \{I_i\}$$

for $\bar{v}$. The linear interpolation on the stencil $\tilde{S}_2(i)$ in (3.10) can be written in the Newton form as

$$P^1(x) = V[x_{i-\frac{1}{2}}] + V[x_{i-\frac{1}{2}}, x_{i+\frac{1}{2}}]\left(x - x_{i-\frac{1}{2}}\right).$$

At the next step, we have only two choices to expand the stencil by adding one point: we can either add the left neighbor $x_{i-\frac{3}{2}}$, resulting in the following quadratic interpolation

$$R(x) = P^1(x) + V[x_{i-\frac{3}{2}}, x_{i-\frac{1}{2}}, x_{i+\frac{1}{2}}]\left(x - x_{i-\frac{1}{2}}\right)\left(x - x_{i+\frac{1}{2}}\right), \qquad (3.11)$$

or add the right neighbor $x_{i+\frac{3}{2}}$, resulting in the following quadratic interpolation

$$S(x) = P^1(x) + V[x_{i-\frac{1}{2}}, x_{i+\frac{1}{2}}, x_{i+\frac{3}{2}}]\left(x - x_{i-\frac{1}{2}}\right)\left(x - x_{i+\frac{1}{2}}\right). \qquad (3.12)$$

We note that the deviations from $P^1(x)$ in (3.11) and (3.12), are the *same* function

$$\left(x - x_{i-\frac{1}{2}}\right)\left(x - x_{i+\frac{1}{2}}\right)$$

multiplied by two different constants

$$V[x_{i-\frac{3}{2}}, x_{i-\frac{1}{2}}, x_{i+\frac{1}{2}}], \text{ and } V[x_{i-\frac{1}{2}}, x_{i+\frac{1}{2}}, x_{i+\frac{3}{2}}]. \qquad (3.13)$$

These two constants are the two second degree divided differences of $V(x)$ in two different stencils. We have already noticed before, in (3.8) and (3.9), that a smaller divided difference implies the function is "smoother" in that stencil. We thus decide upon which point to add to the stencil, by comparing the two relevant divided differences (3.13), and picking the one with a smaller absolute value. Thus, if

$$\left|V[x_{i-\frac{3}{2}}, x_{i-\frac{1}{2}}, x_{i+\frac{1}{2}}]\right| < \left|V[x_{i-\frac{1}{2}}, x_{i+\frac{1}{2}}, x_{i+\frac{3}{2}}]\right|, \qquad (3.14)$$

we will take the 3 point stencil as

$$\tilde{S}_3(i) = \{x_{i-\frac{3}{2}}, x_{i-\frac{1}{2}}, x_{i+\frac{1}{2}}\};$$

otherwise, we will take

$$\tilde{S}_3(i) = \{x_{i-\frac{1}{2}}, x_{i+\frac{1}{2}}, x_{i+\frac{3}{2}}\}.$$

This procedure can be continued, with one point added to the stencil at each step, according to the smaller of the absolute values of the two relevant divided differences, until the desired number of points in the stencil is reached.

We note that, for the uniform grid case $\Delta x_i = \Delta x$, there is no need to compute the divided differences as in (3.2). We should use undivided differences instead:

$$V < x_{i-\frac{1}{2}}, x_{i+\frac{1}{2}} >= V[x_{i-\frac{1}{2}}, x_{i+\frac{1}{2}}] = \bar{v}_i \qquad (3.15)$$

(see (3.5)), and

$$V < x_{i-\frac{1}{2}}, ..., x_{i+j+\frac{1}{2}} > \qquad (3.16)$$
$$\equiv V < x_{i+\frac{1}{2}}, ..., x_{i+j+\frac{1}{2}} > -V < x_{i-\frac{1}{2}}, ..., x_{i+j-\frac{1}{2}} >, \quad j \geq 1.$$

The Newton interpolation formulae (3.6)-(3.7) should also be adjusted accordingly. This both saves computational time and reduces round-off effects.

The FORTRAN program for this ENO choosing process is very simple:

```
* assuming the m-th degree divided (or undivided) differences
* of V(x), with x_i as the left-most point in the arguments,
* are stored in V(i,m), also assuming that "is" is the
* left-most point in the stencil for cell i for a k-th degree
* polynomial

    is=i
    do m=2,k
    if(abs(V(is-1,m)).lt.abs(V(is,m))) is=is-1
    enddo
```

Once the stencil $\tilde{S}(i)$, hence $S(i)$, in (2.8) is found, one could use (2.11), with the prestored values of the constants c_{rj}, (2.20) or (2.21), to compute the reconstructed values at the cell boundary. Or, one could use (2.30) to compute the fluxes. An alternative way is to compute the values or fluxes using the Newton form (3.7) directly. The computational cost is about the same.

We summarize the ENO reconstruction procedure in the following

Algorithm 3.1. 1D ENO reconstruction.

Given the cell averages $\{\bar{v}_i\}$ of a function $v(x)$, we obtain a piecewise polynomial reconstruction, of degree at most $k - 1$, using ENO, in the following way:

1. Compute the divided differences of the primitive function $V(x)$, for degrees 1 to k, using $\bar{v}$, (3.5) and (3.2).
   If the grid is uniform $\Delta x_i = \Delta x$, at this stage, undivided differences (3.15)-(3.16) should be computed instead.
2. In cell I_i, start with a two point stencil

$$\tilde{S}_2(i) = \{x_{i-\frac{1}{2}}, x_{i+\frac{1}{2}}\}$$

for $V(x)$, which is equivalent to a one point stencil,

$$S_1(i) = \{I_i\}$$

for $\bar{v}$.
3. For $l = 2, ..., k$, assuming

$$\tilde{S}_l(i) = \{x_{j+\frac{1}{2}}, ..., x_{j+l-\frac{1}{2}}\}$$

is known, add one of the two neighboring points, $x_{j-\frac{1}{2}}$ or $x_{j+l+\frac{1}{2}}$, to the stencil, following the ENO procedure:
   - If

$$\left|V[x_{j-\frac{1}{2}}, ..., x_{j+l-\frac{1}{2}}]\right| < \left|V[x_{j+\frac{1}{2}}, ..., x_{j+l+\frac{1}{2}}]\right|, \tag{3.17}$$

   add $x_{j-\frac{1}{2}}$ to the stencil $\tilde{S}_l(i)$ to obtain

$$\tilde{S}_{l+1}(i) = \{x_{j-\frac{1}{2}}, ..., x_{j+l-\frac{1}{2}}\};$$

– Otherwise, add $x_{j+l+\frac{1}{2}}$ to the stencil $\tilde{S}_l(i)$ to obtain

$$\tilde{S}_{l+1}(i) = \{x_{j+\frac{1}{2}}, ..., x_{j+l+\frac{1}{2}}\}.$$

4. Use the Lagrange form (2.19) or the Newton form (3.7) to obtain $p_i(x)$, which is a polynomial of degree at most $k-1$ in I_i, satisfying the accuracy condition (2.5), *as long as $v(x)$ is smooth in I_i*.
   We could use $p_i(x)$ to get the approximations at the cell boundaries:

$$v^-_{i+\frac{1}{2}} = p_i(x_{i+\frac{1}{2}}), \qquad v^+_{i-\frac{1}{2}} = p_i(x_{i-\frac{1}{2}}).$$

However, it is usually more convenient, when the stencil is known, to use (2.10), with c_{rj} defined by (2.20) for a nonuniform grid, or by (2.21) and Table 2.1 for a uniform grid, to compute an approximation to $v(x)$ at the cell boundaries.

□

For the same piecewise cubic interpolation to the step function, but this time using the ENO procedure with a two point stencil $\tilde{S}_2(i) = \{x_{i-\frac{1}{2}}, x_{i+\frac{1}{2}}\}$ in the Step 2 of Algorithm 3.1, we obtain a non-oscillatory interpolation, in Fig. 3.1 (right).

For a piecewise smooth function $V(x)$, ENO interpolation starting with a two point stencil $\tilde{S}_2(i) = \{x_{i-\frac{1}{2}}, x_{i+\frac{1}{2}}\}$ in the Step 2 of Algorithm 3.1, as was shown in Fig. 3.1 (right), has the following properties [48]:

1. The accuracy condition

$$P_i(x) = V(x) + O(\Delta x^{k+1}), \qquad x \in I_i$$

   is valid for any cell I_i which does not contain a discontinuity.
   This implies that the ENO interpolation procedure can recover the full high order accuracy right up to the discontinuity.
2. $P_i(x)$ is monotone in any cell I_i which *does* contain a discontinuity of $V(x)$.
3. The reconstruction is TVB (total variation bounded). That is, there exists a function $z(x)$, satisfying

$$z(x) = P_i(x) + O(\Delta x^{k+1}), \qquad x \in I_i$$

   for any cell I_i, including those cells which contain discontinuities, such that

$$TV(z) \le TV(V).$$

Property 3 is clearly a consequence of Properties 1 and 2 (just take $z(x)$ to be $V(x)$ in the smooth cells and take $z(x)$ to be $P_i(x)$ in the cells containing discontinuities). It is quite interesting that Property 2 holds. One would have

expected trouble in those "shocked cells", i.e. cells I_i which contain discontinuities, for ENO would not help for such cases as the stencil starts with two
points already containing a discontinuity. We will give a proof of Property 2
for a simple but illustrative case, i.e. when $V(x)$ is a step function

$$V(x) = \begin{cases} 0, x \le 0; \\ 1, x > 0. \end{cases}$$

and the k-th degree polynomial $P(x)$ interpolates $V(x)$ at $k+1$ points

$$x_{\frac{1}{2}} < x_{\frac{3}{2}} < \ldots < x_{k+\frac{1}{2}}$$

containing the discontinuity

$$x_{j_0-\frac{1}{2}} < 0 < x_{j_0+\frac{1}{2}}$$

for some j_0 between 1 and k. For any interval which does not contain the
discontinuity 0:

$$[x_{j-\frac{1}{2}}, x_{j+\frac{1}{2}}], \qquad j \ne j_0, \tag{3.18}$$

we have

$$P(x_{j-\frac{1}{2}}) = V(x_{j-\frac{1}{2}}) = V(x_{j+\frac{1}{2}}) = P(x_{j+\frac{1}{2}}),$$

hence there is at least one point ξ_j in between, $x_{j-\frac{1}{2}} < \xi_j < x_{j+\frac{1}{2}}$, such
that $P'(\xi_j) = 0$. This way we can find $k - 1$ distinct zeroes for $P'(x)$, as
there are $k - 1$ intervals (3.18) which do not contain the discontinuity 0.
However, $P'(x)$ is a non-zero polynomial of degree at most $k - 1$, hence can
have at most $k - 1$ distinct zeroes. This implies that $P'(x)$ *does not have any
zero inside the shocked interval* $[x_{j_0-\frac{1}{2}}, x_{j_0+\frac{1}{2}}]$, i.e. $P(x)$ is *monotone* in this
shocked interval. This proof can be generalized to a proof for Property 2 [48].

3.2   WENO Approximation

In this subsection we describe the recently developed WENO (weighted ENO)
reconstruction procedure [69,55]. WENO is based on ENO, of course. For
simplicity of presentation, in this subsection we assume the grid is uniform,
i.e. $\Delta x_i = \Delta x$.

   As we can see from Sect. 3.1, ENO reconstruction is uniformly high order
accurate right up to the discontinuity. It achieves this effect by adaptively
choosing the stencil based on the absolute values of divided differences. However, one could make the following remarks about ENO reconstruction, indicating rooms for improvements:

1. The stencil might change even by a round-off error perturbation near
   zeroes of the solution and its derivatives. That is, when both sides of
   (3.17) are near 0, a small change at the round off level would change the
   direction of the inequality and hence the stencil. In smooth regions, this
   "free adaptation" of stencils is clearly not necessary. Moreover, this may
   cause loss of accuracy when applied to a hyperbolic PDE [83,87].

2. The resulting numerical flux (2.23) is not smooth, as the stencil pattern may change at neighboring points.
3. In the stencil choosing process, k candidate stencils are considered, covering $2k - 1$ cells, but only one of the stencils is actually used in forming the reconstruction (2.10) or the flux (2.30), resulting in k-th order accuracy. If all the $2k - 1$ cells in the potential stencils are used, one could get $(2k - 1)$-th order accuracy in smooth regions.
4. ENO stencil choosing procedure involves many logical "if" structures, or equivalent mathematical formulae, which are not very efficient on certain vector computers such as CRAYs (however they are friendly to parallel computers).

There have been attempts in the literature to rectify the first problem, the "free adaptation" of stencils. In [31] and [87], the following "biasing" strategy was proposed. One first identity a "preferred" stencil

$$\tilde{S}_{pref}(i) = \{x_{i-r+\frac{1}{2}}, ..., x_{i-r+k+\frac{1}{2}}\}, \tag{3.19}$$

which might be central or one-point upwind. One then replaces (3.17) by

$$\left| V[x_{j-\frac{1}{2}}, ..., x_{j+l-\frac{1}{2}}] \right| < b \left| V[x_{j+\frac{1}{2}}, ..., x_{j+l+\frac{1}{2}}] \right|,$$

if

$$x_{j+\frac{1}{2}} > x_{i-r+\frac{1}{2}},$$

i.e. if the left-most point $x_{j+\frac{1}{2}}$ in the current stencil $\tilde{S}_l(i)$ has not reached the left-most point $x_{i-r+\frac{1}{2}}$ of the preferred stencil $S_{pref}(i)$ in (3.19) yet; otherwise, if

$$x_{j+\frac{1}{2}} \leq x_{i-r+\frac{1}{2}},$$

one replaces (3.17) by

$$b \left| V[x_{j-\frac{1}{2}}, ..., x_{j+l-\frac{1}{2}}] \right| < \left| V[x_{j+\frac{1}{2}}, ..., x_{j+l+\frac{1}{2}}] \right|.$$

Here, $b > 1$ is the so-called biasing parameter. Analysis in [87] indicates a good choice of the parameter $b = 2$. The philosophy is to stay as close as possible to the preferred stencil, unless the alternative candidate is, roughly speaking, a factor $b > 1$ better in smoothness.

WENO is a more recent attempt to improve upon ENO in these four points. The basic idea is the following: instead of using only one of the candidate stencils to form the reconstruction, one uses a convex combination of all of them. To be more precise, suppose the k candidate stencils

$$S_r(i) = \{x_{i-r}, ..., x_{i-r+k-1}\}, \quad r = 0, ..., k - 1 \tag{3.20}$$

produce k different reconstructions to the value $v_{i+\frac{1}{2}}$, according to (2.11),

$$v_{i+\frac{1}{2}}^{(r)} = \sum_{j=0}^{k-1} c_{rj} \bar{v}_{i-r+j}, \quad r = 0, ..., k - 1, \tag{3.21}$$

WENO reconstruction would take a convex combination of all $v_{i+\frac{1}{2}}^{(r)}$ defined in (3.21) as a new approximation to the cell boundary value $v(x_{i+\frac{1}{2}})$:

$$v_{i+\frac{1}{2}} = \sum_{r=0}^{k-1} w_r v_{i+\frac{1}{2}}^{(r)}. \tag{3.22}$$

Apparently, the key to the success of WENO would be the choice of the weights w_r. We require

$$w_r \geq 0, \qquad \sum_{r=0}^{k-1} w_r = 1 \tag{3.23}$$

for stability and consistency.

If the function $v(x)$ is smooth in all of the candidate stencils (3.20), there are constants d_r such that

$$v_{i+\frac{1}{2}} = \sum_{r=0}^{k-1} d_r v_{i+\frac{1}{2}}^{(r)} = v(x_{i+\frac{1}{2}}) + O(\Delta x^{2k-1}). \tag{3.24}$$

For example, d_r for $1 \leq k \leq 3$ are given by

$$d_0 = 1, \qquad k = 1;$$
$$d_0 = \frac{2}{3}, \quad d_1 = \frac{1}{3}, \qquad k = 2;$$
$$d_0 = \frac{3}{10}, \quad d_1 = \frac{3}{5}, \quad d_2 = \frac{1}{10}, \qquad k = 3.$$

We can see that d_r is always positive and, due to consistency,

$$\sum_{r=0}^{k-1} d_r = 1. \tag{3.25}$$

In this smooth case, we would like to have

$$w_r = d_r + O(\Delta x^{k-1}), \qquad r = 0, ..., k-1, \tag{3.26}$$

which would imply $(2k-1)$-th order accuracy:

$$v_{i+\frac{1}{2}} = \sum_{r=0}^{k-1} w_r v_{i+\frac{1}{2}}^{(r)} = v(x_{i+\frac{1}{2}}) + O(\Delta x^{2k-1}) \tag{3.27}$$

because

$$\sum_{r=0}^{k-1} w_r v_{i+\frac{1}{2}}^{(r)} - \sum_{r=0}^{k-1} d_r v_{i+\frac{1}{2}}^{(r)} = \sum_{r=0}^{k-1} (w_r - d_r) \left(v_{i+\frac{1}{2}}^{(r)} - v(x_{i+\frac{1}{2}}) \right)$$

$$= \sum_{r=0}^{k-1} O(\Delta x^{k-1}) O(\Delta x^k)$$

$$= O(\Delta x^{2k-1})$$

where in the first equality we used (3.23) and (3.25).

When the function $v(x)$ has a discontinuity in one or more of the stencils (3.20), we would hope the corresponding weight(s) w_r to be essentially 0, to emulate the successful ENO idea.

Another consideration is that the weights should be smooth functions of the cell averages involved. In fact, the weights designed in [55] and described below are C^∞.

Finally, we would like to have weights which are computationally efficient. Thus, polynomials or rational functions are preferred over exponential type functions.

All these considerations and ample numerical experiments lead to the following form of weights:

$$w_r = \frac{\alpha_r}{\sum_{s=0}^{k-1} \alpha_s}, \qquad r = 0, ..., k-1 \tag{3.28}$$

with

$$\alpha_r = \frac{d_r}{(\epsilon + \beta_r)^2}. \tag{3.29}$$

Here $\epsilon > 0$ is introduced to avoid the denominator to become 0. We take $\epsilon = 10^{-6}$ in all our numerical tests [55]. β_r are the so-called "smooth indicators" of the stencil $S_r(i)$: if the function $v(x)$ is smooth in the stencil $S_r(i)$, then

$$\beta_r = O(\Delta x^2),$$

but if $v(x)$ has a discontinuity inside the stencil $S_r(i)$, then

$$\beta_r = O(1).$$

Translating into the weights w_r in (3.28), we will have

$$w_r = O(1)$$

when the function $v(x)$ is smooth in the stencil $S_r(i)$, and

$$w_r = O(\Delta x^4)$$

if $v(x)$ has a discontinuity inside the stencil $S_r(i)$. Emulation of ENO near a discontinuity is thus achieved.

One also has to worry about the accuracy requirement (3.26), which must be checked when the specific form of the smooth indicator β_r is given. For any smooth indicator β_r, it is easy to see that the weights defined by (3.28) satisfies (3.23). To satisfy (3.26), it suffices to have, through a Taylor expansion analysis:

$$\beta_r = D\left(1 + O(\Delta x^{k-1})\right), \qquad r = 0, ..., k-1, \tag{3.30}$$

where D is a nonzero quantity independent of r (but may depend on Δx).

As we have seen in Sect. 3.1, the ENO reconstruction procedure chooses the "smoothest" stencil by comparing a hierarchy of divided or undivided differences. This is because these differences can be used to measure the smoothness of the function on a stencil, (3.8)-(3.9). In [55], after extensive experiments, a robust (for third and fifth order at least) choice of smooth indicators β_r is given. As we know, on each stencil $S_r(i)$, we can construct a $(k-1)$-th degree reconstruction polynomial, which if evaluated at $x = x_{i+\frac{1}{2}}$, renders the approximation to the value $v(x_{i+\frac{1}{2}})$ in (3.21). Since the total variation is a good measurement for smoothness, it would be desirable to minimize the total variation for this reconstruction polynomial inside I_i. Consideration for a smooth flux and for the role of higher order variations leads us to the following measurement for smoothness: let the reconstruction polynomial on the stencil $S_r(i)$ be denoted by $p_r(x)$, we define

$$\beta_r = \sum_{l=1}^{k-1} \int_{x_{i-\frac{1}{2}}}^{x_{i+\frac{1}{2}}} \Delta x^{2l-1} \left(\frac{\partial^l p_r(x)}{\partial^l x} \right)^2 dx . \tag{3.31}$$

The right hand side of (3.31) is just a sum of the squares of scaled L^2 norms for all the derivatives of the interpolation polynomial $p_r(x)$ over the interval $(x_{i-\frac{1}{2}}, x_{i+\frac{1}{2}})$. The factor Δx^{2l-1} is introduced to remove any Δx dependency in the derivatives, in order to preserve self-similarity when used to hyperbolic PDEs (Sect. 4).

We remark that (3.31) is similar to but smoother than the total variation measurement based on the L^1 norm. It also renders a more accurate WENO scheme for the case $k = 2$ and 3.

When $k = 2$, (3.31) gives the following smoothness measurement [69,55]:

$$\beta_0 = (\bar{v}_{i+1} - \bar{v}_i)^2 , \qquad \beta_1 = (\bar{v}_i - \bar{v}_{i-1})^2 . \tag{3.32}$$

For $k = 3$, (3.31) gives [55]:

$$\beta_0 = \frac{13}{12}(\bar{v}_i - 2\bar{v}_{i+1} + \bar{v}_{i+2})^2 + \frac{1}{4}(3\bar{v}_i - 4\bar{v}_{i+1} + \bar{v}_{i+2})^2 ,$$

$$\beta_1 = \frac{13}{12}(\bar{v}_{i-1} - 2\bar{v}_i + \bar{v}_{i+1})^2 + \frac{1}{4}(\bar{v}_{i-1} - \bar{v}_{i+1})^2 , \tag{3.33}$$

$$\beta_2 = \frac{13}{12}(\bar{v}_{i-2} - 2\bar{v}_{i-1} + \bar{v}_i)^2 + \frac{1}{4}(\bar{v}_{i-2} - 4\bar{v}_{i-1} + 3\bar{v}_i)^2 .$$

We can easily verify that the accuracy condition (3.30) is satisfied, even near smooth extrema [55]. This indicates that (3.32) gives a third order WENO scheme, and (3.33) gives a fifth order one.

Notice that the discussion here has a one point upwind bias in the optimal linear stencil, suitable for a problem with wind blowing from left to right. If the wind blows the other way, the procedure should be modified symmetrically with respect to $x_{i+\frac{1}{2}}$.

In summary, we have the following WENO reconstruction procedure:

Algorithm 3.2. 1D WENO reconstruction.

Given the cell averages $\{\bar{v}_i\}$ of a function $v(x)$, for each cell I_i, we obtain upwind biased $(2k-1)$-th order approximations to the function $v(x)$ at the cell boundaries, denoted by $v_{i-\frac{1}{2}}^+$ and $v_{i+\frac{1}{2}}^-$, in the following way:

1. Obtain the k reconstructed values $v_{i+\frac{1}{2}}^{(r)}$, of k-th order accuracy, in (3.21), based on the stencils (3.20), for $r = 0, ..., k-1$;

   Also obtain the k reconstructed values $v_{i-\frac{1}{2}}^{(r)}$, of k-th order accuracy, using (2.10), again based on the stencils (3.20), for $r = 0, ..., k-1$;
2. Find the constants d_r and $\tilde{d}_r$, such that (3.24) and

$$v_{i-\frac{1}{2}} = \sum_{r=0}^{k-1} \tilde{d}_r v_{i-\frac{1}{2}}^{(r)} = v(x_{i-\frac{1}{2}}) + O(\Delta x^{2k-1})$$

are valid. By symmetry,

$$\tilde{d}_r = d_{k-1-r}.$$

3. Find the smooth indicators β_r in (3.31), for all $r = 0, ..., k-1$. Explicit formulae for $k = 2$ and $k = 3$ are given in (3.32) and (3.33) respectively.
4. Form the weights ω_r and $\tilde{\omega}_r$ using (3.28)-(3.29) and

$$\tilde{\omega}_r = \frac{\tilde{\alpha}_r}{\sum_{s=0}^{k-1} \tilde{\alpha}_s}, \qquad \tilde{\alpha}_r = \frac{\tilde{d}_r}{(\epsilon + \beta_r)^2}, \qquad r = 0, ..., k-1.$$

5. Find the $(2k-1)$-th order reconstruction

$$v_{i+\frac{1}{2}}^- = \sum_{r=0}^{k-1} \omega_r v_{i+\frac{1}{2}}^{(r)}, \qquad v_{i-\frac{1}{2}}^+ = \sum_{r=0}^{k-1} \tilde{\omega}_r v_{i-\frac{1}{2}}^{(r)}. \tag{3.34}$$

$\square$

We can obtain weights for higher orders of k (corresponding to seventh and higher order WENO schemes) using the same recipe. However, these schemes of seventh and higher order have not been extensively tested yet. Current research of Balsara and Shu [5] addresses this issue.

4   ENO and WENO Schemes in One Dimension

In this section we describe the ENO and WENO schemes for one dimensional conservation laws:

$$u_t(x,t) + f_x(u(x,t)) = 0 \tag{4.1}$$

equipped with suitable initial and boundary conditions.

We will concentrate on the discussion of spatial discretization, and will leave the time variable t continuous (the method-of-lines approach). Time discretization will be discussed in Sect. 9.

Our computational domain is $a \le x \le b$. We have a grid defined by (2.1), with the notations (2.2)-(2.3). Except for in Sect. 4.5, we do not consider boundary conditions. We thus assume that the values of the numerical solution are also available outside the computational domain whenever they are needed. This would be the case for periodic or compactly supported problems.

4.1   Finite Volume Formulation in the Scalar Case

For finite volume schemes, or schemes based on cell averages, we do not solve (4.1) directly, but its integrated version. We integrate (4.1) over the interval I_i to obtain

$$\frac{d\bar{u}(x_i, t)}{dt} = -\frac{1}{\Delta x_i} \left(f(u(x_{i+\frac{1}{2}}, t)) - f(u(x_{i-\frac{1}{2}}, t)) \right), \qquad (4.2)$$

where

$$\bar{u}(x_i, t) \equiv \frac{1}{\Delta x_i} \int_{x_{i-\frac{1}{2}}}^{x_{i+\frac{1}{2}}} u(\xi, t) \, d\xi \qquad (4.3)$$

is the cell average. We approximate (4.2) by the following conservative scheme

$$\frac{d\bar{u}_i(t)}{dt} = -\frac{1}{\Delta x_i} \left(\hat{f}_{i+\frac{1}{2}} - \hat{f}_{i-\frac{1}{2}} \right), \qquad (4.4)$$

where $\bar{u}_i(t)$ is the numerical approximation to the cell average $\bar{u}(x_i, t)$, and the numerical flux $\hat{f}_{i+\frac{1}{2}}$ is defined by

$$\hat{f}_{i+\frac{1}{2}} = h \left(u_{i+\frac{1}{2}}^-, u_{i+\frac{1}{2}}^+ \right) \qquad (4.5)$$

with the values $u_{i+\frac{1}{2}}^{\pm}$ obtained by the ENO reconstruction Algorithm 3.1, or by the WENO reconstruction Algorithm 3.2.

The two argument function h in (4.5) is a monotone flux. It satisfies:

- $h(a, b)$ is a Lipschitz continuous function in both arguments;
- $h(a, b)$ is a nondecreasing function in a and a nonincreasing function in b. Symbolically $h(\uparrow, \downarrow)$;
- $h(a, b)$ is consistent with the physical flux f, that is, $h(a, a) = f(a)$.

Examples of monotone fluxes include:

1. Godunov flux:

$$h(a, b) = \begin{cases} \min_{a \le u \le b} f(u) & \text{if } a \le b \\ \max_{b \le u \le a} f(u) & \text{if } a > b \end{cases}, \qquad (4.6)$$

2. Engquist-Osher flux:

$$h(a, b) = \int_0^a \max(f'(u), 0) du + \int_0^b \min(f'(u), 0) du + f(0). \qquad (4.7)$$

3. Lax-Friedrichs flux:

$$h(a, b) = \frac{1}{2} [f(a) + f(b) - \alpha(b - a)] \tag{4.8}$$

where $\alpha = \max_u |f'(u)|$ is a constant. The maximum is taken over the relevant range of u.

We have listed the monotone fluxes from the least dissipative (less smearing of discontinuities) to the most. For lower order methods (order of reconstruction is 1 or 2), there is a big difference between results obtained by different monotone fluxes. However, this difference becomes much smaller for higher order reconstructions. In Fig. 4.1, we plot the results of a right moving shock for the Burgers' equation ($f(u) = \frac{u^2}{2}$ in (4.1)), with first order reconstruction using Godunov and Lax-Friedrichs monotone fluxes (top), and with fourth order ENO reconstruction using Godunov and Lax-Friedrichs monotone fluxes (bottom). We can clearly see that, while the Godunov flux behaves much better for the first order scheme, the two fourth order ENO schemes behave similarly. We thus use the simple and inexpensive Lax-Friedrichs flux in most of our high order calculations.

We remark that, by the classical Lax-Wendroff theorem [65], the solution to the conservative scheme (4.4), *if converges*, will converge to a weak solution of (4.1).

In summary, to build a finite volume ENO scheme (4.4), given the cell averages $\{\bar{u}_i\}$ (we will often drop the explicit reference to the time variable t), we proceed as follows:

Algorithm 4.1. Finite volume 1D scalar ENO and WENO Schemes.

1. Follow the Algorithm 3.1 in Sect. 3.1 for ENO, or the Algorithm 3.2 in Sect. 3.2 for WENO, to obtain the k-th order reconstructed values $u^-_{i+\frac{1}{2}}$ and $u^+_{i+\frac{1}{2}}$ for all i;
2. Choose a monotone flux (e.g., one of (4.6) to (4.8)), and use (4.5) to compute the flux $\hat{f}_{i+\frac{1}{2}}$ for all i;
3. Form the scheme (4.4).

□

Notice that the finite volume scheme can be applied to arbitrary nonuniform grids.

4.2   Finite Difference Formulation in the Scalar Case

We first assume the grid is uniform and solve (4.1) directly using a conservative approximation to the spatial derivative:

$$\frac{du_i(t)}{dt} = -\frac{1}{\Delta x} \left(\hat{f}_{i+\frac{1}{2}} - \hat{f}_{i-\frac{1}{2}} \right) \tag{4.9}$$

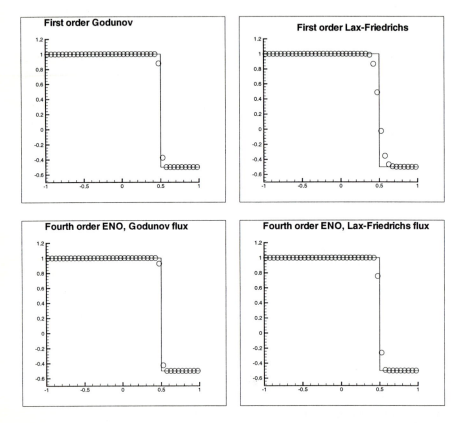

Fig. 4.1. First order (top) and fourth order (bottom) ENO schemes for the Burgers equation, with the Godunov flux (left) and the Lax-Friedrichs flux (right). Solid lines: exact solution; Circles: the computed solution at $t = 4$.

where $u_i(t)$ is the numerical approximation to the point value $u(x_i, t)$, and the numerical flux

$$\hat{f}_{i+\frac{1}{2}} = \hat{f}(u_{i-r}, ..., u_{i+s})$$

satisfies the following conditions:

- $\hat{f}$ is a Lipschitz continuous function in all the arguments;
- $\hat{f}$ is consistent with the physical flux f, that is, $\hat{f}(u, ..., u) = f(u)$.

Again the Lax-Wendroff theorem [65] applies. The solution to the conservative scheme (4.9), *if converges*, will converge to a weak solution of (4.1).

The numerical flux $\hat{f}_{i+\frac{1}{2}}$ is obtained by the ENO or WENO reconstruction procedures, Algorithm 3.1 or 3.2, *with* $\bar{v}(x) = f(u(x, t))$. For stability, it is important that upwinding is used in constructing the flux. The easiest and the most inexpensive way to achieve upwinding is the following: compute the Roe speed

$$\bar{a}_{i+\frac{1}{2}} \equiv \frac{f(u_{i+1}) - f(u_i)}{u_{i+1} - u_i}, \tag{4.10}$$

(when $u_{i+1} = u_i$ one should use $\bar{a}_{i+\frac{1}{2}} \equiv f'(u_i)$) and

- if $\bar{a}_{i+\frac{1}{2}} \geq 0$, then the the wind blows from the left to the right. We would use $v_{i+\frac{1}{2}}^-$ for the numerical flux $\hat{f}_{i+\frac{1}{2}}$;
- if $\bar{a}_{i+\frac{1}{2}} < 0$, then the wind blows from the right to the left. We would use $v_{i+\frac{1}{2}}^+$ for the numerical flux $\hat{f}_{i+\frac{1}{2}}$.

This produces the Roe scheme [82] at the first order level. For this reason, the ENO scheme based on this approach was termed "ENO-Roe" in [90].

In summary, to build a finite difference ENO scheme (4.9) *using the ENO-Roe approach*, given the point values $\{u_i\}$ (we again drop the explicit reference to the time variable t), we proceed as follows:

Algorithm 4.2. Finite difference 1D scalar ENO-Roe and WENO-Roe schemes.

1. Compute the Roe speed $\bar{a}_{i+\frac{1}{2}}$ for all i using (4.10);
2. Identify $\bar{v}_i = f(u_i)$ and use the ENO reconstruction Algorithm 3.1 or the WENO reconstruction Algorithm 3.2, to obtain the cell boundary values $v_{i+\frac{1}{2}}^-$ if $\bar{a}_{i+\frac{1}{2}} \geq 0$, or $v_{i+\frac{1}{2}}^+$ if $\bar{a}_{i+\frac{1}{2}} < 0$;
3. If the Roe speed at $x_{i+\frac{1}{2}}$ is positive

$$\bar{a}_{i+\frac{1}{2}} \geq 0,$$

then take the numerical flux as:

$$\hat{f}_{i+\frac{1}{2}} = v_{i+\frac{1}{2}}^-;$$

otherwise, take the the numerical flux as:

$$\hat{f}_{i+\frac{1}{2}} = v_{i+\frac{1}{2}}^+;$$

4. Form the scheme (4.9).

□

One disadvantage of the ENO-Roe approach is that entropy violating solutions may be obtained, just like in the first order Roe scheme case. For example, if ENO-Roe is applied to the Burgers equation

$$u_t + \left(\frac{u^2}{2}\right)_x = 0$$

with the following initial condition

$$u(x,0) = \begin{cases} -1, & \text{if } x < 0, \\ 1, & \text{if } x \geq 0, \end{cases}$$

it will converge to the entropy violating expansion shock:

$$u(x,t) = \begin{cases} -1, & \text{if } x < 0, \\ 1, & \text{if } x \geq 0. \end{cases}$$

Local entropy correction could be used to rectify this [90]. However, it is usually more robust to use a global "flux splitting":

$$f(u) = f^+(u) + f^-(u) \tag{4.11}$$

where

$$\frac{df^+(u)}{du} \geq 0, \qquad \frac{df^-(u)}{du} \leq 0. \tag{4.12}$$

We would need the positive and negative fluxes $f^\pm(u)$ to have as many derivatives as the order of the scheme. This unfortunately rules out many popular flux splittings (such as those of van Leer [101] and Osher [77]) for high order methods in this framework.

The simplest smooth splitting is the Lax-Friedrichs splitting:

$$f^\pm(u) = \frac{1}{2}(f(u) \pm \alpha u) \tag{4.13}$$

where α is again taken as $\alpha = \max_u |f'(u)|$ over the relevant range of u.

We note that there is a close relationship between a flux splitting (4.11) and a monotone flux (4.5). In fact, for any flux splitting (4.11) satisfying (4.12),

$$h(a,b) = f^+(a) + f^-(b) \tag{4.14}$$

is clearly a monotone flux. However, not every monotone flux can be written in the flux split form (4.11). For example, the Godunov flux (4.6) cannot.

With the flux splitting (4.11), we apply the the ENO or WENO reconstruction procedures, Algorithm 3.1 or 3.2, with $\bar{v}(x) = f^+(u(x,t))$ and

$\bar{v}(x) = f^-(u(x,t))$ separately, to obtain two numerical fluxes $\hat{f}^+_{i+\frac{1}{2}}$ and $\hat{f}^-_{i+\frac{1}{2}}$, and then sum them to get the numerical flux $\hat{f}_{i+\frac{1}{2}}$.

In summary, to build a finite difference ENO or WENO scheme (4.9) *using the flux splitting approach*, given the point values $\{u_i\}$, we proceed as follows:

Algorithm 4.3. Finite difference 1D scalar flux splitting ENO and WENO schemes.

1. Find a smooth flux splitting (4.11), satisfying (4.12);
2. Identify $\bar{v}_i = f^+(u_i)$ and use the ENO or WENO reconstruction procedure, Algorithm 3.1 or 3.2, to obtain the cell boundary values $v^-_{i+\frac{1}{2}}$ for all i;
3. Take the positive numerical flux as

$$\hat{f}^+_{i+\frac{1}{2}} = v^-_{i+\frac{1}{2}};$$

4. Identify $\bar{v}_i = f^-(u_i)$ and use the ENO or WENO reconstruction procedures, Algorithm 3.1 or 3.2, to obtain the cell boundary values $v^+_{i+\frac{1}{2}}$ for all i;
5. Take the negative numerical flux as

$$\hat{f}^-_{i+\frac{1}{2}} = v^+_{i+\frac{1}{2}};$$

6. Form the numerical flux as

$$\hat{f}_{i+\frac{1}{2}} = \hat{f}^+_{i+\frac{1}{2}} + \hat{f}^-_{i+\frac{1}{2}};$$

7. Form the scheme (4.9).

□

We remark that the finite difference scheme in this section and the finite volume scheme in Sect. 4.1 are equivalent for one dimensional, linear PDE with constant coefficients: the only difference is in the initial condition (the finite difference version uses point values and the finite volume version uses cell averages of the exact initial condition). Notice that the schemes are still nonlinear in this case. However, this equivalency does not hold for a nonlinear PDE. Moreover, we will see later that there are significant differences in efficiency of the two approaches for multidimensional problems.

In the following we test the accuracy of the fifth order finite difference WENO schemes on the linear equation:

$$u_t + u_x = 0, \qquad -1 \le x \le 1$$
$$u(x,0) = u_0(x) \qquad \text{periodic.}$$

In Table 4.1, we show the errors of the fifth order WENO scheme given by the weights (3.28)-(3.29) with the smooth indicator (3.33), at time $t = 1$

for the initial condition $u_0(x) = \sin(\pi x)$, and compare them with the errors of the linear 5-th order upstream central scheme (i.e. the scheme with the linear weights d_r as in (3.24)). We can see that fifth order WENO gives the expected order of accuracy starting at about 40 grid points.

Table 4.1. Accuracy on $u_t + u_x = 0$ with $u_0(x) = \sin(\pi x)$.

Fifth order WENO scheme

N	L_∞ error	L_∞ order	L_1 error	L_1 order
10	2.98e-2	-	1.60e-2	-
20	1.45e-3	4.36	7.41e-4	4.43
40	4.58e-5	4.99	2.22e-5	5.06
80	1.48e-6	4.95	6.91e-7	5.01
160	4.41e-8	5.07	2.17e-8	4.99
320	1.35e-9	5.03	6.79e-10	5.00

Fifth order linear upwind-central scheme

N	L_∞ error	L_∞ order	L_1 error	L_1 order
10	4.98e-3	-	3.07e-3	-
20	1.60e-4	4.96	9.92e-5	4.95
40	5.03e-6	4.99	3.14e-6	4.98
80	1.57e-7	5.00	9.90e-8	4.99
160	4.91e-9	5.00	3.11e-9	4.99
320	1.53e-10	5.00	9.73e-11	5.00

In Table 4.2, we show errors for the initial condition $u_0(x) = \sin^4(\pi x)$. The order of accuracy for the fifth order WENO settles down later than in the previous example. Notice that this is the example for which ENO schemes lose their accuracy [83], [87].

We emphasize again that the high order conservative finite difference ENO and WENO schemes of third or higher order accuracy can only be applied to a uniform grid or a smoothly varying grid, i.e. a grid such that a smooth transformation

$$\xi = \xi(x)$$

will result in a uniform grid in the new variable ξ. Here ξ must contain as many derivatives as the order of accuracy of scheme calls for. If this is the case, then (4.1) is transformed to

$$u_t + \xi_x f(u)_\xi = 0$$

Table 4.2. Accuracy on $u_t + u_x = 0$ with $u_0(x) = \sin^4(\pi x)$.

Fifth order WENO scheme

N	L_∞ error	L_∞ order	L_1 error	L_1 order
20	1.08e-1	-	4.91e-2	-
40	8.90e-3	3.60	3.64e-3	3.75
80	1.80e-3	2.31	5.00e-4	2.86
160	1.22e-4	3.88	2.17e-5	4.53
320	4.37e-6	4.80	6.17e-7	5.14
640	9.79e-8	5.48	1.57e-8	5.30

Fifth order linear upwind-central scheme

N	L_∞ error	L_∞ order	L_1 error	L_1 order
20	5.23e-2	-	3.35e-2	-
40	2.47e-3	4.40	1.52e-3	4.46
80	8.32e-5	4.89	5.09e-5	4.90
160	2.65e-6	4.97	1.60e-6	4.99
320	8.31e-8	5.00	4.99e-8	5.00
640	2.60e-9	5.00	1.56e-9	5.00

and the conservative ENO or WENO derivative approximation is then applied to $f(u)_\xi$. It is proven in [77] that this way the scheme is still conservative, i.e. Lax-Wendroff theorem [65] still applies.

4.3 Provable Properties in the Scalar Case

Second order ENO schemes are also TVD (total variation diminishing), hence have at least subsequences which converge to weak solutions. There is no known convergence result for ENO schemes of degree higher than 2, even for smooth solutions.

WENO schemes have better convergence results, mainly because their numerical fluxes are smoother. It is proven [55] that WENO schemes converge for smooth solutions. Also, Jiang and Yu [56] have obtained an existence proof for traveling waves for WENO schemes. This is an important first step towards the proof of convergence for shocked cases.

Even though there are very few theoretical results about ENO or WENO schemes, in practice these schemes are very robust and stable. We caution against any attempts to modify the schemes solely for the purpose of stability or convergence proofs. In [89] we gave a remark about a modification of ENO schemes, which keeps the formal uniform high order accuracy and makes them stable and convergent for general multi dimensional scalar equations.

However it was pointed out there that the modification is not computationally useful, hence the convergence result has little value.

The remark in [89] is illustrative hence we reproduce it here. We start with a flux splitting (4.11) satisfying (4.12), and notice that the first order monotone scheme

$$\frac{du_i}{dt} = -\frac{1}{\Delta x_i}\left(f^+(u_i) - f^+(u_{i-1}) + f^-(u_{i+1}) - f^-(u_i)\right) \equiv R_1(u)_i \quad (4.15)$$

is convergent (also for multi space dimensions). We now construct a high order ENO approximation in the following way: starting from the two point stencil $\{x_{i-1}, x_i\}$, we expand it into a $k+1$ point stencil in an ENO fashion using the divided differences of $f^+(u(x))$. We then build the k-th degree polynomial $P^+(x)$ which interpolates $f^+(u(x))$ in this stencil. $P^-(x)$ is constructed in a similar way, starting from the two point stencil $\{x_i, x_{i+1}\}$. The scheme is finally defined as

$$\frac{du_i}{dt} = -\frac{d}{dx}\left(P^+(x) + P^-(x)\right)\Big|_{x=x_i} \equiv R_k(u)_i \quad (4.16)$$

This scheme is clearly k-th order accurate but is not conservative. We now denote the difference between the high order scheme (4.16) and the first order monotone scheme (4.15) by

$$D(u)_i \equiv R_k(u)_i - R_1(u)_i, \quad (4.17)$$

and limit it by

$$\tilde{D}(u)_i = \overline{m}(D(u)_i, M\Delta x^\alpha), \quad (4.18)$$

where $M > 0$ and $0 < \alpha \le 1$ are constants, and the capping function $\overline{m}$ is defined by

$$\overline{m}(a, b) = \begin{cases} a, & \text{if } |a| \le b; \\ b, & \text{if } a > b; \\ -b, & \text{if } a < -b. \end{cases}$$

The modified ENO scheme is then defined by

$$\frac{du_i}{dt} = \tilde{R}_k(u)_i \equiv R_1(u)_i + \tilde{D}(u)_i. \quad (4.19)$$

We notice that, in smooth regions, the difference between the first order and high order residues, $D(u)_i$, as defined in (4.17), is of the size $O(\Delta x)$, hence the capping (4.18) does not take effect in such regions, if $\alpha < 1$ or if $\alpha = 1$ and M is large enough, when Δx is sufficiently small. This implies that the scheme (4.19) is uniformly accurate. Moreover, since

$$\left|\tilde{R}_k(u)_i - R_1(u)_i\right| \le M\Delta x^\alpha$$

by (4.18), the high order scheme (4.19) shares every good property of the first order monotone scheme (4.15), such as total variation boundedness, entropy conditions, and convergence. From a theoretical point of view, this is the strongest result one could possibly hope for a high order scheme. However, the mesh size dependent limiting (4.18) renders the scheme highly impractical: the quality of the numerical solution will depend strongly on the choice of the parameters M and α, as well as on the mesh size Δx.

4.4   Systems

We only consider hyperbolic $m \times m$ systems, i.e. the Jocobian $f'(u)$ has m real eigenvalues

$$\lambda_1(u) \leq \ldots \leq \lambda_m(u) \tag{4.20}$$

and a complete set of independent eigenvectors

$$r_1(u), \ldots, r_m(u). \tag{4.21}$$

We denote the matrix whose columns are eigenvectors (4.21) by

$$R(u) = (r_1(u), \ldots, r_m(u)) \tag{4.22}$$

Then clearly

$$R^{-1}(u) f'(u) R(u) = \Lambda(u) \tag{4.23}$$

where $\Lambda(u)$ is the diagonal matrix with $\lambda_1(u), \ldots, \lambda_m(u)$ on the diagonal. Notice that the rows of $R^{-1}(u)$, denoted by $l_1(u), \ldots, l_m(u)$ (row vectors), are left eigenvectors of $f'(u)$:

$$l_i(u) f'(u) = \lambda_i(u) l_i(u), \qquad i = 1, \ldots, m. \tag{4.24}$$

There are several ways to generalize scalar ENO or WENO schemes to systems.

The easiest way is to apply the ENO or WENO schemes in a component by component fashion. For the finite volume formulation, this means that we make the reconstruction using ENO or WENO for each of the components of u separately. This produces the left and right values $u_{i+\frac{1}{2}}^{\pm}$ at the cell interface $x_{i+\frac{1}{2}}$. An exact or approximate Riemann solver, $h(u_{i+\frac{1}{2}}^-, u_{i+\frac{1}{2}}^+)$, is then used to build the scheme (4.4)-(4.5). The exact Riemann solver is given by the exact solution of (4.1) with the following step function as initial condition

$$u(x, 0) = \begin{cases} u_{i+\frac{1}{2}}^-, & x \leq 0; \\ u_{i+\frac{1}{2}}^+, & x > 0, \end{cases} \tag{4.25}$$

evaluated at the center $x = 0$. Notice that the solution to (4.1) with the initial condition (4.25) is self-similar, that is, it is a function of the variable $\xi = \frac{x}{t}$, hence is constant along $x = 0$. If we denote this solution by $u_{i+\frac{1}{2}}$, then the flux is taken as

$$h(u_{i+\frac{1}{2}}^-, u_{i+\frac{1}{2}}^+) = f(u_{i+\frac{1}{2}}).$$

In the scalar case, the exact Riemann solver gives the Godunov flux (4.6). Exact Riemann solver can be obtained for many systems including the Euler equations of compressible gas, which is used very often in practice. However, it is usually very costly to get this solution (for Euler equations of compressible gas, an iterative procedure is needed to obtain this solution, see [94]). In practice, approximate Riemann solvers are usually good enough. As in the scalar case, the quality of the solution is usually very sensitive to the choice of approximate Riemann solvers for *lower order* schemes (first or second order), but this sensitivity decreases with an increasing order of accuracy. The simplest approximate Riemann solver (albeit the most dissipative) is again the Lax-Friedrichs solver (4.8), except that now the constant α is taken as

$$\alpha = \max_{u} \max_{1 \le j \le m} |\lambda_j(u)| \tag{4.26}$$

where $\lambda_j(u)$ are the eigenvalues of the Jacobian $f'(u)$, (4.20). The maximum is again taken over the relevant range of u.

We summarize the procedure in the following

Algorithm 4.4. Component-wise finite volume 1D system ENO and WENO schemes.

1. For each component of the solution $\bar{u}$, apply the scalar ENO Algorithm 3.1 or WENO Algorithm 3.2 to reconstruct the corresponding component of the solution at the cell interfaces, $u^{\pm}_{i+\frac{1}{2}}$ for all i;

2. Apply an exact or approximate Riemann solver to compute the flux $\hat{f}_{i+\frac{1}{2}}$ for all i in (4.5);

3. Form the scheme (4.4).

□

For the finite difference formulation, a smooth flux splitting (4.11) is again needed. The condition (4.12) now becomes that the two Jacobians

$$\frac{\partial f^+(u)}{\partial u}, \quad \frac{\partial f^-(u)}{\partial u} \tag{4.27}$$

are still diagonalizable (preferably by the same eigenvectors $R(u)$ as for $f'(u)$), and have only non-negative / non-positive eigenvalues, respectively. We again recommend the Lax-Friedrichs flux splitting (4.13), with α given by (4.26), because of its simplicity and smoothness. A somewhat more complicated Lax-Friedrichs type flux splitting is:

$$f^{\pm}(u) = \frac{1}{2}(f(u) \pm R(u)\,\overline{\Lambda}\,R^{-1}(u)\,u),$$

where $R(u)$ and $R^{-1}(u)$ are defined in (4.22), and

$$\overline{\Lambda} = diag(\overline{\lambda}_1, ..., \overline{\lambda}_m)$$

where $\bar{\lambda}_j = max_u|\lambda_j(u)|$, and the maximum is again taken over the relevant range of u. This way the dissipation is added in each field according to the maximum size of eigenvalues in that field, not globally. One could also use other flux splittings, such as the van Leer splitting for gas dynamics [101]. However, for higher order schemes, the flux splitting must be sufficiently smooth in order to retain the order of accuracy.

With these flux splittings, we can again use the scalar recipes to form the finite difference scheme: just compute the positive and negative fluxes $\hat{f}^+_{i+\frac{1}{2}}$ and $\hat{f}^-_{i+\frac{1}{2}}$ component by component.

We summarize the procedure in the following

Algorithm 4.5. Component-wise finite difference 1D system ENO and WENO schemes.

1. Find a flux splitting (4.11). The simplest example is the Lax-Friedrichs flux splitting (4.13), with α given by (4.26);
2. For each component of the solution u, apply the scalar Algorithm 4.3 to reconstruct the corresponding component of the numerical flux $\hat{f}_{i+\frac{1}{2}}$;
3. Form the scheme (4.9).

□

These component by component versions of ENO and WENO schemes are simple and cost effective. They work reasonably well for many problems, especially when the order of accuracy is not high (second or sometimes third order). However, for more demanding test problems, or when the order of accuracy is high, it is usually advisable to use the following more costly, but much more robust characteristic decompositions.

To explain the characteristic decomposition, we start with a simple example where $f(u) = Au$ in (4.1) is linear and A is a constant matrix. In this situation, the eigenvalues (4.20), the eigenvectors (4.21), and the related matrices R, R^{-1} and Λ (4.22)-(4.23), are all constant matrices. If we define a change of variable

$$v = R^{-1}u, \tag{4.28}$$

then the PDE (4.1) becomes diagonal:

$$v_t + \Lambda v_x = 0 \tag{4.29}$$

that is, the m equations in (4.29) are decoupled and each one is a scalar linear convection equation of the form

$$w_t + \lambda_j w_x = 0. \tag{4.30}$$

We can thus use the reconstruction or flux evaluation techniques for the scalar equations, discussed in Sections 4.1 and 4.2, to handle each of the equations in (4.30). After we obtain the results, we can "come back" to the physical space u by using the inverse of (4.28):

$$u = R v \qquad (4.31)$$

For example, if the reconstructed polynomial for each component j in (4.29) is denoted by $q_j(x)$, then we form

$$q(x) = \begin{pmatrix} q_1(x) \\ \cdot \\ \cdot \\ \cdot \\ q_m(x) \end{pmatrix} \qquad (4.32)$$

and obtain the reconstruction in the physical space by using (4.31):

$$p(x) = R q(x) \qquad (4.33)$$

The flux evaluations for the finite difference schemes can be handled similarly.

We now come to the situation where $f'(u)$ is not constant. The trouble is that now all the matrices $R(u)$, $R^{-1}(u)$ and $\Lambda(u)$ are dependent upon u. We must "freeze" them locally in order to carry out a similar procedure as in the constant coefficient case. Thus, to compute the flux at the cell boundary $x_{i+\frac{1}{2}}$, we would need an approximation to the Jocobian at the middle value $u_{i+\frac{1}{2}}$. This can be simply taken as the arithmetic mean

$$u_{i+\frac{1}{2}} = \frac{1}{2} \left(u_i + u_{i+1} \right), \qquad (4.34)$$

or as a more elaborate average satisfying some nice properties, e.g. the mean value theorem

$$f(u_{i+1}) - f(u_i) = f'(u_{i+\frac{1}{2}})(u_{i+1} - u_i). \qquad (4.35)$$

Roe average [82] is such an example for the compressible Euler equations of gas dynamics and some other physical systems. It is also possible to use two different one-sided Jacobians at a higher computational cost [28].

Once we have this $u_{i+\frac{1}{2}}$, we will use $R(u_{i+\frac{1}{2}})$, $R^{-1}(u_{i+\frac{1}{2}})$ and $\Lambda(u_{i+\frac{1}{2}})$ to help evaluating the numerical flux at $x_{i+\frac{1}{2}}$. We thus omit the notation $i + \frac{1}{2}$ and still denote these matrices by R, R^{-1} and Λ, etc. We then repeat the procedure described above for linear systems. The difference here being, the matrices R, R^{-1} and Λ are different at different locations $x_{i+\frac{1}{2}}$, hence the cost of the operation is greatly increased.

In summary, we have the following procedures:

Algorithm 4.6. Characteristic-wise finite volume 1D ENO and WENO schemes.

1. Compute the divided or undivided differences of the cell averages $\bar{u}$, for all i;
2. At each fixed $x_{i+\frac{1}{2}}$, do the following:
   (a) Compute an average state $u_{i+\frac{1}{2}}$, using either the simple mean (4.34) or a Roe average satisfying (4.35);
   (b) Compute the right eigenvectors, the left eigenvectors, and the eigenvalues of the Jacobian $f'(u_{i+\frac{1}{2}})$, (4.20)-(4.23), and denote them by

   $$R = R(u_{i+\frac{1}{2}}), \qquad R^{-1} = R^{-1}(u_{i+\frac{1}{2}}), \qquad \Lambda = \Lambda(u_{i+\frac{1}{2}});$$

   (c) Transform all those differences computed in Step 1, which are in the potential stencil of the ENO and WENO reconstructions for obtaining $u_{i+\frac{1}{2}}^{\pm}$, to the local characteristic fields by using (4.28). For example,

   $$\bar{v}_j = R^{-1}\bar{u}_j, \qquad j \text{ in a neighborhood of } i;$$

   (d) Perform the scalar ENO or WENO reconstruction Algorithm 4.1, for each component of the characteristic variables $\bar{v}$, to obtain the corresponding component of the reconstruction $v_{i+\frac{1}{2}}^{\pm}$;
   (e) Transform back into physical space by using (4.31):

   $$u_{i+\frac{1}{2}}^{\pm} = R\, v_{i+\frac{1}{2}}^{\pm}$$

3. Apply an exact or approximate Riemann solver to compute the flux $\hat{f}_{i+\frac{1}{2}}$ for all i in (4.5); then form the scheme (4.4).

$\square$

Similarly, the procedure to obtain a finite difference ENO-Roe type scheme using the local characteristic decomposition is:

Algorithm 4.7. Characteristic-wise finite difference 1D system, Roe-type schemes.

1. Compute the undivided differences of the flux $f(u)$ for all i;
2. At each fixed $x_{i+\frac{1}{2}}$, do the following:
   (a) Compute an average state $u_{i+\frac{1}{2}}$, using either the simple mean (4.34) or a Roe average satisfying (4.35);
   (b) Compute the right eigenvectors, the left eigenvectors, and the eigenvalues of the Jacobian $f'(u_{i+\frac{1}{2}})$, (4.20)-(4.23), and denote them by

   $$R = R(u_{i+\frac{1}{2}}), \qquad R^{-1} = R^{-1}(u_{i+\frac{1}{2}}), \qquad \Lambda = \Lambda(u_{i+\frac{1}{2}});$$

   (c) Transform all those differences computed in Step 1, which are in the potential stencil of the ENO and WENO reconstructions for obtaining the flux $\hat{f}_{i+\frac{1}{2}}$, to the local characteristic fields by using (4.28). For example,

   $$v_j = R^{-1} f(u_j), \qquad j \text{ in a neighborhood of } i;$$

(d) Perform the scalar ENO or WENO Roe-type Algorithm 4.2, for each component of the characteristic variables v, to obtain the corresponding component of the flux $\hat{v}_{i+\frac{1}{2}}$. The Roe speed $\bar{a}_{i+\frac{1}{2}}$ is replaced by the eigenvalue $\lambda_l(u_{i+\frac{1}{2}})$ for the l-th component of the characteristic variables v;

(e) Transform back into physical space by using (4.31):

$$\hat{f}_{i+\frac{1}{2}} = R\,\hat{v}_{i+\frac{1}{2}}$$

3. Form the scheme (4.9).

$\square$

Finally, the procedure to obtain a finite difference flux splitting ENO or WENO scheme using the local characteristic decomposition is:

Algorithm 4.8. Characteristic-wise finite difference 1D system, flux splitting schemes.

1. Compute the undivided differences of the flux $f(u)$ and the solution u for all i;
2. At each fixed $x_{i+\frac{1}{2}}$, do the following:
   (a) Compute an average state $u_{i+\frac{1}{2}}$, using either the simple mean (4.34) or a Roe average satisfying (4.35);
   (b) Compute the right eigenvectors, the left eigenvectors, and the eigenvalues of the Jacobian $f'(u_{i+\frac{1}{2}})$, (4.20)-(4.23), and denote them by

$$R = R(u_{i+\frac{1}{2}}), \qquad R^{-1} = R^{-1}(u_{i+\frac{1}{2}}), \qquad \Lambda = \Lambda(u_{i+\frac{1}{2}});$$

   (c) Transform all those differences computed in Step 1, which are in the potential stencil of the ENO and WENO reconstructions for obtaining the flux $\hat{f}_{i+\frac{1}{2}}$, to the local characteristic fields by using (4.28). For example,

$$v_j = R^{-1} u_j, \qquad g_j = R^{-1} f(u_j), \qquad j \text{ in a neighborhood of } i;$$

   (d) Perform the scalar flux splitting ENO or WENO Algorithm 4.3, for each component of the characteristic variables, to obtain the corresponding component of the flux $\hat{g}_{i+\frac{1}{2}}^{\pm}$. For the most commonly used Lax-Friedrichs flux splitting, we can use, for the l-th component of the characteristic variables, the viscosity coefficient

$$\alpha = \max_{1 \leq j \leq N} |\lambda_l(u_j)|;$$

Local Lax Friedrichs flux splitting can also be used here, when α is chosen as a maximum of $|\lambda_l(u_i)|$ and $|\lambda_l(u_{i+1})|$, plus perhaps several other neighbors, rather than as a maximum over the whole domain.

(e) Transform back into physical space by using (4.31):

$$\hat{f}^{\pm}_{i+\frac{1}{2}} = R\,\hat{g}^{\pm}_{i+\frac{1}{2}}$$

3. Form the flux by taking

$$\hat{f}_{i+\frac{1}{2}} = \hat{f}^{+}_{i+\frac{1}{2}} + \hat{f}^{-}_{i+\frac{1}{2}}$$

and then form the scheme (4.9).

□

There are attempts recently to simplify this characteristic decomposition. For example, for the compressible Euler equations of gas dynamics, Jiang and Shu [55] used smooth indicators based on density and pressure to perform the so-called pseudo characteristic decompositions. There are also second and sometimes third order component ENO type schemes [75], [70], with limited success for higher order methods.

4.5   Boundary Conditions

For periodic boundary conditions, or problems with compact support for the entire computation (not just the initial data), there is no difficulty in implementing boundary conditions: one simply set as many ghost points as needed using either the periodicity condition or the compactness of the solution.

Other types of boundary conditions should be handled according to their type: for reflective or symmetry boundary conditions, one would set as many ghost points as needed, then use the symmetry/antisymmetry properties to prescribe solution values at those ghost points. For inflow or partially inflow (e.g. a subsonic outflow where one of the characteristic waves flows in) boundary conditions, one would usually use the physical inflow boundary condition at the exact boundary (for example, if $x_{\frac{1}{2}}$ is the left boundary and a finite volume scheme is used, one would use the given boundary value u_b as $u^{-}_{\frac{1}{2}}$ in the monotone flux at $x_{\frac{1}{2}}$; if x_0 is the left boundary and a finite difference scheme is used, one would use the given boundary value u_b as u_0). Apart from that, the most natural way of treating boundary conditions for the ENO scheme is to *use only the available values inside the computational domain when choosing the stencil.* In other words, only stencils completely contained inside the computational domain is used in the ENO stencil choosing process described in the previous algorithms. In practical implementation, in order to avoid logical structures to distinguish whether a given stencil is completely inside the computational domain, one could set all the ghost values outside the computational domain to be very large with large variations (e.g. setting $u_{-j} = (10j)^{10}$ if x_{-j}, for $j = 1, 2, ...$, are ghost points). This way the ENO stencil choosing procedure will automatically avoid choosing any stencil containing ghost points. Another way of treating boundary conditions is to use extrapolation of suitable order to set the values of the solution in all necessary ghost points. For scalar problems this is actually equivalent to the approach of using only the stencils inside the computational domain in the ENO procedure. WENO can be handled in a similar fashion.

Stability analysis (GKS analysis [39], [98]) can be used to study the linear stability when the boundary treatment described above is applied to a fixed stencil upwind biased scheme. For most practical situations the schemes are linearly stable [3].

5 Reconstruction and Approximation in Multi Dimensions

In this section we describe how the ideas of reconstruction and approximation in Sect. 2 are generalized to multi space dimensions. We will concentrate our discussion in 2D, although things carry over to higher dimensions as well.

In the first two subsections we will consider Cartesian grids, that is, the domain is a rectangle

$$[a, b] \times [c, d] \tag{5.1}$$

covered by cells

$$I_{ij} \equiv [x_{i-\frac{1}{2}}, x_{i+\frac{1}{2}}] \times [y_{j-\frac{1}{2}}, y_{j+\frac{1}{2}}], \qquad 1 \le i \le N_x, \ 1 \le j \le N_y \tag{5.2}$$

where

$$a = x_{\frac{1}{2}} < x_{\frac{3}{2}} < ... < x_{N_x - \frac{1}{2}} < x_{N_x + \frac{1}{2}} = b,$$

and

$$c = y_{\frac{1}{2}} < y_{\frac{3}{2}} < ... < y_{N_y - \frac{1}{2}} < y_{N_y + \frac{1}{2}} = d.$$

The centers of the cells are

$$(x_i, y_j), \quad x_i \equiv \frac{1}{2}\left(x_{i-\frac{1}{2}} + x_{i+\frac{1}{2}}\right), \quad y_j \equiv \frac{1}{2}\left(y_{j-\frac{1}{2}} + y_{j+\frac{1}{2}}\right), \tag{5.3}$$

and we still use

$$\Delta x_i \equiv x_{i+\frac{1}{2}} - x_{i-\frac{1}{2}}, \qquad i = 1, 2, ..., N_x \tag{5.4}$$

and

$$\Delta y_j \equiv y_{j+\frac{1}{2}} - y_{j-\frac{1}{2}}, \qquad j = 1, 2, ..., N_y \tag{5.5}$$

to denote the grid sizes. We denote the maximum grid sizes by

$$\Delta x \equiv \max_{1 \le i \le N_x} \Delta x_i, \qquad \Delta y \equiv \max_{1 \le j \le N_y} \Delta y_j, \tag{5.6}$$

and assume that Δx and Δy are of the same magnitude (their ratio is bounded from above and below during refinement). Finally,

$$\Delta \equiv \max(\Delta x, \Delta y). \tag{5.7}$$

5.1   Reconstruction from Cell Averages — Rectangular Case

The approximation problem we will face, in solving hyperbolic conservation laws using cell averages (finite volume schemes, see Sect. 7.1), is still the following *reconstruction* problem.

Problem 5.1. Two dimensional reconstruction for rectangles.
Given the cell averages of a function $v(x, y)$:

$$\bar{v}_{ij} \equiv \frac{1}{\Delta x_i \Delta y_j} \int_{y_{j-\frac{1}{2}}}^{y_{j+\frac{1}{2}}} \int_{x_{i-\frac{1}{2}}}^{x_{i+\frac{1}{2}}} v(\xi, \eta) \, d\xi \, d\eta, \tag{5.8}$$

$$i = 1, 2, ..., N_x, \quad j = 1, 2, ..., N_y,$$

find a polynomial $p_{ij}(x, y)$, preferably of degree at most $k - 1$, for each cell I_{ij}, such that it is a k-th order accurate approximation to the function $v(x, y)$ inside I_{ij}:

$$p_{ij}(x, y) = v(x, y) + O(\Delta^k), \qquad (x, y) \in I_{ij}, \tag{5.9}$$

$$i = 1, ..., N_x, \quad j = 1, ..., N_y.$$

In particular, this gives approximations to the function $v(x, y)$ at the cell boundaries

$$v_{i+\frac{1}{2},y}^{-} = p_{ij}(x_{i+\frac{1}{2}}, y), \qquad v_{i-\frac{1}{2},y}^{+} = p_{ij}(x_{i-\frac{1}{2}}, y),$$

$$i = 1, ..., N_x, \quad y_{j-\frac{1}{2}} \le y \le y_{j+\frac{1}{2}}$$

$$v_{x,j+\frac{1}{2}}^{-} = p_{ij}(x, y_{j+\frac{1}{2}}), \qquad v_{x,j-\frac{1}{2}}^{+} = p_{ij}(x, y_{j-\frac{1}{2}}),$$

$$j = 1, ..., N_y, \quad x_{i-\frac{1}{2}} \le x \le x_{i+\frac{1}{2}}$$

which are k-th order accurate:

$$v_{i+\frac{1}{2},y}^{\pm} = v(x_{i+\frac{1}{2}}, y) + O(\Delta^k), \quad i = 0, 1, ..., N_x, \ y_{j-\frac{1}{2}} \le y \le y_{j+\frac{1}{2}} \tag{5.10}$$

and

$$v_{x,j+\frac{1}{2}}^{\pm} = v(x, y_{j+\frac{1}{2}}) + O(\Delta^k), \quad j = 0, 1, ..., N_y, \ x_{i-\frac{1}{2}} \le x \le x_{i+\frac{1}{2}}. \tag{5.11}$$

□

Again we will not discuss boundary conditions in this section. We thus assume that $\bar{v}_{ij}$ is also available for $i \le 0$, $i > N_x$ and for $j \le 0$, $j > N_y$ if needed.

In the following we describe a general procedure to solve Problem 5.1.

Given the location I_{ij} and the order of accuracy k, we again first choose a "stencil", based on $\frac{k(k+1)}{2}$ neighboring cells, the collection of these cells still being denoted by $S(i, j)$. We then try to find a polynomial of degree at most

$k - 1$, denoted by $p(x, y)$ (we again drop the subscript ij when it does not cause confusion), whose cell average in each of the cells in $S(i, j)$ agrees with that of $v(x, y)$:

$$\frac{1}{\Delta x_l \Delta y_m} \int_{y_{m-\frac{1}{2}}}^{y_{m+\frac{1}{2}}} \int_{x_{l-\frac{1}{2}}}^{x_{l+\frac{1}{2}}} p(\xi, \eta) \, d\xi \, d\eta = \bar{v}_{lm}, \text{ if } I_{lm} \in S(i, j). \qquad (5.12)$$

We first remark that there are now many more candidate stencils $S(i, j)$ than in the 1D case, More importantly, unlike in the 1D case, here we encounter the following essential difficulties:

- Not all of the candidate stencils can be used to obtain a polynomial $p(x, y)$ of degree at most $k - 1$ satisfying condition (5.12).
  For example, it is an easy exercise to show that neither existence nor uniqueness holds, if one wants to reconstruct a first degree polynomial $p(x, y)$ satisfying (5.12) for the three horizontal cells

$$S(i, j) = \{I_{i-1,j}, I_{ij}, I_{i+1,j}\}.$$

To see this, let's assume that

$$I_{i-1,j} = [-2\Delta, -\Delta] \times [0, \Delta], \, I_{i,j} = [-\Delta, 0] \times [0, \Delta], \, I_{i+1,j} = [0, \Delta] \times [0, \Delta],$$

and the first degree polynomial $p(x, y)$ is given by

$$p(x, y) = \alpha + \beta x + \gamma y$$

then condition (5.12) implies

$$\begin{cases} \alpha - \frac{3}{2}\Delta\beta + \frac{1}{2}\Delta\gamma = \bar{v}_{i-1,j} \\ \alpha - \frac{1}{2}\Delta\beta + \frac{1}{2}\Delta\gamma = \bar{v}_{i,j} \\ \alpha + \frac{1}{2}\Delta\beta + \frac{1}{2}\Delta\gamma = \bar{v}_{i+1,j} \end{cases}$$

which is a singular linear system for α, β and γ.
- Even if one obtains such a polynomial $p(x, y)$, there is no guarantee that the accuracy conditions (5.9) will hold. We again use the same simple example. If we pick the function

$$v(x, y) = 0,$$

then one of the polynomials of degree one satisfying the condition (5.12) is

$$p(x, y) = \Delta - 2y$$

clearly the difference

$$v(x, 0) - p(x, 0) = -\Delta$$

is not at the size of $O(\Delta^2)$ in $x_{i-\frac{1}{2}} \le x \le x_{i+\frac{1}{2}}$, as is required by (5.9).

This difficulty will be more profound for unstructured meshes such as triangles. See, for example, [1], and Sect. 5.3.

For rectangular meshes, if we use the tensor products of 1D polynomials, i.e. use polynomials in Q^{k-1}:

$$p(x,y) = \sum_{m=0}^{k-1} \sum_{l=0}^{k-1} a_{lm} x^l y^m$$

then things can proceed as in 1D. We restrict ourselves in the following tensor product stencils:

$$S_{rs}(i,j) = \{I_{lm} : i - r \le l \le i + k - 1 - r, \; j - s \le m \le j + k - 1 - s\}$$

then we can address Problem 5.1 by introducing the two dimensional primitives:

$$V(x,y) = \int_{-\infty}^{y} \int_{-\infty}^{x} v(\xi,\eta) d\xi d\eta \; .$$

Clearly

$$V(x_{i+\frac{1}{2}}, y_{j+\frac{1}{2}}) = \int_{-\infty}^{y_{j+\frac{1}{2}}} \int_{-\infty}^{x_{i+\frac{1}{2}}} v(\xi,\eta) d\xi d\eta = \sum_{m=-\infty}^{j} \sum_{l=-\infty}^{i} \bar{v}_{lm} \Delta x_l \Delta y_m \; ,$$

hence as in the 1D case, with the knowledge of the cell averages $\bar{v}$ we know the primitive function V exactly at cell corners.

On a tensor product stencil

$$\tilde{S}_{rs}(i,j) = \{(x_{l+\frac{1}{2}}, y_{m+\frac{1}{2}}) : i-r-1 \le l \le i+k-1-r, j-s-1 \le m \le j+k-1-s\}$$

there is a unique polynomial $P(x,y)$ in Q^k which interpolates V at every point in $\tilde{S}_{rs}(i,j)$. We take the mixed derivative of the polynomial P to get:

$$p(x,y) = \frac{\partial^2 P(x,y)}{\partial x \partial y}$$

then $p(x,y)$ is in Q^{k-1}, approximates $v(x,y)$, which is the mixed derivative of $V(x,y)$, to k-th order:

$$v(x,y) - p(x,y) = O(\Delta^k)$$

and also satisfies (5.12):

$$\frac{1}{\Delta x_l \Delta y_m} \int_{y_{m-\frac{1}{2}}}^{y_{m+\frac{1}{2}}} \int_{x_{l-\frac{1}{2}}}^{x_{l+\frac{1}{2}}} p(\xi,\eta) \, d\xi \, d\eta$$

$$= \frac{1}{\Delta x_l \Delta y_m} \int_{y_{m-\frac{1}{2}}}^{y_{m+\frac{1}{2}}} \int_{x_{l-\frac{1}{2}}}^{x_{l+\frac{1}{2}}} \frac{\partial^2 P}{\partial \xi \partial \eta}(\xi,\eta) \, d\xi \, d\eta$$

$$= \frac{1}{\Delta x_l \Delta y_m} \left(P(x_{l+\frac{1}{2}}, y_{m+\frac{1}{2}}) - P(x_{l+\frac{1}{2}}, y_{m-\frac{1}{2}}) \right.$$

$$\left. -P(x_{l-\frac{1}{2}}, y_{m+\frac{1}{2}}) + P(x_{l-\frac{1}{2}}, y_{m-\frac{1}{2}}) \right)$$

$$= \frac{1}{\Delta x_l \Delta y_m} \left(V(x_{l+\frac{1}{2}}, y_{m+\frac{1}{2}}) - V(x_{l+\frac{1}{2}}, y_{m-\frac{1}{2}}) \right.$$

$$\left. -V(x_{l-\frac{1}{2}}, y_{m+\frac{1}{2}}) + V(x_{l-\frac{1}{2}}, y_{m-\frac{1}{2}}) \right)$$

$$= \frac{1}{\Delta x_l \Delta y_m} \int_{y_{m-\frac{1}{2}}}^{y_{m+\frac{1}{2}}} \int_{x_{l-\frac{1}{2}}}^{x_{l+\frac{1}{2}}} v(\xi, \eta) \, d\xi \, d\eta = \bar{v}_{lm},$$

$$i - r \le l \le i + k - 1 - r,$$
$$j - s \le m \le j + k - 1 - s.$$

There is a practical way to perform the reconstruction in 2D. We first perform a one dimensional reconstruction (Problem 2.1), say in the y direction, obtaining one dimensional cell averages of the function v in the other direction (say in the x direction). We then perform a reconstruction in the other direction. Notice that if ENO is used in each direction, the effective two dimensional stencil may not be a tensor product.

It should be remarked that the cost to do this 2D reconstruction is very high: for each grid point, if the cost to perform a one dimensional reconstruction is c, then we need $2c$ per grid point to perform this 2D reconstruction. In general n space dimensions, the cost grows to nc.

We also remark that to use polynomials in Q^{k-1} is a waste: to get the correct order of accuracy only polynomials in P^{k-1} is needed. However, there is no natural way of utilizing polynomials in P^{k-1} (see the comments above, the paper of Abgrall [1], and Sect. 5.3).

The reconstruction problem, Problem 5.1, can also be raised for general, non-Cartesian meshes, such as triangles. However, the solution becomes much more complicated. For discussions, see for example [1] and Sect. 5.3.

5.2   Conservative Approximation to the Derivative from Point Values

The second approximation problem we will face, in solving hyperbolic conservation laws using point values (finite difference schemes, see Sect. 7.2), is again the following problem in obtaining high order conservative approximation to the derivative from point values [89,90]. As in the 1D case, here we also assume that the grid is uniform in each direction. We again ignore the boundary conditions and assume that v_{ij} is available for $i \le 0$ and $i > N_x$, and for $j \le 0$ and $j > N_y$.

Problem 5.2. Two dimensional conservative approximation to the derivatives.

Given the point values of a function $v(x, y)$:

$$v_{ij} \equiv v(x_i, y_j), \qquad i = 1, 2, ..., N_x, \quad j = 1, 2, ..., N_y, \qquad (5.13)$$

find numerical flux functions

$$\hat{v}_{i+\frac{1}{2},j} \equiv \hat{v}(v_{i-r,j}, ..., v_{i+k-1-r,j}), \qquad i = 0, 1, ..., N_x \qquad (5.14)$$

and

$$\hat{v}_{i,j+\frac{1}{2}} \equiv \hat{v}(v_{i,j-s}, ..., v_{i,j+k-1-s}), \qquad j = 0, 1, ..., N_y \qquad (5.15)$$

such that the flux differences approximate the derivatives $v_x(x, y)$ and $v_y(x, y)$ to k-th order accuracy:

$$\frac{1}{\Delta x} \left(\hat{v}_{i+\frac{1}{2},j} - \hat{v}_{i-\frac{1}{2},j} \right) = v_x(x_i, y_j) + O(\Delta x^k), \qquad i = 0, 1, ..., N_x, \quad (5.16)$$

and

$$\frac{1}{\Delta y} \left(\hat{v}_{i,j+\frac{1}{2}} - \hat{v}_{i,j-\frac{1}{2}} \right) = v_y(x_i, y_j) + O(\Delta y^k), \qquad j = 0, 1, ..., N_y, \quad (5.17)$$

$\square$

The solution of this problem is essential for the high order conservative schemes based on point values (finite difference) rather than on cell averages (finite volume).

Having seen the complication of reconstructions in the previous subsection for multi space dimensions, it is a good relieve to see that conservative approximation to the derivative from point values is as simple in multi dimensions as in 1D. In fact, for fixed j, if we take

$$w(x) = v(x, y_j)$$

then to obtain $v_x(x_i, y_j) = w'(x_i)$ we only need to perform the one dimensional procedure in Sect. 2.2, Problem 2.2, to the one dimensional function $w(x)$. Same thing for $v_y(x, y)$.

As in the 1D case, the conservative approximation to derivatives, of third order accuracy or higher, can only be applied to uniform or smoothly varying meshes (curvilinear coordinates). It cannot be applied to general unstructured meshes such as triangles, unless conservation is given up.

5.3   Reconstruction from Cell Averages — Triangular Case

Assuming that we have a triangulation with N triangles

$$\{\triangle_0, \triangle_1, ..., \triangle_N\}, \qquad (5.18)$$

the reconstruction problem similar to Problem 5.1, which we will face, in solving hyperbolic conservation laws using cell averages (finite volume schemes, see Sect. 7.1), is the following:

Problem 5.3. Two dimensional reconstruction for triangles.
Given the cell averages of a function $v(x,y)$:

$$\bar{v}_i \equiv \frac{1}{|\Delta_i|} \int_{\Delta_i} v(\xi, \eta) \, d\xi \, d\eta, \qquad i = 1, 2, ..., N, \qquad (5.19)$$

here $|\Delta_i|$ is the area of the triangle Δ_i, find a polynomial $p_i(x,y)$, of degree at most $k-1$, for each triangle Δ_i, such that it is a k-th order accurate approximation to the function $v(x,y)$ inside Δ_i:

$$p_i(x,y) = v(x,y) + O(\Delta^k), \qquad (x,y) \in \Delta_i, \qquad i = 1, ..., N. \qquad (5.20)$$

Here we again use Δ to denote a typical length of the triangles, for example the longest side of the triangles. ☐

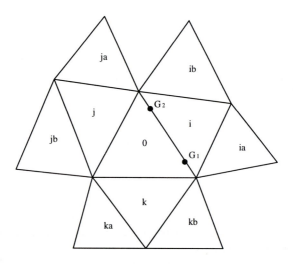

Fig. 5.1. A typical stencil

In particular, (5.20) gives approximations to the function $v(x,y)$ at the triangle boundaries, which are needed in forming the finite volume schemes in Sect. 7.1.

Again we will not discuss boundary conditions in this subsection. We thus assume that $\bar{v}_i$ is also available for triangles Δ_i outside the boundary of the given triangulation if needed.

The following is still a general procedure to solve Problem 5.3.

Given the location $\triangle_i$ and the order of accuracy k, we again first choose a "stencil", based on $m = \frac{k(k+1)}{2}$ neighboring triangles, the collection of these triangles being denoted by $S(i)$. We then try to find a polynomial of degree at most $k - 1$, denoted by $p(x, y)$ (we again drop the subscript i when it does not cause confusion), whose cell average in each of the triangle in $S(i)$ agrees with that of $v(x, y)$:

$$\frac{1}{|\triangle_j|} \int_{\triangle_j} v(\xi, \eta) \, d\xi \, d\eta = \bar{v}_j, \qquad \text{if } \triangle_j \in S(i). \tag{5.21}$$

Notice that (5.21) will give us a $m \times m$ linear system. If this linear system has a unique solution, $S(i)$ is called an *admissible* stencil. Of course, in practice, we also have to worry about any ill conditioned linear system even if it is invertible. For $k = 1$, a stencil formed by $\triangle_i$ itself plus two immediate neighboring triangles is admissible for most triangulations. Thus second order reconstruction is quite easy. We emphasize here that when we talk about order of accuracy in this section it applies only on the approximation level, and also only for "reasonable" triangulations. We will not go into the details of classifying such triangulations.

For a third order reconstruction we need a quadratic polynomial ($k = 2$), which has $m = 6$ degrees of freedom. This time, some of the stencils consisting of $\triangle_i$ and 5 of its neighbors may *not* be admissible. It seems that the most robust way is the least square reconstruction procedure suggested by Barth and Frederickson [7]. For the control volume triangle $\triangle_0$ (see Fig. 5.1), let $\triangle_i, \triangle_j, \triangle_k$ be its three neighbors, and $\triangle_{ia}, \triangle_{ib}$ be the two neighbors (other than $\triangle_0$) of $\triangle_i$, and so on, we determine the quadratic polynomial p^2 by requiring that p^2 has the same cell average as v on $\triangle_0,$, and also p^2 has the same cell average as v on

$$\{\triangle_i, \triangle_{ia}, \triangle_{ib}, \triangle_j, \triangle_{ja}, \triangle_{jb}, \triangle_k, \triangle_{ka}, \triangle_{kb}\},$$

but only in a *least-square* sense (as this is an over-determined system). Notice that some of the neighbors' neighbors ($\triangle_{ia}, \triangle_{ib}, \triangle_{ja}, ...$) may coincide. For example, $\triangle_{ib}$ might be the same as $\triangle_{ja}$. This, however, does not affect the least square procedure to determine p^2.

For a fourth order reconstruction we need a cubic polynomial ($k = 3$), which has $m = 10$ degrees of freedom. If we only consider the case where ia, ib, ja, jb, ka, kb are distinct in the stencil (see Fig. 5.1), it seems that we can construct the cubic polynomial p^3 by requiring that its cell average agrees with that of v on each triangle in the 10-triangle stencil shown in Fig. 5.1, for most triangulations.

6   ENO and WENO Reconstruction and Approximation in Multi Dimensions

For solving hyperbolic conservation laws in multi space dimensions, we are again interested in the class of piecewise smooth functions. We define a piecewise smooth function $v(x, y)$ to be such that, for each fixed y, the one dimensional function $w(x) = v(x, y)$ is piecewise smooth in the sense described in Sect. 3. Likewise, for each fixed x, the one dimensional function $w(y) = v(x, y)$ is also assumed to be piecewise smooth. Such functions are again "generic" for solutions to multi dimensional hyperbolic conservation laws in practice.

In the previous section, we have already discussed the problems of reconstruction and conservative approximations to derivatives in multi space dimensions. For structured meshes, both the reconstruction and the conservative approximation can be obtained from one dimensional procedures. For unstructured meshes, the procedure has to be truly two dimensional.

6.1   Structured Meshes

For a rectangular mesh, we can proceed using the one dimensional results. For the reconstruction, we first use a one dimensional ENO or WENO reconstruction procedure, Algorithm 3.1 or 3.2, on the two dimensional cell averages, say in the y direction, to obtain one dimensional cell averages in x only. Then, another one dimensional reconstruction in the remaining direction, say in the x direction, is performed to recover the function itself, again using the one dimensional ENO or WENO methodology, Algorithm 3.1 or 3.2.

For the conservative approximation to derivatives, since they are already formulated in a dimension by dimension fashion, one dimensional ENO and WENO procedures can be trivially applied. In effect, the FORTRAN program for the 2D problem is the same as the one for the 1D problem, with an outside "do loop".

What happens to general geometry which cannot be covered by a Cartesian grid?

If the domain is smooth enough, it usually can be mapped *smoothly* to a rectangle (or at least to a union of non-overlapping rectangles). That is, the transformation

$$\xi = \xi(x, y), \qquad \eta = \eta(x, y) \tag{6.1}$$

maps the physical domain Ω where (x, y) belongs, to a rectangular computational domain

$$a \leq \xi \leq b, \qquad c \leq \eta \leq d. \tag{6.2}$$

We require the transformation functions (6.1) to be smooth (i.e. it has as many derivatives as the accuracy of the scheme calls for). Using chain rule, we could write, for example,

$$v_x = \xi_x v_\xi + \eta_x v_\eta \tag{6.3}$$

We can then use our ENO or WENO approximations on v_ξ and v_η, as they are now defined in rectangular domains. The smoothness of ξ_x and η_x will guarantee that this leads to a high order approximation to v_x as well through

(6.3). It is proven in [77] that this way the scheme is still conservative, i.e. Lax-Wendroff theorem [65] still applies. For Euler equations of gas dynamics or other homogeneous of degree zero systems, it is also possible to write the system in the new ξ and η variables as a strongly conservative system, see [77].

If the domain is really ugly, or if one wants to use unstructured meshes for other purposes (e.g. for adaptivity), then ENO and WENO approximations for unstructured meshes must be studied. This will be discussed briefly in the next subsection.

6.2   Unstructured Meshes

For unstructured meshes a truly two dimensional ENO or WENO reconstruction must be carried out. We will present here one approach, adopted by Hu and Shu in [49], [50], for third and fourth order WENO reconstructions. Alternative (lower order) WENO reconstruction procedures can also be found in [32]. For an ENO reconstruction procedure, we refer the readers to [1] and [97].

We start with the third order reconstruction. A key step in building a high order WENO scheme based on lower order polynomials is carried out in the following. We want to construct several linear polynomials whose weighted average will give the same result as the quadratic reconstruction p^2 at each quadrature point (the weights are different for different quadrature points). Referring to Fig. 5.1, we can build the following 9 linear polynomials by agreeing with the cell averages of v on the following stencils: p_1 on triangles $0, j, k$, p_2 on triangles $0, k, i$, p_3 on triangles $0, i, j$, p_4 on triangles $0, i, ia$, p_5 on triangles $0, i, ib$, p_6 on triangles $0, j, ja$, p_7 on triangles $0, j, jb$, p_8 on triangles $0, k, ka$, and p_9 on triangles $0, k, kb$. For each quadrature point (x^G, y^G), we want to find the linear weights γ_s, such that the linear polynomial obtained from a linear combination of these p_s

$$R(x, y) = \sum_{s=1}^{9} \gamma_s p_s(x, y) \tag{6.4}$$

satisfies

$$R(x^G, y^G) = p^2(x^G, y^G) \tag{6.5}$$

where p^2 is defined before in Sect. 5.3 using the least squares procedure, for arbitrary choices of cell averages

$$\{\bar{u}_0, \bar{u}_i, \bar{u}_j, \bar{u}_k, \bar{u}_{ia}, \bar{u}_{ib}, \bar{u}_{ja}, \bar{u}_{jb}, \bar{u}_{ka}, \bar{u}_{kb}\}. \tag{6.6}$$

Since both the left side and the right side of the equality (6.5) are linear in the cell averages (6.6), for the equality to hold for arbitrary $\bar{u}$'s in (6.6) one must have all 10 coefficients of the $\bar{u}$'s to be identically zero (when all terms are moved to one side of the equality), which leads to 10 linear equations for the nine weights γ_s. This looks like an over-determined system, but is in fact under-determined of rank 8, allowing for one degree of freedom in the choice of the nine γ_s.

Before explaining this, we first look at a simpler but illustrative one dimensional example. Let us denote I_j, $j = 0, 1, 2$, as three equal sized consecutive intervals. The two linear polynomials p_s, where p_1 agrees with u on cell averages in the intervals I_0 and I_1, and p_2 agrees with u on cell averages in the intervals I_1 and I_2, give the following two second order approximations to the value of u at the point $x_{\frac{3}{2}}$ (the boundary of I_1 and I_2):

$$-\frac{1}{2}\bar{u}_0 + \frac{3}{2}\bar{u}_1, \qquad \frac{1}{2}\bar{u}_1 + \frac{1}{2}\bar{u}_2. \tag{6.7}$$

The quadratic polynomial p^2, which agrees with u on cell averages in the intervals I_0, I_1 and I_2, gives the following third order approximation to the value of u at the point $x_{\frac{3}{2}}$:

$$-\frac{1}{6}\bar{u}_0 + \frac{5}{6}\bar{u}_1 + \frac{1}{3}\bar{u}_1. \tag{6.8}$$

We would like to find γ_s such that

$$\gamma_1 \left(-\frac{1}{2}\bar{u}_0 + \frac{3}{2}\bar{u}_1\right) + \gamma_2 \left(\frac{1}{2}\bar{u}_1 + \frac{1}{2}\bar{u}_2\right) = -\frac{1}{6}\bar{u}_0 + \frac{5}{6}\bar{u}_1 + \frac{1}{3}\bar{u}_1 \tag{6.9}$$

for arbitrary $\bar{u}$'s. This leads to the following three equations:

$$-\frac{1}{2}\gamma_1 = -\frac{1}{6}, \qquad \frac{3}{2}\gamma_1 + \frac{1}{2}\gamma_2 = \frac{5}{6}, \qquad \frac{1}{2}\gamma_2 = \frac{1}{3},$$

for the two unknowns γ_1 and γ_2. It looks like an over-determined system but is in fact rank 2 and has a unique solution

$$\gamma_1 = \frac{1}{3}, \qquad \gamma_2 = \frac{2}{3}.$$

The reason can be understood if we ask for the validity of the equality (6.9) in the cases of $u = 1$, $u = x$ and $u = x^2$. Clearly if (6.9) holds in these three cases then it holds for arbitrary choices of $\bar{u}$'s. The crucial observation is that (6.9) holds for both $u = 1$ and $u = x$ as long as $\gamma_1 + \gamma_2 = 1$, as all three expressions in (6.7) and (6.8) reproduce linear functions exactly. Hence the equality (6.9) is valid for all the three cases $u = 1$, $u = x$ and $u = x^2$ with only two conditions: $\gamma_1 + \gamma_2 = 1$ and another one obtained when $u = x^2$, resulting in a solvable 2×2 system for γ_s.

The same argument can be applied in the current two dimensional case. Although there are 10 linear equations for the nine weights γ_s resulting from the equality (6.5), we should notice that the equality (6.5) is valid for all three cases $u = 1$, $u = x$ and $u = y$ under only one constraint on γ_s, namely

$\sum_{s=1}^{9} \gamma_s = 1$, again because $p_s(x)$ and $p^2(x)$ all reproduce linear functions exactly. Thus we can eliminate two equations from the ten, resulting in a rank 8 system with one degree of freedom in the solution for γ_s. In practice, we obtain the solution γ_s for $s \geq 2$ with γ_1 as the degree of freedom.

Note that there are situations when ia, ib, ja, jb, ka, kb might not be distinct, in these cases, we simply discard some of the p_s, or just set the corresponding coefficient γ_s to zero. For example, if $ib = ja$, we will just use $p_1, p_2, p_3, p_4, p_5, p_7, p_8, p_9$ and discard p_6. In this case there is one fewer coefficient but also one fewer condition to satisfy for (6.5), as there is one fewer triangle in the stencil. The discussion carried out above still applies.

The first effort we would like to make is to use this degree of freedom to obtain a set of non-negative γ_s, which is important for the WENO procedure. Unfortunately, it turns out that, for many triangulations, this is impossible. Some grouping is needed and is discussed next. We want to group these 9 linear polynomials into 3 groups:

$$\sum_{s=1}^{9} \gamma_s p_s(x, y) = \sum_{s=1}^{3} \tilde{\gamma}_s \tilde{p}_s(x, y),$$

each $\tilde{p}_s(x, y)$ being still a linear polynomial and a second order approximation to u, with positive coefficients $\tilde{\gamma}_s \geq 0$. We also require the stencils corresponding to the three new linear polynomials $\tilde{p}_s(x, y)$ to be reasonably separated, so that when shocks are present, not all stencils will contain the shock under normal situations.

The grouping we will introduce in the following works for most triangulations. There are however cases when it does give some negative coefficients, especially when one is doing adaptive meshing and is near the adaptively refined regions where triangle sizes are changing very abruptly. In such cases one would need to use a Lax-Friedrichs like procedure, namely breaking each coefficient $\tilde{\gamma}_s = 2\tilde{\gamma}_s - \tilde{\gamma}_s$ and collecting the three positive terms and the three negative terms separately to obtain WENO weights. This procedure is currently being developed by Hu and Shu and have been performing well numerically in our preliminary tests. It will appear in a future publication. In the following we will only consider those triangulations when our grouping strategy will produce positive weights.

For the first quadrature point on side i (G_1 in Fig. 5.1), Group 1 contains $p_2(0, k, i)$, $p_4(0, i, ia)$, and $p_5(0, i, ib)$,

$$\tilde{p}_1 = (\gamma_2 p_2 + \gamma_4 p_4 + \gamma_5 p_5)/(\gamma_2 + \gamma_4 + \gamma_5), \qquad \tilde{\gamma}_1 = \gamma_2 + \gamma_4 + \gamma_5,$$

Group 2 contains $p_3(0, i, j)$, $p_6(0, j, ja)$, and $p_7(0, j, jb)$,

$$\tilde{p}_2 = (\gamma_3 p_3 + \gamma_6 p_6 + \gamma_7 p_7)/(\gamma_3 + \gamma_6 + \gamma_7), \qquad \tilde{\gamma}_2 = \gamma_3 + \gamma_6 + \gamma_7,$$

Group 3 contains $p_1(0, j, k)$, $p_8(0, k, ka)$, and $p_9(0, k, kb)$,

$$\tilde{p}_3 = (\gamma_1 p_1 + \gamma_8 p_8 + \gamma_9 p_9)/(\gamma_1 + \gamma_8 + \gamma_9), \qquad \tilde{\gamma}_3 = \gamma_1 + \gamma_8 + \gamma_9.$$

The resulting linear polynomial

$$\tilde{R}(x,y) = \sum_{s=1}^{3} \tilde{\gamma}_s \tilde{p}_s(x,y) \tag{6.10}$$

is identical to $R(x,y)$ in (6.4) and in most cases the coefficients $\tilde{\gamma}_s$ can be made non-negative by suitably choosing the value of the degree of freedom γ_1, through the solution of a group of 3 linear inequalities for γ_1.

We remark that for practical implementation, it is the 5 constants a_i, which depend on the local geometry only, such that

$$\tilde{p}_1(x^{G^1}, y^{G^1}) = a_1 \bar{u}_0 + a_2 \bar{u}_i + a_3 \bar{u}_k + a_4 \bar{u}_{ia} + a_5 \bar{u}_{ib}, \tag{6.11}$$

that have to be precomputed and stored once the mesh is generated. We do not need to store any information about the polynomial $\tilde{p}_1$ itself.

For the second quadrature point on side i, (G_2 in Fig. 5.1), Group 1 contains $p_3\,(0, i, j)$, $p_4\,(0, i, ia)$, and $p_5\,(0, i, ib)$, with the combination coefficient $\tilde{\gamma}_1 = \gamma_3 + \gamma_4 + \gamma_5$; Group 2 contains $p_2\,(0, k, i)$, $p_8\,(0, k, ka)$, and $p_9\,(0, k, kb)$; with the combination coefficient $\tilde{\gamma}_2 = \gamma_2 + \gamma_8 + \gamma_9$; Group 3 contains $p_1\,(0, j, k)$, $p_6\,(0, j, ja)$, and $p_7\,(0, j, jb)$; with combination coefficient $\tilde{\gamma}_3 = \gamma_1 + \gamma_6 + \gamma_7$. We can do the same thing for the other two sides (j, k).

Next we describe the fourth order reconstruction. Again, the key step to build a high order WENO scheme based on lower order polynomials is carried out in the following. We would like to construct several quadratic polynomials whose weighted average will give the same result as the cubic reconstruction p^3, which was described in Sect. 6.2, at each quadrature point (the weights are different for different quadrature points). The following 6 quadratic polynomials are constructed by having the same cell averages as u on the corresponding triangles:
q_1 (on triangles: $0, i, ia, ib, k, kb$), q_2 (on triangles: $0, i, ia, ib, j, ja$), q_3 (on triangles: $0, j, ja, jb, i, ib$), q_4 (on triangles: $0, j, ja, jb, k, ka$), q_5 (on triangles: $0, k, ka, kb, j, jb$), q_6 (on triangles: $0, k, ka, kb, i, ia$).

For each quadrature point (x^G, y^G), we would like to find the linear weights such that the linear combination of these q_s

$$Q(x,y) = \sum_{s=1}^{6} \gamma_s q_s(x,y) \tag{6.12}$$

satisfies

$$Q(x^G, y^G) = p^3(x^G, y^G) \tag{6.13}$$

for all $\bar{u}$'s.

As before, (6.13) results in 10 linear equations for the 6 unknowns γ_s, which are the coefficients of the 10 cell averages $\bar{u}$'s in (6.6). This looks like a grossly over-determined system, but it is in fact under-determined with rank 5, thus allowing a solution for γ_s with one degree of freedom. A crucial observation is again that (6.13) is valid for all the 6 cases $u = 1, x, y, x^2, xy, y^2$ under just one constraint on the γ_s, namely $\sum_{s=1}^{9} \gamma_s = 1$, because $q_s(x)$ and $p^3(x)$ all reproduce quadratic functions exactly. We can thus eliminate 5 equations from the 10, resulting in a rank 5 system with one degree of freedom in the solution for γ_s. In practice, we obtain the solution γ_s for $s \geq 2$ with γ_1 as the degree of freedom.

Again, the first effort we would like to make is to use this degree of freedom to obtain a set of non-negative γ_s, through the solution of a group of 5 linear inequalities for γ_1. This is important for the WENO procedure. Positivity seems achievable for the mostly near-uniform meshes used in the numerical examples. For general triangulations negative coefficients do appear, and the investigation of using the Lax-Friedrichs like procedure mentioned above for the third order case is currently undertaken.

We finally come to the point of smooth indicators and nonlinear weights. For this we follow exactly as in Jiang and Shu [55], see Sect. 3.2. For a polynomial $p(x,y)$ with degree up to n, we define the following measurement for smoothness

$$S = \sum_{1 \leq |\alpha| \leq n} \int_{\triangle} |\triangle|^{|\alpha|-1} (D^{\alpha} p(x,y))^2 \, dx dy \qquad (6.14)$$

where α is a multi-index and D is the derivative operator, for example, when $\alpha = (1,2)$ then $|\alpha| = 3$ and $D^{\alpha} p(x,y) = \frac{\partial p^3(x,y)}{\partial x \partial y^2}$. The non-linear weights are then defined as:

$$\omega_j = \frac{\tilde{\omega}_j}{\sum_i \tilde{\omega}_i}, \qquad \tilde{\omega}_i = \frac{\gamma_i}{(\epsilon + S_i)^2} \qquad (6.15)$$

where γ_i is the i-th coefficient in the linear combination of polynomials (i.e. the $\tilde{\gamma}_s$ in (6.10) for the third order case and the γ_s in (6.12) for the fourth order case), S_i is the measurement of smoothness of the i-th polynomial $p_i(x,y)$ (i.e. the $\tilde{p}_s$ in (6.10) for the third order case and the q_s in (6.12) for the fourth order case), and ϵ is a small positive number which we take as $\epsilon = 10^{-3}$ for all the numerical experiments for triangles. The numerical results are not very sensitive to the choice of ϵ in a range from 10^{-2} to 10^{-6}. In general, larger ϵ gives better accuracy for smooth problems but may generate small oscillations for shocks. Smaller ϵ is more friendly to shocks. The nonlinear weights ω_j in (6.15) would then replace the linear weights γ_j to form a WENO reconstruction.

We emphasize that the smoothness measurements (6.14) are quadratic functions of the cell averages in the stencil. For example, it is the 10 constants b_i and c_i, which depend on the local geometry only, such that

$$S = (b_1 \bar{u}_0 + b_2 \bar{u}_i + b_3 \bar{u}_k + b_4 \bar{u}_{ia} + b_5 \bar{u}_{ib})^2 + (c_1 \bar{u}_0 + c_2 \bar{u}_i + c_3 \bar{u}_k + c_4 \bar{u}_{ia} + c_5 \bar{u}_{ib})^2 \qquad (6.16)$$

for the smoothness measurements (6.14) of $\tilde{p}_1$ in (6.10), that have to be precomputed and stored once the mesh is generated. We do not need to store any information about the polynomial $\tilde{p}_1$ itself.

7   ENO and WENO Schemes in Multi Dimensions

In this section we describe the ENO and WENO schemes for 2D conservation laws:

$$u_t(x, y, t) + f_x(u(x, y, t)) + g_y(u(x, y, t)) = 0 \qquad (7.1)$$

again equipped with suitable initial and boundary conditions.

Although we present everything in 2D, most of the discussion is also valid for higher dimensions.

We again concentrate on the discussion of spatial discretizations, and will leave the time variable t continuous (the method-of-lines approach). Time discretization will be discussed in Sect. 9.

For structured meshes, our computational domain is rectangular, given by (5.1). In such cases our grids will be Cartesian, given by (5.2) and (5.3). For unstructured meshes, we assume a triangulation consisting of triangles (5.18).

We do not discuss boundary conditions in this section. We thus assume that the values of the numerical solution are also available outside the computational domain whenever they are needed. This would be the case for periodic or compactly supported problems. Two dimensional boundary condition treatments are similar to the one dimensional case discussed in Sect. 4.5.

7.1   Finite Volume Formulation in the Scalar Case

For finite volume schemes, or schemes based on cell averages, we do not solve (7.1) directly, but its integrated version. For a structured mesh, we integrate (7.1) over the cell I_{ij} to obtain

$$\frac{d\bar{u}_{ij}(t)}{dt} = -\frac{1}{\Delta x_i \Delta y_j} \left(\int_{y_{j-\frac{1}{2}}}^{y_{j+\frac{1}{2}}} f(u(x_{i+\frac{1}{2}}, y, t)) \, dy - \int_{y_{j-\frac{1}{2}}}^{y_{j+\frac{1}{2}}} f(u(x_{i-\frac{1}{2}}, y, t)) \, dy \right.$$

$$\left. + \int_{x_{i-\frac{1}{2}}}^{x_{i+\frac{1}{2}}} g(u(x, y_{j+\frac{1}{2}}, t)) \, dx - \int_{x_{i-\frac{1}{2}}}^{x_{i+\frac{1}{2}}} g(u(x, y_{j-\frac{1}{2}}, t)) \, dx \right) \qquad (7.2)$$

where

$$\bar{u}_{ij}(t) \equiv \frac{1}{\Delta x_i \Delta y_j} \int_{y_{j-\frac{1}{2}}}^{y_{j+\frac{1}{2}}} \int_{x_{i-\frac{1}{2}}}^{x_{i+\frac{1}{2}}} u(\xi, \eta, t) \, d\xi \, d\eta \qquad (7.3)$$

is the cell average. We approximate (7.2) by the following conservative scheme

$$\frac{d\bar{u}_{ij}(t)}{dt} = -\frac{1}{\Delta x_i} \left(\hat{f}_{i+\frac{1}{2},j} - \hat{f}_{i-\frac{1}{2},j} \right) - \frac{1}{\Delta y_j} \left(\hat{g}_{i,j+\frac{1}{2}} - \hat{g}_{i,j-\frac{1}{2}} \right), \qquad (7.4)$$

where the numerical flux $\hat{f}_{i+\frac{1}{2},j}$ is defined by

$$\hat{f}_{i+\frac{1}{2},j} = \sum_{\alpha} w_\alpha h \left(u^-_{i+\frac{1}{2}, y_j + \beta_\alpha \Delta y_j}, u^+_{i+\frac{1}{2}, y_j + \beta_\alpha \Delta y_j} \right), \qquad (7.5)$$

where β_α and w_α are Gaussian quadrature nodes and weights, for approximating the integration in y:

$$\frac{1}{\Delta y_j} \int_{y_{j-\frac{1}{2}}}^{y_{j+\frac{1}{2}}} f(u(x_{i+\frac{1}{2}}, y, t))\, dy$$

inside the integral form of the PDE (7.2), and $u_{i+\frac{1}{2},y}^{\pm}$ are the k-th order accurate reconstructed values obtained by ENO or WENO reconstruction described in the previous section. As before, the superscripts $\pm$ imply the values are obtained within the cell I_{ij} (for the superscript -) and the cell $I_{i+1,j}$ (for the superscript +), respectively. The flux $\hat{g}_{i,j+\frac{1}{2}}$ is defined similarly by

$$\hat{g}_{i,j+\frac{1}{2}} = \sum_\alpha w_\alpha h\left(u_{x_i+\beta_\alpha \Delta x_i, j+\frac{1}{2}}^{-}, u_{x_i+\beta_\alpha \Delta x_i, j+\frac{1}{2}}^{+}\right), \tag{7.6}$$

for approximating the integration in x:

$$\frac{1}{\Delta x_i} \int_{x_{i-\frac{1}{2}}}^{x_{i+\frac{1}{2}}} g(u(x, y_{j+\frac{1}{2}}, t))\, dx$$

inside the integral form of the PDE (7.2). $u_{x,j+\frac{1}{2}}^{\pm}$ are again the k-th order accurate reconstructed values obtained by ENO or WENO reconstruction described in the previous section. h is again a one dimensional monotone flux, examples being given in (4.6)-(4.8).

We summarize the procedure to build a finite volume ENO or WENO 2D scheme (7.4) on structured mesh, given the cell averages $\{\bar{u}_{ij}\}$ (we again drop the explicit reference to the time variable t), and a one dimensional monotone flux h, as follows:

Algorithm 7.1. Finite volume 2D scalar ENO and WENO schemes for a rectangular mesh.

1. Follow the procedures described in Sect. 6.1, to obtain ENO or WENO reconstructed values at the Gaussian points,

$$u_{i+\frac{1}{2}, y_j+\beta_\alpha \Delta y_j}^{\pm} \quad \text{and} \quad u_{x_i+\beta_\alpha \Delta x_i, j+\frac{1}{2}}^{\pm}.$$

   Notice that this step involves two one dimensional reconstructions, each one to remove a one dimensional cell average in one of the two directions. Also notice that the optimal weights used in the WENO reconstruction procedure are different for different Gaussian points indexed by α;
2. Compute the flux $\hat{f}_{i+\frac{1}{2},j}$ and $\hat{g}_{i,j+\frac{1}{2}}$ using (7.5) and (7.6);
3. Form the scheme (7.4).

$\square$

We remark that the finite volume scheme in 2D, as described above, is very expensive due to the following reasons:

- A two dimensional reconstruction, at the cost of two one dimensional reconstructions per grid point, is needed. For general n space dimensions, the cost becomes n one dimensional reconstructions per grid point;
- More than one quadrature points are needed in formulating the flux (7.5)-(7.6), for order of accuracy higher than two. Thus, for ENO, although the stencil choosing process needs to be done only once, the reconstruction (2.10) has to be done for each quadrature point used in the flux formulation. For WENO, the optimal weights are also different for each quadrature point. This becomes much more costly for $n > 2$ dimension, as then the fluxes are defined by integrals in $n - 1$ dimension and a $n - 1$ dimensional quadrature rule must be used.

This is why multidimensional finite volume schemes of order of accuracy higher than 2 are rarely used for structured mesh. For 2D, based on [43], Casper [14] has coded up a fourth order finite volume ENO scheme for Cartesian grids, see also [15]. 3D finite volume ENO code of order of accuracy higher than 2 for a rectangular mesh does not exist yet, to the author's knowledge. A finite difference version to be described in Sect. 7.2 is much more economical for a multidimensional structured mesh.

At the second order level, the cost is greatly reduced because:

- There is no need to perform a reconstruction, as the cell average $\bar{u}_{ij}$ agrees with the point value at the center $u(x_i, y_j)$ to second order $O(\Delta^2)$;
- The quadrature rule in defining the flux (7.5)-(7.6) needs only one (mid) point.

One advantage of finite volume ENO or WENO schemes is that they can be defined on arbitrary meshes, provided that an ENO or WENO reconstruction on that mesh is available. This is described below. See also [1].

Taking the triangle Δ_i as our control volume, we formulate the semi-discrete finite volume scheme for equation (7.1) as:

$$\frac{d}{dt}\bar{u}_i(t) + \frac{1}{|\Delta_i|} \int_{\partial\Delta_i} F \cdot n \, ds = 0 \qquad (7.7)$$

where $\bar{u}_i(t)$ is the cell average of u on the cell Δ_i, $F = (f, g)^T$, n is the outward unit normal of the triangle boundary $\partial\Delta_i$.

The line integral in (7.7) is discretized by a q-point Gaussian integration formula,

$$\int_{\Gamma_k} F \cdot n \, ds \approx |\Gamma_k| \sum_{j=1}^{q} \omega_j F(u(G_j, t)) \cdot n \qquad (7.8)$$

and $F(u(G_j, t)) \cdot n$ is replaced by a one dimensional numerical flux in the n direction. We can for example use any one of (4.6)-(4.8). The simple Lax-Friedrichs flux is for example given by

$$F(u(G_j, t)) \cdot n \qquad (7.9)$$

$$\approx \frac{1}{2} \left[\left(F(u^-(G_j, t)) + F(u^+(G_j, t)) \right) \cdot n - \alpha \left(u^+(G_j, t) - u^-(G_j, t) \right) \right]$$

where α is taken as an upper bound for $|F'(u) \cdot n|$. Here, u^- and u^+ are the values of the reconstructed values of u inside the triangle and outside the triangle (inside the neighboring triangle) at the Gaussian point, see Sect. 6.2.

Since we are constructing schemes up to fourth order accuracy, two point Gaussian $q = 2$ is used, which has $G_1 = cP_1 + (1 - c)P_2$, $G_2 = cP_2 + (1 - c)P_1$, $c = \frac{1}{2} + \frac{\sqrt{3}}{6}$ and $\omega_1 = \omega_2 = \frac{1}{2}$ for the line with end points P_1 and P_2.

We now give some test results about accuracy for the third and fourth order WENO schemes constructed on triangulations above.

The first example is the two-dimensional linear equation:

$$u_t + u_x + u_y = 0 \tag{7.10}$$

with the initial condition $u_0(x, y) = \sin(\frac{\pi}{2}(x+y))$, $-2 \leq x \leq 2$, $-2 \leq y \leq 2$, and periodic boundary conditions.

We first use uniform triangular meshes which are obtained by adding one diagonal line in each rectangle, shown in Fig. 7.1 for the coarsest case $h = \frac{2}{5}$. The accuracy results are shown for both the third order scheme (from the combination of linear polynomials) and the fourth order scheme (from the combination of quadratic polynomials), for both the linear constant weights in Table 7.1 and the WENO weights in Table 7.2. Here h is the length of the rectangles. The results shown are at $t = 2.0$. The errors presented are those of the cell averages of u.

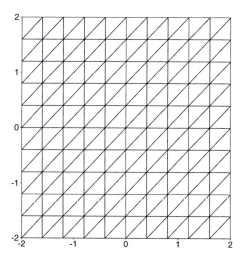

Fig. 7.1. Uniform mesh with $h = \frac{2}{5}$ for the accuracy test.

Table 7.1. Accuracy for the 2D linear equation, uniform meshes, linear schemes.

	P^1 (3rd order)				P^2 (4th order)			
h	L^1 error	order	L^∞ error	order	L^1 error	order	L^∞ error	order
2/5	1.80E-01	—	2.79E-01	—	1.40E-02	—	2.17E-02	—
1/5	2.81E-02	2.68	4.37E-02	2.68	9.11E-04	3.94	1.41E-03	3.94
1/10	3.65E-03	2.95	5.72E-03	2.93	5.57E-05	4.03	8.72E-05	4.02
1/20	4.60E-04	2.99	7.22E-04	2.99	3.43E-06	4.02	5.39E-06	4.02
1/40	5.76E-05	3.00	9.05E-05	3.00	2.12E-07	4.02	3.34E-07	4.01
1/80	7.21E-06	3.00	1.13E-05	3.00	1.32E-08	4.01	2.07E-08	4.01

Table 7.2. Accuracy for the 2D linear equation, uniform meshes, WENO schemes.

	P^1 (3rd order)				P^2 (4th order)			
h	L^1 error	order	L^∞ error	order	L^1 error	order	L^∞ error	order
2/5	2.66E-01	—	4.30E-01	—	1.38E-02	—	2.94E-02	—
1/5	8.11E-02	1.71	1.93E-01	1.16	1.80E-03	2.94	2.74E-03	3.42
1/10	2.65E-02	1.62	6.16E-02	1.65	8.87E-05	4.34	1.46E-04	4.23
1/20	2.68E-03	3.31	8.77E-03	2.81	4.34E-06	4.35	7.11E-06	4.36
1/40	1.44E-04	4.22	4.88E-04	4.17	2.30E-07	4.24	3.71E-07	4.26
1/80	8.05E-06	4.16	2.40E-05	4.35	1.34E-08	4.10	2.12E-08	4.13

We then use non-uniform meshes, shown in Fig. 7.2 for the coarsest case $h = \frac{2}{5}$, where h is just an average mesh size. The refinement of the meshes is done in a uniform way, namely by cutting each triangle into 4 smaller similar ones. The accuracy result is shown in Table 7.3 for the linear constant weights case and in Table 7.4 for the WENO case.

The second example is the two-dimensional Burgers' equation:

$$u_t + \left(\frac{u^2}{2}\right)_x + \left(\frac{u^2}{2}\right)_y = 0 \qquad (7.11)$$

with the initial condition $u_0(x, y) = 0.3 + 0.7 \sin(\frac{\pi}{2}(x + y))$, $-2 \leq x \leq 2$, $-2 \leq y \leq 2$, and periodic boundary conditions.

We first use the same uniform triangular meshes as in the previous example, shown in Fig. 7.1 for the coarsest case $h = \frac{2}{5}$. In Table 7.5, the accuracy results for the linear schemes are shown for both the third order scheme and

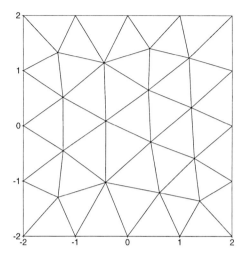

Fig. 7.2. Non-uniform mesh with $h = \frac{2}{5}$ for accuracy test.

Table 7.3. Accuracy for the 2D linear equation, non-uniform meshes, linear schemes.

h	P^1 (3rd order)				P^2 (4th order)			
	L^1 error	order	L^∞ error	order	L^1 error	order	L^∞ error	order
$h_0/2$	1.21E-01	—	2.25E-01	—	4.95E-03	—	1.73E-02	—
$h_0/4$	1.81E-02	2.74	3.74E-02	2.59	2.90E-04	4.09	1.42E-03	3.61
$h_0/8$	2.36E-03	2.94	5.39E-03	2.80	2.21E-05	3.71	8.32E-05	4.09
$h_0/16$	3.00E-04	2.98	7.19E-04	2.91	1.29E-06	4.10	5.09E-06	4.03
$h_0/32$	3.78E-05	2.99	9.40E-05	2.94	7.76E-08	4.06	3.16E-07	4.01
$h_0/64$	4.75E-06	2.99	1.22E-05	2.95	4.75E-09	4.03	1.95E-08	4.02

Table 7.4. Accuracy for the 2D linear equation, non-uniform meshes, WENO schemes.

h	P^1 (3rd order)				P^2 (4th order)			
	L^1 error	order	L^∞ error	order	L^1 error	order	L^∞ error	order
$h_0/2$	2.79E-01	—	5.28E-01	—	1.77E-02	—	6.41E-02	—
$h_0/4$	8.43E-02	1.73	2.32E-01	1.19	8.85E-04	4.32	3.07E-03	4.38
$h_0/8$	2.53E-02	1.74	7.47E-02	1.64	4.08E-05	4.44	1.43E-04	4.42
$h_0/16$	2.24E-03	3.50	1.14E-02	2.71	1.82E-06	4.49	6.37E-06	4.49
$h_0/32$	1.18E-04	4.25	6.83E-04	4.06	8.95E-08	4.35	3.36E-07	4.25
$h_0/64$	6.21E-06	4.25	3.15E-05	4.44	4.92E-09	4.19	2.00E-08	4.07

the fourth order scheme, at $t = 0.5/\pi^2$ when the solution is still smooth. The errors presented are those of the point values at the 6 quadrature points of each triangle. In Table 7.6, the same accuracy results for the WENO schemes are shown.

Table 7.5. Accuracy for 2D Burgers' equation, uniform meshes, linear schemes.

	P^1 (3rd order)				P^2 (4th order)			
h	L^1 error	order	L^∞ error	order	L^1 error	order	L^∞ error	order
2/5	2.67E-02	—	7.75E-02	—	8.63E-03	—	2.18E-02	—
1/5	3.65E-03	2.87	1.16E-02	2.74	6.08E-04	3.83	1.70E-03	3.68
1/10	4.60E-04	2.99	1.52E-03	2.93	3.97E-05	3.94	1.16E-04	3.87
1/20	5.75E-05	3.00	1.91E-04	2.99	2.51E-06	3.98	7.37E-06	3.98
1/40	7.18E-06	3.01	2.38E-05	3.01	1.57E-07	4.00	4.62E-07	4.00
1/80	8.96E-07	3.00	2.97E-06	3.00	9.83E-09	4.00	2.89E-08	4.00

Table 7.6. Accuracy for 2D Burgers' equation, uniform meshes, WENO schemes.

	P^1 (3rd order)				P^2 (4th order)			
h	L^1 error	order	L^∞ error	order	L^1 error	order	L^∞ error	order
2/5	2.76E-02	—	8.18E-02	—	8.64E-03	—	2.106-02	—
1/5	4.63E-03	2.58	1.20E-02	2.77	6.05E-04	3.84	1.73E-03	3.60
1/10	6.97E-04	2.73	2.16E-03	2.47	3.94E-05	3.94	1.18E-04	3.87
1/20	7.12E-05	3.29	1.90E-04	3.51	2.50E-06	3.98	7.42E-06	3.99
1/40	7.63E-06	3.22	2.36E-05	3.01	1.57E-07	3.99	4.63E-07	4.00
1/80	9.08E-07	3.07	2.96E-06	3.00	9.83E-09	4.00	2.89E-08	4.00

We then use the same non-uniform meshes as in the previous example, shown in Fig. 7.2 for the coarsest case. The accuracy result is shown in Table 7.7 for the linear constant weights case and in Table 7.8 for the WENO case.

To demonstrate the application for shock computation, we continue the the WENO calculation to $t = 5/\pi^2$ when discontinuities develop. Fig. 7.3 is the result for $h = 1/20$ of a uniform mesh. Fig. 7.4 is the result for $h = h_0/16$ of a non-uniform mesh. We can see that the shock transitions are sharp and non-oscillatory.

Table 7.7. Accuracy for 2D Burgers' equation, non-uniform meshes, linear schemes.

	P^1 (3rd order)				P^2 (4th order)			
h	L^1 error	order	L^∞ error	order	L^1 error	order	L^∞ error	order
$h_0/2$	1.69E-02	—	7.95E-01	—	3.96E-03	—	1.88E-02	—
$h_0/4$	2.23E-03	2.92	1.23E-02	2.69	2.87E-04	3.79	2.17E-03	3.12
$h_0/8$	2.84E-04	2.97	1.69E-03	2.86	1.90E-05	3.92	1.81E-04	3.58
$h_0/16$	3.57E-05	2.99	2.22E-04	2.93	1.20E-06	3.99	1.34E-05	3.77
$h_0/32$	4.48E-06	2.99	3.00E-05	2.89	7.57E-08	3.99	1.00E-06	3.74
$h_0/64$	5.63E-07	2.99	4.26E-06	2.82	4.75E-09	4.00	7.57E-08	3.72

Table 7.8. Accuracy for 2D Burgers' equation, non-uniform meshes, WENO schemes.

	P^1 (3rd order)				P^2 (4th order)			
h	L^1 error	order	L^∞ error	order	L^1 error	order	L^∞ error	order
$h_0/2$	2.01E-02	—	9.16E-02	—	4.18E-03	—	2.376-02	—
$h_0/4$	3.85E-03	2.38	1.80E-02	2.35	2.90E-04	3.85	2.61E-03	3.18
$h_0/8$	5.79E-04	2.73	3.39E-03	2.41	1.85E-05	3.97	1.92E-04	3.77
$h_0/16$	5.34E-05	3.44	3.55E-04	3.26	1.18E-06	3.97	1.35E-05	3.83
$h_0/32$	5.12E-06	3.38	2.95E-05	3.59	7.45E-08	3.99	9.99E-07	3.76
$h_0/64$	5.82E-07	3.14	4.23E-06	2.80	4.67E-09	4.00	7.56E-08	3.72

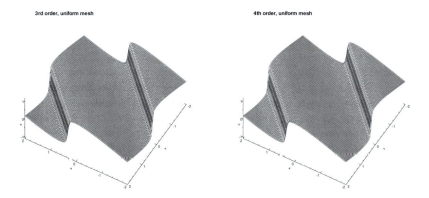

3rd order, uniform mesh   4th order, uniform mesh

Fig. 7.3. 2D Burgers' equation: $t = 5/\pi^2$, uniform mesh. Left: third order WENO; Right: fourth order WENO.

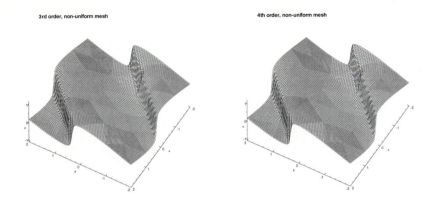

3rd order, non-uniform mesh   4th order, non-uniform mesh

Fig. 7.4. 2D Burgers' equation: $t = 5/\pi^2$, non-uniform mesh. Left: third order WENO; Right: fourth order WENO.

7.2   Finite Difference Formulation in the Scalar Case

Here we assume a uniform grid and solve (7.1) directly using a conservative approximation to the spatial derivative:

$$\frac{du_{ij}(t)}{dt} = -\frac{1}{\Delta x}\left(\hat{f}_{i+\frac{1}{2},j} - \hat{f}_{i-\frac{1}{2},j}\right) - \frac{1}{\Delta y}\left(\hat{g}_{i,j+\frac{1}{2}} - \hat{g}_{i,j-\frac{1}{2}}\right) \qquad (7.12)$$

where $u_{ij}(t)$ is the numerical approximation to the point value $u(x_i, y_j, t)$.

The numerical flux $\hat{f}_{i+\frac{1}{2},j}$ is obtained by the one dimensional ENO or WENO approximation procedure, Algorithm 3.1 or 3.2, with $v(x) = f(u(x, y_j, t))$ and with j fixed. Likewise, the numerical flux $\hat{g}_{i,j+\frac{1}{2}}$ is obtained by the one dimensional ENO or WENO approximation procedure, with $v(y) = f(u(x_i, y, t))$ and with i fixed.

All the one dimensional discussions in Sect. 4.2, such as upwinding, ENO-Roe, flux splitting, etc., can be applied here dimension by dimension.

The discussion here is also valid for higher spatial dimension n. In effect, it is the same one dimensional conservative derivative approximation applied to each space dimension.

It is a straight forward exercise [16] to show that, in terms of operation count, the finite difference ENO or WENO schemes are about a factor of 4 less than the finite volume counterpart of the same order. In 3D this factor becomes about 9.

We thus strongly recommend the usage of the finite difference version of ENO and WENO schemes (also called ENO and WENO schemes based on point values), whenever possible.

7.3   Provable Properties in the Scalar Case

Second order ENO schemes are also maximum norm non-increasing for multi-dimensions. Of course, this stability is too weak to imply any convergence. As was mentioned before, there is no known convergence result for ENO schemes of order higher than 2, even for smooth solutions.

WENO schemes have better convergence results also in the current multi-D case, mainly because their numerical fluxes are smoother. It is proven [55] that WENO schemes converge for smooth solutions.

We again emphasize that, even though there are very few theoretical results about ENO or WENO schemes, in practice they are very robust and stable. We once again caution against any attempts to modify the schemes solely for the purpose of stability or convergence proofs. In fact the modification of ENO schemes in [89], presented in Sect. 4.3, which keeps the formal uniform high order accuracy, actually produces schemes which are convergent to entropy solutions for general multi dimensional scalar equations. However it was pointed out there that the modification is not computationally useful, hence the convergence result has little practical value.

7.4   Systems

The advice here is that, when the fluxes are computed along a cell boundary, a one dimensional local characteristic decomposition normal to the boundary is performed. Also, the monotone flux is replaced with a one dimensional exact or approximate Riemann solver. Thus, the discussion in Sect. 4.4 can be applied here. For second and some third order schemes, a componentwise ENO or WENO scheme usually gives satisfactory results for most test problems, with a significantly lower computational cost than the characteristic decompositions.

There are discussions in the literature about truly multi-dimensional recipes. However, these tend to become extremely complicated for order of accuracy higher than two, so they have not been used extensively in practice for higher order schemes. Another reason to suggest against using such complicated truly multidimensional recipes for order of accuracy higher than two is that, while dimension by dimension schemes as advocated in these lecture notes are not rotationally invariant, the direction related non-symmetry actually diminishes with increased order [16].

8   Further Topics in ENO and WENO Schemes

In this section we discuss some miscellaneous (but not necessarily unimportant!) topics in ENO and WENO schemes.

8.1   Subcell Resolution

This idea was first raised by Harten [44]. The observation is that, since in interpolating the primitive V, *two* points must be included in the initial stencil (see Algorithm 3.1), one cannot avoid having at least one cell for each discontinuity, inside which the reconstructed polynomial is not accurate ($O(1)$ error there). We can clearly see this $O(1)$ error in the ENO interpolation in Fig. 3.1. The reconstruction in this shocked cell, although inaccurate, will always be monotone (Property 2 in Sect. 3.1), so stability will not be a problem. However, it does cause a smearing of the discontinuity (over one cell, initially).

If we are solving a truly nonlinear shock, then characteristics flow into the shock, thus any error one makes during time evolution tends to be absorbed into the shock (we also say that the shock has a self sharpening mechanism). However, we are less lucky with a linear discontinuity, such as a discontinuity carried by the linear equation $u_t + u_x = 0$. Such linear discontinuities are also called contact discontinuities in gas dynamics. The characteristics for such cases are parallel to the discontinuity, hence any numerical smearing tends to accumulate and the discontinuity becomes progressively more smeared with time. Harten argues that the smearing of the discontinuity is at the rate of $O(\Delta x^{1-\frac{1}{k+1}})$ where k is the order of the scheme. Although higher order schemes have less smearing, when time is large the smearing is still very significant.

Harten [44] makes the following simple observation: in the shocked cell I_i, instead of using the reconstruction polynomial $p_i(x)$, which is highly inaccurate (the only useful information it carries is the cell average in the cell), one could try to find the location of the discontinuity inside the cell I_i, say at x_s, and then use the neighboring reconstructions $p_{i-1}(x)$ extended to x_s from left and $p_{i+1}(x)$ extended to x_s from right. To find the shock location, one could argue that $p_{i-1}(x)$ is a very accurate approximation to $v(x)$ up to the discontinuity x_s from left, and $p_{i+1}(x)$ is a very accurate approximation to $v(x)$ up to the discontinuity x_s from right. We thus extend $p_{i-1}(x)$ from the left into the cell I_i, and extend $p_{i+1}(x)$ from the right into the cell I_i, and require that the cell average $\bar{v}_i$ be preserved:

$$\int_{x_{i-\frac{1}{2}}}^{x_s} p_{i-1}(x)\,dx + \int_{x_s}^{x_{i+\frac{1}{2}}} p_{i+1}(x)\,dx = \Delta x_i \bar{v}_i. \qquad (8.1)$$

It can be proven that under very general conditions, (8.1) has only one root x_s inside the cell I_i, hence one could use Newton iterations to find this root.

Subcell resolution can be applied to both finite volume and finite difference ENO and WENO schemes [44], [90], However, it should be applied only to sharpen contact discontinuities. It is quite dangerous to apply the subcell resolution to a shock, since it might generate entropy violating expansion shocks in the numerical solution.

Another very serious restriction about subcell resolution is that it is very difficult to be applied to 2D. However, see Siddiqi, Kimia and Shu [93], where a geometrical ENO is used to extend the subcell resolution idea to 2D for image processing problems (we termed it geometric ENO, or GENO).

8.2   Artificial Compression

Another very useful idea to sharpen a contact discontinuity is the artificial compression, first developed by Harten [41] and further improved by Yang [105]. The idea is to *increase* the magnitude of the slope of a reconstruction, of course subject to certain monotonicity restrictions, near such a discontinuity. Notice that this goes against the idea of limiting, which typically *decreases* the magnitude of the slope of a reconstruction.

Artificial compression can be applied both to finite volume and to finite difference ENO and WENO schemes [105], [90], [55]. Unlike subcell resolution, artificial compression can also be applied easily to multi space dimensions, at least in principle.

8.3   Other Building Blocks

It is not necessary to stay within polynomial building blocks, although polynomials are the most natural functions to work with. For some applications, other building blocks, such as rational functions, trigonometric polynomials, exponential functions, radial functions, etc., may be more appropriate. The idea of ENO or WENO can be applied also in such situations. The key idea is to find suitable "smooth indicators", similar to the Newton divided differences for the polynomial case, for applying the ENO or WENO idea. See [17] and [52] for some examples.

9   Time Discretization

Up to now we have only considered spatial discretizations, leaving the time variable continuous (method of lines). In this section we consider the issue of time discretization. The techniques discussed in this section can also be applied to other types of spatial discretizations using the method of lines approach, such as various TVD and TVB schemes [66,100,85] and discontinuous Galerkin methods [18–21].

9.1   TVD Runge-Kutta Methods

A class of TVD (total variation diminishing) high order Runge-Kutta methods is developed in [89] and further in [36].

These Runge-Kutta methods are used to solve a system of initial value problems of ODEs written as:

$$u_t = L(u), \tag{9.1}$$

resulting from a method of lines spatial approximation to a PDE such as:

$$u_t = -f(u)_x. \tag{9.2}$$

We have written the equation in (9.2) as a 1D conservation law, but the discussion which follows apply to general initial value problems of PDEs in any spatial dimensions. Clearly, $L(u)$ in (9.1) is an approximation (e.g. ENO or WENO approximation in these lecture notes), to the derivative $-f(u)_x$ in the PDE (9.2).

If we *assume* that a first order Euler forward time stepping:

$$u^{n+1} = u^n + \Delta t L(u^n) \tag{9.3}$$

is stable in a certain norm:

$$||u^{n+1}|| \leq ||u^n|| \tag{9.4}$$

under a suitable restriction on Δt:

$$\Delta t \leq \Delta t_1, \tag{9.5}$$

then we look for higher order in time Runge-Kutta methods such that the same stability result (9.4) holds, under a perhaps different restriction on Δt:

$$\Delta t \leq c \, \Delta t_1. \tag{9.6}$$

where c is termed *the CFL coefficient* for the high order time discretization.

We remark that the stability condition (9.4) for the first order Euler forward in time (9.3) is easy to obtain in many cases, such as various TVD and TVB schemes in 1D (where the norm is the total variation norm) and in multi dimensions (where the norm is the L^∞ norm), see, e.g. [66,100,85].

Originally in [89,86] the norm in (9.4) was chosen to be the total variation norm, hence the terminology "TVD time discretization".

As it stands, the TVD high order time discretization defined above maintains stability in whatever norm, of the Euler forward first order time stepping, for the high order time discretization, under the time step restriction (9.6). For example, if it is used for multi dimensional scalar conservation laws, for which TVD is not possible but maximum norm stability can be maintained for high order spatial discretizations plus forward Euler time stepping (e.g. [20]), then the same maximum norm stability can be maintained if TVD high order time discretization is used. As another example, if an entropy inequality can be proved for the Euler forward, then the same entropy inequality is valid under a high order TVD time discretization.

In [89], a general Runge-Kutta method for (9.1) is written in the form:

$$u^{(i)} = \sum_{k=0}^{i-1} \left(\alpha_{ik} u^{(k)} + \Delta t \beta_{ik} L(u^{(k)}) \right), \qquad i = 1, ..., m \tag{9.7}$$

$$u^{(0)} = u^n, \qquad u^{(m)} = u^{n+1}.$$

Clearly, if all the coefficients are nonnegative $\alpha_{ik} \geq 0$, $\beta_{ik} \geq 0$, then (9.7) is just a convex combination of the Euler forward operators, with Δt replaced by $\frac{\beta_{ik}}{\alpha_{ik}} \Delta t$, since by consistency $\sum_{k=0}^{i-1} \alpha_{ik} = 1$. We thus have

Lemma 9.1. [89] The Runge-Kutta method (9.7) is TVD under the CFL coefficient (9.6):

$$c = \min_{i,k} \frac{\alpha_{ik}}{\beta_{ik}}, \tag{9.8}$$

provided that $\alpha_{ik} \geq 0$, $\beta_{ik} \geq 0$. □

In [89], schemes up to third order were found to satisfy the conditions in Lemma 9.1 with CFL coefficient equal to 1.

The optimal second order TVD Runge-Kutta method is given by [89,36]:

$$u^{(1)} = u^n + \Delta t L(u^n) \tag{9.9}$$
$$u^{n+1} = \frac{1}{2}u^n + \frac{1}{2}u^{(1)} + \frac{1}{2}\Delta t L(u^{(1)}),$$

with a CFL coefficient $c = 1$ in (9.8).

The optimal third order TVD Runge-Kutta method is given by [89,36]:

$$u^{(1)} = u^n + \Delta t L(u^n)$$
$$u^{(2)} = \frac{3}{4}u^n + \frac{1}{4}u^{(1)} + \frac{1}{4}\Delta t L(u^{(1)}) \tag{9.10}$$
$$u^{n+1} = \frac{1}{3}u^n + \frac{2}{3}u^{(2)} + \frac{2}{3}\Delta t L(u^{(2)}),$$

with a CFL coefficient $c = 1$ in (9.8).

It can be shown that for any order of accuracy, $c = 1$ is the best one can get for a CFL coefficient. We have also found, for a linear spatial operator L, optimal TVD Runge-Kutta methods for arbitrary order of accuracy with a CFL coefficient $c = 1$. These results will appear in a forthcoming paper [37].

Unfortunately, if L is nonlinear, it is proven in [36] that no four stage, fourth order TVD Runge-Kutta method exists with nonnegative α_{ik} and β_{ik}. We thus have to consider the situation where $\alpha_{ik} \geq 0$ but β_{ik} might be negative. In such situations we need to introduce an adjoint operator $\tilde{L}$. The requirement for $\tilde{L}$ is that it approximates the same spatial derivative(s) as L, but is TVD (or stable in another relevant norm) for first order Euler, backward in time:

$$u^{n+1} = u^n - \Delta t \tilde{L}(u^n) \tag{9.11}$$

This can be achieved, for hyperbolic conservation laws, by solving the backward in time version of (9.2):

$$u_t = f(u)_x. \tag{9.12}$$

Numerically, the only difference is the change of upwind direction. Clearly, $\tilde{L}$ can be computed with the same cost as that of computing L. We then have the following lemma:

Lemma 9.2. [89] The Runge-Kutta method (9.7) is TVD under the CFL coefficient (9.6):

$$c = \min_{i,k} \frac{\alpha_{ik}}{|\beta_{ik}|}, \tag{9.13}$$

provided that $\alpha_{ik} \geq 0$, and L is replaced by $\tilde{L}$ for negative β_{ik}. $\square$

Notice that, if for the same k, both $L(u^{(k)})$ and $\tilde{L}(u^{(k)})$ must be computed, the cost as well as storage requirement for this k is doubled. For this reason, we would like to avoid negative β_{ik} as much as possible.

An extensive search performed in [36] gives the following preferred four stage, fourth order TVD Runge-Kutta method:

$$u^{(1)} = u^n + \frac{1}{2}\Delta t L(u^n)$$

$$u^{(2)} = \frac{649}{1600}u^{(0)} - \frac{10890423}{25193600}\Delta t \tilde{L}(u^n) + \frac{951}{1600}u^{(1)} + \frac{5000}{7873}\Delta t L(u^{(1)})$$

$$u^{(3)} = \frac{53989}{2500000}u^n - \frac{102261}{5000000}\Delta t \tilde{L}(u^n) + \frac{4806213}{20000000}u^{(1)}$$

$$- \frac{5121}{20000}\Delta t \tilde{L}(u^{(1)}) + \frac{23619}{32000}u^{(2)} + \frac{7873}{10000}\Delta t L(u^{(2)}) \tag{9.14}$$

$$u^{n+1} = \frac{1}{5}u^n + \frac{1}{10}\Delta t L(u^n) + \frac{6127}{30000}u^{(1)} + \frac{1}{6}\Delta t L(u^{(1)})$$

$$+ \frac{7873}{30000}u^{(2)} + \frac{1}{3}u^{(3)} + \frac{1}{6}\Delta t L(u^{(3)})$$

with a CFL coefficient $c = 0.936$ in (9.13). Notice that two $\tilde{L}$'s must be computed. The effective CFL coefficient, comparing with an ideal case without $\tilde{L}$'s, is $0.936 \times \frac{4}{6} = 0.624$. Since it is difficult to solve the global optimization problem, we do not claim that (9.14) is the optimal 4 stage, 4th order TVD Runge-Kutta method.

A fifth order TVD Runge-Kutta method is also given in [89].

For large scale scientific computing in three space dimensions, storage is usually a paramount consideration. There are therefore discussions about low storage Runge-Kutta methods [103], [13], which only require 2 storage units per ODE equation. In [36], we considered the TVD properties among such low storage Runge-Kutta methods and found third order low storage TVD Runge-Kutta methods.

The general low-storage Runge-Kutta schemes can be written in the form [103], [13]:

$$du^{(i)} = A_i du^{(i-1)} + \Delta t L(u^{(i-1)}) \tag{9.15}$$

$$u^{(i)} = u^{(i-1)} + B_i du^{(i)}, \qquad i = 1, ..., m$$

$$u^{(0)} = u^n, \quad u^{(m)} = u^{n+1}, \quad A_0 = 0$$

Only u and du must be stored, resulting in two storage units for each variable.

Carpenter and Kennedy [13] have classified all the three stage, third order (m=3) low storage Runge-Kutta methods, obtaining the following one parameter family:

$$z_1 = \sqrt{36c_2^4 + 36c_2^3 - 135c_2^2 + 84c_2 - 12}$$
$$z_2 = 2c_2^2 + c_2 - 2$$
$$z_3 = 12c_2^4 - 18c_2^3 + 18c_2^2 - 11c_2 + 2$$
$$z_4 = 36c_2^4 - 36c_2^3 + 13c_2^2 - 8c_2 + 4$$
$$z_5 = 69c_2^3 - 62c_2^2 + 28c_2 - 8$$
$$z_6 = 34c_2^4 - 46c_2^3 + 34c_2^2 - 13c_2 + 2$$
$$B_1 = c_2 \tag{9.16}$$
$$B_2 = \frac{12c_2(c_2 - 1)(3z_2 - z_1) - (3z_2 - z_1)^2}{144c_2(3c_2 - 2)(c_2 - 1)^2}$$
$$B_3 = \frac{-24(3c_2 - 2)(c_2 - 1)^2}{(3z_2 - z_1)^2 - 12c_2(c_2 - 1)(3z_2 - z_1)}$$
$$A_2 = \frac{-z_1(6c_2^2 - 4c_2 + 1) + 3z_3}{(2c_2 + 1)z_1 - 3(c_2 + 2)(2c_2 - 1)^2}$$
$$A_3 = \frac{-z_4 z_1 + 108(2c_2 - 1)c_2^5 - 3(2c_2 - 1)z_5}{24z_1 c_2(c_2 - 1)^4 + 72c_2 z_6 + 72c_2^6(2c_2 - 13)}$$

In [36] we converted this form into the form (9.7), by introducing three new parameters. Then we searched for values of these parameters that would maximize the CFL restriction, by a computer program. The result seems to indicate that

$$c_2 = 0.924574 \tag{9.17}$$

gives an almost best choice, with CFL coefficient $c = 0.32$ in (9.8). This is of course less optimal than (9.10) in terms of CFL coefficients, however the low storage form is useful for large scale calculations.

We end this subsection by quoting the following numerical example [36], which shows that, even with a very nice second order TVD spatial discretization, if the time discretization is by a non-TVD but linearly stable Runge-Kutta method, the result may be oscillatory. Thus it would always be safer to use TVD Runge-Kutta methods for hyperbolic problems.

The numerical example uses the standard minmod based MUSCL second order spatial discretization [101]. We will compare the results of a TVD versus a non-TVD second order Runge-Kutta time discretizations. The PDE is the simple Burgers equation

$$u_t + \left(\frac{1}{2}u^2\right)_x = 0 \tag{9.18}$$

with a Riemann initial data:

$$u(x,0) = \begin{cases} 1, & \text{if } x \le 0 \\ -0.5, & \text{if } x > 0. \end{cases} \tag{9.19}$$

The nonlinear flux $\left(\frac{1}{2}u^2\right)_x$ in (9.18) is approximated by the conservative difference

$$\frac{1}{\Delta x}\left(\hat{f}_{i+\frac{1}{2}} - \hat{f}_{i-\frac{1}{2}}\right),$$

where the numerical flux $\hat{f}_{i+\frac{1}{2}}$ is defined by

$$\hat{f}_{i+\frac{1}{2}} = h\left(u_{i+\frac{1}{2}}^-, u_{i+\frac{1}{2}}^+\right)$$

with

$$u_{i+\frac{1}{2}}^- = u_i + \frac{1}{2}minmod(u_{i+1} - u_i, u_i - u_{i-1}),$$

$$u_{i+\frac{1}{2}}^+ = u_{i+1} - \frac{1}{2}minmod(u_{i+2} - u_{i+1}, u_{i+1} - u_i)$$

The monotone flux h is the Godunov flux defined by (4.6), and the *minmod* function is given by

$$minmod(a, b) = \frac{sign(a) + sign(b)}{2} \min(|a|, |b|).$$

It is easy to prove, by using Harten's Lemma [42], that the Euler forward time discretization with this second order MUSCL spatial operator is TVD under the CFL condition (9.5):

$$\Delta t \le \frac{\Delta x}{2\max_j |u_j^n|} \tag{9.20}$$

Thus $\Delta t = \frac{\Delta x}{2\max_j |u_j^n|}$ will be used in all our calculations. Actually, apart from a slight difference (the *minmod* function is replaced by a minimum-in-absolute-value function), this MUSCL scheme is the same as the second order ENO scheme discussed in Sect. 4.1.

The TVD second order Runge-Kutta method we consider is the optimal one (9.9). The non-TVD method we use is:

$$u^{(1)} = u^n - 20\Delta t L(u^n) \tag{9.21}$$

$$u^{n+1} = u^n + \frac{41}{40}\Delta t L(u^n) - \frac{1}{40}\Delta t L(u^{(1)}).$$

It is easy to verify that both methods are second order accurate in time. The second one (9.21) is however clearly non-TVD, since it has negative β's in both stages (i.e. it partially simulates backward in time with wrong upwinding).

If the operator L is linear (for example the first order upwind scheme applied to a linear PDE), then both Runge-Kutta methods (actually all the two stage, second order Runge-Kutta methods) yield identical results (the two stage, second order Runge-Kutta method for a linear ODE is unique). However, since our L is nonlinear, we may and do observe different results when the two Runge-Kutta methods are used.

In Fig. 9.1 we show the result of the TVD Runge-Kutta method (9.9) and the non-TVD method (9.21), after the shock moves about 50 grids (400 time steps for the TVD method, 528 time steps for the non-TVD method). We can clearly see that the non-TVD result is oscillatory (there is an overshoot).

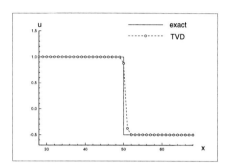 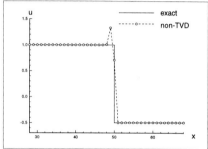

Fig. 9.1. Second order TVD MUSCL spatial discretization. Solution after the shock moves 50 grids. Left: with TVD time discretization (9.9); Right: with non-TVD time discretization (9.21).

Such oscillations are also observed when the non-TVD Runge-Kutta method coupled with a second order TVD MUSCL spatial discretization is applied to a linear PDE ($u_t + u_x = 0$) (the scheme is still nonlinear due to the *minmod* functions). Moreover, for some Runge-Kutta methods, if one looks at the intermediate stages, i.e. $u^{(i)}$ for $1 \leq i < m$ in (9.7), one observes even bigger oscillations. Such oscillations may render difficulties when physical problems are solved, such as the appearance of negative density and pressure for Euler equations of gas dynamics. On the other hand, TVD Runge-Kutta method guarantees that each middle stage solution is also TVD.

This simple numerical test convinces us that it is much safer to use a TVD Runge-Kutta method for solving hyperbolic problems.

9.2 TVD Multi-Step Methods

If one prefers multi-step methods rather than Runge-Kutta methods, one can use the TVD high order multi-step methods developed in [86]. The philosophy is very similar to the TVD Runge-Kutta methods discussed in the previous subsection. One starts with a method of lines approximation (9.1) to the

PDE (9.2), and an assumption that the first order Euler forward in time discretization (9.3) is stable under a certain norm (9.4), with the time step restriction (9.5). One then looks for higher order in time multi-step methods such that the same stability result (9.4) holds, under a perhaps different restriction on Δt in (9.6), where c is again termed *the CFL coefficient* for the high order time discretization.

The general form of the multi-step methods studied in [86] is:

$$u^{n+1} = \sum_{k=0}^{m} \left(\alpha_k u^{n-k} + \Delta t \beta_k L(u^{n-k}) \right), \tag{9.22}$$

Similar to the Runge-Kutta methods in the previous subsection, if all the coefficients are nonnegative $\alpha_k \geq 0$, $\beta_k \geq 0$, then (9.22) is just a convex combination of the Euler forward operators, with Δt replaced by $\frac{\beta_k}{\alpha_k} \Delta t$, since by consistency $\sum_{k=0}^{m} \alpha_k = 1$. We thus have

Lemma 9.3. [86] The multi-step method (9.22) is TVD under the CFL coefficient (9.6):

$$c = \min_{k} \frac{\alpha_k}{\beta_k}, \tag{9.23}$$

provided that $\alpha_k \geq 0$, $\beta_k \geq 0$. □

In [86], schemes up to third order were found to satisfy the conditions in Lemma 9.3. Here we list a few examples.

The following three step ($m = 2$) scheme is second order and TVD

$$u^{n+1} = \frac{3}{4} u^n + \frac{3}{2} \Delta t L(u^n) + \frac{1}{4} u^{n-2} \tag{9.24}$$

with a CFL coefficient $c = 0.5$ in (9.23). This translates to the same efficiency as the optimal second order TVD Runge-Kutta scheme (9.9), as here only one residue evaluation is needed per time step. Of course, the storage requirement is bigger here. There is also the problem of the starting values u^1 and u^2.

The following five step ($m = 4$) scheme is third order and TVD

$$u^{n+1} = \frac{25}{32} u^n + \frac{25}{16} \Delta t L(u^n) + \frac{7}{32} u^{n-4} + \frac{5}{16} \Delta t L(u^{n-4}) \tag{9.25}$$

with a CFL coefficient $c = 0.5$ in (9.23). This translates to a better efficiency than the optimal third order TVD Runge-Kutta scheme (9.10), as here only one residue evaluation is needed per time step. Of course, the storage requirement is much bigger here. There is also the problem of the starting values u^1, u^2, u^3 and u^4.

There are many other TVD multi-step methods satisfying the conditions in Lemma 9.3 listed in [86]. It seems that if one uses more storage (larger m) one could get better CFL coefficients.

In [86] we have been unable to find multi-step schemes of order four or higher satisfying the condition of Lemma 9.3. As in the Runge-Kutta case, we can relax the condition $\beta_k \geq 0$ by introducing the adjoint operator $\tilde{L}$. We thus have

Lemma 9.4. [86] The multi-step method (9.22) is TVD under the CFL coefficient (9.6):

$$c = \min_k \frac{\alpha_k}{|\beta_k|}, \tag{9.26}$$

provided that $\alpha_k \geq 0$, and L is replaced by $\tilde{L}$ for negative β_k.   □

Again, notice that, if we have both positive and negative β_k's, then both $L(u^n)$ and $\tilde{L}(u^n)$ must be computed, the cost as well as storage requirement will thus be doubled.

We list here a six step ($m = 5$), fourth order multi-step method which is TVD with a CFL coefficient $c = 0.245$ in (9.23) [86]:

$$u^{n+1} = \frac{747}{1280}u^n + \frac{237}{128}\Delta t L(u^n) + \frac{81}{256}u^{n-4} + \frac{165}{128}\Delta t L(u^{n-4})$$
$$+ \frac{1}{10}u^{n-5} - \frac{3}{8}\Delta t \tilde{L}(u^{n-5}) \tag{9.27}$$

9.3   The Lax-Wendroff Procedure

Another way to discretize the time variable is by the Lax-Wendroff procedure [65]. This is also referred to as the Taylor series method for discretizing the ODE (9.1). We will again use the simple 1D scalar conservation law (9.2) as an example to illustrate the procedure, however it applies to more general multidimensional systems.

Starting from a Taylor series expansion in time:

$$u(x, t + \Delta t) = u(x, t) + u_t(x, t)\Delta t + u_{tt}(x, t)\frac{\Delta t^2}{2} + \dots \tag{9.28}$$

The expansion is carried out to the desired order of accuracy in time. For example, a second order in time would need the three terms written out in (9.28). We then use the PDE (9.2) to replace the time derivatives by the spatial derivatives:

$$u_t(x, t) = -f(u(x, t))_x = -f'(u(x, t))\, u_x(x, t);$$
$$u_{tt}(x, t) = -(f(u(x, t))_{tx}$$
$$= -(f'(u(x, t)\, u_t(x, t))_x \tag{9.29}$$
$$= ((f'(u(x, t))^2 u_x(x, t))_x$$
$$= 2f'(u(x, t))\, f''(u(x, t)\, (u_x(x, t))^2 + (f'(u(x, t)))^2\, u_{xx}(x, t);$$

This little exercise in (9.29) should convince us that it is always possible to write all the time derivatives as functions of the $u(x, t)$ and its spatial derivatives. But the expression could be terribly complicated, especially for multidimensional systems.

Once this is done, we substitute (9.29) into (9.28), and then discretize the spatial derivatives of $u(x, t)$ by whatever methods we use. For example, in the cell averaged (finite volume) ENO schemes discussed in Sect. 4.1, we proceed as follows. We first integrate the PDE (4.1) in space-time over the region $[x_{i-\frac{1}{2}}, x_{i+\frac{1}{2}}] \times [t^n, t^{n+1}]$ to obtain

$$\bar{u}_i^{n+1} = \bar{u}_i^n - \frac{1}{\Delta x_i} \left(\int_{t^n}^{t^{n+1}} f(u(x_{i+\frac{1}{2}}, t)) dt - \int_{t^n}^{t^{n+1}} f(u(x_{i-\frac{1}{2}}, t)) dt \right) \quad (9.30)$$

Then, we use a suitable Gaussian quadrature to discretize the time integration for the flux in (9.30):

$$\frac{1}{\Delta t} \int_{t^n}^{t^{n+1}} f(u(x_{i+\frac{1}{2}}, t)) dt \approx \sum_\alpha w_\alpha f(u(x_{i+\frac{1}{2}}, t^n + \beta_\alpha \Delta t), \quad (9.31)$$

where β_α and w_α are Gaussian quadrature nodes and weights. Next we replace each

$$f(u(x_{i+\frac{1}{2}}, t^n + \beta_\alpha \Delta t)$$

by a monotone flux:

$$f(u(x_{i+\frac{1}{2}}, t^n + \beta_\alpha \Delta t) \approx h\left(u(x_{i+\frac{1}{2}}^-, t^n + \beta_\alpha \Delta t), u(x_{i+\frac{1}{2}}^+, t^n + \beta_\alpha \Delta t)\right), \quad (9.32)$$

and use the Lax-Wendroff procedure (9.28)-(9.29) to convert

$$u(x_{i+\frac{1}{2}}^\pm, t^n + \beta_\alpha \Delta t)$$

to $u(x_{i+\frac{1}{2}}^\pm, t^n)$ and its spatial derivatives also at t^n, which can then be obtained by the reconstructions $p(x)$ inside I_i and I_{i+1}. Notice that the accuracy is just enough in this procedure, as each derivative of the reconstruction $p(x)$ will be one order lower in accuracy, but this is compensated by the Δt in front of it in (9.28).

This Lax-Wendroff procedure, comparing with the method of lines approach coupled with TVD Runge-Kutta or multi-step time discretizations, has the following advantages and disadvantages.

Advantages:

1. This is a truly one step method, hence it is quite compact (a second order method in space and time uses only three cells on time level n to advance to time level $n + 1$ for one cell), and there are no complications such as boundary conditions needed in middle stages;

2. It utilizes the PDE more extensively than the method of lines approach. This is also one reason that it can be so compact.

Disadvantages:

1. The algebra is very, very complicated for multi dimensional systems. This also increases operation counts for complicated nonlinear systems;
2. It is more difficult to prove stability properties (e.g. TVD) for higher order methods in this framework;
3. It is difficult and costly to apply this procedure to the conservative finite difference framework established in Sections 4.2 and 7.2.

10 Formulation of the ENO and WENO Schemes for the Hamilton-Jacobi Equations

In this section we describe high order ENO and WENO approximations to the Hamilton-Jacobi equation:

$$\begin{cases} \phi_t + H(\phi_x, \phi_y) = 0 \\ \phi(x, y, 0) = \phi^0(x, y) \end{cases} \tag{10.1}$$

where H is a locally Lipschitz continuous Hamiltonian and the initial condition $\phi^0(x, y)$ is locally Lipschitz continuous. We have written the equation (10.1) in two space dimensions, but the discussion is valid for other space dimensions as well.

As is well known, solutions to (10.1) are Lipschitz continuous but may have discontinuous derivatives, regardless of the smoothness of $\phi^0(x, y)$. The non-uniqueness of such generalized solutions also necessitates the definition of viscosity solutions, to single out a unique, practically relevant solution. The viscosity solution to (10.1) is a locally Lipschitz continuous function $\phi(x, y, t)$, which satisfies the initial condition and the following property: for any smooth function $\psi(x, y, t)$, if (x_0, y_0, t_0) is a local maximum point of $\phi - \psi$, then

$$\psi_t(x_0, y_0, t_0) + H(\psi_x(x_0, y_0, t_0) + \psi_y(x_0, y_0, t_0)) \le 0,$$

and, if (x_0, y_0, t_0) is a local minimum point of $\phi - \psi$, then

$$\psi_t(x_0, y_0, t_0) + H(\psi_x(x_0, y_0, t_0) + \psi_y(x_0, y_0, t_0)) \ge 0.$$

Of course, the above definition means that whenever $\phi(x, y, t)$ is differentiable, (10.1) is satisfied in the classical sense. Viscosity solution defined this way exists and is unique. For details and equivalent definitions of viscosity solutions, see Crandall and Lions [24].

Hamilton-Jacobi equations are actually easier to solve than conservation laws, because the solutions are typically continuous (only the derivatives are discontinuous).

As before, given mesh sizes Δx, Δy and Δt, we denote the mesh points as $(x_i, y_j, t_n) = (i\Delta x, j\Delta y, n\Delta t)$. The numerical approximation to the viscosity solution $\phi(x_i, y_j, t_n)$ of (10.1) at the mesh point (x_i, y_j, t_n) is denoted by ϕ_{ij}^n. We again use a semi-discrete (discrete in the spatial variables only) formulation as a middle step in designing algorithms. In such cases, the numerical approximation to the viscosity solution $\phi(x_i, y_j, t)$ of (10.1) at the mesh point (x_i, y_j, t) is denoted by $\phi_{ij}(t)$, the temporal variable t is not discretized. We will also use the notations $D_{\pm}^x \phi_{ij} = \frac{\pm(\phi_{i\pm1,j} - \phi_{ij})}{\Delta x}$ and $D_{\pm}^y \phi_{ij} = \frac{\pm(\phi_{i,j\pm1} - \phi_{ij})}{\Delta y}$ to denote the first order forward/backward difference approximations to the left and right derivatives of $\phi(x, y)$ at the location (x_i, y_j).

Since the viscosity solution to (10.1) is usually only Lipschitz continuous but not everywhere differentiable, the *formal* order of accuracy of a numerical scheme is again defined as that determined by the local truncation error in the smooth regions of the solution. Thus, a monotone scheme of the form

$$\phi_{ij}^{n+1} = G(\phi_{i-p,j-r}^n, \cdots, \phi_{i+q,j+s}^n) \tag{10.2}$$

where G is a non-decreasing function of each argument, is called a first order scheme, although the provable order of accuracy in the L_∞ norm is just $\frac{1}{2}$ [25]. In the semi-discrete formulation, a five point monotone scheme (it does not pay to use more points for a monotone scheme because the order of accuracy of a monotone scheme is at most one [45]) is of the form

$$\frac{d}{dt}\phi_{ij}(t) = -\hat{H}(D_+^x \phi_{ij}(t), D_-^x \phi_{ij}(t), D_+^y \phi_{ij}(t), D_-^y \phi_{ij}(t)). \tag{10.3}$$

The numerical Hamiltonian $\hat{H}$ is assumed to be locally Lipschitz continuous, consistent with H: $\hat{H}(u, u, v, v) = H(u, v)$, and is non-increasing in its first and third arguments and non-decreasing in the other two. Symbolically $\hat{H}(\downarrow, \uparrow, \downarrow, \uparrow)$. It is easy to see that, if the time derivative in (10.3) is discretized by Euler forward differencing, the resulting fully discrete scheme, in the form of (10.2), will be monotone when Δt is suitably small. We have chosen the semi-discrete formulation (10.3) in order to apply suitable nonlinearly stable high order Runge-Kutta type time discretization, see Sect. 9.

Semi-discrete or fully discrete monotone schemes (10.3) and (10.2) are both convergent towards the viscosity solution of (10.1) [25]. However, monotone schemes are at most first order accurate. As before, we will use the monotone schemes as building blocks for higher order ENO and WENO schemes.

ENO schemes were adapted to the Hamilton-Jacobi equations (10.1) by Osher and Sethian [78] and Osher and Shu [79]. As we know now, the key feature of the ENO algorithm is an adaptive stencil high order interpolation which tries to avoid shocks or high gradient regions whenever possible. Since the Hamilton-Jacobi equation (10.1) is closely related to the conservation law (7.1), in fact in one space dimension they are exactly the same if one takes $u = \phi_x$, it is not surprising that successful numerical schemes for the conservation laws (7.1), such as ENO and WENO, can be applied to the Hamilton-Jacobi equation (10.1). ENO and WENO schemes, when applied to Hamilton-Jacobi equations (10.1), can produce high order accuracy in the smooth regions of the solution, and sharp, non-oscillatory corners (discontinuities in derivatives).

There are many monotone Hamiltonians [25], [78], [79]. In this section we mainly discuss the following two:

1. For the special case $H(u,v) = f(u^2, v^2)$ where f is a monotone function of both arguments, such as the example $H(u,v) = \sqrt{u^2 + v^2}$, we can use the Osher-Sethian monotone Hamiltonian [78]:

$$\hat{H}^{OS}(u^+, u^-, v^+, v^-) = f(u^2, v^2) \qquad (10.4)$$

where, if f is a non-increasing function of u^2, u^2 is implemented by

$$u^2 = (\min(u^-, 0))^2 + (\max(u^+, 0))^2 \qquad (10.5)$$

and, if f is a non-decreasing function of u^2, u^2 is implemented by

$$u^2 = (\min(u^+, 0))^2 + (\max(u^-, 0))^2 \qquad (10.6)$$

Similarly for v^2. This Hamiltonian is purely upwind (i.e. when $H(u,v)$ is monotone in u in the relevant domain $[u^-, u^+] \times [v^-, v^+]$, only u^- or u^+ is used in the numerical Hamiltonian according to the wind direction), and simple to program. Whenever applicable it should be used. This flux is similar to the Engquist-Osher monotone flux (4.7) for the conservation laws.

2. For the general H we can always use the Godunov type Hamiltonian [6], [79]:

$$\hat{H}^G(u^+, u^-, v^+, v^-) = ext_{u \in I(u^-, u^+)} \, ext_{v \in I(v^-, v^+)} \, H(u, v) \qquad (10.7)$$

where the extrema are defined by

$$ext_{u \in I(a,b)} = \begin{cases} \min_{a \le u \le b} & \text{if } a \le b \\ \max_{b \le u \le a} & \text{if } a > b \end{cases} \qquad (10.8)$$

Godunov Hamiltonian is obtained by attempting to solve the Riemann problem of the equation (10.1) exactly with piecewise linear initial condition determined by $u^\pm$ and $v^\pm$. It is in general not unique, because in general $\min_u \max_v H(u,v) \ne \max_v \min_u H(u,v)$ and interchanging the order of the two ext's in (10.7) can produce a different monotone Hamiltonian.

Godunov Hamiltonian is purely upwind and is the least dissipative among all monotone Hamiltonians [76]. However, it might be extremely difficult to program, since in general analytical expressions for things like $\min_u \max_v H(u,v)$ can be quite complicated. The readers will be convinced by doing the exercise of obtaining the analytical expression and programming H^G for the ellipse in ellipse case in image processing where $H(u,v) = \sqrt{au^2 + 2buv + cv^2}$. For this case the Osher-Sethian Hamiltonian H^{OS} does not apply.

We are now ready to discuss about higher order ENO or WENO schemes for (10.1). The framework is quite simple: we simply replace the first order scheme (10.3) by:

$$\frac{d}{dt}\phi_{ij}(t) = -\hat{H}(u_{ij}^+(t), u_{ij}^-(t), v_{ij}^+(t), v_{ij}^-(t)) \tag{10.9}$$

where $u_{ij}^\pm(t)$ are high order approximations to the left and right x-derivatives of $\phi(x, y, t)$ at (x_i, y_j, t):

$$u_{ij}^\pm(t) = \frac{\partial \phi}{\partial x}(x_i^\pm, y_j, t) + O(\Delta x^r) \tag{10.10}$$

Similarly for $v_{ij}^\pm(t)$. Notice that there is no cell-averaged version now.

The key feature of ENO to avoid numerical oscillations is through the following interpolation procedure to obtain $u_{ij}^\pm(t)$ and $v_{ij}^\pm(t)$. These are just the same ENO procedure we discussed before in Sect. 3. We repeat it here with its own notations:

ENO Interpolation Algorithm: Given point values $f(x_j)$, $j = 0, \pm 1, \pm 2, \cdots$ of a (usually piecewise smooth) function $f(x)$ at discrete nodes x_j, we associate an r-th degree polynomial $P_{j+1/2}^{f,r}(x)$ with each interval $[x_j, x_{j+1}]$, with the left-most point in the stencil as $x_{k_{min}^{(r)}}$, constructed inductively as follows:

(1) $P_{j+1/2}^{f,1}(x) = f[x_j] + f[x_j, x_{j+1}](x - x_j)$, $k_{min}^{(1)} = j$;

(2) If $k_{min}^{(l-1)}$ and $P_{j+1/2}^{f,l-1}(x)$ are both defined, then let

$$a^{(l)} = f[x_{k_{min}^{(l-1)}}, \cdots, x_{k_{min}^{(l-1)}+l}] \quad b^{(l)} = f[x_{k_{min}^{(l-1)}-1}, \cdots, x_{k_{min}^{(l-1)}+l-1}]$$

and

(i) If $|a^{(l)}| \geq |b^{(l)}|$, then $c^{(l)} = b^{(l)}$ and $k_{min}^{(l)} = k_{min}^{(l-1)} - 1$; otherwise $c^{(l)} = a^{(l)}$ and $k_{min}^{(l)} = k_{min}^{(l-1)}$;

(ii) $P_{j+1/2}^{f,l}(x) = P_{j+1/2}^{f,l-1}(x) + c^{(l)} \prod_{i=k_{min}^{(l-1)}}^{k_{min}^{(l-1)}+l-1}(x - x_i)$.

□

In the above procedure $f[\cdot, \cdots, \cdot]$ are the standard Newton divided differences, inductively defined as $f[x_1, x_2, \cdots, x_{k+1}] = \frac{f[x_2, \cdots, x_{k+1}] - f[x_1, \cdots, x_k]}{x_{k+1} - x_1}$ with $f[x_1] = f(x_1)$.

ENO Interpolation Algorithm starts with a first degree polynomial $P_{j+1/2}^{f,1}(x)$ interpolating the function $f(x)$ at the two grid points x_j and x_{j+1}. If we stop here, we would obtain the first order monotone scheme. When higher order is desired, we will in each step add just one point to the existing stencil, chosen from the two immediate neighbors by the size of the two relevant divided differences, which measures the local smoothness of the function $f(x)$.

The approximations to the left and right x-derivatives of ϕ are then taken as

$$u_{ij}^{\pm} = \frac{\partial}{\partial x} P_{i\pm 1/2,j}^{\phi,r}(x_i). \tag{10.11}$$

where $P_{i\pm 1/2,j}^{\phi,r}(x)$ is obtained by the ENO Interpolation Algorithm in the x-direction, with $y = y_j$ and t both fixed. $v_{ij}^{\pm}$ are obtained in a similar fashion. The resulting ODE (10.9) is then discretized by an r-th order TVD Runge-Kutta time discretization in Sect. 9 to guarantee nonlinear stability. More specifically, the high order Runge-Kutta method we use in Sect. 9 will maintain TVD (total-variation-diminishing) or other stability properties, if these properties are valid for the simple first order Euler forward time discretization of the ODE (10.9). Notice that this is different from the usual linear stability requirement for the ODE solver. We thus obtain both nonlinear stability and high order accuracy in time. The second order ($r = 2$) and third order ($r = 3$) methods we use which has this stability property are given by (9.9) and (9.10), respectively.

Time step restriction is taken as

$$\Delta t \left(\frac{1}{\Delta x} \max_{u,v} \left| \frac{\partial}{\partial u} H(u,v) \right| + \frac{1}{\Delta y} \max_{u,v} \left| \frac{\partial}{\partial v} H(u,v) \right| \right) \leq 0.6$$

where the maximum is taken over the relevant ranges of u, v. Here 0.6 is just a convenient number used in practice. This number should be chosen between 0.5 and 0.7 according to our numerical experience.

WENO schemes can be used in a similar fashion for Hamilton-Jacobi equations [57]. We will not present the details here.

11   Applications to Compressible Gas Dynamics I: Structured Mesh for Polytropic Gas

One of the main application areas of ENO and WENO schemes is compressible gas dynamics. In this section we describe the applications of ENO and WENO schemes in structured mesh for polytropic gas dynamics.

In 3D, the Euler equations of a polytropic gas are written as

$$U_t + f(U)_x + g(U)_y + h(U)_z = 0 \tag{11.1}$$

where

$$U = (\rho, \rho u, \rho v, \rho w, E),$$

$$f(U) = (\rho u, \rho u^2 + P, \rho uv, \rho uw, u(E + P)),$$

$$g(U) = (\rho v, \rho uv, \rho v^2 + P, \rho vw, v(E + P)),$$

$$h(U) = (\rho w, \rho uw, \rho vw, \rho w^2 + P, w(E + P)).$$

Here ρ is density, (u, v, w) is the velocity, E is the total energy, P is the pressure, related to the total energy E by

$$E = \frac{P}{\gamma - 1} + \frac{1}{2}\rho(u^2 + v^2 + w^2)$$

with $\gamma = 1.4$ for air.

In two space dimensions, there is one fewer equation with the w component of the velocity eliminated; in one space dimension, there are two fewer equations with the v and w components of the velocity eliminated.

For the form of the Navier-Stokes equations, for the eigenvalues and eigenvectors needed for the characteristic-wise ENO and WENO schemes, and for those equations appearing in curvilinear coordinates, see, e.g. [91].

Example 11.1. Shock tube problem. This is a standard problem for testing codes for one dimensional shock calculations. However, it is not the best test case for high order methods, as the solution structure is relatively simple (basically piecewise linear). The set-up is a Riemann type initial data:

$$U(x, 0) = \begin{cases} U_L \text{ if } x \leq 0 \\ U_R \text{ if } x > 0 \end{cases}$$

The two standard test cases are the Sod's problem [95]:

$$(\rho_L, q_L, P_L) = (1, 0, 1); \tag{11.2}$$
$$(\rho_R, q_R, P_R) = (0.125, 0, 0.1)$$

and the Lax's problem [64]:

$$(\rho_L, q_L, P_L) = (0.445, 0.698, 3.528); \tag{11.3}$$
$$(\rho_R, q_R, P_R) = (0.5, 0, 0.571)$$

We show the results of the finite difference WENO (third order and fifth order) schemes for the Lax problem, in Fig. 11.1. Notice that "PS" in the pictures means a way of treating the system cheaper than the local characteristic decompositions (for details, see [55]). "A" stands for Yang's artificial compression [105] applied to these cases [55].

We can see from Fig. 11.1 that WENO perform reasonably well for these shock tube problem. The contact discontinuity is smeared more than the shock, as expected. Artificial compression helps sharpening contacts. For this problem, which is not the most demanding, the less expensive "PS" version of WENO work quite well.

ENO schemes on this test case perform similarly. We will not give the pictures here. See [90].

Example 11.2. Shock entropy wave interactions. This problem is very suitable for high order ENO and WENO schemes, because both shocks and

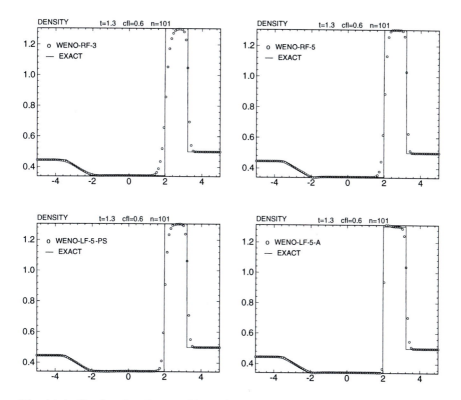

Fig. 11.1. Shock tube, Lax problem, density. Top left: third order WENO; Top right: fifth order WENO; Bottom left: fifth order WENO with a "cheaper" characteristic decomposition; Bottom right: fifth order WENO with artificial compression.

complicated smooth flow feature co-exist. In this example, a moving shock interacting with an entropy wave of small amplitude. On a domain $[0, 5]$, the initial condition is:

$$\rho = 3.85714; \quad u = 2.629369; \quad P = 10.33333;$$

when $x < 0.5$, and

$$\rho = e^{-\epsilon \sin(kx)}; \quad u = 0; \quad P = 1;$$

when $x \geq 0.5$, where ϵ and k are the amplitude and wave number of the entropy wave, respectively. The mean flow is a pure right moving Mach 3 shock. If ϵ is small compared to the shock strength, the shock will march to the right at approximately the non-perturbed shock speed and generate a sound wave which travels along with the flow behind the shock. At the same time, the perturbing entropy wave, after "going through" the shock, is compressed and amplified and travels approximately at the speed of $u + c$ where u and c are the velocity and speed of the sound of the mean flow left to the shock. The amplification factor for the entropy wave can be obtained by linear analysis.

Since the entropy wave here is set to be very weak relative to the shock, any numerical oscillation might pollute the generated waves (e.g. the sound waves) and the amplified entropy waves. In our tests, we take $\epsilon = 0.01$ and $k = 13$. The amplitude of the amplified entropy waves predicted by the linear analysis is 0.08690716 (shown in the following figures as horizontal solid lines).

In Fig. 11.2, we show the result (entropy) when 12 waves have passed through the shock. It is clear that a lower order method (more dissipative) damps the magnitude of the transmitted wave more seriously, especially when the waves are traveling more and more away from the shock. We can see that, while fifth order WENO with 800 points already resolves the passing waves well, and with 1200 points resolves the waves excellently, a second order TVD scheme (which is a good one among second order schemes) with 2000 points still shows excessive dissipation downstream. If we agree that fifth order WENO with 800 points behaves similarly as second order TVD with 2000 points, then there is a saving of a factor of 2.5 in grid points. This factor is *per dimension*, hence for a 3D time dependent problem the saving of the number of space-time grids will be a factor of $2.5^4 \approx 40$, a significant saving even after factoring in the extra cost per grid point for the higher order WENO method.

ENO schemes behave similarly for this problem.

There is a two dimensional version of this problem, when the entropy wave can make an angle with the shock. The simulation results again show an advantage in using a higher order method, in Fig. 11.3. Several curves are clustered in Fig. 11.3 around the exact solution, belonging to various fourth and fifth order ENO or WENO schemes. The circles correspond to a second order TVD scheme, which dissipates the amplitude of the transmitted entropy wave much more rapidly.

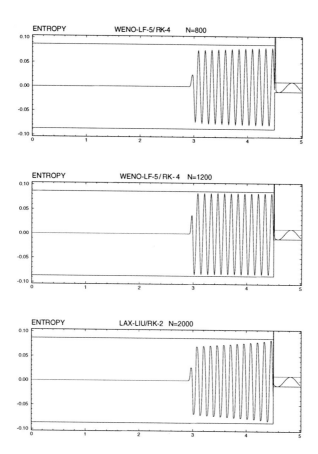

Fig. 11.2. 1D shock entropy wave interaction. Entropy. Top: fifth order WENO with 800 points; middle: fifth order WENO with 1200 points; bottom: second order TVD with 2000 points.

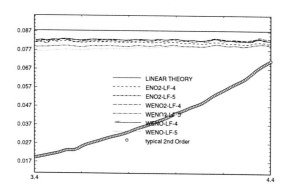

Fig. 11.3. 2D shock entropy wave interaction. Amplitude of amplified entropy waves. 800 points (about 20 points per entropy wave length).

Example 11.3. Steady state calculations. This is important both in gas dynamics and in other fields of applications, such as in semiconductor device simulation. For ENO or TVD schemes, the residue does not settle down to machine zero during the time evolution. It will decay first and then hang at the level of the local truncation errors. Presumably this is due to the fact that the numerical flux is not smooth enough (it is only Lipschitz continuous but not C^1). Although this is not satisfactory, it does not seem to affect the final solution (up to the truncation error level, which is how accurate the solution will be anyway).

WENO schemes are much better in getting the residues to settle down to machine zeroes, due to the smoothness of their fluxes.

In Fig. 11.4 we show the result of a one dimensional nozzle calculation. The residue in this case settles down nicely to machine zeros. Both fourth and fifth order WENO results are shown.

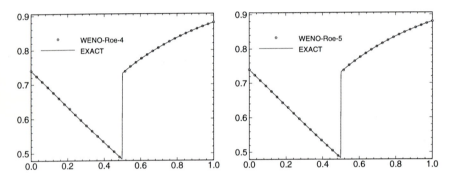

Fig. 11.4. Density. Steady quasi-1D nozzle flow. 34 points. Left: fourth order WENO; Right: fifth order WENO.

Example 11.4. Forward facing step problem. This is a standard test problem for high resolution schemes [104]. However, second order methods usually already work well. High order methods might have some advantage in resolving the slip lines. We refer the readers to [21] for an illustration of such advantages of high order schemes.

The set up of the problem is the following: the wind tunnel is 1 length unit wide and 3 length units long. The step is 0.2 length units high and is located 0.6 length units from the left-hand end of the tunnel. The problem is initialized by a right-going Mach 3 flow. Reflective boundary conditions are applied along the walls of the tunnel and in-flow and out-flow boundary conditions are applied at the entrance (left-hand end) and the exit (right-hand end). For the treatment of the singularity at the corner of the step, we adopt the same technique used in [104], which is based on the assumption of a nearly steady flow in the region near the corner.

In Fig. 11.5 we present the results of fifth order WENO and fourth order ENO with 242×79 grid points.

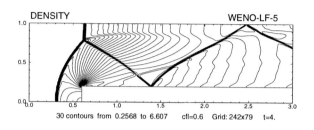

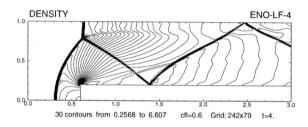

Fig. 11.5. Flow past a forward facing step. Density: 242×79 grid points. Top: fifth order WENO; bottom: fourth order ENO.

Example 11.5. Double Mach reflection. This is again a standard test problem for high resolution schemes [104]. However, second order methods usually again already work well. High order methods have some advantage in resolving the flow below the Mach stem. We again refer the readers to [21] for an illustration of such advantages of high order schemes.

The computational domain for this problem is chosen to be $[0, 4] \times [0, 1]$, although only part of it, $[0, 3] \times [0, 1]$, is shown [104]. The reflecting wall lies at the bottom of the computational domain starting from $x = \frac{1}{6}$. Initially a right-moving Mach 10 shock is positioned at $x = \frac{1}{6}, y = 0$ and makes a $60°$ angle with the x-axis. For the bottom boundary, the exact post-shock condition is imposed for the part from $x = 0$ to $x = \frac{1}{6}$ and a reflective boundary condition is used for the rest. At the top boundary of our computational domain, the flow values are set to describe the exact motion of the Mach 10 shock. See [104] for a detailed description of this problem.

In Fig. 11.6 we present the results of fifth order WENO and fourth order ENO with 480×119 grid points.

In Fig. 11.7 we present the result of fifth order WENO with a more refined mesh, 1920×479 grid points. and a "blow-up" portion of the picture near

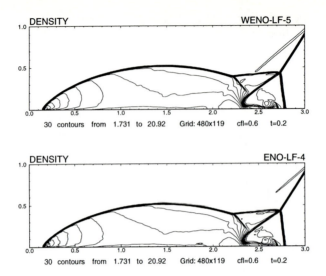

Fig. 11.6. Double Mach reflection. Density: 480×119 grid points. Top: fifth order WENO; bottom: fourth order ENO.

the Mach stem. We can see the complicated structures being captured by the scheme.

Example 11.6. 2D shock vortex interactions. High order methods have some advantages in this case, as it resolves the vortex and the interaction better.

The model problem we use describes the interaction between a stationary shock and a vortex. The computational domain is taken to be $[0, 2] \times [0, 1]$. A stationary Mach 1.1 shock is positioned at $x = 0.5$ and normal to the x-axis. Its left state is $(\rho, u, v, P) = (1, \sqrt{\gamma}, 0, 1)$. A small vortex is superposed to the flow left to the shock and centers at $(x_c, y_c) = (0.25, 0.5)$. We describe the vortex as a perturbation to the velocity (u, v), temperature $(T = \frac{P}{\rho})$ and entropy $(S = \ln \frac{P}{\rho^\gamma})$ of the mean flow and denote it by the tilde values:

$$\tilde{u} = \epsilon \tau e^{\alpha(1-\tau^2)} \sin \theta$$
$$\tilde{v} = -\epsilon \tau e^{\alpha(1-\tau^2)} \cos \theta$$
$$\tilde{T} = -\frac{(\gamma - 1)\epsilon^2 e^{2\alpha(1-\tau^2)}}{4\alpha\gamma}$$
$$\tilde{S} = 0$$

where $\tau = \frac{r}{r_c}$ and $r = \sqrt{(x - x_c)^2 + (y - y_c)^2}$. Here ϵ indicates the strength of the vortex, α controls the decay rate of the vortex and r_c is the critical radius for which the vortex has the maximum strength. In our tests, we choose

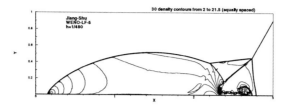

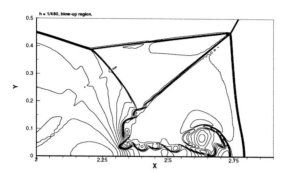

Fig. 11.7. Double Mach reflection. Density: 1920 × 479 grid points. Fifth order WENO. Top: the whole region; bottom: the blow-up region near the Mach stem.

$\epsilon = 0.3, r_c = 0.05$ and $\alpha = 0.204$. The above defined vortex is a steady state solution to the 2D Euler equation.

We use a grid of 251×100 which is uniform in y but refined in x around the shock. The upper and lower boundaries are intentionally set to be reflective. The results (pressure contours) are shown in Fig. 11.8 for a fifth order WENO with the cheap "PS" way of treating characteristic decomposition for the system.

In [30], interaction of a shock with a longitudinal vortex is also investigated by the ENO method.

Example 11.7. 2D bow shock. How does the finite difference version of ENO and WENO handle non-rectangular domain? As we mentioned before, as long as the domain can be *smoothly* transformed to a rectangle, the schemes can be handily applied.

We consider, as an example, the problem of a supersonic flow past a cylinder. In the physical space, a cylinder of unit radius is positioned at the origin on a x–y plane. The computational domain is chosen to be $[0, 1] \times [0, 1]$ on $\xi - \eta$ plane. The mapping between the computational domain and the physical domain is:

$$x = (R_x - (R_x - 1)\xi) \cos(\theta(2\eta - 1)) \tag{11.4}$$
$$y = (R_y - (R_y - 1)\xi) \sin(\theta(2\eta - 1)) \tag{11.5}$$

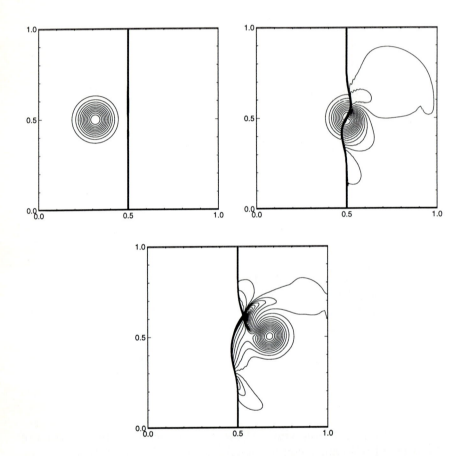

Fig. 11.8. 2D shock vortex interaction. Pressure. Fifth order WENO-LF-5-PS. 30 contours. Top left: t=0.05, Top right: t=0.20, Bottom: t=0.35.

where we take $R_x = 3, R_y = 6$ and $\theta = \frac{5\pi}{12}$. Fifth order WENO and a uniform mesh of 60×80 in the computational domain are used.

The problem is initialized by a Mach 3 shock moving toward the cylinder from the left. Reflective boundary condition is imposed at the surface of the cylinder, i.e. $\xi = 1$, inflow boundary condition is applied at $\xi = 0$ and outflow boundary condition is applied at $\eta = 0, 1$,

We present an illustration of the mesh in the physical space (drawing every other grid line), and the pressure contour, in Fig. 11.9. Similar results are obtained by the ENO schemes but are not shown here.

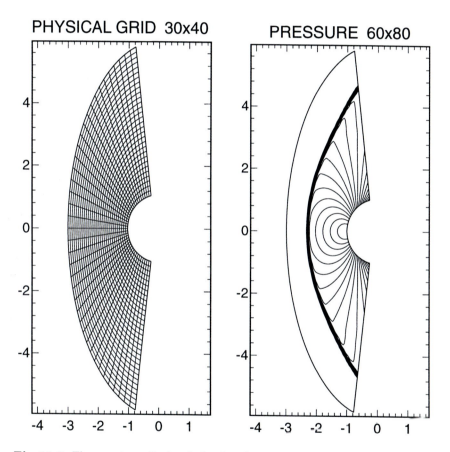

Fig. 11.9. Flow past a cylinder. Left: the physical grid, Right: pressure. WENO-LF-5. 20 contours.

Example 11.8. Vortex evolution. Finally, we use the following problem to illustrate more clearly the power of high order methods. Consider the following idealized problem for the Euler equations in 2D: the mean flow is $\rho = 1$, $P = 1$, and $(u,v) = (1,1)$ (diagonal flow). We add, to this mean flow, an isentropic vortex (perturbations in (u,v) and the temperature $T = \frac{P}{\rho}$, no perturbation in the entropy $S = \frac{P}{\rho^\gamma}$):

$$(\delta u, \delta v) = \frac{\epsilon}{2\pi} e^{0.5(1-r^2)} (-\bar{y}, \bar{x})$$

$$\delta T = -\frac{(\gamma-1)\epsilon^2}{8\gamma\pi^2} e^{1-r^2}, \qquad \delta S = 0,$$

where $(\bar{x}, \bar{y}) = (x-5, y-5)$, $r^2 = \bar{x}^2 + \bar{y}^2$, and the vortex strength $\epsilon = 5$.

Since the mean flow is in the diagonal direction, the vortex movement is not aligned with the mesh direction.

The computational domain is taken as $[0,10] \times [0,10]$, *extended periodically in both directions*. This allows us to perform long time simulation without having to deal with a large domain. As we will see, the advantage of the high order methods are more obvious for long time simulations.

It is clear that the exact solution of the Euler equation with the above initial and boundary conditions is just the passive convection of the vortex with the mean velocity.

A grid of 80^2 points is used. The simulation is performed until $t = 100$ (10 periods in time). As can be seen from Fig. 11.10, fifth order WENO has a much better resolution than a second order TVD scheme, especially for the larger time $t = 100$.

12   Applications to Compressible Gas Dynamics II: Unstructured Mesh for Polytropic Gas

In this section we describe the application of the third and fourth order WENO schemes in Sect. 6.2 and Sect. 7.1, [49,50] to the two dimensional Euler equations of a polytropic gas in general triangulations. The equations are given by (11.1) without the third dimension.

As was mentioned in Sect. 7.4, there are two ways to extend the scalar schemes to systems. One is to do so component by component. This is easy to implement and cost effective, and it seems to work well for the third order scheme. We will use component-wise methods for all numerical examples with the third order WENO scheme in this section. Another extension method is by the characteristic decomposition. We will give a brief description in the following.

Let us take one side of the triangle which has the outward unit normal (n_x, n_y). Let A be some average Jacobian at one quadrature point,

$$A = n_x \frac{\partial f}{\partial u} + n_y \frac{\partial g}{\partial u}. \qquad (12.1)$$

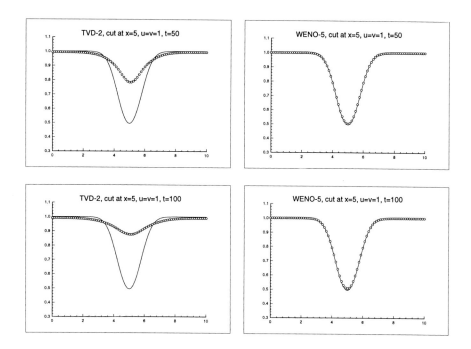

Fig. 11.10. Vortex evolution. Cut at $x = 5$. Solid: exact solution; circles: computed solution. Top: $t = 50$ (after 5 time periods); Bottom: $t = 100$ (after 10 time periods). Left: second order TVD scheme; Right: fifth order WENO scheme.

For Euler systems, the Roe's mean matrix [82] is used. Denote by R the matrix of right eigenvectors and L the matrix of left eigenvectors of A. Then the scalar triangular WENO scheme can be applied to each of the characteristic fields, i.e. to each component of the vector $v = L u$. With the reconstructed point values v, we define our reconstructed point values u by $u = R v$.

Example 12.1. Vortex evolution. This is the same test case as in Example 11.8.

The reconstruction procedure is applied to each component of the solution U. We first compute the solution to $t = 2.0$ for the accuracy test. The meshes are the same as those used in the accuracy tests in Sect. 7.1 for the scalar linear and Burgers equations, suitably scaled for the new spatial domain. The accuracy results for the linear schemes are shown in Table 12.1 for the uniform meshes and Table 12.2 for the non-uniform meshes. The errors presented are those of the cell averages of ρ. The accuracy results for the WENO schemes are shown in Table 12.3 for the uniform meshes and Table 12.4 for the non-uniform meshes.

Table 12.1. Accuracy for 2D Euler equation of smooth vortex evolution, uniform meshes, linear schemes.

h	P^1 (3rd order)				P^2 (4th order)			
	L^1 error	order	L^∞ error	order	L^1 error	order	L^∞ error	order
1	1.65E-02	—	2.60E-01	—	5.26E-03	—	7.89E-02	—
1/2	6.31E-03	1.39	1.21E-01	1.10	7.36E-04	2.84	1.62E-02	2.28
1/4	1.31E-03	2.27	2.53E-02	2.26	5.40E-05	3.77	1.03E-03	3.98
1/8	2.21E-04	2.57	4.66E-03	2.44	2.32E-06	4.54	5.36E-05	4.26
1/16	2.98E-05	2.89	6.44E-04	2.86	1.10E-07	4.40	2.48E-06	4.43
1/32	3.77E-06	2.98	8.23E-05	2.97	6.37E-09	4.11	1.25E-07	4.31

We then fix the mesh at $h = \frac{1}{8}$ (uniform) and compute the long time evolution of the vortex. Fig. 12.1 is the result by the third order scheme at $t = 0$ and after 1, 5 and 10 time periods, and Fig. 12.2 is the result by the fourth order scheme. We show the line cut through the center of the vortex for the density ρ. It is easy to see the difference between the third and fourth order schemes. The fourth order scheme gives almost no dissipation even after 10 periods, while the dissipation is quite noticeable for the long time results of the third order scheme.

Table 12.2. Accuracy for 2D Euler equation of smooth vortex evolution, non-uniform meshes, linear schemes.

h	P^1 (3rd order)				P^2 (4th order)			
	L^1 error	order	L^∞ error	order	L^1 error	order	L^∞ error	order
$h_0/2$	1.81E-02	—	2.98E-01	—	7.00E-03	—	8.16E-02	—
$h_0/4$	7.74E-03	1.28	1.44E-01	1.05	1.18E-03	2.57	1.61E-02	2.34
$h_0/8$	1.67E-03	2.21	2.47E-02	2.54	8.17E-05	3.85	1.31E-03	3.62
$h_0/16$	2.86E-04	2.55	4.79E-03	2.37	4.70E-06	4.12	1.10E-04	3.57
$h_0/32$	3.94E-05	2.86	7.95E-04	2.59	2.68E-07	4.13	7.73E-06	3.83
$h_0/64$	5.07E-06	2.96	1.25E-04	2.67	1.56E-08	4.10	5.99E-07	3.69

Table 12.3. Accuracy for 2D Euler equation of smooth vortex evolution, uniform meshes, WENO schemes.

h	P^1 (3rd order)				P^2 (4th order)			
	L^1 error	order	L^∞ error	order	L^1 error	order	L^∞ error	order
1	1.87E-02	—	2.95E-01	—	1.30E-02	—	2.05E-01	—
1/2	1.01E-02	0.89	2.09E-01	0.50	2.50E-03	2.38	4.45E-02	2.49
1/4	2.78E-03	1.86	6.37E-02	1.71	1.79E-04	3.80	3.29E-03	3.76
1/8	6.47E-04	2.10	3.05E-02	1.06	6.92E-06	4.69	1.96E-04	4.07
1/16	8.74E-05	2.89	8.14E-03	1.91	2.03E-07	5.09	4.95E-06	5.31
1/32	7.10E-06	3.62	5.66E-04	3.85	7.83E-09	4.70	1.96E-07	4.66

Table 12.4. Accuracy for 2D Euler equation of smooth vortex evolution, non-uniform meshes, WENO schemes.

h	P^1 (3rd order)				P^2 (4th order)			
	L^1 error	order	L^∞ error	order	L^1 error	order	L^∞ error	order
$h_0/2$	2.12E-02	—	3.33E-01	—	1.84E-02	—	2.14E-01	—
$h_0/4$	1.28E-02	0.73	2.27E-01	0.55	2.80E-03	2.69	3.43E-02	2.64
$h_0/8$	3.84E-03	1.74	6.85E-02	1.73	2.12E-04	3.72	6.57E-03	2.38
$h_0/16$	8.32E-04	2.21	3.02E-02	1.18	1.09E-05	4.28	5.91E-04	3.48
$h_0/32$	1.26E-04	2.72	5.64E-03	2.42	3.76E-07	4.86	1.97E-05	4.91
$h_0/64$	1.16E-05	3.44	6.19E-04	3.19	1.66E-08	4.50	6.78E-07	4.86

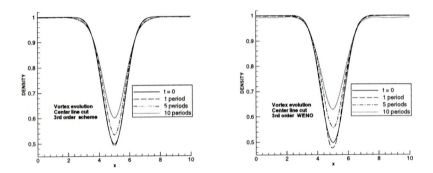

Fig. 12.1. 2D vortex evolution: third order schemes. Left: linear scheme; Right: WENO scheme.

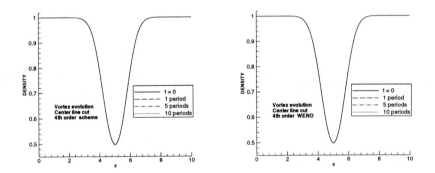

Fig. 12.2. 2D vortex evolution: fourth order schemes. Left: linear scheme; Right: WENO scheme.

Example 12.2. Shock tube problem. This is the same test case as in Example 11.1, except that we compute the problem in two dimensions. We consider the solution of the Euler equations in a domain of $[-1, 1] \times [0, 0.2]$ with a triangulation of 101 vertices in the x-direction and 11 vertices in the y-direction. The velocity in the y-direction is zero, and periodic boundary condition is used in the y-direction. A portion of the mesh is shown in Fig. 12.3. The pictures shown below are obtained by extracting the data along the central cut line for 101 equally spaced points.

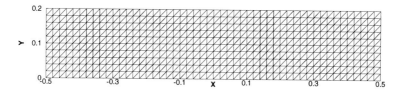

Fig. 12.3. A portion of the mesh for the Riemann problems.

The first test case is Sod's problem (11.2). Density at $t = 0.40$ is shown in Fig. 12.4, left.

The second test case is the Riemann problem proposed by Lax (11.3). Density at $t = 0.26$ is shown in Fig. 12.4, right.

We can observe a better resolution of the fourth order scheme over the third order one, and also a less oscillatory result from the characteristic version of the fourth order scheme over the component version.

Example 12.3. Forward facing step problem. This is the same test case as in Example 11.4. However, for the corner singularity, instead of adopting the same technique used in [104] and in Example 11.4, which is based on the assumption of a nearly steady flow in the region near the corner, we do not modify our method near the corner, instead we adopt the same technique as the one used in [21], namely refining the mesh near the corner and using the same scheme in the whole domain.

We use the third order scheme for this problem. Four meshes have been used, see Fig. 12.5. For the first mesh, the triangle size away from the corner is roughly equal to a rectangular element case of $\Delta x = \Delta y = \frac{1}{40}$, while it is onc-quarter of that near the corner. For the second mesh, the triangle size away from the corner is the same as in the first mesh, but it is one-eighth of that near the corner. The third mesh has a triangle size of $\Delta x = \Delta y = \frac{1}{80}$ away from the corner, and it is one-quarter of that near the corner. The last mesh has a triangle size of $\Delta x = \Delta y = \frac{1}{160}$ away from the corner, and it is one-half of that near the corner. Fig. 12.6 is the contour picture for the density at time $t = 4.0$. It is clear that with more triangles near the corner the artifacts from the singularity decrease significantly.

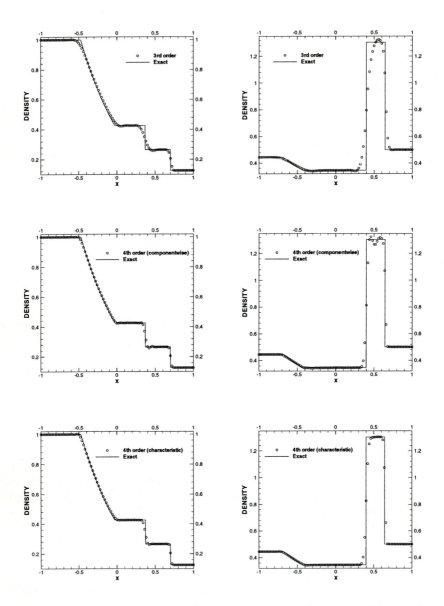

Fig. 12.4. Riemann problems of Euler equations. Density. Left: Sod's problem; Right: Lax's problem. Top: third order componentwise WENO; Middle: fourth order componentwise WENO; Bottom: fourth order characteristicwise WENO.

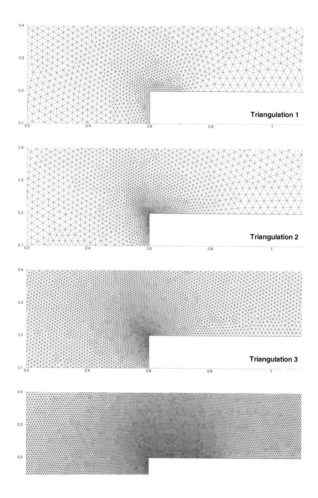

Fig. 12.5. Triangulations for the forward step problem: part near the corner.

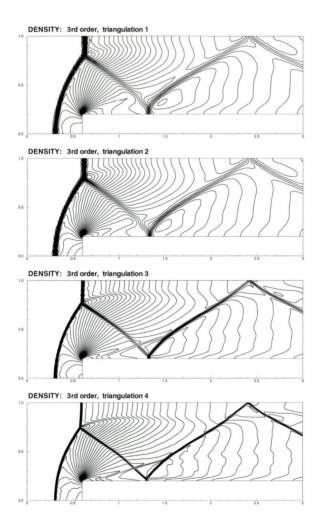

Fig. 12.6. Forward step problem: 30 contours from 0.32 to 6.15.

Example 12.4. Double Mach reflection. This is the same test case as in Example 11.5.

We test both the third and the fourth order schemes. Four triangle sizes are used, they are roughly equal to rectangular element cases of $\Delta x = \Delta y = \frac{1}{50}$, $\Delta x = \Delta y = \frac{1}{100}$, $\Delta x = \Delta y = \frac{1}{200}$, and $\Delta x = \Delta y = \frac{1}{400}$ respectively. For the third order scheme, we use both uniform triangular mesh (equilateral triangles) and locally refined triangular mesh (the refined region has the above triangle sizes, Fig. 12.7 shows the region $[0, 2] \times [0, 1]$ of such a mesh of $\Delta x = \Delta y = \frac{1}{50}$ locally). For the fourth order, we use uniform triangular mesh only. For the cases of $\Delta x = \Delta y = \frac{1}{200}$ and $\Delta x = \Delta y = \frac{1}{400}$, we present both the picture of whole region ($[0, 3] \times [0, 1]$) and a blow-up region around the double Mach stems. All pictures are the density contours with 30 equally spaced contour lines from 1.5 to 21.5. We can clearly see that the fourth order scheme captures the complicated flow structure under the triple Mach stem much better than the third order scheme. We refer to [21] for similar results obtained with discontinuous Galerkin methods.

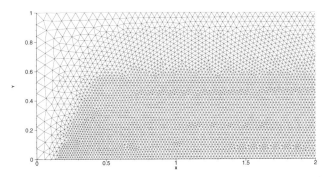

Fig. 12.7. Triangulation for the double Mach reflection.

13   Applications to Compressible Gas Dynamics III: Structured Mesh for Real Gas

In this section we describe the application of the fifth order WENO scheme on a structured mesh in Sect. 4.4 and Sect. 7.4 to solve the Euler equations of a real gas [74].

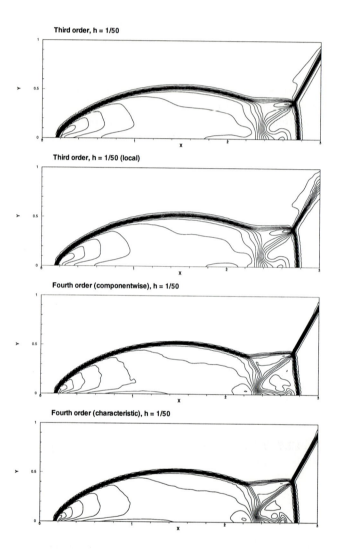

Fig. 12.8. Double Mach reflection: $h = \frac{1}{50}$, $t = 0.2$.

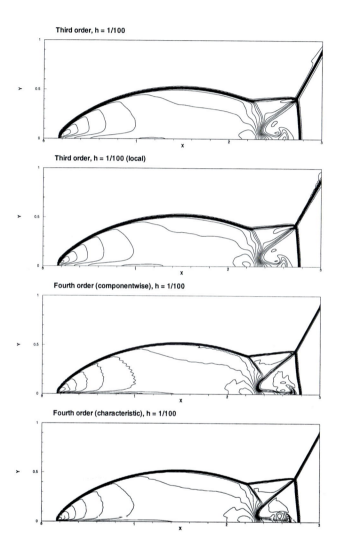

Fig. 12.9. Double Mach reflection: $h = \frac{1}{100}$, $t = 0.2$.

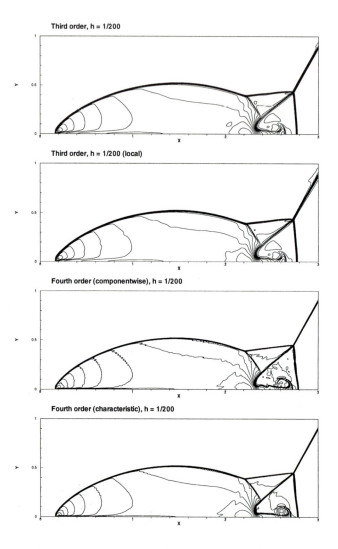

Fig. 12.10. Double Mach reflection: $h = \frac{1}{200}$, $t = 0.2$.

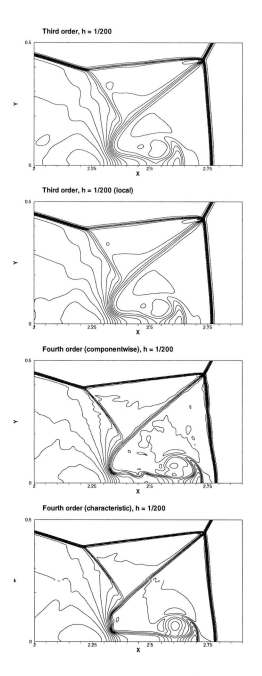

Fig. 12.11. Double Mach reflection: $h = \frac{1}{200}$, $t = 0.2$ (blow-up).

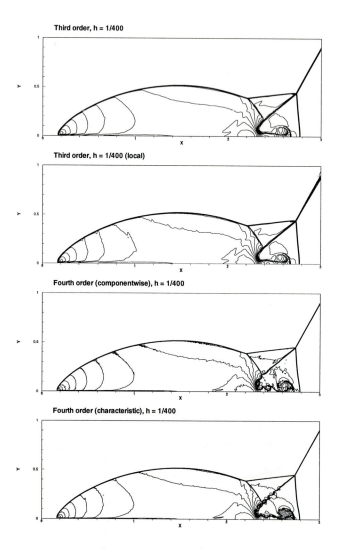

Fig. 12.12. Double Mach reflection: $h = \frac{1}{400}$, $t = 0.2$.

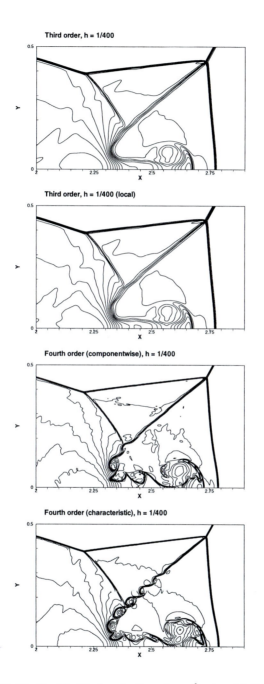

Fig. 12.13. Double Mach reflection: $h = \frac{1}{400}$, $t = 0.2$ (blow-up).

We consider the Euler equations for a real compressible inviscid fluid,

$$\partial_t \rho + \mathrm{div}\,(\rho u) = 0, \quad t \geq 0, x \in \mathbb{R}^d,$$
$$\partial_t \rho u + \mathrm{div}\,(\rho u \otimes u + p) = 0,$$
$$\partial_t E + \mathrm{div}\,((E + p)u) = 0, \tag{13.1}$$
$$E = \frac{1}{2}\rho|u|^2 + \rho\varepsilon,$$

where the quantities ρ, u, p, E and ε represent the density, velocity, pressure, total energy and specific internal energy, respectively. In addition, there is an equation of state (EOS) of the form $p = p(\rho, \varepsilon)$ associated with a strictly convex entropy $\rho s(\rho, \varepsilon)$ which satisfies the following entropy inequalities

$$\partial_t \rho s + \mathrm{div}\,(\rho s u) \leq 0. \tag{13.2}$$

The pressure law is furthermore assumed to satisfy

$$p_{,\varepsilon}(\rho, \varepsilon) > 0, \tag{13.3}$$
$$p(\rho, 0) = 0 \quad \text{and} \quad p(\rho, \infty) = \infty.$$

In the literature research has been done in order to extend classical schemes designed for perfect gas to real gases. Collela and Glaz [22] extended the numerical procedure for obtaining the exact Riemann solution to a real-gas case, Grossman and Walters [38], Liou, van Leer and Shuen [68] extended the method of flux-vector splitting and flux-difference splitting, Montagné, Yee and Vinokur [73] developed second-order explicit shock-capturing schemes for real gas, Glaister [34] presented an extension of approximate linearized Riemann solver with different averaged matrices, while Loh and Liou [71] used the generalization of their Lagrangian approach (originally proposed for perfect gas) to obtain the real gas Riemann solution.

Most of the previous proposed methods would require a computation of the pressure law and its derivatives, or a Riemann solver. This is not only costly but also problematic when there is no analytical expressions of the pressure law (for example if we have only table values).

Recently Coquel and Perthame [23] have introduced an energy relaxation theory for Euler equations of real gas. The main idea is to introduce a relaxation of the nonlinear pressure law by considering an energy decomposition under the form $\varepsilon = \varepsilon_1 + \varepsilon_2$. The internal energy ε_1 is associated with a simpler pressure law p_1 (which is taken as the γ-law in this section), while ε_2 stands for the nonlinear perturbation and is simply convected by the flow. These two energies are also subject to a relaxation process and in the limit of an infinite relaxation rate, one recovers the initial pressure law p.

From this general framework, Coquel and Perthame have also deduced the extension to general pressure laws of classical schemes for polytropic gases, which only uses a single call to the pressure law per grid point and time step. No derivatives of the pressure law or any Riemann solvers need to be computed. Another advantage of their approach is that its implementation does not depend on the particular expression of the equation of states. For the first order Godunov scheme, they have shown that this extension satisfies stability, entropy and accuracy conditions. Numerical examples have been provided using first order schemes by A. In [51].

The aim of this section is to study the implementation of this relaxation method with high order WENO schemes [55] for real gases. One and two dimensional numerical examples will be given.

In Sect. 13.1 we provide the general framework of the energy relaxation theory of [23]. We then give the details of the construction of the relaxed WENO schemes for general gases. In Sect. 13.2 numerical examples are given. We start with a description of the different equations of states used in this section, followed by one dimensional shock tube test problems. Two dimensional test cases of a smooth vortex, to test the accuracy of the schemes, and of the double Mach reflection problem, are then presented.

13.1 Implementation of the Energy Relaxation Method with WENO

The principle of the energy relaxation theory developed by Coquel and Perthame [23] is to find a pressure law $p_1(\rho^\lambda, \varepsilon_1^\lambda)$ (simpler than p, typically a polytropic law) and an internal energy $\phi(\rho^\lambda, \varepsilon_1^\lambda)$ so that the system (13.1) and the entropy inequality (13.2) can be recovered, in the limit of an infinite relaxation rate λ (called the *equilibrium limit*), from the following system (called the *relaxation system*):

$$
\begin{aligned}
&\partial_t \rho^\lambda + \mathrm{div}\left(\rho^\lambda u^\lambda\right) = 0, \quad t \geq 0, x \in \mathbb{R}^d, \\
&\partial_t \rho^\lambda u^\lambda + \mathrm{div}\left(\rho^\lambda u^\lambda \otimes u^\lambda + p_1^\lambda\right) = 0, \\
&\partial_t E_1^\lambda + \mathrm{div}\left((E_1^\lambda + p_1^\lambda)u^\lambda\right) = \lambda \rho^\lambda \left(\varepsilon_2^\lambda - \phi(\rho^\lambda, \varepsilon_1^\lambda)\right), \\
&\partial_t \rho^\lambda \varepsilon_2^\lambda + \mathrm{div}(\rho^\lambda u^\lambda \varepsilon_2^\lambda) = -\lambda \rho^\lambda \left(\varepsilon_2^\lambda - \phi(\rho^\lambda, \varepsilon_1^\lambda)\right), \\
&E_1^\lambda = \frac{1}{2}\rho^\lambda |u^\lambda|^2 + \rho^\lambda \varepsilon_1^\lambda,
\end{aligned}
\tag{13.4}
$$

where $p_1(\rho^\lambda, \varepsilon_1^\lambda) = (\gamma_1 - 1)\rho^\lambda \varepsilon_1^\lambda$ with γ_1 a given constant greater than 1. One can prove [23] that the relaxation system (13.4) can be supplemented by entropy inequalities under the form

$$
\partial_t \rho^\lambda \Sigma + \mathrm{div}(\rho^\lambda \Sigma u^\lambda) \leq \mathrm{RED}^\lambda := -\lambda \rho^\lambda (\Sigma_{,s_1} s_{1,\varepsilon_1} - \Sigma_{,\varepsilon_2})(\varepsilon_2 - \phi(\rho^\lambda, \varepsilon_1^\lambda))
$$

where $s_1(\rho, \varepsilon_1) = \rho^{\gamma_1 - 1}/\varepsilon_1$ and the specific entropy Σ denotes an arbitrary function in $C^1(\mathbb{R}_+^2)$ such that $\rho \Sigma$ is convex in $(\rho, \rho\varepsilon_1, \rho\varepsilon_2)$ and that can be written under the form $\Sigma = \Sigma(s_1(\rho, \varepsilon_1), \varepsilon_2)$. RED^λ represents the Rate of Entropy Dissipation.

Formally, the original Euler system (13.1) will be recovered at $\lambda \to +\infty$ with

$$
\varepsilon = \varepsilon_1 + \varepsilon_2 = \varepsilon_1 + \phi(\rho, \varepsilon_1),
\tag{13.5}
$$

provided that we have the following condition (called the *consistency condition*)

$$p\left(\rho, \varepsilon_1 + \phi(\rho, \varepsilon_1)\right) = p_1(\rho, \varepsilon_1) = (\gamma_1 - 1)\rho\varepsilon_1. \tag{13.6}$$

This last condition can be fulfilled for any given choice of $\gamma_1 > 1$.

But in addition to the conservative system (13.1), one also wants to recover at the limit the entropy inequality (13.2). The following result, due to Coquel and Perthame [23], gives this last condition under a characterization of the admissible γ_1.

Theorem 13.1. Assuming that γ_1 satisfies

$$\gamma_1 > \sup_{\rho,\varepsilon} \Gamma(\rho, \varepsilon), \ \Gamma(\rho, \varepsilon) = 1 + \frac{p_{,\varepsilon}}{\rho}, \\ \gamma_1 > \sup_{\rho,\varepsilon} \gamma(\rho, \varepsilon), \ \gamma(\rho, \varepsilon) = \frac{\rho}{p}p_{,\varepsilon} + \frac{p_{,\varepsilon}}{\rho}, \tag{13.7}$$

provided that γ_1 is finite, we then have
(i) there exists a (unique) specific entropy $\Sigma(s_1, \varepsilon_2)$ such that at equilibrium $(\varepsilon = \varepsilon_1 + \phi(\rho, \varepsilon_1))$

$$s(\rho, \varepsilon) = \Sigma(s_1(\rho, \varepsilon_1), \phi(\rho, \varepsilon_1)),$$

(ii) this entropy is uniformly compatible with the relaxation procedure, i.e.:

$$\text{RED}^\lambda \leq 0, \ \text{for all } \lambda > 0.$$

$\square$

The procedure to solve the Euler system (13.1) within the framework of the energy relaxation theory is the following. Given the numerical equilibrium solution at the time level t^n

$$\rho(x, t^n), u(x, t^n), \varepsilon(x, t^n), \tag{13.8}$$

this approximation is advanced to the next time level $t^{n+1} = t^n + \Delta t$ in two steps.

- First step: relaxation. The two internal energies $\varepsilon_1(x, t^n)$ and $\varepsilon_2(x, t^n)$ are obtained by (13.5) and the consistency condition (13.6):

$$\varepsilon_1(x, t^n) = \frac{p\left(\rho(x, t^n), \varepsilon(x, t^n)\right)}{(\gamma_1 - 1)\rho(x, t^n)}, \\ \varepsilon_2(x, t^n) = \varepsilon(x, t^n) - \varepsilon_1(x, t^n). \tag{13.9}$$

Notice that this step involves just one call to the pressure law per grid point and does not involve any derivatives of the pressure law or any iterations.

– Second step: evolution in time. For $t^n \leq t \leq t^{n+1}$, we solve the Cauchy problem for the relaxation system (13.4), with zero on the right side:

$$
\begin{aligned}
&\partial_t \rho^\lambda + \operatorname{div}\left(\rho^\lambda u^\lambda\right) = 0, \quad t \geq 0, x \in \mathbb{R}^d, \\
&\partial_t \rho^\lambda u^\lambda + \operatorname{div}\left(\rho^\lambda u^\lambda \otimes u^\lambda + p_1^\lambda\right) = 0, \\
&\partial_t E_1^\lambda + \operatorname{div}\left((E_1^\lambda + p_1^\lambda)u^\lambda\right) = 0, \\
&\partial_t \rho^\lambda \varepsilon_2^\lambda + \operatorname{div}(\rho^\lambda u^\lambda \varepsilon_2^\lambda) = 0, \\
&E_1^\lambda = \frac{1}{2}\rho^\lambda |u^\lambda|^2 + \rho^\lambda \varepsilon_1^\lambda,
\end{aligned}
\tag{13.10}
$$

and the initial data

$$
\rho(x, t^n), u(x, t^n), \varepsilon_1(x, t^n), \varepsilon_2(x, t^n),
\tag{13.11}
$$

and we obtain at time t^{n+1-}

$$
\rho(x, t^{n+1-}), \ u(x, t^{n+1-}), \ \varepsilon_1(x, t^{n+1-}), \ \varepsilon_2(x, t^{n+1-}).
\tag{13.12}
$$

At last, we compute the equilibrium solution at time t^{n+1} by

$$
\begin{aligned}
\rho(x, t^{n+1}) &= \rho(x, t^{n+1-}), \\
u(x, t^{n+1}) &= u(x, t^{n+1-}), \\
\varepsilon(x, t^{n+1}) &= \varepsilon_1(x, t^{n+1-}) + \varepsilon_2(x, t^{n+1-}).
\end{aligned}
\tag{13.13}
$$

Remark 13.1. The first step is clearly a relaxation phase, as it is equivalent to the solution of the following ODE problem for $t \geq t^n$

$$
\left|
\begin{aligned}
&d_t \rho^\lambda = 0, \\
&d_t \rho^\lambda u^\lambda = 0, \\
&d_t E_1^\lambda = \lambda \rho^\lambda \left(\varepsilon_2^\lambda - \phi(\rho^\lambda, \varepsilon_1^\lambda)\right), \\
&d_t \rho^\lambda \varepsilon_2^\lambda = -\lambda \rho^\lambda \left(\varepsilon_2^\lambda - \phi(\rho^\lambda, \varepsilon_1^\lambda)\right),
\end{aligned}
\right.
\tag{13.14}
$$

with initial data at time level t^n

$$
\rho(x, t^{n-}), u(x, t^{n-}), \varepsilon_1(x, t^{n-}), \varepsilon_2(x, t^{n-}).
\tag{13.15}
$$

and to let $\lambda \to +\infty$.   □

We now describe the numerical method we will use for the step of evolution in time. Although our numerical results concern both one and two dimensional problems, for simplicity of presentations we shall restrict our description to one space dimension. As we are using the finite difference version of WENO schemes in [55], extensions to two and more spatial dimensions are simply done dimension by dimension. Essentially, the two dimensional code is the one dimensional code with an outside "do loop".

We have to solve for $t^n \leq t < t^{n+1}$ the following system of four equations

$$
\begin{aligned}
&\partial_t U + \partial_x F(U) = 0, \\
&+ \text{ initial conditions given by (13.11),}
\end{aligned}
\tag{13.16}
$$

where
$$U = (\rho, \rho u, E_1, \rho \varepsilon_2)^T,$$
$$F(U) = (\rho u, \rho u^2 + p_1, (E_1 + p_1)u, \rho u \varepsilon_2)^T. \tag{13.17}$$

In order to solve the ordinary differential equation

$$\frac{d}{dt}U = L(U), \tag{13.18}$$

where $L(U)$ is a discretization of the spatial operator, we use a third-order TVD Runge-Kutta scheme (9.10), [89].

Remark 13.2. We have two possibilities for the placement of the relaxation step: each Runge-Kutta inner stage or each time step. With Example 13.3 below we show that the two approaches give nearly identical results in accuracy. Of course the second approach is less costly. We thus perform all our calculations using the second approach. □

We now discretize the space into uniform intervals of size Δx and denote $x_j = j \Delta x$. Various quantities at x_j will be identified by the subscript j.

We use the WENO procedure described in Sect. 4.1 to obtain the spatial operator $L_j(U)$ which approximates $-\partial_x F(U)$ at x_j. We have tested several possibilities for the definition of $L(U)$ based on WENO schemes. The first one is to use a WENO Lax-Friedrichs scheme with a full characteristic decomposition. For this purpose we need to compute a Roe matrix for the system (13.16) and its eigenvalues and eigenvectors. The details of this derivation can be found in [74].

The other possibility is to compute the first three components of the numerical flux $\hat{F}_{j+\frac{1}{2}}^1, \hat{F}_{j+\frac{1}{2}}^2, \hat{F}_{j+\frac{1}{2}}^3$ by using a WENO Lax-Friedrichs scheme with a decomposition on the Euler system characteristics and to obtain the last numerical flux $\hat{F}_{j+\frac{1}{2}}^4$ with a scalar WENO Lax-Friedrichs scheme. This is possible because the first three equations of system (13.16) are independent from the last one.

Remark 13.3. We have also tried to compute the last numerical flux by using a first order scheme specially designed in order to preserve the maximum principle for ε_2 [63]. But with this approach, we lose the accuracy of the high-order WENO scheme also for the other variables. □

Remark 13.4. In order to make comparisons in the numerical results we have also implemented a WENO Lax-Friedrichs scheme with a full characteristic decomposition for a two molecular vibrating gas (see next subsection for a description of the related EOS). For this purpose we need a definition of the corresponding Roe average matrix, see [74]. For the numerical comparisons for the other real gases we use a component-wise WENO Lax-Friedrichs scheme which requires only the computation of the sound velocity

$$c = \sqrt{p_{,\rho} + p \frac{p_{,\varepsilon}}{\rho^2}}. \tag{13.19}$$

□

13.2   Numerical Results

We present here several equations of states which we will use in the computation. We find the second one in the paper of In [51], while the third one comes from Glaister [33]).

- *Polytropic ideal gas.* The equation of states for a polytropic ideal gas (also called perfect gas) is the following

$$p(\rho, \varepsilon) = (\gamma - 1)\rho\varepsilon. \tag{13.20}$$

Then we have

$$p_{,\rho} = (\gamma - 1)\varepsilon, \quad p_{,\varepsilon} = (\gamma - 1)\rho. \tag{13.21}$$

Air under normal conditions (p and T moderate enough) can be considered as a perfect gas with $\gamma = 7/5 = 1.4$ (approximately a mixture of two diatomic molecular species: 20% of O_2, 80% of N_2).

- *Two molecular vibrating gas.* When the temperature increases the vibrational motion of oxygen and nitrogen molecules in air becomes important, and specific heats vary with temperatures. So that one must consider the following thermally perfect, calorically imperfect model for two molecular vibrating gas

$$p(\rho, \varepsilon) = r\rho T(\varepsilon) \tag{13.22}$$

where the temperature T is given by the implicit expression

$$\rho\varepsilon = c_v^{tr}T + \rho\frac{\alpha\Theta_{vib}}{\exp\left(\frac{\Theta_{vib}}{T}\right) - 1}, \tag{13.23}$$

with $r = 287.086 \, J \cdot kg^{-1} \cdot K^{-1}$, $C_v^{tr} = r/(\gamma_{tr} - 1)$, $\gamma_{tr} = 1.4$, $\Theta_{vib} = 10^3 \, K$, $\alpha = r$. Then we have

$$p_{,\rho} = rT(\varepsilon), \quad p_{,\varepsilon} = \frac{r\rho}{\varepsilon'(T(\varepsilon))}. \tag{13.24}$$

- *Osborne model* R. K. Osborne from the Los Alamos Scientific Laboratory has developed a quite general equation of states in the following form [81]

$$p(\rho, \varepsilon) = \frac{1}{E + \phi_0} \left(\zeta(a_1 + a_2\zeta) + E(b_0 + \zeta(b_1 + b_2\zeta) + E(c_0 + c_1\zeta))\right) \tag{13.25}$$

where $E = \rho_0\varepsilon$ and $\zeta = \frac{\rho}{\rho_0} - 1$ and the constants ρ_0, a_1, a_2, b_0, b_1, b_2, c_0, c_1, ϕ_0 depend on the material in question. The typical values for water are $\rho_0 = 10^{-2}$, $a_1 = 3.84 \times 10^{-4}$, $a_2 = 1.756 \times 10^{-3}$, $b_0 = 1.312 \times 10^{-2}$, $b_1 = 6.265 \times 10^{-2}$, $b_2 = 0.2133$, $c_0 = 0.5132$, $c_1 = 0.6761$ and $\phi_0 = 2. \times 10^{-2}$. Then we have

$$p_{,\rho} = \frac{1}{\rho_0(E + \phi_0)}\left((a_1 + 2a_2\zeta) + E(b_1 + 2b_2\zeta + Ec_1)\right),$$

$$p_{,\varepsilon} = -\frac{\rho_0}{E + \phi_0}p + \frac{\rho_0}{E + \phi_0}(b_0 + \zeta(b_1 + b_2\zeta) + 2E(c_0 + c_1\zeta)).$$

Example 13.1. Shock tube problem. This is the one dimensional Riemann problem test case with perfect gas, already used in Example 11.1. Of course for this perfect gas situation there is no need to use the relaxation model in practice. The purpose of this test problem is to test the behavior of different relaxation models (different γ_1's) and different ways of treating the relaxed system (fully characteristic and partially characteristic for the first three equations only).

For this example, a uniform grid of 100 points are used and every 2 points are drawn in the figures.

We first give, in Table 13.1, a CPU time comparison among the traditional WENO characteristic scheme for the perfect gas, and the WENO scheme applied to the relaxation system, both with a fully characteristic decomposition and with a partially characteristic decomposition for the first three equations only. The calculation is done on a SUN Ultra1 workstation. We can see that while a fully characteristic decomposition is significantly more costly, the partially characteristic decomposition is only slightly more costly than the WENO scheme applied to the original perfect gas Euler equations.

Table 13.1. CPU time (in seconds) of different schemes for the Sod and Lax shock tube problems for a perfect gas.

Case	WENO with characteristic	Relaxed WENO with full characteristic	Relaxed WENO with partial characteristic
Sod Shock	2.28	3.49	2.91
Lax Shock	3.32	4.93	4.08

In Figures 13.1 and 13.3, we present the comparison for the Sod's and Lax's shock tube problems, of the fifth order WENO schemes, applied directly to the perfect gas Euler equations using a characteristic decomposition, and applied to the relaxation model with $\gamma_1 = 3$ using only partial characteristic decomposition of the first 3 equations. We can see that the results are very close, except for the slight over- and under-shoots in entropy for the relaxation model calculation. This indicates the feasibility of using the relaxation model.

In Figures 13.2 and 13.4, we present the comparison for the Sod's and Lax's shock tube problems, of the fifth order WENO schemes. The top left figure compares the full characteristic decomposition for the relaxation model, with a partial characteristic decomposition for the first 3 equations only, for $\gamma_1 = 3$. We can see that the results are quite close, again indicating the feasibility of using the less costly partial characteristic decomposition for the relaxation model. The top right figure compares the effect of different γ_1's in the relaxation model. Apparently bigger γ_1 corresponds to larger numerical

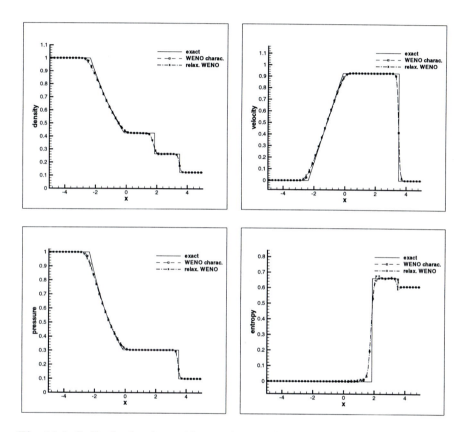

Fig. 13.1. Sod's shock tube problem with WENO-LF-5 characteristic and relaxed WENO-LF-5 partial characteristic with $\gamma_1 = 3.0$. Top left: density; Top right: velocity; Bottom left: pressure; Bottom right: entropy.

dissipation. This indicates that one should always choose the smallest possible γ_1 subject to stability considerations. The bottom figure compares the relaxation WENO results for $\gamma_1 = 3$ and a partial characteristic decomposition, with a component-wise WENO scheme applied directly on the original perfect gas Euler equations. Although neither uses the correct characteristic information, apparently the relaxation model results are better than the component-wise results, especially for the Lax's problem in Figure 13.4.

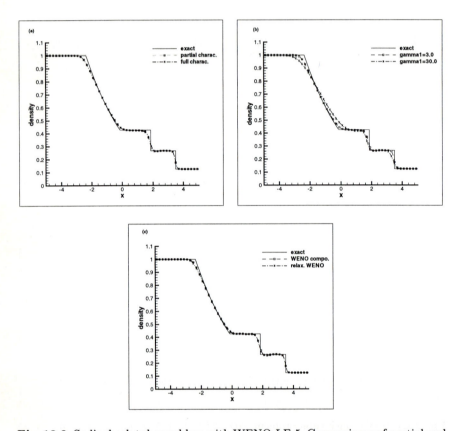

Fig. 13.2. Sod's shock tube problem with WENO-LF-5. Comparisons of partial and full characteristic decompositions for the relaxation model with $\gamma_1 = 3$ (top, left); $\gamma_1 = 3$ and $\gamma_1 = 30$ for the relaxation model with partial characteristic decomposition (top, right); and the relaxation model with partial characteristic decomposition with $\gamma_1 = 3$ versus the component-wise WENO applied to the original perfect gas Euler equations (bottom).

Example 13.2. Shock tube problem for real gas. In this example we compute the solutions to the Riemann shock tube problem, for the two molec-

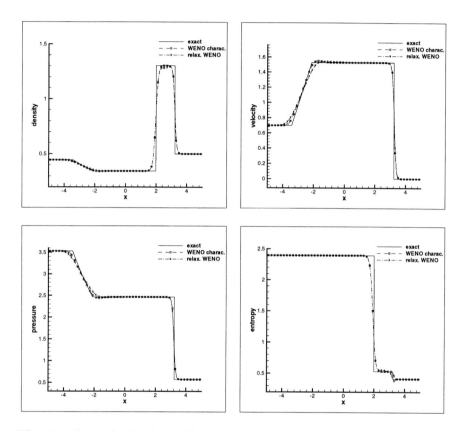

Fig. 13.3. Lax's shock tube problem with WENO-LF-5 characteristic and relaxed WENO-LF-5 partial characteristic with $\gamma_1 = 3.0$. Top left: density; Top right: velocity; Bottom left: pressure; Bottom right: entropy.

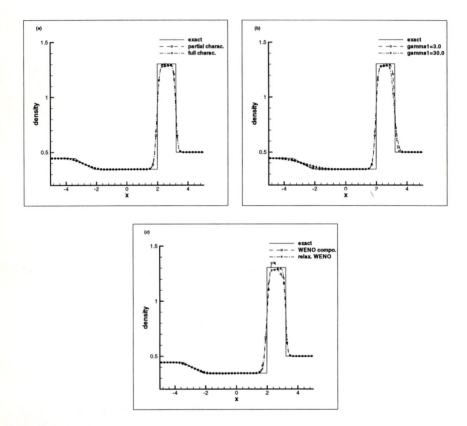

Fig. 13.4. Lax's shock tube problem with WENO-LF-5. Comparisons of partial and full characteristic decompositions for the relaxation model with $\gamma_1 = 3$ (top, left); $\gamma_1 = 3$ and $\gamma_1 = 30$ for the relaxation model with partial characteristic decomposition (top, right); and the relaxation model with partial characteristic decomposition with $\gamma_1 = 3$ versus the component-wise WENO applied to the original perfect gas Euler equations (bottom).

ular vibrating gas (13.22)-(13.24) and the Osborne model (13.25), with the following initial conditions in Table 13.2.

Table 13.2. Initial conditions for the test cases for real gases.

Case	State	ρ	u	ε
A	Left	0.066	0.0	7.22e6
	Right	0.030	0.0	1.44e6
B	Left	1.40	0.0	2.22e6
	Right	0.14	0.0	2.24e6
C	Left	1.2900	0.0	1.95e6
	Right	0.0129	0.0	2.75e6
D	Left	1.00	0.0	2.00e6
	Right	0.01	0.0	2.50e5
E	Left	0.01	2200.0	1.44e5
	Right	0.14	0.0	4.00e5

For this example, a uniform grid of 200 points are used and every 4 points are drawn in the figures. Also, the "exact solution" in the figures are obtained with the best scheme using 2000 points.

We first give a CPU time comparison between the full characteristic decomposition for the original model and the partial characteristic decomposition using only the first three equations of the relaxation model, for the two molecular vibrating gas model, in Table 13.3. We can see that the partial characteristic decomposition for the relaxed model is usually more than twice less costly than the full characteristic version for the original system. Although the relaxed model has one more equation, it does not require the computation of the complicated derivatives of the EOS.

In Figure 13.5 we show the comparison of the full characteristic decomposition for the original model and the partial characteristic decomposition using only the first three equations of the relaxation model, for the two molecular vibrating gas model, with case A initial condition. The results are almost identical, indicating that the relaxation model with a partial characteristic decomposition works well with a much reduced cost.

In Figure 13.6 we show the comparison of the component WENO scheme on the original system, and the partially characteristic WENO scheme on the relaxed system with $\gamma_1 = 2.0$, for the Osborne gas model with case A initial condition. We can see that the result of the relaxed model is much better, especially for the density. This indicates that the relaxation model is a good one for the computation of real gases.

Table 13.3. CPU time (in seconds) depending on full or partial characteristic decomposition with a two vibrating molecular gas.

Case	WENO with characteristic	Relaxed WENO with partial characteristic
A	12.68	5.21
B	4.8	2.63
C	12.53	4.87
D	15.0	5.35
E	15.0	7.84

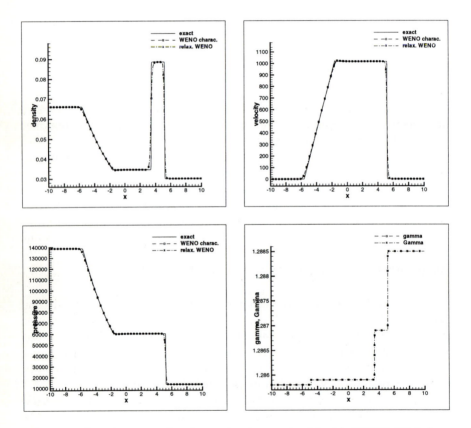

Fig. 13.5. Case A + two vibrating molecular gas model with WENO-LF-5 characteristic and relaxed WENO-LF-5 partial characteristic with $\gamma_1 = 1.5$. Top left: density; Top right: velocity; Bottom left: pressure; Bottom right: γ and Γ.

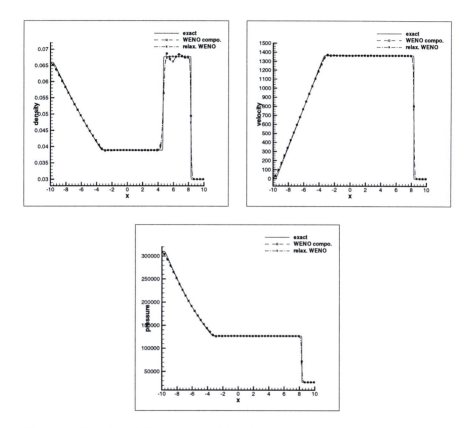

Fig. 13.6. Case A + Osborne gas model with component-wise WENO-LF-5 for the original system and relaxed WENO-LF-5 partial characteristic with $\gamma_1 = 2.0$. Top left: density; Top right: velocity; Bottom: pressure.

In Figure 13.7 we show the comparison of taking $\gamma_1 = 10$, which satisfies the stability condition (13.7), and $\gamma_1 = 2$, which satisfies only the second inequality in the stability condition (13.7), for the partial characteristic decomposition using only the first three equations of the relaxation model, and the Osborne gas model with case A initial condition. We can see that the $\gamma_1 = 2$ results are stable and less dissipative, indicating that in practice one does not always have to choose γ_1 satisfying both inequalities in condition (13.7).

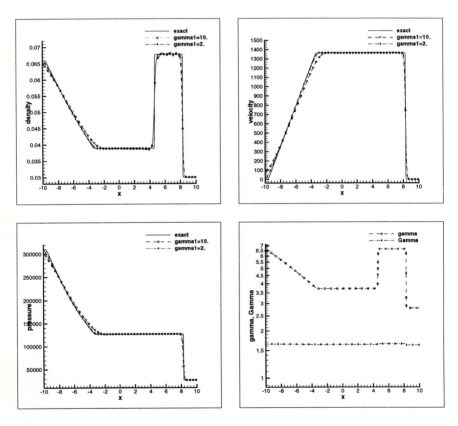

Fig. 13.7. Case A + Osborne gas model with the relaxed WENO-LF-5 partial characteristic with $\gamma_1 = 10.0$ and $\gamma_1 = 2.0$.

We have also tested the same problems for the other initial condition cases B, C, D and E. The results are mostly similar qualitatively as in case A. To save space we will not present the results here.

Example 13.3. Vortex evolution. This is the same case as in Example 11.8, the purpose here being to verify the accuracy of the relaxation approach, especially the placement of the relaxation steps during time stepping. The gas is ideal but we still use the relaxation model.

In Table 13.4 we show the accuracy result at $t = 10$ (one time period). We can see that WENO for the relaxed model with $\gamma_1 = 3$ gives a somewhat larger error than WENO applied directly to the original system, but the order of accuracy is correct. Moreover, to place the relaxation step for each Runge-Kutta inner stage or just for each time step seems to give almost identical results. We have thus used the less costly version of putting the relaxation step for every time step in all the numerical examples in this section.

Table 13.4. L1 error and order of accuracy at $t = 10$ (1 period)

Nb. points	WENO	
	error	order
20 × 20	1.07e-2	
40 × 40	1.06e-3	3.3
80 × 80	6.50e-5	4.0
160 × 160	2.09e-6	4.9

Nb. points	Relaxed WENO each time step		Relaxed WENO each R-K step	
	error	order	error	order
20 × 20	1.22e-2		1.22e-2	
40 × 40	2.16e-3	2.5	2.17e-3	2.5
80 × 80	1.77e-4	3.6	1.78e-4	3.6
160 × 160	7.57e-6	4.6	7.60e-6	4.6

Example 13.4. Double Mach reflection. First we present the results for a perfect gas, which is the same as the case in Example 11.5. We compare the results using WENO directly on the original system [55], and using it on the relaxed model with $\gamma_1 = 1.5$ and $\gamma_1 = 3.0$, in Fig. 13.8 for a mesh of 480 × 120 points and Fig. 13.9 for a mesh of 960 × 240 points. We can see that the relaxed model results are quite satisfactory, although a bigger γ_1 results in some small oscillations.

Next, we show the results of the same problem with the two vibrating molecular gas. The purpose here is to show that the relaxation model based algorithm does work, rather than on the details of the flow with more physical models. The results with both a 480 × 120 grid and a 960 × 240 grid are shown in Fig. 13.10. Comparing with the results in [26], we can see that the main

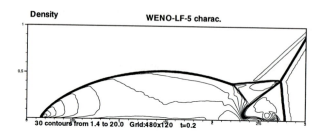

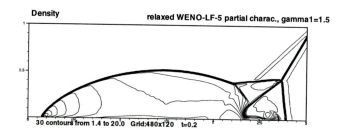

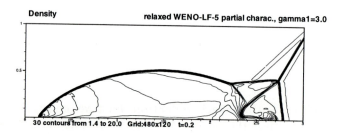

Fig. 13.8. Double-Mach reflection, perfect gas, 480×120 grid points.

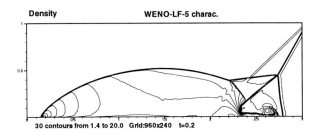

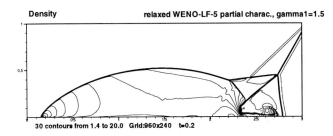

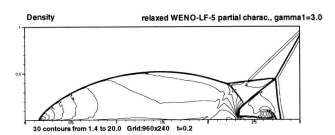

Fig. 13.9. Double-Mach reflection, perfect gas, 960×240 grid points.

features such as the main shock being closer to the bottom boundary, and the shock below the triple point being bent, are also observed here.

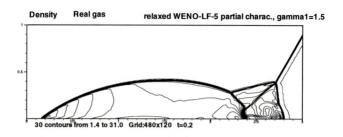

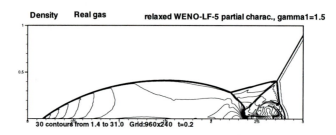

Fig. 13.10. Double-Mach reflection, two vibrating molecular gas.

14   Applications to Incompressible Flows

In this section we consider numerically solving the incompressible Navier-Stokes or Euler equations

$$
\begin{aligned}
u_t + uu_x + vu_y &= \mu(u_{xx} + u_{yy}) - p_x \\
v_t + uv_x + vv_y &= \mu(v_{xx} + v_{yy}) - p_y \\
u_x + v_y &= 0
\end{aligned}
\tag{14.1}
$$

or their equivalent conservative form

$$
\begin{aligned}
u_t + (u^2)_x + (uv)_y &= \mu(u_{xx} + u_{yy}) - p_x \\
v_t + (uv)_x + (v^2)_y &= \mu(v_{xx} + v_{yy}) - p_y \\
u_x + v_y &= 0
\end{aligned}
\tag{14.2}
$$

where (u, v) is the velocity vector, p is the pressure, $\mu > 0$ for the Navier-Stokes equations and $\mu = 0$ for the Euler equations, using ENO and WENO schemes. We do not discuss the issue of boundary conditions here, thus the equation is defined on the box $[0, 2\pi] \times [0, 2\pi]$ with periodic boundary conditions in both directions. We choose two space dimensions for easy presentation, although our method is also applicable for three space dimensions.

In some sense equations (14.1) are easier to solve numerically than their compressible counter-parts in the previous three sections, because the latter have solutions containing possible discontinuities (for example shocks and contact discontinuities). However, the solution to (14.1), even if for most cases smooth mathematically, may evolve rather rapidly with time t and may easily become too complicated to be fully resolved on a feasible grid. Traditional linearly stable schemes, such as spectral methods and high-order central difference methods, are suitable for the cases where the solution can be fully resolved, but typically produce signs of instability such as oscillations when small scale features of the flow, such as shears and roll-ups, cannot be adequately resolved on the computational grid. Although in principle one can always overcome this difficulty by refining the grid, today's computer capacity seriously restricts the largest possible grid size.

As we know, the high resolution "shock capturing" schemes such as ENO and WENO are based on the philosophy of giving up fully resolving rapid transition regions or shocks, just to "capture" them in a stable and somehow globally correct fashion (e.g., with correct shock speed), but at the same time to require a high resolution for the smooth part of the solution. The success of such an approach for the conservation laws is documented by many examples in these lecture notes and the references. One example is the one and two dimensional shock interaction with vorticity or entropy waves [90], [91]. The shock is captured sharply and certain key quantities related to the interaction between the shock and the smooth part of the flow, such as the amplification and generation factors when a wave passes through a shock, are well resolved. Another example is the homogeneous turbulence for compressible Navier-Stokes equations studied in [91]. In one of the test cases, the spectral method can resolve all the scales using a 256^2 grid, while third order ENO with just 64^2 points can adequately resolve certain interesting quantities although it cannot resolve local quantities achieved inside the rapid transition region such as the minimum divergence. The conclusion seems to be that, when fully resolving the flow is either impossible or too costly, a "capturing" scheme such as ENO can be used on a coarse grid to obtain at least some partial information about the flow.

We thus expect that, also for the incompressible flow, we can use high-order ENO or WENO schemes on a coarse grid, without fully resolving the flow, but still get back some useful information.

A pioneer work in applying shock capturing compressible flow techniques to incompressible flow is by Bell, Colella and Glaz [8], in which they considered a second order Godunov type discretization, investigated the projection into divergence-free velocity fields for general boundary conditions, and discussed accuracy of time discretizations. Higher order ENO and WENO schemes for incompressible flows are extensions of such methods.

We solve (14.2) in its equivalent projection form

$$\begin{pmatrix} u \\ v \end{pmatrix}_t = \mathbf{P}\left[-\begin{pmatrix} u^2 \\ uv \end{pmatrix}_x - \begin{pmatrix} uv \\ v^2 \end{pmatrix}_y + \mu\left(\begin{pmatrix} u \\ v \end{pmatrix}_{xx} + \begin{pmatrix} u \\ v \end{pmatrix}_{yy} \right) \right] \qquad (14.3)$$

where $\mathbf{P}$ is the Hodge projection into divergence-free fields, i.e., if $\begin{pmatrix} \tilde{u} \\ \tilde{v} \end{pmatrix} = \mathbf{P}\begin{pmatrix} u \\ v \end{pmatrix}$, then $\tilde{u}_x + \tilde{v}_y = 0$ and $\tilde{v}_y - \tilde{u}_x = v_y - u_x$. See, e.g., [8]. For the current periodic case the additional condition to obtain a unique projection $\mathbf{P}$ is that the mean values of u and v are preserved, i.e., $\int_0^{2\pi} \int_0^{2\pi} \tilde{u}(x,y)dxdy = \int_0^{2\pi} \int_0^{2\pi} u(x,y)dxdy$ and $\int_0^{2\pi} \int_0^{2\pi} \tilde{v}(x,y)dxdy = \int_0^{2\pi} \int_0^{2\pi} v(x,y)dxdy$.

We use N_x and N_y (even numbers) equally spaced grid points in x and y, respectively. The grid sizes are denoted by $\Delta x = \frac{N_x}{2\pi}$ and $\Delta y = \frac{N_y}{2\pi}$, and the grid points are denoted by $x_i = i\Delta x$ and $y_j = j\Delta y$. The approximated numerical values of u and v at the grid point (x_i, y_j) are denoted by u_{ij} and v_{ij}.

We first describe the numerical implementation of the projection $\mathbf{P}$. In the periodic case this is easily achieved in the Fourier space. We first expand u and v using Fourier collocation:

$$u_N(x,y) = \sum_{l=-\frac{N_y}{2}}^{\frac{N_y}{2}} \sum_{k=-\frac{N_x}{2}}^{\frac{N_x}{2}} \hat{u}_{kl} e^{I(kx+ly)}, \qquad (14.4)$$

$$v_N(x,y) = \sum_{l=-\frac{N_y}{2}}^{\frac{N_y}{2}} \sum_{k=-\frac{N_x}{2}}^{\frac{N_x}{2}} \hat{v}_{kl} e^{I(kx+ly)}$$

where $I = \sqrt{-1}$, $\hat{u}_{kl}$ and $\hat{v}_{kl}$ are the Fourier collocation coefficients which can be computed from the point values u_{ij} and v_{ij}, using either FFT or matrix-vector multiplications. The detail can be found in, e.g., [12]. Derivatives, either by spectral method or by central differences, involve only multiplications by factors d_k^x or d_l^y in (14.4) because $e^{I(kx+ly)}$ are eigenfunctions of such derivative operators. For example,

$$d_k^x = Ik, \qquad d_l^y = Il \qquad (14.5)$$

for spectral derivatives;

$$d_k^x = \frac{2I\sin(\frac{k\Delta x}{2})}{\Delta x}, \qquad d_l^y = \frac{2I\sin(\frac{l\Delta y}{2})}{\Delta y} \qquad (14.6)$$

for the second order central differences which, when used twice, will produce the second order central difference approximation $\frac{w_{i+1}-2w_i+w_{i-1}}{\Delta x^2}$ for w_{xx}, and

$$d_k^x = \frac{2I\sqrt{(1-\cos(k\Delta x))(7-\cos(k\Delta x))}}{\Delta x},$$

$$d_l^y = \frac{2I\sqrt{(1 - \cos(l\Delta y))(7 - \cos(l\Delta y))}}{\Delta y} \tag{14.7}$$

for the fourth order central differences which, when used twice, will produce the fourth order central difference approximation

$$\frac{16(w_{i+1} + w_{i-1}) - (w_{i+2} + w_{i-2}) - 30w_i}{12\Delta x^2}$$

for w_{xx}. High order filters, such as the exponential filter [72], [58]:

$$\sigma_k^x = e^{-\alpha(\frac{k}{N_x})^{2p}}, \qquad \sigma_l^y = e^{-\alpha(\frac{l}{N_y})^{2p}} \tag{14.8}$$

where $2p$ is the order of the filter and α is chosen so that $e^{-\alpha}$ is machine zero, can be used to enhance the stability while keeping at least $2p$-th order of accuracy. This is especially helpful when the projection $\mathbf{P}$ is used for the under-resolved coarse grid with ENO methods. We use the fourth order projection (14.7) and the filter (14.8) with $2p = 8$ in our calculations. This will guarantee third order accuracy (fourth order in L_1) of the ENO scheme. We will denote this combination (the fourth order projection plus the eighth order filtering) by $\mathbf{P}_4$. To be precise, if $\begin{pmatrix} \tilde{u} \\ \tilde{v} \end{pmatrix} = \mathbf{P}_4 \begin{pmatrix} u \\ v \end{pmatrix}$ and $\hat{u}_{kl}$ and $\hat{v}_{kl}$ are Fourier collocation coefficients of u and v, then the Fourier collocation coefficients of $\tilde{u}$ and $\tilde{v}$ are given by

$$\hat{\tilde{u}} = \sigma_k^x \sigma_l^y \frac{d_l^y(d_l^y \hat{u} - d_k^x \hat{v})}{(d_k^x)^2 + (d_l^y)^2}, \qquad \hat{\tilde{v}} = \sigma_k^x \sigma_l^y \frac{-d_k^x(d_l^y \hat{u} - d_k^x \hat{v})}{(d_k^x)^2 + (d_l^y)^2} \tag{14.9}$$

where σ_k^x and σ_l^y are defined by (14.8) with $2p = 8$, and d_k^x and d_l^y are defined by (14.7).

Next we shall describe the ENO scheme for (14.2). Since (14.2) is equivalent to the non-conservative form (14.1), it is natural to implement upwinding by the signs of u and v, and to implement ENO equation by equation (the component version described in Sect. 4.4). The r-th order ENO approximation of, e.g., $(u^2)_x$ is thus carried out using the ENO Algorithm 4.2. We mention a couple of facts needing attention:

1. Take $f(x) = u^2(x, y)$ with y fixed. We start with the point values $f_i = f(x_i)$;
2. The stencil of the reconstruction is determined adaptively by upwinding and smoothness of $f(x)$. It starts with either x_j or x_{j+1} according to whether $u > 0$ or $u < 0$.

There are two ways to handle the second derivative terms for the Navier-Stokes equations. One can absorb them into the convection part and treat them using ENO. For example, $f(x) = u^2(x, y)$ can be replaced by $f(x) = u^2(x, y) - \mu u(x, y)_x$, where $u(x, y)_x$ itself can be obtained using either ENO or central difference of a suitable order. The remaining procedure for computing $f(x)_x$ would be the same as described above. Another simpler possibility is just to use standard central differences (of suitable order) to compute the double derivative terms. Our experience with compressible flow is that there is little difference between the two approaches, especially when the viscosity μ is small.

In the above we have described the discretization for the spatial derivatives

$$L_{ij} \approx \left[-\left(\frac{u^2}{uv}\right)_x - \left(\frac{uv}{v^2}\right)_y + \mu\left(\left(\frac{u}{v}\right)_{xx} + \left(\frac{u}{v}\right)_{yy}\right)\right]_{\substack{x = x_i \\ y = y_j}} . \qquad (14.10)$$

We then use the third order TVD (total variation diminishing) Runge-Kutta method (9.10) to discretize the resulting ODE:

$$\left(\frac{u}{v}\right)_t = \mathbf{P}_4 L_{ij} \qquad (14.11)$$

obtaining:

$$\left(\frac{u}{v}\right)^{(1)} = \mathbf{P}_4\left[\left(\frac{u}{v}\right)^n + \Delta t L_{ij}^n\right]$$

$$\left(\frac{u}{v}\right)^{(2)} = \mathbf{P}_4\left[\frac{3}{4}\left(\frac{u}{v}\right)^n + \frac{1}{4}\left(\frac{u}{v}\right)^{(1)} + \frac{1}{4}\Delta t L_{ij}^{(1)}\right] \qquad (14.12)$$

$$\left(\frac{u}{v}\right)^{n+1} = \mathbf{P}_4\left[\frac{1}{3}\left(\frac{u}{v}\right)^n + \frac{2}{3}\left(\frac{u}{v}\right)^{(2)} + \frac{2}{3}\Delta t L_{ij}^{(2)}\right]$$

Notice that we have used the property $\mathbf{P}_4 \circ \mathbf{P}_4 = \mathbf{P}_4$ in obtaining the discretization (14.12) from (14.11).

This explicit time discretization is expected to be nonlinearly stable under the CFL condition

$$\Delta t\left[\max_{i,j}\left(\frac{|u_{ij}|}{\Delta x} + \frac{|v_{ij}|}{\Delta y}\right) + 2\mu\left(\frac{1}{\Delta x^2} + \frac{1}{\Delta y^2}\right)\right] \leq 1 \qquad (14.13)$$

For small μ (which is the case we are interested in) this is not a serious restriction on Δt.

We present some numerical examples in the following.

Example 14.1. Accuracy test. This example is used to check the third order accuracy of our ENO scheme for smooth solutions. We first take the initial condition as

$$u(x, y, 0) = -\cos(x)\sin(y), \qquad (14.14)$$
$$v(x, y, 0) = \sin(x)\cos(y)$$

which was used in [8]. The exact solution for this case is known:

$$u(x, y, t) = -\cos(x)\sin(y)e^{-2\mu t}, \qquad (14.15)$$
$$v(x, y, t) = \sin(x)\cos(y)e^{-2\mu t}$$

We take $\Delta x = \Delta y = \frac{1}{N}$ with $N = 32, 64, 128$ and 256. The solution is computed up to $t = 2$ and the L_2 error and numerical order of accuracy are listed in Table 14.1. For the $\mu = 0.05$ case, we list results both with fourth order central approximation to the double derivative terms (central) and with ENO to handle the double derivative terms by absorbing them into the convection part (ENO). We can clearly observe fully third order accuracy (actually better in many cases because the spatial ENO is fourth order in the L_1 sense) in this table.

Table 14.1. Accuracy of ENO Schemes for (14.2).

N	$\mu = 0$		$\mu = 0.05$, central		$\mu = 0.05$, ENO	
	L_2 error	order	L_2 error	order	L_2 error	order
32	9.10(-4)		5.28(-4)		4.87(-4)	
64	5.73(-5)	3.99	3.20(-5)	4.04	3.09(-5)	3.98
128	3.62(-6)	3.98	1.93(-6)	4.05	1.89(-6)	4.03
256	2.28(-7)	3.99	1.18(-7)	4.03	1.16(-7)	4.03

Example 14.2. Double shear layer. This is our test example to study resolution of ENO schemes when the grid is coarse. It is a double shear layer taken from [8]:

$$u(x, y, 0) = \begin{cases} \tanh((y - \pi/2)/\rho) & y \leq \pi \\ \tanh((3\pi/2 - y)/\rho) & y > \pi \end{cases}$$
$$v(x, y, 0) = \delta \sin(x) \tag{14.16}$$

where we take $\rho = \pi/15$ and $\delta = 0.05$. The Euler equations ($\mu = 0$) are used for this example. The solution quickly develops into roll-ups with smaller and smaller scales, so on any fixed grid the full resolution is lost eventually. For example, the expensive run we performed using 512^2 points for the spectral collocation code (with a 18-th order filter (14.8)) is able to resolve the solution fully up to $t = 8$, Fig. 14.1, top left, as verified by the spectrum of the solution (not shown here), but begins to lose resolution as indicated by the wriggles in the vorticity contour at $t = 10$ (not shown here). On the other hand, the ENO runs with 64^2 (top right) and 128^2 points (bottom left) produces smooth, stable results Fig. 14.1. In Fig. 14.1, bottom right, we show a cut at $x = \pi$ for v at $t = 8$. This gives a better feeling about the resolution in physical space. Apparently with these coarse grids the full structure of the roll-up is not resolved. However, when we compute the total circulation

$$c_\Omega = \int_\Omega \omega(x, y)dxdy = \int_{\partial\Omega} udx + vdy \tag{14.17}$$

around the roll-up by taking $\Omega = [\frac{\pi}{2}, \frac{3\pi}{2}] \times [0, 2\pi]$ and using the rectangular rule (which is infinite order accurate for the periodic case) on the line integrals at the right-hand-side of (14.7), we can see that this number is resolved much better than the roll-up itself, Table 14.2.

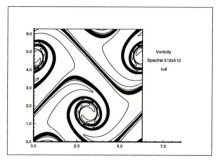
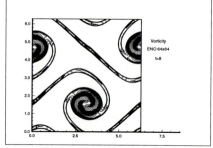
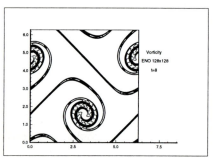
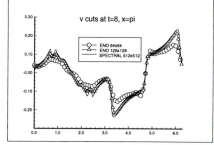

Fig. 14.1. Double shear layer. Contours of vorticity. $t = 8$. Top left: spectral with 512^2 points; Top right: ENO with 64^2 points; Bottom left: ENO with 128^2 points; Bottom right: the cut at $x = \pi$ of v, spectral method with 512^2 points, ENO method with 64^2 and with 128^2 points.

Example 14.3. Level set formulation and vortex sheet. As an application of ENO scheme for incompressible flow, we consider the motion of an incompressible fluid, in two and three dimensions, in which the vorticity is concentrated on a lower dimensional set [40]. Prominent examples are vortex sheets and vortex filaments in three dimensions, and vortex sheets, vortex dipole sheets and point vortices in two dimensions.

In three dimensions, the equations are written in the form

$$\xi_t + v\nabla\xi - \nabla v\, \xi = 0$$
$$\nabla \times v = \xi \qquad (14.18)$$
$$\nabla \cdot v = 0$$

where $\xi(x, y, z, t)$ is the vorticity vector, and $v(x, y, z, t)$ is the velocity vector.

Table 14.2. Resolution of the Total Circulation.

t	2	4	6	8	10
ENO 64^2	0.87300	3.07100	7.16889	9.88063	10.90122
ENO 128^2	0.87452	2.97810	7.30999	10.34414	11.79418
spectral 512^2	0.87433	2.98029	7.28308	10.46212	11.85875

In a vortex sheet, ξ is a singular measure concentrated on a two dimensional surface, while in a vortex filament, ξ is a function concentrated on a tubular neighborhood of a curve.

We use an Eulerian, fixed grid, approach, that works in general in two and three dimensions. In the particular case of the two dimensional vortex sheet problem in which the vorticity does not change sign, the approach yields a very simple and elegant formulation.

The basic observation involves a variant of the level set method for capturing fronts, developed in [78].

The formulation we use here regularizes general ill-posed problems via the level set approach, using the idea that a simple closed curve which is the level set of a function cannot change its index, i.e. there is an automatic topological regularization. This is very helpful for numerical calculations. The regularization is automatically accomplished through the use of dissipative schemes, which has the effect of adding a small curvature term (which vanishes as the grid size goes to zero) to the evolution of the interface. The formulation allows for topological changes, such as merging of surfaces.

The main idea is to decompose ξ into a product of the form

$$\xi = P(\varphi)\eta \tag{14.19}$$

where P is a scalar function, typically an approximate δ function. The variable φ is a scalar function whose zero level set represents the points where vorticity concentrates, and η represents the vorticity strength vector. This decomposition is performed at time zero and is of course not unique.

The observation is that once a decomposition is found, the following system of equations yields a solution to the Euler equations, replacing the original set of equations (14.18).

$$\begin{aligned} \varphi_t + v\nabla\varphi &= 0 \\ \eta_t + v\nabla\eta - \nabla v\,\eta &= 0 \\ \nabla \times v &= P(\varphi)\eta \\ \nabla \cdot v &= 0 \end{aligned} \tag{14.20}$$

These equations have initial conditions

$$\varphi(0, \cdot) = \varphi_0$$

$$\eta(0, \cdot) = \eta_0$$

where φ_0, η_0 and P are chosen so that (14.19) holds at time $t = 0$. Notice that (14.19) and (14.20) imply that $\nabla\varphi$ is orthogonal to η, and $div(\eta) = 0$. This is enforced in the initial condition and is maintained automatically by (14.19) and (14.20).

When P is a distribution, such as a δ function, approaching P with a sequence of smooth mollifiers P_ϵ yields a sequence of approximating solutions. This is the approach used in numerical calculations, since the δ function can only be represented approximately on a finite grid. The parameter ϵ is usually chosen to be proportional to the mesh size.

The advantage of this formulation, is that it replaces a possibly singular and unbounded vorticity function ξ, by bounded, smooth (at least uniformly Lipschitz) functions φ and η. Therefore, while it is not feasible to compute solutions of (14.18) directly, it is very easy to compute solutions of (14.20).

In two dimensions, the vorticity is given by

$$\xi = \begin{pmatrix} 0 \\ 0 \\ w(t, x, y) \end{pmatrix}$$

and hence the Euler equations are given by

$$\begin{aligned} w_t + v\nabla w &= 0 \\ curl(v) &= w \\ div(v) &= 0 \end{aligned} \tag{14.21}$$

Our formulation (14.20), becomes

$$\begin{aligned} \varphi_t + v\nabla\varphi &= 0 \\ \eta_t + v\nabla\eta &= 0 \\ curl(v) &= P(\varphi)\eta \\ div(v) &= 0 \end{aligned} \tag{14.22}$$

where η is now a scalar.

If the vortex sheet strength η does not change sign along the curve, it can be normalized to $\eta \equiv 1$ and the equations take on a particularly simple and elegant form:

$$\varphi_t + v(\varphi)\nabla\varphi = 0 \tag{14.23}$$

where the velocity $v(\varphi)$ is given by

$$v = - \begin{pmatrix} -\partial_y \\ \partial_x \end{pmatrix} \Delta^{-1} P(\varphi) \tag{14.24}$$

In this case, the vortex sheet strength along the curve is given by $\frac{1}{|\nabla\varphi|}$ (see (14.26)).

We first consider the periodic vortex sheet in two dimensions, i.e. $P(\varphi) = \delta(\varphi)$ in (14.24). The three dimensional case is defined in detail later. The evolution of the vortex sheet in the Lagrangian framework has been considered by various authors. Krasny [59], [60] has computed vortex sheet roll-up using vortex blobs and point vortices with filtering. Baker and Shelley [4] have approximated the vortex sheet by a layer of constant vorticity which they computed by Lagrangian methods. In the context of our approach, their approximation corresponds to approximating the δ function by a step function.

In our framework, we use a fixed Eulerian grid, and approximate (14.23) by the third order upwind ENO finite difference scheme with a third order TVD Runge-Kutta time stepping. At every time step, the velocity v is first obtained by solving the Poisson equation for the stream function Ψ:

$$\Delta\Psi = -P(\varphi)$$

with boundary conditions

$$\Psi(x, \pm 1) = 0$$

and periodic in x. This is done by using a second order elliptic solver FISH-PAK. Once Ψ is obtained, the velocity is recovered by $v = (-\Psi_y, \Psi_x)$ by using either ENO or central difference approximations (we do not observe major difference among the two: the results shown are those obtained by central difference). Once v is obtained, upwind biased ENO is easily applied to (14.23).

The initial conditions are similar to the ones in [60], i.e given by a sinusoidal perturbation of a flat sheet:

$$\varphi_0(x, y) = y + 0.05 \sin(\pi x)$$

The boundary condition for φ are periodic, of the form:

$$\varphi(t, -1, y) = \varphi(t, 1, y)$$

$$\varphi(t, x, -1) = \varphi(t, x, 1) - 2$$

The δ function is approximated as in [80],[99] by

$$\delta_\epsilon(\phi) = \begin{cases} \frac{1}{2\epsilon}\left(1 + \cos\left(\frac{\pi\phi}{\epsilon}\right)\right) & \text{if } |\varphi| < \epsilon \\ 0 & \text{otherwise} \end{cases} \tag{14.25}$$

For fixed ϵ, there is convergence as $\Delta x \to 0$ to a smooth solution. One can then take $\epsilon \to 0$. This two step limit is very costly to implement numerically. Our numerical results show that one can take ϵ to be proportional to Δx, but convergence is difficult to establish theoretically.

In Fig. 14.2, top left, we present the result at $t = 4$, of using ENO with 128^2 grid points with the parameter ϵ in the approximate δ function chosen as $\epsilon = 12\Delta x$. We use the graphic package TECPLOT to draw the level curve of $\varphi = 0$. Next, we keep $\epsilon = 12\Delta x$ but double the grid points in each direction to 256^2, the result of $t = 4$ is shown in Fig. 14.2, top right. Comparing with Fig. 14.2, top left, we can see that there are more turns in the core at the same physical time when the grid size is reduced and the δ function width ϵ is kept proportional to Δx. One might wonder whether the core structure of Fig. 14.2, top right, is distorted by numerical error. To verify that this is not the case, we keep $\epsilon = 12 \times \frac{2}{256} = \frac{3}{32}$ *fixed*, and reduce Δx, Fig. 14.2, bottom two. The three pictures overlay very well, the bottom two pictures in Fig. 14.2 are indistinguishable, indicating that the core structure is a resolved solution to the problem and convergence is obtained with fixed ϵ. By reducing ϵ for the more refined grids, more turns in the core can be obtained in shorter time (pictures not shown).

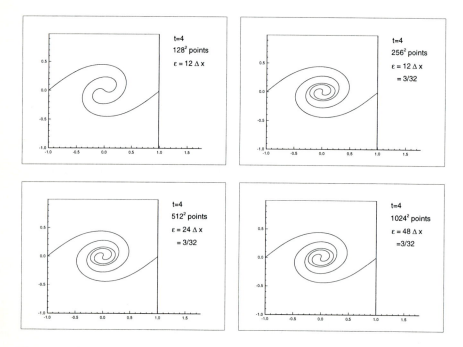

Fig. 14.2. Two dimensional vortex sheet simulation. $t = 4$. Top left: ENO with 128^2 points, δ function width $\epsilon = 12\Delta x = \frac{3}{16}$; Top right: ENO with 256^2 points, δ function width $\epsilon = 12\Delta x = \frac{3}{32}$; Bottom left: ENO with 512^2 points, δ function width $\epsilon = 24\Delta x = \frac{3}{32}$; Bottom right; ENO with 1024^2 points, δ function width $\epsilon = 48\Delta x = \frac{3}{32}$.

The smoothing of the δ function, and the third order truncation error in the advection step and the second order error in the inverse Laplacian are the only smoothing steps in our method.

We now give the same example in three dimensions. We first sketch the algorithm for initializing and computing a periodic 3D vortex sheet, using (14.20).

We let $P(\varphi) = \delta(\varphi)$ (in practice δ is replaced by an approximation). The zero level set of φ is the vortex sheet $\Gamma(s)$, parameterized by surface area s. The variable η_0 is chosen to fit the initial vortex sheet strength. For instance, given any smooth test function g

$$\langle \xi, g \rangle = \langle \eta_0 \delta(\varphi_0), g \rangle$$
$$= \int \eta_0(\Gamma_0(s)) g(\Gamma_0(s)) \frac{1}{|\nabla \varphi_0|} ds$$

Thus, the initial vortex sheet strength is given by

$$\frac{\eta_0}{|\nabla \varphi_0|} \tag{14.26}$$

To obtain the velocity vector, one introduces the vector potential A, where

$$v = \nabla \times A, \quad div(A) = 0$$

and solves the Poisson equation

$$\triangle A = -P(\varphi)\eta \tag{14.27}$$

To ensure that $div(A) = 0$, we require that $div(\eta) = 0$ and that $\nabla \varphi \cdot \eta = 0$ initially. It is easy to see that these equalities are maintained as t increases.

The boundary conditions for the velocity are $v_2(x, \pm 1, z) = 0$ and periodic in x and z. To obtain the boundary conditions for $A = (A_1, A_2, A_3)$, we use the divergence free condition on A in addition to the velocity boundary condition. Thus,

$$A_1(x, \pm 1, z) = A_3(x, \pm 1, z) = 0 \tag{14.28}$$
$$\partial_y A_2(x, \pm 1, z) = 0$$

and periodic in x, z. The Neumann condition requires the following compatibility condition

$$\int \xi_2(x, y, z, 0) dx dy dz = 0$$

Three dimensional runs are much more expensive than two dimensional runs, not only because the number of grid points increases, but also because there are now four evolution equations (for φ and η), and three potential equations. We still use the third order ENO scheme coupled with the second order elliptic solver FISHPAK, with 64^3 grid points, and ϵ is chosen as $6\triangle x$, which is the same in magnitude as that used in Fig. 14.2 of the two dimensional example. The boundary conditions for φ are similar to the ones in two dimensions: periodic in all directions (module the linear term in y). The vortex sheet strength vector η is periodic in all directions.

We first verify whether we can recover the two dimensional results with the three dimensional setting. We use the initial condition

$$\varphi_0(x, y, z) = y + 0.05 \sin(\pi x)$$

which is the same as that for the two dimensional example, and choose a constant initial condition for η as $\eta_0(x, y, z) = (0, 0, 1)$. We observe exact agreement with our two dimensional results in Fig. 14.2. Next, we consider the truly three dimensional problem with the initial condition chosen as

$$\varphi_0(x, y, z) = y + 0.05 \sin(\pi x) + 0.1 \sin(\pi z)$$

and η is chosen as $\eta_0(x, y, z) = (0, -0.1\pi \cos(\pi z), 1)$ which satisfies the divergence free condition as well as the condition to be orthogonal to $\nabla \varphi$. In Fig. 14.3, left, we show the level set of $\varphi = 0$ for $t = 5$. We can clearly see the roll up process and the three dimensional features. The cut at the constants $z = 0$ plane is shown in Fig. 14.3, right.

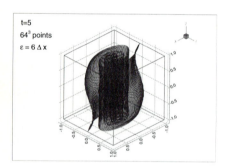 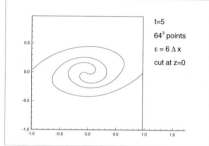

Fig. 14.3. Three dimensional vortex sheet simulation. $t = 5$. ENO with 64^3 points. δ function width $\epsilon = 6\Delta x$. Left: three dimensional level surface; Right: $z = 0$ plane cut.

References

1. R. Abgrall, *On essentially non-oscillatory schemes on unstructured meshes: analysis and implementation*, Journal of Computational Physics, v114 (1994), pp.45–58.
2. N. Adams and K. Shariff, *A high-resolution hybrid compact-ENO scheme for shock-turbulence interaction problems*, Journal of Computational Physics, v127 (1996), pp.27–51.
3. H. Atkins and C.-W. Shu, *GKS and eigenvalue stability analysis of high order upwind scheme*, in preparation.

4. G. R. Baker and M. J. Shelley, *On the connection between thin vortex layers and vortex sheets*, Journal of Fluid Mechanics, v215 (1990), pp.161–194.

5. D. Balsara and C.-W. Shu, *Monotonicity preserving weighted essentially non-oscillatory schemes with increasingly high order of accuracy*, preprint.

6. M. Bardi and S. Osher, *The nonconvex multi-dimensional Riemann problem for Hamilton-Jacobi equations*, SIAM Journal on Mathematical Analysis, v22 (1991), pp.344–351.

7. T. Barth and P. Frederickson, *High order solution of the Euler equations on unstructured grids using quadratic reconstruction*, AIAA Paper No. 90-0013.

8. J. Bell, P. Colella and H. Glaz, *A Second Order Projection Method for the Incompressible Navier-Stokes Equations*, Journal of Computational Physics, v85, 1989, pp.257–283.

9. F. Bianco, G. Puppo and G. Russo, *High order central schemes for hyperbolic systems of conservation laws*, SIAM Journal on Scientific Computing, to appear.

10. B. Bihari and A. Harten, *Application of generalized wavelets: an adaptive multiresolution scheme*, Journal of Computational and Applied Mathematics, v61 (1995), pp.275–321.

11. W. Cai and C.-W. Shu, *Uniform high-order spectral methods for one- and two-dimensional Euler equations*, Journal of Computational Physics, v104 (1993), pp.427–443.

12. C. Canuto, M.Y. Hussaini, A. Quarteroni and T. A. Zang, *Spectral Methods in Fluid Dynamics*, Springer-Verlag, 1988.

13. M. Carpenter and C. Kennedy, *Fourth-order 2N-storage Runge-Kutta schemes*, NASA TM 109112, NASA Langley Research Center, June 1994.

14. J. Casper, *Finite-volume implementation of high-order essentially nonoscillatory schemes in two dimensions*, AIAA Journal, v30 (1992), pp.2829–2835.

15. J. Casper and H. Atkins, *A finite-volume high-order ENO scheme for two dimensional hyperbolic systems*, Journal of Computational Physics, v106 (1993), pp.62–76.

16. J. Casper, C.-W. Shu and H. Atkins, *Comparison of two formulations for high-order accurate essentially nonoscillatory schemes*, AIAA Journal, v32 (1994), pp.1970–1977.

17. S. Christofi, *The study of building blocks for ENO schemes*, Ph.D. thesis, Division of Applied Mathematics, Brown University, September 1995.

18. B. Cockburn and C.-W. Shu, *TVB Runge-Kutta local projection discontinuous Galerkin finite element method for conservation laws II: general framework*, Mathematics of Computation, v52 (1989), pp.411–435.

19. B. Cockburn, S.-Y. Lin and C.-W. Shu, *TVB Runge-Kutta local projection discontinuous Galerkin finite element method for conservation laws III: one dimensional systems*, Journal of Computational Physics, v84 (1989), pp.90–113.

20. B. Cockburn, S. Hou and C.-W. Shu, *The Runge-Kutta local projection discontinuous Galerkin finite element method for conservation laws IV: the multidimensional case*, Mathematics of Computation, v54 (1990), pp.545–581.

21. B. Cockburn and C.-W. Shu, *The Runge-Kutta discontinuous Galerkin method for conservation laws V: multidimensional systems*, Journal of Computational Physics, v141 (1998), pp.199–224.

22. P. Colella and H.M. Glaz, *Efficient solution algorithms for the Riemann problem for real gases*, Journal of Computational Physics, v59 (1985), pp.264–289.

23. F. Coquel and B. Perthame, *Relaxation of energy and approximate Riemann solvers for general pressure laws in fluid dynamics equations*, SIAM Journal on Numerical Analysis, v35 (1998), pp.2223–2249.

24. M. Crandall and P. Lions, *Viscosity solutions of Hamilton-Jacobi equations*, Transactions of the American Mathematical Society, v277 (1983), pp.1–42.

25. M. Crandall and P. Lions, *Two approximations of solutions of Hamilton-Jacobi equations*, Mathematics of Computation, v43 (1984), pp.1–19.

26. R.L. Deschambault and I.I. Glass, *An update on non-stationary oblique shock-wave reflections: actual isopycnics and numerical experiments*, Journal of Fluid Mechanics, v131 (1983), pp.27–57.

27. A. Dolezal and S. Wong, *Relativistic hydrodynamics and essentially non-oscillatory shock capturing schemes*, Journal of Computational Physics, v120 (1995), pp.266–277.

28. R. Donat and A. Marquina, *Capturing shock reflections: an improved flux formula*, Journal of Computational Physics, v125 (1996), pp.42–58.

29. W. E and C.-W. Shu, *A numerical resolution study of high order essentially non-oscillatory schemes applied to incompressible flow*, Journal of Computational Physics, v110 (1994), pp.39–46.

30. G. Erlebacher, Y. Hussaini and C.-W. Shu, *Interaction of a shock with a longitudinal vortex*, Journal of Fluid Mechanics, v337 (1997), pp.129–153.

31. E. Fatemi, J. Jerome and S. Osher, *Solution of the hydrodynamic device model using high order non-oscillatory shock capturing algorithms*, IEEE Transactions on Computer-Aided Design of Integrated Circuits and Systems, v10 (1991), pp.232–244.

32. O. Friedrich, *Weighted essentially non-oscillatory schemes for the interpolation of mean values on unstructured grids*, Journal of Computational Physics, v144 (1998), pp.194–212.

33. P. Glaister, *An efficient numerical method for compressible flows of a real gas using arithmetic averaging*, Computers and Mathematics with Applications, v28 (1994), pp.97–113.

34. P. Glaister, *An analysis of averaging procedures in a Riemann solver for compressible flows of a real gas*, Computers and Mathematics with Applications, v33 (1997), pp.105–119.

35. E. Godlewski and P.-A. Raviart, *Numerical approximation of hyperbolic systems of conservation laws*, Springer, 1996.

36. S. Gottlieb and C.-W. Shu, *Total variation diminishing Runge-Kutta schemes*, Mathematics of Computation, v67 (1998), pp.73-85.

37. S. Gottlieb, C.-W. Shu and E. Tadmor, *Norm preserving time discretizations*, in preparation.

38. B. Grossman and R.W. Walters, *Analysis of flux-split algorithms for Euler's equations with real gases*, AIAA Journal, v27 (1989), pp.524–531.

39. B. Gustafsson, H.-O. Kreiss and A. Sundstrom, *Stability theory of difference approximations for mixed initial boundary value problems, II*, Mathematics of Computation, v26 (1972), pp.649–686.

40. E. Harabetian, S. Osher and C.-W. Shu, *An Eulerian approach for vortex motion using a level set regularization procedure*, Journal of Computational Physics, v127 (1996), pp.15–26.

41. A. Harten, *The artificial compression method for computation of shocks and contact discontinuities III: self-adjusting hybrid schemes*, Mathematics of Computation, v32 (1978), pp.363–389.

42. A. Harten, *High resolution schemes for hyperbolic conservation laws*, Journal of Computational Physics, v49 (1983), pp.357–393.

43. A. Harten, *Preliminary results on the extension of ENO schemes to two dimensional problems*, in Proceedings of the International Conference on Hyperbolic Problems, Saint-Etienne, 1986.

44. A. Harten, *ENO schemes with subcell resolution*, Journal of Computational Physics, v83 (1989), pp.148–184.

45. A. Harten, J. Hyman and P. Lax, *On finite difference approximations and entropy conditions for shocks*, Communications in Pure and Applied Mathematics, v29 (1976), pp.297–322.

46. A. Harten and S. Osher, *Uniformly high-order accurate non-oscillatory schemes, I*, SIAM Journal on Numerical Analysis, v24 (1987), pp.279–309.

47. A. Harten, B. Engquist, S. Osher and S. Chakravarthy, *Uniformly high order essentially non-oscillatory schemes, III*, Journal of Computational Physics, v71 (1987), pp.231–303.

48. A. Harten, S. Osher, B. Engquist and S. Chakravarthy, *Some results on uniformly high order accurate essentially non-oscillatory schemes*, Applied Numerical Mathematics, v2 (1986), pp.347–377.

49. C. Hu and C.-W. Shu, *High order weighted ENO schemes for unstructured meshes: preliminary results*, Computational Fluid Dynamics 98, Invited Lectures, Minisymposia and Special Technological Sessions of the Fourth European Computational Fluid Dynamics Conference, K. Papailiou, D. Tsahalis, J. Periaux and D. Knorzer, Editors, John Wiley and Sons, v2, September 1998, pp.356-362.

50. C. Hu and C.-W. Shu, *Weighted essentially non-oscillatory schemes on triangular meshes*, Journal of Computational Physics, to appear.

51. A. In, *Numerical evaluation of an energy relaxation method for inviscid real fluids*, SIAM Journal on Scientific Computing, to appear.

52. A. Iske and T. Soner, *On the structure of function spaces in optimal recovery of point functionals for ENO-schemes by radial basis functions*, Numerische Mathematik, v74 (1996), pp.177–201.

53. J. Jerome and C.-W. Shu, *Energy models for one-carrier transport in semiconductor devices*, in IMA Volumes in Mathematics and Its Applications, v59, W. Coughran, J. Cole, P. Lloyd and J. White, editors, Springer-Verlag, 1994, pp.185–207.

54. J. Jerome and C.-W. Shu, *Transport effects and characteristic modes in the modeling and simulation of submicron devices*, IEEE Transactions on Computer-Aided Design of Integrated Circuits and Systems, v14 (1995), pp.917–923.

55. G. Jiang and C.-W. Shu, *Efficient implementation of weighted ENO schemes*, Journal of Computational Physics, v126 (1996), pp.202–228.

56. G. Jiang and S.-H. Yu, *Discrete shocks for finite difference approximations to scalar conservation laws*, SIAM Journal on Numerical Analysis, v35 (1998), pp.749–772.

57. G. Jiang and D. Peng, *Weighted ENO schemes for Hamilton-Jacobi equations*, SIAM Journal on Scientific Computing, to appear.

58. D. A. Kopriva, *A Practical Assessment of Spectral Accuracy for Hyperbolic Problems with Discontinuities*, Journal of Scientific Computing, v2, 1987, pp.249–262.

59. R. Krasny, *A study of singularity formation in a vortex sheet by the point-vortex approximation*, Journal of Fluid Mechanics, v167 (1986), pp.65–93.

60. R. Krasny, *Desingularization of periodic vortex sheet roll-up*, Journal of Computational Physics, v65 (1986), pp.292–313.

61. F. Ladeinde, E. O'Brien, X. Cai and W. Liu, *Advection by polytropic compressible turbulence*, Physics of Fluids, v7 (1995), pp.2848–2857.

62. F. Lafon and S. Osher, *High-order 2-dimensional nonoscillatory methods for solving Hamilton-Jacobi scalar equations*, Journal of Computational Physics, v123 (1996), pp.235–253.

63. B. Larrouturou, *How to preserve the mass fractions positivity when computing compressible multi-component flows*, Journal of Computational Physics, v95 (1991), pp.59–84.

64. P. D. Lax, *Weak solutions of non-linear hyperbolic equations and their numerical computations*, Communications in Pure and Applied Mathematics, v7 (1954), pp.159–193.

65. P. D. Lax and B. Wendroff, *Systems of conservation laws*, Communications in Pure and Applied Mathematics, v13 (1960), pp.217–237.

66. R. J. LeVeque, *Numerical Methods for Conservation Laws*, Birkhauser Verlag, Basel, 1990.

67. D. Levy, G. Puppo and G. Russo, *Central WENO schemes for hyperbolic systems of conservation laws*, Mathematical Modelling and Numerical Analysis, to appear.

68. M.S. Liou, B. van Leer, and J.-S. Shuen, *Splitting of inviscid fluxes for real gases*, Journal of Computational Physics, v87 (1990), pp.1–24.

69. X.-D. Liu, S. Osher and T. Chan, *Weighted essentially nonoscillatory schemes*, Journal of Computational Physics, v115 (1994), pp.200–212.

70. X.-D. Liu and S. Osher, *Convex ENO high order multi-dimensional schemes without field by field decomposition or staggered grids*, Journal of Computational Physics, v142 (1998), pp.304–330.

71. C.-Y. Loh and M.S. Liou, *Lagrangian solution of supersonic real gas flows*, Journal of Computational Physics, v104 (1993), pp.150–161.

72. A. Majda, J. McDonough and S. Osher, *The Fourier Method for Nonsmooth Initial Data*, Mathematics of Computation, v32, 1978, pp.1041–1081.

73. J.-L. Montagné, H.C. Yee, and M. Vinokur, *Comparative study of high-resolution shock-capturing schemes for a real gas*, AIAA Journal, v27 (1989), pp.1332–1346.

74. P. Montarnal and C.-W. Shu, *Real gas computation using an energy relaxation method and high order WENO schemes*, Journal of Computational Physics, to appear.

75. H. Nessyahu and E. Tadmor, *Non-oscillatory central differencing for hyperbolic conservation laws*, Journal of Computational Physics, v87 (1990), pp.408–463.

76. S. Osher, *Riemann solvers, the entropy condition, and difference approximations*, SIAM Journal on Numerical Analysis, v21 (1984), pp.217–235.

77. S. Osher and S. Chakravarthy, *Upwind schemes and boundary conditions with applications to Euler equations in general geometries*, Journal of Computational Physics, v50 (1983), pp.447–481.

78. S. Osher and J. Sethian, *Fronts propagating with curvature-dependent speed: algorithms based on Hamilton-Jacobi formulation*, Journal of Computational Physics, v79 (1988), pp.12–49.

79. S. Osher and C.-W. Shu, *High-order essentially nonoscillatory schemes for Hamilton-Jacobi equations*, SIAM Journal on Numerical Analysis, v28 (1991), pp.907–922.

80. C.S. Peskin, *Numerical analysis of blood flow in the heart*, Journal of Computational Physics, v25 (1977), pp.220–252.

81. T.D. Riney, *Numerical evaluation of hypervelocity impact phenomena*, in High-velocity impact phenomena, R. Kinslow, ed., Academic Press, 1970, ch. V, pp.158–212.

82. P. L. Roe, *Approximate Riemann solvers, parameter vectors, and difference schemes*, Journal of Computational Physics, v43 (1981), pp.357–372.

83. A. Rogerson and E. Meiberg, *A numerical study of the convergence properties of ENO schemes.* Journal of Scientific Computing, v5 (1990), pp.151–167.

84. J. Sethian, *Level Set Methods: Evolving Interfaces in Geometry, Fluid Dynamics, Computer Vision, and Material Science*, Cambridge Monographs on Applied and Computational Mathematics, Cambridge University Press, New York, New York, 1996.

85. C.-W. Shu, *TVB uniformly high order schemes for conservation laws*, Mathematics of Computation, v49 (1987), pp.105–121.

86. C.-W. Shu, *Total-Variation-Diminishing time discretizations*, SIAM Journal on Scientific and Statistical Computing, v9 (1988), pp.1073–1084.

87. C.-W. Shu, *Numerical experiments on the accuracy of ENO and modified ENO schemes*, Journal of Scientific Computing, v5 (1990), pp.127–149.

88. C.-W. Shu, *Preface to the republication of "Uniform high order essentially non-oscillatory schemes, III," by Harten, Engquist, Osher, and Chakravarthy*, Journal of Computational Physics, v131 (1997), pp.1–2.

89. C.-W. Shu and S. Osher, *Efficient implementation of essentially non-oscillatory shock capturing schemes*, Journal of Computational Physics, v77 (1988), pp.439–471.

90. C.-W. Shu and S. Osher, *Efficient implementation of essentially non-oscillatory shock capturing schemes II*, Journal of Computational Physics, v83 (1989), pp.32–78.

91. C.-W. Shu, T.A. Zang, G. Erlebacher, D. Whitaker, and S. Osher, *High order ENO schemes applied to two- and three- dimensional compressible flow*, Applied Numerical Mathematics, v9 (1992), pp.45–71.

92. C.-W. Shu and Y. Zeng, *High order essentially non-oscillatory scheme for viscoelasticity with fading memory*, Quarterly of Applied Mathematics, v55 (1997), pp.459–484.

93. K. Siddiqi, B. Kimia and C.-W. Shu, *Geometric shock-capturing ENO schemes for subpixel interpolation, computation and curve evolution*, Computer Vision Graphics and Image Processing: Graphical Models and Image Processing (CVGIP:GMIP), v59 (1997), pp.278–301.

94. J. Smoller, *Shock Waves and Reaction-Diffusion Equations*, Springer-Verlag, New York, 1983.

95. G. A. Sod, *A survey of several finite difference methods for systems of non-linear hyperbolic conservation laws*, Journal of Computational Physics, v27 (1978), pp.1–32.

96. G. A. Sod, *Numerical Methods in Fluid Dynamics*, Cambridge University Press, Cambridge, 1985.

97. T. Sonar, *On the construction of essentially non-oscillatory finite volume approximations to hyperbolic conservation laws on general triangulations: polynomial recovery, accuracy and stencil selection*, Computer Methods in Applied Mechanics and Engineering, v140 (1997), pp.157–181.

98. J. Strikwerda, *Initial boundary value problems for the method of lines*, Journal of Computational Physics, v34 (1980), pp.94–107.

99. M. Sussman, P. Smereka, S. Osher, *A level set approach for computing solutions to incompressible two phase flow*, Journal of Computational Physics, v114 (1994), pp.146–159.

100. P. K. Sweby, *High resolution schemes using flux limiters for hyperbolic conservation laws*, SIAM Journal on Numerical Analysis, v21 (1984), pp.995–1011.

101. B. van Leer, *Towards the ultimate conservative difference scheme V. A second order sequel to Godunov's method*, Journal of Computational Physics, v32 (1979), pp.101–136.

102. F. Walsteijn, *Robust numerical methods for 2D turbulence*, Journal of Computational Physics, v114 (1994), pp.129–145.

103. J.H. Williamson, *Low-storage Runge-Kutta schemes*, Journal of Computational Physics, v35 (1980), pp.48–56.

104. P. Woodward and P. Colella, *The numerical simulation of two-dimensional fluid flow with strong shocks*, Journal of Computational Physics, v54 (1984), pp.115–173.

105. H. Yang, *An artificial compression method for ENO schemes, the slope modification method*, Journal of Computational Physics, v89 (1990), pp.125–160.

General Remarks

Lecture Notes are printed by photo-offset from the master-copy delivered in camera-ready form by the authors. For this purpose Springer-Verlag provides technical instructions for the preparation of manuscripts. See also *Editorial Policy*.

Careful preparation of manuscripts will help keep production time short and ensure a satisfactory appearance of the finished book. The actual production of a Lecture Notes volume normally takes approximately 12 weeks.

Authors receive 50 free copies of their book. No royalty is paid on Lecture Notes volumes.

For conference proceedings, editors receive a total of 50 free copies of their volume for distribution to the contributing authors.

Authors are entitled to purchase further copies of their book and other Springer mathematics books for their personal use, at a discount of 33,3 % directly from Springer-Verlag.

Commitment to publish is made by letter of intent rather than by signing a formal contract. Springer-Verlag secures the copyright for each volume.

Addresses:

Professor M. Griebel
Institut für Angewandte Mathematik
der Universität Bonn
Wegelerstr. 6
D-53115 Bonn, Germany
e-mail: griebel@iam.uni-bonn.de

Professor D. E. Keyes
Computer Science Department
Old Dominion University
Norfolk, VA 23529–0162, USA
e-mail: keyes@cs.odu.edu

Professor R. M. Nieminen
Laboratory of Physics
Helsinki University of Technology
02150 Espoo, Finland
e-mail: rniemine@csc.fi

Professor D. Roose
Department of Computer Science
Katholieke Universiteit Leuven
Celestijnenlaan 200A
3001 Leuven-Heverlee, Belgium
e-mail: dirk.roose@cs.kuleuven.ac.be

Professor T. Schlick
Department of Chemistry and
Courant Institute of Mathematical
Sciences
New York University
and Howard Hughes Medical Institute
251 Mercer Street, Rm 509
New York, NY 10012-1548, USA
e-mail: schlick@nyu.edu

Springer-Verlag, Mathematics Editorial
Tiergartenstrasse 17
D-69121 Heidelberg, Germany
Tel.: *49 (6221) 487-185
e-mail: peters@springer.de
http://www.springer.de/math/
peters.html

Editorial Policy

§1. Submissions are invited in the following categories:

i)   Research monographs
ii) Lecture and seminar notes
iii) Reports of meetings

Those considering a project which might be suitable for the series are strongly advised to contact the publisher or the series editors at an early stage.

§2. Categories i) and ii). These categories will be emphasized by Lecture Notes in Computational Science and Engineering. **Submissions by interdisciplinary teams of authors are encouraged.** The goal is to report new developments – quickly, informally, and in a way that will make them accessible to non-specialists. In the evaluation of submissions timeliness of the work is an important criterion. Texts should be well-rounded and reasonably self-contained. In most cases the work will contain results of others as well as those of the authors. In each case the author(s) should provide sufficient motivation, examples, and applications. In this respect, articles intended for a journal and Ph.D. theses will usually be deemed unsuitable for the Lecture Notes series. Proposals for volumes in this category should be submitted either to one of the series editors or to Springer-Verlag, Heidelberg, and will be refereed. A pro-visional judgment on the acceptability of a project can be based on partial information about the work: a detailed outline describing the contents of each chapter, the estimated length, a bibliography, and one or two sample chapters – or a first draft. A final decision whether to accept will rest on an evaluation of the completed work which should include

– at least 100 pages of text;
– a table of contents;
– an informative introduction perhaps with some historical remarks which should be accessible to readers unfamiliar with the topic treated;
– a subject index.

§3. Category iii). Reports of meetings will be considered for publication provided that they are both of exceptional interest and devoted to a single topic. In exceptional cases some other multi-authored volumes may be considered in this category. One (or more) expert participants will act as the scientific editor(s) of the volume. They select the papers which are suitable for inclusion and have them individually refereed as for a journal. Papers not closely related to the central topic are to be excluded. Organizers should contact Lecture Notes in Computational Science and Engineering at the planning stage.

§4. Format. Only works in English are considered. They should be submitted in camera-ready form according to Springer-Verlag's specifications. Electronic material can be included if appropriate. Please contact the publisher. Technical instructions and/or TEX macros are available on http://www.springer.de/author/tex/help-tex.html; the name of the macro package is "LNCSE – LaTEX2e class for Lecture Notes in Computational Science and Engineering". The macros can also be sent on request.

Lecture Notes
in Computational Science and Engineering

For further information on these books please have a look at our mathematics catalogue at the following URL: http://www.springer.de/math/index.html

Springer
and the
environment

At Springer we firmly believe that an international science publisher has a special obligation to the environment, and our corporate policies consistently reflect this conviction.
We also expect our business partners – paper mills, printers, packaging manufacturers, etc. – to commit themselves to using materials and production processes that do not harm the environment. The paper in this book is made from low- or no-chlorine pulp and is acid free, in conformance with international standards for paper permanency.

 Springer

Printing: Mercedes-Druck, Berlin
Binding: Stürtz AG, Würzburg